THE MEKONG
BIOPHYSICAL ENVIRONMENT OF AN
INTERNATIONAL RIVER BASIN

AQUATIC ECOLOGY Series

Series Editor

Prof. James H. Thorp
Kansas Biological Survey, and
Department of Ecology and Evolutionary Biology,
University of Kansas,
Lawrence, Kansas, USA

Groundwater Ecology
Janine Gilbert, Dan L. Danielopol, Jack A. Stanford

Algal Ecology
R. Jan Stevenson, Max L. Bothwell, Rex L. Lowe

Streams and Ground Waters
Jeremy B. Jones, Patrick J. Mulholland

Ecology and Classification of North American Freshwater Invertebrates,
Second Edition
James H. Thorp, Alan P. Covich

Freshwater Ecology
Walter K. Dodds

Aquatic Ecosystems
Stuart E.G. Findlay, Robert L. Sinsabaugh

Tropical Stream Ecology
David Dudgeon

The Riverine Ecosystem Synthesis
James H. Thorp, Martin C. Thoms, Michael D. Delong

Ecology and Classification of North American Freshwater Invertebrates,
Third Edition
James H. Thorp, Alan P. Covich

THE MEKONG
BIOPHYSICAL
ENVIRONMENT
OF AN
INTERNATIONAL
RIVER BASIN

Edited by

IAN C. CAMPBELL

Principal Scientist, River Health Group,
GHD,
Melbourne, Australia
and
Adjunct Research Associate, School of Biological Sciences,
Monash University

AMSTERDAM • BOSTON • HEIDELBERG • LONDON
NEW YORK • OXFORD • PARIS • SAN DIEGO
SAN FRANCISCO • SINGAPORE • SYDNEY • TOKYO

Academic Press is an imprint of Elsevier

Academic Press is an imprint of Elsevier

360 Park Avenue South, New York, NY 10010–1710
30 Corporate Drive, Suite 400, Burlington, MA 01803, USA
525 B Street, Suite 1900, San Diego, CA 92101-4495, USA
32 Jamestown Road, London NW1 7BY, UK
Radarweg 29, PO Box 211, 1000 AE Amsterdam, The Netherlands

First edition 2009

Library of Congress Cataloging-in-Publication Data
The Mekong : biophysical environment of an international river basin / edited by Ian Campbell. – 1st ed.
    p. cm. – (Aquatic ecology series)
  ISBN 978-0-12-374026-7
  1. Environmental management–Mekong River Watershed–International cooperation.
2. Mekong River Watershed–Environmental conditions. 3. Natural history–Mekong River
Watershed. 4. Physical geology–Mekong River Watershed. 5. Aquatic ecology–Mekong River
Watershed. I. Campbell, Ian C., 1950-GE320.M47M45 2009
  333.91'62130959–dc22

2009022377

British Library Cataloguing in Publication Data
A catalogue record for this book is available from the British Library

ISBN 978-0-12-374026-7

For information on all Academic Press publications
visit our web site at elsevierdirect.com

Printed and bound in USA
09 10 10 9 8 7 6 5 4 3 2 1

Dedication

For Sai and Allan—who love the river and had to live with the book.

Contents

Preface

The Mekong River has been at the center of human conflicts of one sort or another for hundreds of years. Over the next few years to decades the conflicts will intensify, focussed on the way the river should be managed and the consequences of management decisions for those who use it. The six riparian countries—China, Myanmar, Lao PDR, Thailand, Cambodia, and Viet Nam—are at differing stages of development, and each has a different goal for the river. Within each riparian country there are a variety of different interest groups including subsistence farmers and fishers, small and large scale boat operators, electricity supply and construction companies, environmentalists and NGOs, all of whom have their own views on the desirable future of the river.

Interest in the Mekong extends well beyond the six riparian countries. If large scale hydropower developments are to occur, it is likely that many of the companies involved in the development will be based in countries outside the Mekong Basin, either elsewhere in Southeast Asia or in Europe or North America. Several developed countries have long-running aid programs in the Mekong Basin, some of which have been assisting government and intergovernment agencies within the basin to plan for future development. The Mekong Basin has also been a focus for many international NGOs including groups concerned with river conservation, environmental protection, and social justice. Finally, the management of the Mekong has sometimes been cited as a model, both good and bad, for river basin development in developing regions.

Decisions about the future of the Mekong are ultimately political and must be made by the people of the region. However, those decisions should be informed as much as possible by an understanding of the biophysical nature of the system and the likely consequences of various development trajectories on the river and its ecosystems, and ultimately on themselves.

Over the past 10 years or so there has been a substantial increase in the amount of biophysical research carried out on the Mekong Basin, partly through the activities of the Mekong River Commission (MRC) and the aid agencies supporting it, some through agencies such as the ADB and the World Bank, and some through the work of independent researchers. But much of that work has never been widely published and consequently has been difficult to access. This book is an attempt to synthesize and evaluate that work, to set it in context, and to make it more widely available, increasing its potential impact on the decision making processes.

The original idea for the book arose out of discussions with a number of MRC secretariat staff members including Joern Kristensen, then the CEO of the secretariat, Anders Thuren, Chris Barlow, Robyn Johnston, and Hans Guttman. I am grateful to the contributors who have contributed a huge amount of time, effort, and data over multiple chapter revisions to bring this volume to fruition. I am also grateful to Paul Carling, Kent Hortle, Vince Resh,

Robyn Johnston, Ed Ongley, and Chris Barlow who reviewed chapters.

Finally, I thank Andy Richford for his initial support and Emily McCloskey and the production staff at Elsevier for their patience as the book slowly coalesced, and for their skill and care with the production.

Ian Campbell

Introduction

Ian C. Campbell

Principal Scientist, River Health, GHD, 180 Lonsdale Street, Melbourne 3000, Australia & Adjunct Research Associate, School of Biological Sciences, Monash University, 3800 Australia

1. SIGNIFICANCE OF THE MEKONG RIVER

The Mekong is one of the world's most significant rivers. Its importance lies in its size as one of the world's largest rivers and the economic and ecological resources it contains, as well as its location, which has made, and continues to make, it politically significant. The political significance of this river will grow in future.

The river supports a huge population, many of whom are totally dependant on it for their subsistence, but it is also a significant ecological system supporting some of the world's highest diversities of fish and snails, as well as a number of critically endangered charismatic species such as the Irrawaddy dolphin and the giant Mekong catfish. Finally, the Mekong has been the focus of major international river management efforts for over 50 years, efforts that have drawn both international praise and condemnation.

The English language name of the river, Mekong, comes from the Thai and Lao languages. Mekong is an abbreviation commonly used in Thai and Lao for maenam kong, which means kong river. The word for river is a compound word from the words mother and water, but the components of the word are not viewed as being separable in this context any more than the components of the English word "indeed." So, although it sounds romantic, the name of the river does not mean "mother of waters" as many English-speaking writers have suggested (e.g., Luard, 2001; Mekong Descent Foundation, undated; Yamashita, 1996).

One measure of a river's significance is its size, and almost every paper published on the Mekong, including many of the chapters in this volume, begins by comparing the size of the Mekong with other rivers. The two most common measures of size are the mean annual discharge and length. In fact, comparisons based on either measure are far from precise. For example, various authors have provided rankings of the world's rivers based on discharge. Dai and Trenberth (2002) rank the Mekong as 10th, but several other authors have ranked the river somewhat lower (e.g., Milliman and Meade (1983), Van der Leeden (1975), and Van der Leeden *et al.* (1990) all ranked the Mekong below the top 10).

There are a number of difficulties faced by those wanting to rank rivers based on discharge data. Many of the world's largest rivers are located in developing countries that may have limited capacity for river gauging, and large floodplain rivers are very difficult to gauge, particularly near the mouth where there is often a large flat delta subject to overland flows during high flow periods. As a result, the flow estimates often have large associated errors. It is generally considered that Kratie, in central Cambodia, is the most downstream site at which flow estimates for the Mekong are reliable. However, gauging at Kratie excludes almost 20% of the catchment area.

Estimates of river lengths are even less reliable, because river lengths are fractal dimensions. Fractal dimensions are scale dependant, so their value depends on the scale at which they are measured. It is clearly true of river lengths that, the larger the scale of the map on which they are measured, the longer they appear because the larger scale maps have more details of the twists and turns in the river channel. For example, measuring the length of the Mekong using a planimeter on maps of 1:12,900,000, 1:3,600,000, and 1:500,000 gave river lengths of 3510, 4440 and 4770 km, respectively. Thus, to compare the lengths of the world's rivers would need lengths to be measured from maps of identical scale. Shaochuang *et al.* (2007) claim to have measured the "exact" length of the river at 4909 km using remote-sensing tools, but obviously they have simply measured the river at a larger scale than other authors.

The only really meaningful linear dimension for rivers is the length of the catchment measured as a straight line. This was the dimension used by the Mekong River Commission in the 2003 State of the Basin Report (MRC, 2003a). Catchment areas, while subject to larger estimation errors, can also provide useful comparative measures. The Mekong catchment is estimated to be about 795,000 km^2 (MRC, 2003a), about the 22nd largest river basin worldwide (van der Leeden *et al.*, 1990).

2. GEOGRAPHY AND BIOLOGICAL RESOURCES

The Mekong rises in the Tibetan plateau (Fig. 1.1); the exact location can probably never be precisely defined, but it appears to have become become controversial. French, Japanese, American, and Chinese explorers, adventurers, and geographers have all participated in expeditions that claimed to have located the source of the river. A useful summary of some recent claims has been published online (Shangri La, 2006) which also provides a reference list. Clearly, personal and national pride and institutional prestige appear to play as important arole in locating the source as any scientific factors. But the source appears to be around 94°41′ E longitude and 33°46′ N latitude, at an altitude of 5200 m or more.

The annual Mekong flood pulse provides the driving force for a very large and productive ecological system that now supports an extraordinary aquatic fauna. When first proposed, the flood pulse was thought to enhance floodplain river productivity by submerging terrestrial vegetation that had grown through the low flow period, and was then either consumed directly or through the detritus food chain by invertebrates and fish. What has become evident more recently is that much of the boost to productivity occurs through algal production occurring in the relatively warm, shallow nutrient-rich water on the floodplain rather than through direct or

FIGURE 1.1 Map of the Mekong River basin.

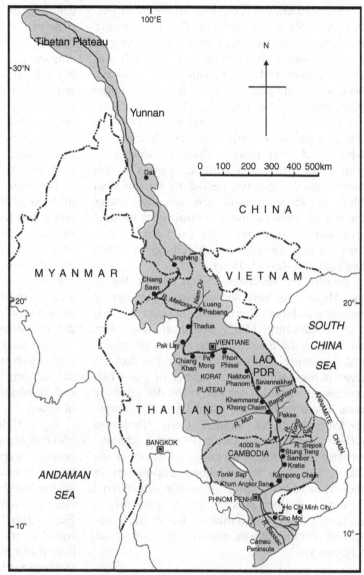

indirect consumption of dry season terrestrial organic material. This pattern has been found in large South American river systems (e.g., Wantzen *et al.*, 2002), as well as in the Tonle Sap system in the Mekong (Campbell *et al.*, Chapter 10).

At the family level, the fish diversity of the Mekong River is exceptional. Valbo-Jorgensen

et al. (this volume) list 87 known families of indigenous fish species from the Mekong. In contrast, Géry (1984) lists only 47 for the Amazon, certainly an underestimate, but at the family level, the Mekong appears to be one of the planet's rivers most diverse in fish. The flood pulse is not the cause of the diversity, but does play a key role in supporting it.

Less well known is the high diversity of gastropod snails in the Mekong. The area around the junction of the Mun and the Mekong is the only area to have been well studied, and 121 species of freshwater gastropod snails have been recorded there, 111 of which are endemic (Groombridge and Jenkins, 2002). Bird diversity is also high (e.g., Campbell *et al.*, 2006) with one species known only from the Mekong, the white-eyed river martin (*Pseudochelidon sirintarae*) now presumed extinct. Species which were thought to have nested in burrows on Mekong River islands are coming under intense pressure as human populations expand and islands are increasingly used for agriculture during the dry season (Dudgeon, 2002; Lekagul and Round, 1991).

Besides the diversity of fish and molluscs, the Mekong is best known for a number of threatened large charismatic species of fish and a dolphin. Much attention has been focused on the giant Mekong catfish (*Pangasianodon gigas*) (Hogan *et al.*, 2004), but in fact the river is the habitat for a number of giant fish species, all of which are becoming disturbingly rare. These include the giant barb (*Catlocarpio siamensis*), Julien's Golden Carp (*Probarbus jullieni*), and two probarbids (*Probarbus labeamajor* and *Probarbus labeaminor*) (Mattson *et al.*, 2002). The now rare Mekong population of the Irrawaddy dolphin (*Orcaella brevirostris*) is not a distinct species but is probably a distinct genetic population of the more widespread Pacific Ocean species (Beasley *et al.* Chapter 15).

3. HISTORY OF THE MEKONG BASIN

The productive riverine ecological system which supports the unusual Mekong fauna has also supported a succession of human civilizations and empires. The Khmer civilization centered on Angkor extended throughout the present day Cambodian lowlands and a good deal of modern Thailand (Higham, 2001). Angkor city at its peak in the 1400s was probably at least 1000 km^2 but with a reasonably low population density, perhaps about 100 people per hectare in Angkor Thom, the central city area, and a lower density outside this center (Coe, 2003). In area, it was by far the largest low-density preindustrial city known (Evans *et al.*, 2007).

The Mekong also played an important role in most of the other and subsequent empires and states that developed in the region. Luang Prabang, downstream of the junction of the Mekong and the Nam Ou, Vientiane, Champassak, and Phnom Penh are all cities located along the river, which were at various times the administrative centers of states. Today Vientiane and Phnom Penh are capitals of modern-day countries. The river provides a convenient trading and transportation route, while the adjacent floodplain provides abundant, flat, potential farming land that can provide agricultural and fisheries surplus to support the growth of cities.

In the 19th century the apparent riches of the Mekong attracted the attention of the European powers, especially France. The French had begun to establish a colonial presence in what is now Viet Nam from 1862, when they first forced the Vietnamese to cede three provinces to France (Hall, 1981). In 1866, French explorers traveled up the Mekong in part seeking a new trade route to China (Osborne, 1996), and throughout the rest of the century the French expanded their colonial influence through the establishment of protectorates. Between 1893 and 1907, Thailand was forced to accept a series of treaties ceding territory on the east bank of the Mekong and in northwest Cambodia to what became French Indochina (Wyatt, 1991). In the north, areas formerly under the control of the Siamese Government were lost to Britain.

The European powers were expelled from the region by the Japanese during World War II, following which the region was torn by a series of wars of independence as the colonial powers returned and were resisted. These in turn merged into the cold war with the Mekong countries forming part of the "hot" front line, and a series of what were essentially civil wars, with various outside countries supporting favored participants with weapons, supplies, training, and refuge.

Unrest in the Mekong countries, including conflicts between them, was an important factor retarding development in the Mekong region until the 1990s. Following the departure of the western powers and the fall of Saigon in 1975, lower level conflict persisted in Lao PDR and continues to some extent to this day. In addition there have been small scale border conflicts between Lao PDR and Thailand over islands within the Mekong that have been claimed by both countries. There is still no agreement of the precise location of the border—whether it occurs in midstream or closer to one bank. The problem is exacerbated because the river itself, like many large floodplain rivers, slowly moves laterally.

Cambodia suffered under the control of the Khmer Rouge who took power in Phnom Penh in 1975 and were largely displaced by Vietnamese forces in 1979 although combat continued in some areas for at least a further 10 years (Chandler, 1996). Under the Khmer Rouge, at least 2 million Cambodians died and Cambodia declined from having one of the highest levels of literacy in southeast Asia to being among the least literate. Many of the best-educated Cambodians, including most of the teachers and academics, were killed or fled the country during the Khmer Rouge period.

Past conflicts between the Mekong countries have left a residual suspicion that will need to be overcome if they are to cooperate fully. The Thai government supported the US during the American wars in Viet Nam, Laos, and Cambodia, and since that time there have been smaller scale conflicts over border demarcation and religious sites between the Thais and the Laos and the Thais and the Cambodians. Similarly, there have been disagreements between the Cambodians and Vietnamese over border regions and the Vietnamese minority living in Cambodia. Incitement of anti-Vietnamese and anti-Thai feelings have been common election strategies in Cambodia.

The Mekong region continues to be of great political interest. China, a superpower and one of the world's most significant and fastest growing economies, is obviously a major influence in the region politically, economically and, since the Mekong arises in Chinese territory, hydrologically. China is a major trading partner for all the lower Mekong countries, and many of their citizens have Chinese ancestry, which creates strong cultural links as well. Chinese investment in the lower Mekong countries is substantial and has been growing rapidly as the Chinese economy grows (e.g., Bangkok Post, 2007). Clearly, the political future of the Mekong region will be linked more and more closely to China.

The Mekong basin is home to some of the world's poorest people. Much of the rapid economic development in China has occurred around the cities of the east and along the coast, resulting in a widening of the economic disparity between the poor agricultural western provinces and the comparatively wealthy cities such as Beijing and Shanghai. Imperatives which include the demand for increased electricity in the east, and the need to increase economic opportunities in the west, and the physical suitability of the southwestern rivers for hydropower development have provided the incentives for China to propose a number of massive hydropower schemes on rivers including the Mekong or Lancang, the Nu, and the Jinsha or Yangtze (Dore and Xiaogang, 2004).

Rapid population growth and poverty in the lower Mekong countries have made development a high priority, partly as a poverty-alleviation strategy. Both Laos and Cambodia have per capita GDP levels below USD $2000 (measured as purchasing power parity) (CIA, 2008). Viet Nam and Thailand are wealthier, Viet Nam having a per capita GDP of USD $2000 and Thailand USD $8000, but the area of Thailand within the Mekong basin, the northeast area of Isaan, is a poor area of the country with incomes substantially lower than is the case in the central regions.

Thus, there has been a strong focus by various national and multinational development agencies on development within the Mekong region. The World Bank, Asian Development Bank, and UNDP have substantial programs within the Mekong region, as do a number of national development aid agencies including, most notably, those from Denmark, Sweden, Australia, and Germany.

In turn, there is an active debate within the region, often involving outside players, about the appropriate trajectory for development for the Mekong basin. Rollason (1996) raised concerns that the development models being pursued would fail to deliver benefits to minorities and marginalized poor communities. Öjendal (2000) contrasted a "mainstream" development approach involving large-scale hydropower and irrigation schemes with an "alternative" small scale approach defending traditional livelihoods.

4. MANAGING THE MEKONG

The Mekong is an international transboundary river with a relatively long, modern history of attempts at basinwide, or at least large scale, management. Hinton (1996) questioned whether large-scale river management of the sort advocated or practiced in developed, politically stable countries using sophisticated modeling

approaches was applicable in regions like the Mekong. There have been several summary histories of development in the Mekong produced from various points of view (Browder and Ortolano, 2000; Hirsch and Jensen, 2006; Mekong Secretariat, 1989).

Shortly after the end of World War II, the United Nations established the Economic Commission for Asia and the Far East (ECAFE) to focus on development issues within the Asian region generally. ECAFE established a Bureau of Flood Control in 1949, and in 1951 requested the Bureau to conduct a study on international rivers. The Bureau responded by suggesting a study of the lower Mekong, resulting in an 18-page report produced in 1952, the conclusion of which called for more intensive studies. Following the signing of the Geneva accords in 1954, the US Bureau of Reclamation conducted a reconnaissance study of the river in 1955 and recommended further studies to establish a more extensive understanding of the river as a basis for planning future development (Mekong Secretariat, 1989).

In 1956 a team of seven experts recruited by ECAFE conducted a further reconnaissance to establish the potential of the basin for hydropower irrigation and flood control. Their report, produced in 1957, advocated an international approach to the development of the lower Mekong River system with the four riparian countries cooperating. It recommended the establishment of an international clearing agency, standing committee, or commission to promote the exchange of information and plans and coordination of projects.

ECAFE accepted the report of the expert team and established a Committee for Coordination of Investigations of the Lower Mekong Basin, often referred to as the Mekong Committee, in September 1957. The committee was a development organization that saw its role as developing plans to harness the resources of the Mekong, and mobilizing capital to implement them.

At the preparatory meeting which established the Committee, the UN Technical Assistance Administration was requested to provide a mission to review the existing Mekong development studies. This was conducted under the leadership of Raymond Wheeler, a retired Lieutenant General from the US Army Corps of Engineers, and has become known as the-Wheeler mission. Much of this early work on the Mekong appears to have been strongly influenced by the Tennessee Valley Authority model of river basin development (Hirsch and Jensen, 2006) developed within the US, although not widely adopted there (Campbell, 2007a).

In April and May 1956, a team organized by ECAFE investigated the potential of the basin for hydropower, irrigation, and flood control (Mekong Secretariat, 1989). The team identified five potential dam sites at Pa Mong and Khemerat between Laos and Thailand, at Khone Falls in Laos, and at Sambor and Tonle Sap in Cambodia.

The Mekong Committee reviewed the proposals of the ECAFE team and developed a 5-year program of investigations to provide a basis for development (Mekong Secretariat, 1989). In 1970 they developed the "1970 Indicative Basin Plan," identifying 180 possible projects which would promote basinwide development, including a cascade of seven mainstream dams. However, a review by the World Bank concluded that the objectives of the plan were "ambitious, unrealistic, and inconsistent with the needs of the countries" (Kirmani and Le Moigne, 1997, p. 8), and suggested that a number of pilot projects be developed instead. None of the main- stream projects proposed by the Mekong Committee has been constructed although the committee did play a role in assisting several projects to be developed on tributaries.

Geopolitical events overtook the Mekong Committee in 1975 with the previously mentioned wars impacting Laos, Viet Nam, and Cambodia, and in 1976 three member countries failed to appoint representatives to the committee. When Laos and Viet Nam did reappoint members in 1977, the committee continued without Cambodian participation as the Interim Mekong Committee.

The Interim Mekong Committee operated until 1995, and successfully initiated a number of activities. Basinwide water quality monitoring was initiated in 1985 (but without sites in Cambodia), and a revised Indicative Basin Plan was released in 1987. The new indicative plan identified 29 projects, including a cascade of eight mainstream dams, the Pak Mun dam in Thailand and the Yali Falls dam in Viet Nam. The mainstream dams have not been constructed but Pak Mun and Yali Falls have been built, and both have been controversial (Fisheries Office, 2000; Roberts, 1993).

In 1992, with Cambodia wishing to rejoin the Mekong Committee, the four member countries negotiated a new agreement with a different emphasis. The 1995 Agreement (MRC, undated, 2000) was now an agreement to cooperate on *sustainable* development, ..., management, and conservation of the water and related resources of the Mekong Basin (Article 1) (my emphasis). Other articles dealt with environmental protection (Article 3), equitable utilization of the water (Article 5), maintenance of mainstream flows (Article 6), and prevention of harmful effects (Article 7) (Table 1.1).

The 1995 agreement established the Mekong River Commission as it now exists. The structure consists of a ministerial council on which the four member countries are represented, and which meets annually. Below that is the Joint Committee of senior bureaucrats meeting three times a year, which is effectively the board of the organization. A secretariat headed by a Chief Executive Officer, with four riparian directors, provides technical and logistics support to the Council and Joint Committee.

The new agreement established a river basin organization with both development and basin management roles explicitly included. However,

TABLE 1.1 List of the articles included under Chapter 3 (Objectives and Principles of Cooperation) of the Agreement on the Cooperation for the Sustainable Development of the Mekong River Basin, 5 April 1995

Article number	Topic
1	Areas of cooperation
2	Projects, programs, and planning
3	Protection of the environment and ecological balance (from pollution or harmful effects of development)
4	Sovereign equality and territorial integrity
5	Reasonable and equitable utilization (of the waters)
6	Maintenance of flows on the mainstream
7	Prevention and cessation of harmful effects
8	State responsibility for damages
9	Freedom of navigation
10	Emergency situations

as several authors have noted previously (e.g., see Varis *et al.*, 2008), the two roles are not compatible. An organization promoting development is not accepted as an "honest broker" when providing advice on basin management, and an honest basin manager is liable to be criticized by the development lobby for insufficiently promoting development projects. The MRC has resolved this contradiction between the two roles by swinging from one to the other, identifying itself as a basin management organization under some chief executive officers, and as a development agency under others.

Joern Kristensen, CEO from 2000 to 2003, identified the MRC's mission as ensuring the sustainable management of water and related resources (Kristensen, 2003). Under his leadership the organization produced and released a series of products intended to provide information to decision makers and civil society in the basin. These included the first State of the Basin Report (MRC, 2003a), the social atlas (Hook *et al.*, 2003), and the river awareness kit (MRC, 2003b).

In contrast, Olivier Cogels, CEO from 2004 to 2007, saw the Commission as taking on a development role that he believed was a new direction for the organization (Cogels, 2005). He believed that the MRC was an organization that was too "green," more interested in environment protection than economic development (Cogels, 2006) which was an orientation that he wanted to change. He stated that the negative impacts of dams were often exaggerated, and that, although there may be concerns about their impacts on fish migration and the flow regime, they are beneficial by increasing the flow of water in the dry season and mitigating the devastating effects of floods (Cogels, 2005, 2007).

Consistent with Cogels' vision of the MRC as a development agency, the mission statement of the organization was changed, deleting reference to the organization "providing scientific information and policy advice." A proposal to include a commitment to the impartial development of data, information, and knowledge along with the free and open distribution of that data, information, and knowledge in the Information and Knowledge Management Programme was rejected by Dr Cogels, who felt that the organization should only release information which supported the views of the council (Clews, personal communication). Under his leadership there was a rapid loss of technical staff from the MRC secretariat, the flow of information products declined, and MRC studies, for example, that failed to demonstrate benefits of large-scale water resources developments, or identified environmental problems in the basin, were suppressed. Examples included reports of an economic analysis of water resources development and a basin environment report card. Dr Cogels left the MRC in 2007, at the conclusion of his 3-year contract.

In fact, historically, the MRC and its predecessor committees have not been successful as development agencies. Few of the projects included in the various basin plans have been implemented. Most of those that have been implemented were done so largely by national agencies, and a number of them have failed to provide the benefits predicted. Moreover, when they have been implemented they have often come at large, unpredicted, social, and environmental cost. Yali Falls and Pak Mun are the most obvious, but by no means the only, examples.

The problems of the Mekong Basin are rapidly becoming more urgent. Osborne (2004) described it as a river at risk, and Varis *et al.* (2008) described it as being at the crossroads. The increasing demand for electricity in the downstream countries and China is increasing the pressure for the rapid development of hydropower projects in China, down the Anamite Range in Laos and Viet Nam, and even on the mainstream of the lower river.

The next decade will see growing pressures within and between the riparian states of the Mekong. China, Thailand, and Cambodia have already seen the growth of a number of local activist groups with interests in Mekong River management, and there are international groups with conservation concerns about the river, such as the World Wide Fund for Nature (WWF) and International Union for the Conservation of Nature (IUCN), active in those three countries as well as Laos and Viet Nam. As changes in the river increasingly affect farmers and fishers, there is potential for increasing conflict between interest groups within each country and between people living in different geographical parts of a particular country.

There is also the potential for increasing conflict between countries, especially between upstream and downstream countries. The discussions that have taken place so far have essentially been limited to bilateral discussions between governments. The Chinese Government is pursuing hydropower development without consideration of the impacts on downstream countries as several authors have pointed out (Dore and Xiaogang, 2004; Hirsch and Jensen, 2006). Downstream governments are wary of offending the Chinese by protesting these actions. However, during the 2004 drought it was evident that the people in downstream countries are quick to blame upstream countries, when they perceive that the river and their livelihoods are under threat. Low flows in Thailand, Laos, and Cambodia on that occasion were not caused by the operation of dams in China (Campbell and Manusthiparom, 2004) but there were numerous stories in the print media and on television incorrectly attributing the problems to the Chinese, and there was widespread public anger at China (e.g., see Osborne, 2004). How much more strongly will public anger in these countries grow when there are real grounds for attributing environmental impacts to Chinese activities upstream?

This book is intended to provide mainly technical, background information to inform the debate about the future of the Mekong. Information about the biophysical systems in the Mekong Basin has been diffuse and often inaccessible. Sometimes consultants' reports were not released, but more often they were simply not widely distributed. Often information was widely scattered with no synthesis or overview available, and occasionally data were available but had not been analyzed.

Those making decisions about the future of the Mekong must base their decisions on the information available, and their own perceptions about the basin and how it will be affected by future developments. Besides the information available never being complete, the perceptions of those making decisions may be at variance with the data that are available, a situation which has been documented for the Mekong several times recently (Campbell, 2007b; Kummu *et al.*, 2008). It is my hope that this book will make a useful contribution to addressing that problem.

References

Bangkok Post. (2007). China top Laos investor last year. Bangkok Post 3 October 2007. <http://www.bangkok-post.com/breaking_news/breakingnews.php?id=122253>. Accessed 13 October 08.

Browder, G. and Ortolano, L. (2000). The evolution of an international water resources management regime in the Mekong River Basin. *Natural Resources Journal* **40**, 499-532.

Campbell, I. C. (2007a). The management of large rivers: Technical and political challenges. *In* "Large Rivers. Geomorphology and Management," Gupta, A. (ed.), pp 571-585. Wiley, New York, 689pp.

Campbell, I. C. (2007b). Perceptions, data and river management: Lessons from the Mekong River. *Water Resources Research* **43**, W02407.

Campbell, I. C. and Manusthiparom, C. (2004). "Technical Report on Rainfall and Discharge in the Lower Mekong Basin in 2003-2004," Mekong River Commission Secretariat, Vientiane, 7pp.

Campbell, I. C., Poole, C., Giesen, W., and Valbo-Jorgensen, J. (2006). Species diversity and ecology of the Tonle Sap Great Lake, Cambodia. *Aquatic Sciences* **68**, 355-373.

Chandler, D. (1996). "A History of Cambodia," 2nd edition, Silkworm Books, Chiang Mai, Thailand, 288pp.

CIA. (2008). The world fact book. <https://www.cia.gov/library/publications/the-world-factbook>. Accessed 15 August 2008.

Coe, M. D. (2003). "Angkor and the Khmer Civilization," Thames and Hudson, London, 240pp.

Cogels, O. (2005). River commission takes on development role in the Lower Mekong Basin. *Mekong Update and Dialogue* **8**(2), 2-3.

Cogels, O. (2006). Opening address to Asia 2006—The international symposium on water resources and renewable energy development in Asia. <www.mrcmekong.org/MRC_news/speeches/30-nov-06_open_htm>. Accessed 10 January 2006.

Cogels, O. (2007). Dam's impact "often exaggerated"—Cogels. Letter to the Bangkok Post 9 January 2007. <www.newsmekong.org/dams_impact_often_exaggerated_-_cogels>. Accessed 22 March 2007.

Dai A. and Trenberth, K. E. (2002). Estimates of freshwater discharge from continents: Latitudinal and seasonal variations. *Journal of Hydrometeorology* **3**(6), 660-687.

Dore, J. and Xiaogang, Yu. (2004). "Yuunan Hydropower Expansion. Update on China's Energy Industry Reforms and the Nu Lancang and Jinsha Hydropower Dams," Chiang Mai University Social and Environmental Research Unit, and Green Watershed, Chiang Mai, Thailand, 30pp.

Dudgeon, D. (2002b). The most endangered ecosystems in the world? Conservation of riverine biodiversity in Asia. *Verhandlungen Internationale Vereinigung Limnologie* **28**, 59-68.

Evans, D., Pottier, C., Fletcher, R., Hensley, S., Tapley, I., and Milne, A. (2007). A comprehensive archaeological map of the world's largest pre-industrial settlement complex at Angkor, Cambodia. *Proceedings of the National Academy of Sciences* **104**, 14277-14282.

Fisheries Office. (2000). "A Study of the Downstream Impacts of the Yali Falls Dam in the Se San River Basin in Ratanakiri Province, Northeast Cambodia," The Fisheries Office, Ratanakiri Province and the Non-Timber Forest Products Project. Ratanakiri Province, Cambodia, 40pp.

Géry, J. (1984). The fishes of Amazonia. *In* "The Amazon. Limnology and Landscape Ecology of a Mighty Tropical River and its Basin," Sioli, H. (ed.), pp 353-370. Dr W. Junk, Dordrecht, The Netherlands, 763pp.

Groombridge, B. and Jenkins, M. D. (2002). "World Atlas of Biodiversity," University of California Press, Berkeley, USA, 340pp.

Hall, D. G. E. (1981). "A History of South-East Asia," 4th edition, Macmillan Education, London, 1070pp.

Higham, C. (2001). "The Civilization of Angkor," Weidenfeld and Nicholson, London, 192pp.

Hinton, P. (1996). Is it possible to "manage" a river? Reflections from the Mekong. *In* "Development Dilemmas in the Mekong Subregion," Stensholt, B. (ed.), pp 49-56. Monash Asia Institute, Clayton, Australia, 277pp.

Hirsch, P. and Jensen, K. M. (2006). "National Interests and Transboundary Water Governance in the Mekong," Australian Mekong Resource Centre, University of Sydney, Australia, 171pp.

Hogan, Z., Moyle, P., May, B., Zanden, J. V., and Baird, I. (2004). The imperiled giants dof the Mekong. *American Scientist* **92**, 228-237.

Hook, J., Novak, S., and Johnston, R. (2003). "Social Atlas of the Lower Mekong Basin," Mekong River Commission, Phnom Penh, 154pp.

Kirmani, S. and Le Moigne, G. (1997). Fostering riparian cooperation in international river basins. The World Bank at its best in development diplomacy. World Bank Technical Paper 335, World Bank, Washington, DC, 42pp.

Kristensen, J. (2003). Foreword. *In* "Social Atlas of the Lower Mekong Basin," Hook *et al.* (eds.). Mekong River Commission, Phnom Penh, 154pp.

Kummu, M., Keskinen, M., and Varis, O. (eds.). (2008). "Modern Myths of the Mekong," Water & Development Publications TKK-WD-01, Helsinki University of Technology, Finland, 187pp.

Lekagul, B. and Round, P. D. (1991). "A Guide to the Birds of Thailand," Saha Karn Bhaet, Bangkok, 457pp.

Luard, T. (2001). Mekong: Mother of rivers. <http://news.bbc.co.uk/1/hi/world/asia-pacific/1506086.stm> Accessed 15 October 2008.

Mattson, N. S., Kongpheng, B., Naruepon, S., Nguyen, T., and Ouk, V. (2002). Cambodia Mekong giant fish species: On their management and biology. MRC Technical Paper No. 3, Mekong River Commission, Phnom Penh, 29pp.

Mekong Descent Foundation. (undated). Exploring the mother of waters. <www.mekongdescentfoundation.org/mother_of_waters.html>. Accessed 15 October 2008.

Mekong Secretariat. (1989). "The Mekong Committee. A Historical Account (1957-89)," Secretariat of the Interim Committee for Coordination of Investigations of the Lower Mekong Basin, Bangkok, 84pp.

Milliman, J. D. and Meade, R. H. (1983). World-wide delivery of river sediments to the oceans. *Journal of Geology* **91**, 1-21.

MRC. (2000). "Agreement on Cooperation for the Sustainable Development of the Mekong River Basin. 5 April 1995," 3rd reprint, MRC Secretariat, Phnom Penh, 19pp.

MRC. (2003a). "State of the Basin Report 2003," Mekong River Commission, Phnom Penh, 300pp.

MRC. (2003b). "Mekong River Awareness Kit. Interactive Self-Study CD-ROM." Mekong River Commission, Phnom Penh.

MRC. (undated). "1995 Mekong Agreement and Procedural Rules," Mekong River Commission, Vientiane, 54pp.

Öjendal, J. (2000). Sharing the good: Modes of managing water resources in the lower Mekong river basin, Ph.D. dissertation. Göteborg University, Sweden, 325pp.

Osborne, M. (1996). "River Road to China. The Search for the Source of the Mekong, 1866-73," Archipelago Press, Singapore, 248pp.

Osborne, M. (2004). "River at Risk. The Mekong and the Water Politics of China and Southeast Asia," Lowy Institute, Sydney, Australia, 56pp.

Roberts, T. (1993). Just another dammed river? Negative impacts of Pak Mun dam on the fishes of the Mekong Basin. *Natural History Bulletin of the Siam Society* **41**, 105-133.

Rollason, R. (1996). Poverty and participation in the Mekong subregion. *In* "Development Dilemmas in the Mekong Subregion," Stensholt, B. (ed.), pp 163-171. Monash Asia Institute, Clayton, Australia, 277pp.

Shangri La. (2006). The source of the Mekong River, Qinghai, China. <http://www.shangri-la-river-expeditions.com/1stdes/mekong/mekongsource/mekongsource.html>. Accessed 4 August 2008.

Shaochuang, L., Pingli, L., Donghui, L., and Peidong, J. (2007). Pinpointing source of Mekong and measuring its length through analysis of satellite imagery and field investigations. *Geo-Spatial Information Science* **10**, 51-56.

Van der Leeden, F. (1975). "Water Resources of the World," Water Information Center, Plainview, NY, 568pp.

Van der Leeden, F., Troise, F. L., and Todd, D. K. (1990). "The Water Encyclopedia," 2nd edition, Lewis Publishers, Boca Raton, FL, 808pp.

Varis, O., Keskenin, M., and Kummu, M. (2008). Mekong at the Crossroads. *Ambio* **37**, 146-149.

Wantzen, K. M., Machado, F. A., Voss, M., Boriss, H., and Junk, W. J. (2002). Seasonal isotopic changes in fish of the Pantanal wetland, Brazil. *Aquatic Sciences* **64**, 239-251.

Wyatt, D. K. (1991). "Thailand, A Short History," Silkworm Books, Chiang Mai, Thailand, 351pp.

Yamashita, M. (1996). "Mekong: A Journey on the Mother of Waters," Takarajima Books, New York, 223pp.

2

The Geology of the Lower Mekong River

Paul A. Carling

School of Geography, Highfield, University of Southampton, Southampton SO17 1BJ

OUTLINE

1. INTRODUCTION

The purpose of this chapter is to provide a general overview of the tectonic and geological setting of the Mekong River south of the international border between The Peoples' Republic of China to the north and the Lao P.D.R. (Laos) to the south. To the north lies the Upper Mekong Basin within which the river has various local names, but is usually termed the Lancang in China. The division of the two basins is not purely arbitrary or political but is physiographic and useful in the present context, demarking the rapid broadening of the basin from its northern extended curvilinear form as it crosses the Greater Himalaya and debouches from the confines of the great mountain ranges and is joined by numerous and sizable tributaries (Fig. 2.1). However, brief details are provided concerning the tectonic control on the development of the Mekong River within China as this is pertinent to the evolution of the Lower Mekong River.

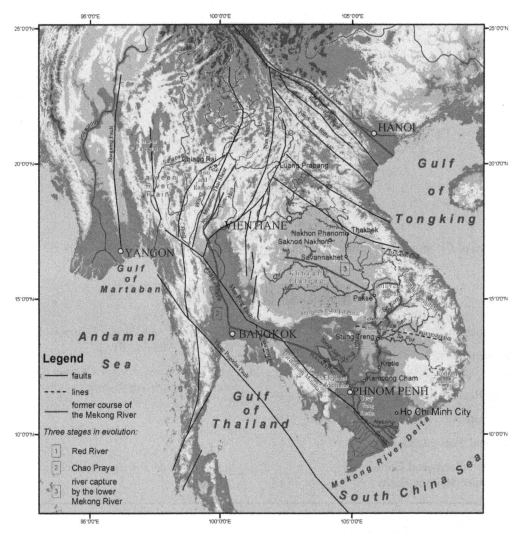

FIGURE 2.1 Simplified tectonic map of Indochina with information drawn from various sources cited in the text but chiefly from Mouret (1994). The Indochina block sits between the South China Block to the north-east and the Central-Sunda Plate to the south-west. The boundaries between these three blocks are, respectively, the Red River Shear and the Danang Line in the north-east and the Rovieng, Pursat, and Sakeo Lines in the south-west. The possible former courses of the Mekong River are discussed in the text.

The geology is significant to an understanding of the nature of the modern Mekong River in many respects. At the regional scale, the movement of cratonic blocks constituting SE Asia conditioned the prior and present courses of the river although unraveling the earlier (largely pre-Tertiary) history of the river is not the focus of this overview. The modern river largely follows a course linking Tertiary basins mediated by Cenozoic tectonism (Brookfield, 1998) and is now set often within autogenic Quaternary and Holocene fluvial sediments. However, unusually for such a large river, much of the course within Thailand and Laos and parts of northern Cambodia is bedrock confined and the course is conditioned by the regional tectonic history. However, the controls on the reach-scale channel form are often authigenic; that is, the immediate local geology of many reaches of the river influences the bank-line character and the nature of the bed, as well as being a major influence on the channel planform and cross section. The usual major gradient and directional changes in the course of the river are tectonically controlled, whereas some rapids and local detail of channel planform are lithologically as well as tectonically controlled. Examples are (i) the spectacular Khone Falls located at the Rovieng terrane boundary that forms the Laos:Cambodian border, and (ii) the course of the Mekong River in Cambodia between Stung Treng and Kratie.

Nevertheless, beyond generalizations from aerial photography, the influence of the local geology on the river planform can only be deduced in a few cases where unpublished reports of potential run-of-river dam sites contain detailed surveys, such as for Pa Mong (north of Vientiane) and for Stung Treng and Sabor (near Kratie, Fig. 2.1) in Cambodia. Within, for example, the Siphandone region of southern Laos, unpublished photogeological maps, related to hydrocarbon exploration, clearly demonstrate that the local structure has major influences on the river gradient, course, and planform (Fig. 2.2).

2. PERSPECTIVE

Detailed study of channel planform patterns (e.g., Twidale, 2004), river profiles (Brookfield, 1998; Seeber and Gornitz, 1983), and the early

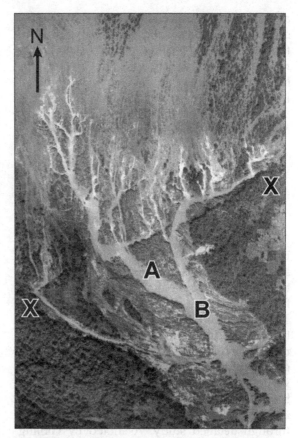

FIGURE 2.2 Vertical view of anabranch of the Mekong River close to Kon Phapheng in Siphandone area, Laos. Structural lineations, aligned from top to center of the image and curving to the right in the lower portion of the image, control the alignment of small-scale rivulets and the avulsive channel within the forest on the far left. Major channels may conform to the lineation (A) or cut through the lineation (B) but are disrupted by a distinct fault line running obliquely across the image (X to X). From air photograph, 1993, 1:15,000. Horizontal field of view ~1.5 km.

Quaternary and older alluvial deposits of a river can shed some light on the tectonic and sea-level adjustments within a given region. For example, rivers are often drawn into subsiding basins (e.g., Gawthorpe and Hurst, 1993) and evidence of preferential lateral channel migration may indicate differential subsidence. Sudden changes in river course alignment and especially "straight" channel sections can indicate the persistent control of fault lines on the channel course (e.g., Gaudemer et al., 1989). Such reflexive relationships between river styles and geology are examined in the thought-provoking review of Gibling (2006) and some examples pertaining to the Mekong River are described below.

3. CLIMATE CONTROLLED CHANGE TO RIVER REGIME

Adding a complexity to the Quaternary interpretation of the interaction of the river and the geology, is the large supply of fluvially transported fine sediment; largely sourced from within the Himalayan massif and the Tibetan Plateau mediated by temporal variations in precipitation, glaciation, evapotranspiration; and thus, river discharge through geological time. The river today seems to carry very little coarse sand and gravel, the load being dominated by fine sands (Conlan, 2008), although this assertion requires verification through detailed study. As noted by Gibling (2006), across the Indian subcontinent, the Late Quaternary Indian monsoon precipitation may have varied from the modern value by at least 30% during periods as short as a few thousands or tens of thousands of years and the same maybe true of the east-Asian monsoon (e.g., Kale et al., 2003; Wang et al., 2005) as the monsoon strengthened from the Early to Late Pleistocene (Zhisheng et al., 1990) and during the Holocene (Porter and Weijian, 2006). Such adjustments are in response to orbital insolation and glacial boundary conditions (Overpeck et al., 1996; Prell and Kutzbach, 1987) as well as longer term tectonic forcing by the rising Himalaya massif (An et al., 2001). Knox (1993) noted that for the well-studied upper Mississippi River, USA, lesser changes in precipitation than the 30% noted above could

induce noticeable changes in river hydrology and changes in sedimentation style (Knox, 2000, 2006). The effects of such changes during the Cenozoic to the monsoon in Indochina on the palaeohydrology of the Mekong River are not known for certain but will be reflected in the sedimentary archives of the floodplains and in the shallow marine environment (Clift et al., 2006a). Goodbred and Kuehl (2000), for example, have claimed an enhanced sediment discharge occurred from the Ganges-Brahmaputra system owing to a strengthened Holocene monsoon. However, for the Mekong, studies at the subepoch time scale are few, and various studies have suggested a fairly stable sediment yield throughout the Neogene (Métivier and Gaudemer, 1999).

4. THE TECTONIC AND GEOLOGICAL DATABASE

There is authoritative and general consistent literature from which the tectonic history of the upper Mekong basin can be adduced. In addition, there is a large literature of specialized papers on the stratigraphic geology of the region but few authoritative and detailed summations that are useful in respect to understanding the history of the Mekong River. The geology of Lower Mekong River Basin has been mapped variously between 1:250,000 and 1:50,000 scales, although in certain regions and districts there is no complete map coverage. Of the five countries bordering the Mekong River in the lower basin, only the geology of Thailand and Vietnam is reasonably well known with Cambodia, Laos, and Myanmar generally being poorly or partially mapped (ESCAP, 1990a,b, 1993, 2001; Fontaine and Workman, 1997; Moores and Fairbridge, 1997; Steinshouer et al., 1997; Workman, 1997a, b,c). Little is known of the geology of Myanmar (Bender, 1983; ESCAP, 1996; Goossens, 1978) and, as the length of the Mekong River along the Myanmar border is short, no further consideration is given to this issue, except to note the presence here of gneisses and granites of unknown age (Moores and Fairbridge, 1997).

Before the 1950s, French geologists predominated in the Basin along with the occasional presence of other Europeans including USSR geologists (Gerasimov, 1964). From 1923 to 1927, Fromaget compiled a 1:500,000 scale geological map of Indochina which he synthesized in 1937 as the 1:2,000,000 scale geological map of Indochina, which was then reedited and published for the second time in 1952 (http://eusoils.jrc.it/esdb_archive/EuDASM/asia/maps/XA2005_GE.htm). At the same time, he compiled a 1:5,000,000 scale geostructural map of Indochina. In 1963, the South Vietnam Department of Geography published a 1:500,000 scale geological map of Indochina with Saurin as the editor, which basically reproduced the results of Fromaget. Subsequent investigations have been local, or have been in respect to mineral prospecting, such that the best regional geology maps available for Laos and Cambodia are still those of Fromaget, and of Steinshouer et al. (1997) at 1:5,000,000. Details of other studies conducted after 1975 of regional rather than local context are summarized by Tran (2000) with a focus on Laos. When comparing different maps and accounts of the regional geology, it should be noted that different authorities may ascribe the geological units within any given area to different epochs or periods and thus some caution is required in interpreting the geological history of the region.

Since 1975, there has been an increasing contribution to geological studies by other nations including riparian nationals. However, as a result of this multifaceted approach, different mapping conventions have been used along with a plethora of "incompatible" terminologies to record stratigraphic successions. The Mekong River is also the international border between Myanmar and Laos and more significantly between Laos and Thailand. The geology mapped by national agencies stops at the borders and maps available for each side of the river are not always compatible (Fontaine, 2007). A detailed summary of Vietnamese geology is available (Tong and Vu, 2006) but this consists of descriptions of formations with few supporting maps and diagrams, such that local knowledge is required to use this volume. Although it might be surmised that the Mekong would have contributed to sedimentation within the Cenozoic, and within the Quaternary period especially (a period particularly relevant to readers of this chapter), often the terrestrial Quaternary is the least-well documented period within the region (Thiramongkol, 1989) with offshore Cenozoic sedimentation being better documented (e.g., Clift et al., 2003). Despite these provisos, it is possible to use existing geological maps and literature to outline the geological context of the Mekong River and a few, more detailed accounts, are available for some reaches of the river especially with respect to the Quaternary and Holocene history of the delta (e.g., Kitazawa et al., 2006).

5. REGIONAL TECTONIC CONTEXT

The Mekong River today has its source in Tibet and flows in a great arc around the northeastern syntaxis of the Himalayan massif exiting to the south of the mountain belt in the "Three Rivers" region between Myanmar and the Yunnan province of China where the Yangtze, the Mekong, and the Salween rivers are closely aligned; the course of each largely being confined by specific sutures. In SW Yunnan province, the Mekong closely follows a line of faults (Heppe et al., 2007, their figs. 1 and 2) to the east of the late Triassic Litian-Jinsha suture through a complex tectonic region where the individual blocks of India, South

China, and Indochina meet (Moores and Fairbridge, 1997). Today, seismic activity is evident in this area (Socquet and Pubellier, 2005). Intense crustal deformation in this region has resulted in well-defined structural and topographic trends, most notably a strong north-south alignment of the valleys of the great rivers and intervening mountain chains.

To the south of Yunnan, the mountains of northern Laos and Vietnam represent a major shear zone between the South China Plate (to the east of Indochina) and the Indochina plate, which is strongly demarcated in northern Vietnam by the Red River Shear (Fig. 2.1). A further major shear zone, marked out by the Mai Ping and the Three Pagodas Fault Zones, delimits the western extent of the Indochina plate, where it abuts the Sunda Plate (Fig. 2.1). The latter shear zone extends from Myanmar south-east across western mainland Thailand past Bangkok and extends into southern Cambodia. The Indochina plate has been subject to major cross plate compression which results in distinctive cross plate tensional faulting approximately trending perpendicular to the major bounding shear zones.

These faults, although anastomosed and complex, are orientated approximately northeast to north-north-east (e.g., Morley, his fig. 6), are often deep half graben or full graben rifts and define, for example, the characteristic structure of the Basin and Ranges of Thailand south-east of the Shan Plateau. Thus, the general structural pattern of the mountains of northern Vietnam and Laos, as well as much of Thailand and Cambodia exhibits a sigmoidal form (Fig. 2.1), bounded by the Red River shear and the Myanmar-Thai shear, which shape is conditioned by a component of torque to the compression of the Indochina plate with rotation largely active from the Oligocene to Pliocene times (Wood, 1985). This regional and local tectonic pattern affects the modern and ancient course of the Mekong River as is explained below.

6. REGIONAL GEOLOGICAL CONTEXT OF THE LOWER MEKONG BASIN

The geological evolution of the region is summarized by Hutchinson (1989). Additional information is drawn from Fontaine and Workman (1978), the summaries of the geology of the individual nations of SE Asia are provided by Workman (in Moores and Fairbridge, 1997), the ESCAP reports, and the pre-Mesozoic history is summarized by Liere and McNeely (2005). The generalized distribution of rocks associated with the geological periods and the boundaries of the main geological provinces are shown in Fig. 2.3.

The Lower Mekong River traverses the Indosinia cratonic block which has been relatively stable since the Jurassic period. Extensive areas were covered by an inland sea during the Upper Mesozoic which accounts for much of the red-bed sandstones and evaporates in the region. Vigorous uplift and igneous activity during the late Cretaceous and early Tertiary (Paleogene) in northern Vietnam developed the distinctive Red River rift along the northeast margin of the Fan Si Pan range of hills, which rift is now filled with ca. 1000 m of Tertiary and Quaternary sediments. At the same time, Indosinia was warped into broad domes and basins along southeasterly tending alignments with minor secondary axes at right angles. Northeast Thailand and adjacent parts of Laos (especially the Khorat Plateau) remained low lying, such that the Mesozoic succession has been preserved in the Mun-Chi and Sakhon Nakhon basins, with deposits of salts that are 250 m thick. In contrast, throughout much of Cambodia, the Mesozoic sandstones were gradually domed up during the Tertiary and, as a result, have been mainly leveled by erosion, causing intrusions of the basement complex to become exposed as granite outcrops. In northern Thailand, block

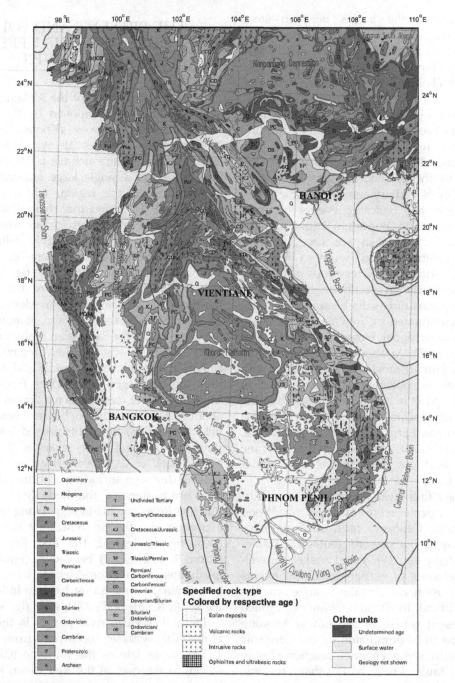

FIGURE 2.3 Geological map of Indochina. Boundaries of the main geological provinces are shown as red lines. Scale: 1:5,000,000. Redrawn from Steinshouer *et al.* (1997).

faulting formed flat basins between steep mountain ranges (the Basin and Range structure noted above) which later filled with Quaternary alluvial sediments including possibly proto-Mekong River deposits as described below.

The Cenozoic, especially the Quaternary period, has been most important in forming the present fluvial landforms and establishing the Mekong River in its present course mediated by the presence of older geological units notably of Tertiary age. During the Cenozoic, tectonic and volcanic events occurred in the region, and important rapid climatic changes and changes in sea level, affected southern Cambodia and southern Vietnam especially. Although Cretaceous volcanics occur extensively in southern Laos, widespread eruptions of basalts began in the Miocene and culminated in the Quaternary being related to extensional tectonics (Moores and Fairbridge, 1997). At about 600,000 years ago, vulcanism formed most of the basalt areas in the Annamites, northeast Thailand, and in Cambodia and it is outcrops of these basaltic rocks which influence the course of the Mekong in southern Laos and in Cambodia, and in part constitute the Rovieng terrane boundary and give rise to the spectacular Khone Falls.

The major fault systems of the region during the Cenozoic were mapped by Wood (1985). These Cenozoic fault motions have largely shaped the present geological framework (Hutchinson, 1989) and make unraveling the influence of earlier geological events and plate motions on the course of the Mekong difficult to elucidate. However, much of the present course of the Lower Mekong River seems to be dominated by the Cenozoic motions with some channel reaches closely following fault lines and other tectonic lineaments.

7. THE PRESENT AND FORMER COURSES OF THE UPPER MEKONG RIVER

Although the course of the Mekong River in Yunnan today is conditioned by the tectonics of the eastern syntaxis (Koons, 1995), the Mekong River has not always had such an extended headwater catchment. Rather, 40 million years ago a single large antecedent river system drained the region, which is known today as SE Tibet and this river flowed south into an ocean equivalent to the South China Sea along a course now roughly delimited by the Red and/or Black Rivers in northern Vietnam (Brookfield, 1998, his fig. 23; Clark et al., 2004, their fig. 7) before Late Tertiary deformation. As Tibet rose in elevation from ca. 50 million years (Ma) ago, it initiated the development of the East Asian monsoon system. Later as Tibet rose further, regional tilting caused rivers to change course and to disconnect the drainage of SE Tibet from that of Indochina. Tibet reached its maximum elevation about 15 Ma (Harris, 2006) but as it reached sufficient elevation, the uplift caused widespread river capture and flow reversals (Brookfield, 1998; Clark et al., 2004) diverting much SE Tibetan drainage, including a portion of the former headwaters of the Red River, into the Yangtse River ca. 20 Ma ago (Wang, 2004) and a portion into the proto-Mekong (Clark et al., 2004, fig. 8) leaving the Red River system (curve 1 in Fig. 2.1) beheaded. Similarly, the headwaters of the proto-Mekong were captured by a drainage system directed into Indochina to the south (Clark et al., 2004, their fig. 3c).

Incision of steep river gorges in the eastern Himalaya syntaxis occurred from around 13 Ma, or as late as the Pliocene (Clift et al., 2006b) in the case of the Red River, such that the present pattern of rivers was established by about 8 Ma (Clift *et al.*, 2003). Through this process, a portion of the former palaeo-river catchment of the proto-Red River became an extended narrow headwater for the Mekong, from the Miocene or later (Clift *et al.*, 2006b).

8. THE PRESENT AND FORMER COURSES OF THE LOWER MEKONG RIVER

8.1. The Modern River: China Border to Vientiane

The course of the Mekong in northern Laos is complex and is directly influenced by the Himalayan orogenic history as well intra-plate warping and faulting. Flowing at first to the south-west between zones of lesser slip-faults the river is constrained within a rugged relief that ranges from 500 to 300 m in elevation. Lacassin *et al.* (1998) have argued that the tectonic evidence shows that the Mekong River in this area has kept nearly the same trace for at least several million years. Relatively straight reaches of the river are linked by hair-pin loops that reflect the control of the local faults with left-lateral offsets of up to 30 km during the Quaternary but with some excursions of the river course out of the active fault zones (Lacassin *et al.*, 1998). For example, immediately north of Chang Rai, the river is a straight bedrock canyon 600 m deep that forms the Laos-Myanmar border. At 20° N, it is diverted abruptly to the east, broadens to an alluvial plain about 4 km wide at the confluence with the tributary Kok River (Fenton *et al.*, 2003; Lacassin *et al.*, 1998) where it crosses a tectonic pull-apart basin associated with the Mae Chan fault, before turning to the southeast again between Luang Prabang and Vientiane where the zig-zag course of "straight" canyon-like reaches is strongly influenced by the major Lay Chau fault, and the alignment of the Late Triassic-Early Jurassic, Nan (or

Uttaradit-Luang Prabang) suture. In the vicinity of Luang Prabang and north-west of Vientiane, the river traverses a complex of rock types (the Truong Son Fold Belt: Fig. 2.3), including Permian andesites, dacites, and some basalts as well as Devonian shales, including sandy and calcareous shales and limestones. The presence of curvilinear Quaternary deposits immediately to the west and north-west of Vientiane (Fig. 2.3) might readily be associated with the Mekong River. However, the offset of the plotted position of the river in Fig. 2.3 and the location of these Quaternary deposits is not due to incorrect plotting of the river course and requires further investigation.

As is found in the case of other major rivers fronting major orogenic belts (Gupta, 1997; Hovius, 1996), there are few laterally restricted points at which the Mekong River can exit the montane massif in the north and debouche into the piedmont to the south. In this respect, Hutchinson (1989) has argued that, prior to Late Cenozoic faulting, the Mekong originally flowed directly to the south, west of the Cretaceous platform (Fig. 2.3) of the Khorat Plateau, through the Basin and Range region of Thailand, into the northern central plains of Thailand (curve 2 in Fig. 2.1), to join the proto Chao Phraya via a series of small linear Eocene to Miocene basins between 18° and 20° N which contain extensive and thick fluvial sediments of Pleistocene age (Geological Map of Thailand, 1:1,500,000). In each basin, the fluvial infill has no immediately associated major river source. The alignment of these Quaternary fills often accords with a series of larger scale Tertiary basins (grabens) that run down through central Thailand (Chaodumrong *et al.*, 1983; Moores and Fairbridge, 1997; Morley, 2001, fig. 1; Nichols and Uttamo, 2005, fig. 2; Polachan *et al.*, 1991). These basins are mainly intermontane rift basins with graben/half graben geometry, oriented in approximately N-S direction. Most of the basins are located to the west of the Nan

Suture zone. The thick Tertiary sediments (< 2000 m thick) are unconformably overlain by the Quaternary sediments. Workman (1997c) ascribes this Quaternary infill to local erosion induced by major uplift of northern Thailand in the Plio-Pleistocene or Miocene (Fenton *et al.*, 2003), and Sinsakul *et al.* (2002) ascribe these Quaternary sediments specifically to tributaries of the Chao Phraya. However, there may be coalesced and superimposed channel deposits in Thailand related to both the Chao Phraya tributaries and former courses of the Mekong River but this possibility has not been explicitly noted previously. Nichols and Uttamo (2005) consider and dismiss sedimentary evidence for the former presence of a large river flowing north to south through one of these Tertiary basins (the Li Basin) during the mid-late Oligocene to early Miocene. Brookfield (1998) has argued that the most likely former course of the Mekong River was via the Yom River in Thailand, a major tributary of the Chao Phraya River (Fig. 2.1) but that headward extension of a tributary of an ancestral Lower Mekong River from the Khorat plateau resulted in river capture and diversion of the flow from the Yom into the present course of the Mekong. Brookfield (1998) argued that this river capture occurred as recently as the Late Quaternary or still later. Workman (1997b) suggested that the Mekong may have departed its present course from a point upstream of Vientiane, as noted earlier but then flowed across the Khorat Plateau, into southern Laos, close to Pakse. The reasoning is based, in part, on the thin and intermittent presence of Mekong alluvium (generally mapped as Quaternary: Fig. 2.3) for the river course south of Vientiane, some of which might only be of Holocene age, thus the river may not have been in its present course for long.

8.2. The Modern River: Vientiane to Southern Laos

In the vicinity of Vientiane, the Mekong first turns to the north-east and then turns to the south-east in accord with a major fault zone, flowing past the towns of Nakhon Phanom and Mukdahan to Thaket, thus cutting across the eastern side of the so-called Khorat Plateau (Fig. 2.1); the geological Khorat Province (Fig. 2.3). The term "plateau" here is somewhat misleading, as much of the area is only 100-200 m above sea level. The Khorat Plateau is rimmed by cuestas some 600-1000 m above sea level on its western and southern margins but elsewhere it slopes gently towards the Mekong River. The plateau is, in fact, an epicontinental basin, a major Tertiary cratonic basin, the basement of which is predominately Mesozoic sandstones (Fig. 2.3) with localized Quaternary or Holocene floodplain fill associated with the course of the Mekong and two major tributaries: the Chi and Mun Rivers. However, this recent fill is not very extensive laterally, nor thick, throughout the course of the Mekong in Laos. The easterly extension of the Khorat basin east of the Mekong River is called the Sakhon Nakhon Basin, and extends from Vientiane through Savannaket to Pakse, in which latter region it is often called the Savannaket Basin.

Workman (1997b) has suggested that, in southern Laos, the Mekong during the Quaternary flowed east at latitude 16° N, roughly along the course of the present Don River (see curve 3 in Fig. 2.1), toward the town of Saravane, to then flow southward on the eastern side of the Bolovens Plateau following structural alignments and the Kong River to join the present-day Mekong tributary, the Se Kong River, to rejoin the modern course south of the Khone Falls in Cambodia. This supposition is based on the general topography, the mapped substantial deposits of Quaternary river alluvium close to Saravane, and the presumed blocking and diversion of the Mekong by Quaternary basalt flows that occur extensively south of Saravane as far as Champasak (Workman, personal communication, 2007).

The Quaternary deposits of the Mekong are poorly studied. Those deposits within the Nakhon Province of Thailand bordering the Mekong River were investigated by Wongsomsak (1992) and these observations summarized by Sinsakul *et al.* (2002). A high-level terrace 18 m above the modern river consists largely of well-rounded gravel with a sandy-clay matrix and tentatively has been dated using tektites to 600,000-700,000 years BP (Bunopas *et al.* 1999; Sinsakul *et al.*, 2002). A further terrace at 13 m also consists of gravel with admixtures of sand, silt, and clay. A lower elevation at 8 m often accords with the modern Mekong River bank-top and consists of stiff sandy clays with orange and red mottles. Concretions which appear to be iron and/or manganese are widespread below about 1 m depth indicating an oxidizing environment. Laterite soils occur in the upper meter where flooding is infrequent and, where flooding occurs, around 1 m of alluvium is recorded devoid of mottling and concretions. Extensive reaches in Laos are not inundated annually which implies a degree (i.e., several meters) of Late Quaternary-Holocene incision has occurred (see Carling, Chapter 5).

The Cenozoic history of alluviation and incision along the Mekong is not known but must be related to tectonics and, as in the case of the Indian subcontinent (Kale, 2007), it may correlate with Late Pleistocene-Holocene changes in the monsoon intensity (e.g., Bookhagen *et al.*, 2005), although the trends in monsoon rainfall may not be the same for Indian and SE Asian monsoons (e.g., Staubwasser, 2006).

The Cenozoic structural history of southern Laos is not well known but the distribution of Quaternary fluvial deposits and the drainage pattern indicate that epeirogenic movements were pronounced in the late Cenozoic (Pliocene-Early Pleistocene?) (Moores and Fairbridge, 1997). In particular, uplift has caused the Khorat basin to be referred to as a plateau owing to it having a mesa-like appearance when viewed from the south and west due to strongly up-tilted western and southern margins (Hutchinson, 1989). To the south it is delimited by the Dangrek escarpment (Fig. 2.1), which cuesta approximately defines the international borders between Thailand, Laos, and Cambodia in this region. The Khorat plateau lies entirely within the drainage basin of the Mekong. In southern Laos, the river demarcates the western flank of the Savannaket Mesozoic basin and skirts the western flank of Neogene and Pleistocene basalts of the Bolovens Plateau. South of Champasak, the river closely follows the western limit of Early Cretaceous volcanics before cutting through Triassic and Jurassic volcanic rocks which constrain the complex zone of anastomosed bedrock-confined channels of the Siphandone area on the southern Laos border.

8.3. The Modern River: Cambodia to the Delta

Leaving the Khorat Plateau the Mekong plunges over the Dangrek escarpment, which here consists of Mesozoic rhyolites, tuffs, and sandstones (Brambati and Carulli, 2001) and more local Quaternary basalts at the Khone Falls, into Cambodia. The Mekong south of Khone Falls, and near Stung Treng is constrained at first primarily by Mesozoic and some Eocene bedrock as well as Lower and Upper Paleozoic rocks. The former consists of sandstones and arenaceous slates but the river passes Carboniferous rocks and local outcrops of andesites close to the Sekong confluence. Cambrian quartz slates have also been recorded here (Geofiska, 1968). Thus here, the river displays bedrock-confined channel features south of Stung Treng and an associated narrow band of Lower Quaternary river-fill, 4-5 km wide, is about 10-20 m thick. To the south, near Kratie, the river has a more pronounced alluvial character flowing across a broader zone of Quaternary alluvium (ca. 12 km wide) that overlies Lower to Middle Jurassic and Triassic continental red beds ("terraine rouge") which consist predominately of calcareous shales and sands. Massifs of continental Pliocene-Pleistocene basalts occur just south of the town of Kratie but only locally impinge on the channel. The river marks its entry into the Prey Veng Basin of late Quaternary fill in southern Cambodia by executing a sharp turn to the west, after which it follows a straight course for some 25 km that is consistent with the strike of a conjectured major local "fault" that is a eastward extension of the regional Chao Praya strike-slip fault system noted below. This local "fault" might equally be related to the local disruption of the Quaternary fill by the injection of the Pliocene-Pleistocene basalts noted earlier which crop-out between Kratie and Kampong Cham. Nonetheless, in similar manner, the river executes sharp turns at Kompong Cham and at Phnom Penh that may be controlled by mapped suspected fault alignments buried by alluvium. From ca. 20 km south of Kratie there is a laterally extensive Quaternary fill associated with the Prey Veng Basin and the Tonle-Sap (Phnom Penh) basin (Figs. 2.1 and 2.3). These thick, extensive deposits are poorly studied (Carbonnel, 1965) but include Quaternary river terraces. Many of the surfaces are modern floodplains, up to 100 km in lateral extent, that are inundated annually and are ornamented by numerous nonincised or little-incised palaeochannels and scrollbars which reflect repeated avulsion and lateral channel shifts of the Mekong River which in this region is often anastomosed and anabranching (*sensu* Carling, Chapter 5). Thus, the geomorphological evidence here indicates a long Quaternary history of persistent anastomosed channel behavior with vertical accretion and the deposition of laterally amalgamated fluvial segments predominating into the Holocene until today. Such anastomosed behavior is often the response of a large river close to the coast where rapid sea-level rise has occurred (Aslan and Autin, 1999; Törnqvist *et al.*, 1993) but in this case, the system is too far inland to have been affected by base-level changes associated with Quaternary sea-level changes. Rather the anastomosed behavior may be conditioned by the extremely low river gradient and the complex recent history of subsidence to the west of the river, within the Tonle Sap Basin (Carling, Chapter 5).

Although a number of investigators have suggested that the Tonlé Sap lake basin might be an asteroid impact crater (e.g., Son and Koeberl, 2005), traditionally the lake has been associated with subsidence and the development of the Quaternary basin as influenced by the alignment of the Tonle Sap-Mekong shear zone, along the major Chao Praya strike-slip fault system and the Pursat Line, demarking the boundary between the Indochina plate and the Central-Sunda subplate which runs north-west to south-east influencing the alignment of the Tonle Sap lake and the eponymous river. The subsidence necessary to form the lake was only slight and may have occurred as recently as the Holocene as the lake is only around 5000 years old (Carbonnel, 1965). A slight bedrock ridge near the town of Kampong Cham prevents complete drainage of the lake. Workman (1977) and Hutchinson (1989) believe that a Tertiary basin (mapped in Tong and Vu, 2006) some times known as the Siagon-Vung Tau Basin (Fontaine and Workman, 1997) may be concealed by the Quaternary deposits and that the Tonlé-Sap basin may have provided a former course for the lower proto-Mekong (curve 2 in Fig. 2.1) from the Chao Phraya Basin (Bangkok) in Thailand to the south-east until Late Cenozoic diversion. This scenario indicates that the lower proto-Mekong, when still flowing south through Thailand, might have been drawn eastward into a more rapidly subsiding extensional basin in southern Cambodia. Extensional tectonics along the Chao Phraya fault also are responsible for the basalt province noted earlier and possibly for the numerous small igneous outcrops that punctuate the Prey Veng Quaternary infill to the west of the Mekong River from south of Kratie to the coast south of Phom Penh.

The lower parts of the river underwent changes associated with fluctuating Pleistocene and Holocene sea levels (Hanebuth *et al.*, 2000) but the effect of base-level change extends upstream only a short distance and largely influenced deltaic processes. The delta is situated in a horst-graben that has a trend parallel to the regional NW to SE structural trend but imposed on this tendency are NE to SE swells and faults with lateral throws up to 1-2 km which result in a series of minor basins and blocks (Xang, 1998). This structure influences the local topography, most noticeably constraining the straight course of the Bassac River which follows a boundary between a horst and a graben (Takaya, 1974; Xang, 1998). The sediments of the delta are generally Eocene to Pleistocene, with a Holocene carapace, and the total fill above the basement has been reported from boreholes to vary between ca. 800 and 200 m, although recent unpublished aeromagnetic and aerogravity surveys now include a 3.6 km thick sedimentary section that includes the basement. The basement is exposed, in the north-west of the delta, as isolated granite and limestone hills which are extensions of the Cardamom and Elephant ranges of SW Cambodia. The boundary between the Holocene and the Quaternary is shallower to the north of the delta such that the Pleistocene crops out in the north, through the Holocene units, as river terraces, which delimit the northerly extent of the Holocene deposits of the delta (Morgan, 1970). The evolution of the delta during the Holocene is considered by Nguyen *et al.* (1997, 2005; Ta *et al.*, 2005) and summarized by Hashimoto (2001). Finally, the Mekong debouches into the Vung Tau graben, offshore of the delta. Beyond this, an extensive palaeo-river system exists offshore on the Sunda Shelf associated with former lower sea levels (Voris, 2000).

ACKNOWLEDGMENT

Thanks are extended to Lyubov Meshkova who drew Figs. 2.1 and 2.3.

References

An, Z. S., Kutzbach, J. E., Prell, W. L., and Porter, S. C. (2001). Evolution of Asian monsoon and phased uplift of the Himalaya-Tibetan Plateau since late Miocene times. *Nature* **411**, 62-66.

Aslan, A. and Autin, W. J. (1999). Evolution of the Holocene Mississippi River floodplain, Ferriday, Louisiana: Insights on the origin of fine-grained floodplains. *Journal of Sedimentary Research* **69**, 800-815.

Bender, F. K. (1983). "Geology of Burma," Gebruder Borntraeger Verlagsbuchhandlung, Berlin-Stuttgart, 293pp.

Bookhagen, B., Thiede, R. C., and Strecker, M. R. (2005). Late Quaternary intensified monsoon phases control landscape evolution in the northwest Himalaya. *Geology* **33**, 149-152.

Brambati, A. and Carulli, G. B. (2001). Geology, geomorphology and hydrogeology. *In* "Siphandone Wetlands," Daconto, G. (ed.), pp 24-38. CESVI, Cooperation and Development, Bergamo, Italy, 192pp.

Brookfield, M. E. (1998). The evolution of the great river systems of southern Asia during the Cenozoic India-Asia coillison: Rivers draining southwards. *Geomorphology* **22**, 285-312.

Bunopas, S., Wasson, J. T., Vella, P., Fontaine, H., Hada, S. H., Burrett, C., Suphajunpa, T., and Khositanont, S. (1999). Catastrophic loess, mass mortality and forest fires suggest that a Pleistocene cometary impact in Thailand caused the Australarian Tektite field. *Journal of the Geological Society of Thailand* **1**, 1-17.

Carbonnel, J. P. (1965). Essai d'interpretation morphotectonique de la cuvette cambodgienne. *Revue de Géologie Dynamique et Geographie Physique* 7(3), 277-281.

Chaodumrong, P., Ukakimapan, Y., Snansieng, S., Janmaha, S., Pradidtan, S., and Sae Leow, N. (1983). A review of the Tertiary sedimentary rocks of Thailand. *In* "Proceedings of the Workshop on Stratigraphic Correlation of Thailand and Malaysia," Haad Yai, Thailand, September 8-10, Nutalaya, P. (ed.) Chiang Mai University, pp 159-187.

Clark, M. K., Schoenbohm, L. M., Royden, L. H., Whipple, K. X., Burchfield, B. C., Zhang, X., Tang, W., Wang, E., and Chen, L. (2004). Surface uplift, tectonics, and erosion of eastern Tibet from large-scale drainage patterns. *Tectonics* 23, TC1006, doi:10.1029/2002TC001402.

Clift, P. D., Wang, P., Kuhnt, W., Hall, R., and Tada, R. (2003). Continental-ocean interactions within the East Asian marginal seas. *Eos* 84(15), 139-141.

Clift, P. D., Carter, A., Cambell, I. H., Pringle, M. S., Nguyen, V. L., Allen, C. M., Hodges, K. V., and Mai, T. T. (2006a). Thermochronology of mineral grains in the Red and Mekong Rivers, Vietnam: Provenance and exhumation implications for Southeast Asia. *Geochemistry, Geophysics and Geosystems* 7, Q10005, doi:10.1029/2006GC001336.

Clift, P. D., Blusztajn, J., and Nguyen, A. D. (2006b). Large-scale drainage capture and surface uplift in eastern Tibet-SW China before 24 ma inferred from sediments of the Hanoi Basin, Vietnam. *Geophysical Research Letters* 33, L19403, doi:1029/2006GL027772.

Conlan, I. (2008). The geomorphology of deep pools on the lower-Mekong River: Controls on pool spacing and dimensions, processes of pool maintenance and potential future changes to pool morphology. Unpublished report to MRC, 98pp.

ESCAP (1990a). "Atlas of Mineral Resources of the ESCAP Region," Volume 7—Lao PDR, Economic and Social Commission for Asia and the Pacific and Department of Mineral Resources of Thailand, United Nations, New York, 19pp.

ESCAP (1990b). "Atlas of Mineral Resources of the ESCAP Region," Volume 6—Viet Nam, Economic and Social Commission for Asia and the Pacific and Department of Mineral Resources of Thailand, United Nations, New York, 124pp.

ESCAP (1993). "Atlas of Mineral Resources of the ESCAP Region," Volume 10—Cambodia, Economic and Social Commission for Asia and the Pacific and Department of Mineral Resources of Thailand, United Nations, New York, 87pp.

ESCAP (1996). "Atlas of Mineral Resources of the ESCAP Region," Volume 12—Geology and Mineral Resources of Myanmar, Economic and Social Commission for Asia and the Pacific and Department of Mineral Resources of Thailand, United Nations, New York, 193pp.

ESCAP (2001). "Atlas of Mineral Resources of the ESCAP Region," Volume 16—Mineral Resources of Thailand, Economic and Social Commission for Asia and the Pacific and Department of Mineral Resources of Thailand, United Nations, New York, 239pp.

Fenton, C. H., Charusiri, P., and Wood, S. H. (2003). Recent paleoseismic investigations in Northern and Western Thailand. *Annals of Geophysics* 46, 957-981.

Fontaine, H. (2007). Book review. *Journal of Asian Earth Sciences* 29, 183-184.

Fontaine, H. and Workman, D. R. (1978). Review of the geology and mineral resources of Kampuchea, Laos and Vietnam. *In* "Geology and Mineral Resources of Southeast Asia," Nutalaya, P. (ed.), pp 538-603. Asian Institute of Technology, Bangkok.

Fontaine, H. and Workman, D. R. (1997). Vietnam. *In*: "Encyclopedia of European and Asian Regional Geology," Moores, E. M. and Fairbridge, R. W. (eds.), pp 774-782. Chapman & Hall, London, 804pp.

Gaudemer, Y., Tapponnier, P., and Turcotte, D. L. (1989). River offsets across active strike-slip faults. *Annales Tectonicae* 3, 55-76.

Gawthorpe, R. L. and Hurst, J. M. (1993). Transfer zones in extensional basins: Their structural style and influence on drainage development and stratigraphy. *Journal of the Geological Society of London* 150, 1137-1152.

Geofiska (1968). Final report on seismic investigation of Stung Treng damsite, Lower Mekong Basin—Cambodia, Enterprise for Applied Geophysics, Zagreb, Yugoslavia, 6pp.

Gerasimov, I. P. (1964). "Physico-Geographical World Atlas," Scientific Academy of USSR and Cartographic and Geodesic Central Committee, Moscow, 298pp.

Gibling, M. R. (2006). Width and thickness of fluvial channel bodies and valley fills in the geological record: A literature compilation and classification. *Journal of Sedimentary Research* 76, 731-770.

Goodbred, S. L., Jr. and Kuehl, S. A. (2000). Enormous Ganges-Brahmaputra sediment discharge during strengthened early Holocene monsoon. *Geology* 28, 1083-1086.

Goossens, P. J. (1978). Earth sciences bibliography of Burma, Yunnan and Andaman Islands. *In* "Third Regional Conference on Geology and Mineral Resources of Southeast Asia," Bangkok, Thailand, 14-17 November, 1978, Nutalaya, P. (ed.), pp 495-523. Asian Institute of Technology, Bangkok.

Gupta, S. (1997). Himalayan drainage patterns and the origin of fluvial megafans in the Ganges foreland basin. *Geology* 25, 11-14.

Hanebuth, T., Stattegger, K., and Grootes, P. M. (2000). Rapid flooding of the Sunda Shelf: A late-glacial sea-level record. *Science* 288, 1033-1035.

Harris, N. B. W. (2006). The elevation history of the Tibetan Plateau and its implications for the Asian monsoon.

Palaeogeography, Palaeoclimatology and Palaeoecology **241**, 4-15.

Hashimoto, T. (2001). "Environmental Issues and Recent Infrastructure Development in the Mekong Delta: Review, Analysis and Recommendations with Particular Reference to Large-Scale Water Control Projects and the Development of Coastal Areas." Australian Mekong Resource Centre, Working Paper No. 4. University of Sydney, Australia.

Heppe, K., Helmcke, D., and Wemmer, K. (2007). The Lancang River Zone of southwestern Yunnan, China: A questionable location for the active continental margin of Paleotethys. *Journal of Asian Earth Sciences* **30**, 706-720.

Hovius, N. (1996). Regular spacing of drainage outlets from linear mountain belts. *Basin Research* **8**, 29-44.

Hutchinson, C. S. (1989). "Geological Evolution of South-East Asia," Oxford Scientific Publications, Oxford, UK, 368pp.

Kale, V. S. (2007). Fluvio-sedimentary response of the monsoon-fed Indian rivers to Late Pleistocene-Holocene changes in monsoon strength: Reconstruction based on existing ^{14}C dates. *Quaternary Science Reviews* **26**, 1610-1620.

Kale, V. S., Gupta, A., and Singhvi, A. K. (2003). Late Pleistocene-Holocene palaeohydrology of monsoon Asia. *In* "Palaeohydrology: Understanding Global Change," Gregory, K. J. and Benito, G. (eds.), pp 213-232. Wiley, Chichester, UK, 396pp.

Kitazawa, T., Nakagawa, T., Hashimoto, T., and Tateishi, M. (2006). Stratigraphy and optically stimulated luminescence (OSL) dating of a Quaternary sequence along the Dong Nai River, southern Vietnam. *Journal of Asian Earth Sciences* **27**(6), 788-804.

Knox, J. C. (1993). Large increases in flood magnitude in response to modest changes in climate. *Nature* **361**, 430-432.

Knox, J. C. (2000). Sensitivity of modern and Holocene floods to climate change. *Quaternary Science Reviews* **19**, 439-457.

Knox, J. C. (2006). Floodplain sedimentation in the Upper Mississippi Valley: Natural versus human accelerated. *Geomorphology* **79**, 286-310.

Koons, P. O. (1995). Modelling the topographic evolution of collisional mountain belts. *Annual Reviews of Earth and Planetary Sciences* **23**, 375-408.

Lacassin, R., Replumaz, A., and Leloup, P. H. (1998). Hairpin river loops and slip-sense inversion on Southeast Asian strike-slip faults. *Geology* **26**, 703-706.

van Liere, W. and McNeely, J. A. (2005). "Agriculture in the Lower Mekong Basin: Experience from the Critical Decade of 1966-1976." IUCN, Gland, Switzerland and Cambridge, UK. doi:10.2305/IUCN.CH.2005.5.en. ISBN: 2-8317-0891-5.

Métivier, F. and Gaudemer, Y. (1999). Stability of output fluxes of large rivers in South and East Asia during the last 2 million years: Implications on floodplain processes. *Basin Research* **11**, 293-303.

Moores, E. M. and Fairbridge, R. W. (1997). "Encyclopedia of European and Asian Regional Geology," Chapman & Hall, London, 804pp.

Morgan, J. P. (1970). Depositional processes and products in the deltaic environment. *In* "Deltaic Sediment: Modern and Ancient," Morgan, J. P. and Shaver, R. H. (eds.), pp 31-47. Society of Economic Paleontologists and Mineralogists Special Publication No. 15, Tulsa, OK.

Morley, C. K. (2001). Combined escape tectonics and subduction rollback-back arc extension: A model for the evolution of Tertiary rift basins in Thailand, Malaysia and Laos. *Journal of the Geological Society, London* **158**, 461-474.

Mouret, C. (1994). Geological history of northeastern Thailand since the Carboniferous, relations with Indochina and Carboniferous to Early Cenozoic evolution model. *In* "Proceedings of the International Symposium on Stratigraphic Correlation of Southeast Asia," Angsuwathana, P. (ed.), pp 132-158. Department of Mineral Resources, Bangkok, Thailand.

Nguyen, V. L., Ta, T. K. O., and Tateishi, M. (1997). Late Holocene depositional environments and coastal evolution of the Mekong River Delta, southern Vietnam. *In* "Program and Abstracts, International Symposium on Quaternary Environmental Change in the Asia and Western Pacific Region," October 14-17, University of Tokyo, Japan, Yonekura, N. and Kayanne, H. (eds.), p 68. M&J International, Yokohama, Japan.

Nguyen, V. L., Ta, T. K. O., Tateishi, M., Kobayashi, I., Umitsu, M., and Saito, Y. (2005). Late Quaternary depositional sequences in the Mekong River Delta, Vietnam. *In* "Mega-Deltas of Asia: Geological Evolution and Human Impact," Chen, Z. Y., Saito, Y., and Goodbred, S. L., Jr. (eds.), pp 121-127. China Ocean Press, Beijing.

Nichols, G. and Uttamo, W. (2005). Sedimentation in a humid, interior, extensional basin: The Cenozoic Li Basin, northern Thailand. *Journal of the Geological Society, London* **162**, 333-347.

Overpeck, J., Anderson, D., Trumbore, S., and Prell, W. (1996). The Southwest Indian monsoon over the last 18000 years. *Climate Dynamics* **12**, 213-225.

Polachan, S., Pradidtan, S., Tongtaow, C., Janmaha, S., Intarawijitr, K., and Sangsuwan, C. (1991). Development of Cenozoic basins in Thailand. *Marine and Petroleum Geology* **8**, 85-97.

Porter, S. C. and Weijian, Z. (2006). Synchronism of Holocene East Asian monsoon variations and North Atlantic drift-ice tracers. *Quaternary Research* **65**, 443-449.

Prell, W. L. and Kutzbach, J. E. (1987). Monsoon variability over the past 150,000 years. *Journal of Geophysical Research* **92**, 8411-8425.

Seeber, L. and Gornitz, V. (1983). River profiles along the Himalayan arc as indicators of active tectonics. *Tectonophysics* **92**, 335-367.

Sinsakul, S., Chaimane, N., and Tiyapairach, S. (2002). Quaternary geology of Thailand. *In* "Proceedings of the Symposium on Geology of Thailand," 26-31 August 2002, Bangkok, Thailand. http://www.dmr.go.th/dmr_data/DMR_eng/geology/queternary/queternary.htm. Accessed on March 2009.

Socquet, A. and Pubellier, M. (2005). Cenozoic deformation in western Yunnan (China-Myanmar border). *Journal of Asian Earth Sciences* **24**, 495-515.

Son, T. H. and Koeberl, C. (2005). Chemical variation within fragments of Australasian tektites. *Meteoritics & Planetary Science* **40**(6), 805-815.

Staubwasser, M. F. (2006). An overview of Holocene south Asian monsoon records—Monsoon domains and regional contrasts. *Journal of the Geological Society of India* **68**, 433-446.

Steinshouer, D. W., Qiang, J., McCabe, P. J. M., and Ryder, R. T. (1997). Maps showing geology, oil and gas fields, and geological provinces of the Asia Pacific Region. U.S. Geological Survey Open-File Report 97-470F.

Ta, T. K. O., Nguyen, V. L., Tateishi, M., Kobayashi, I., and Saito, Y. (2005). Sediment facies change and delta evolution during Holocene in the Mekong River Delta, Vietnam. *In* "Mega-Deltas of Asia: Geological Evolution and Human Impact," Chen, Z. Y., Saito, Y., and Goodbred, S. L., Jr. (eds.), pp 107-112. China Ocean Press, Beijing.

Takaya, Y. (1974). A physiographic classification of rice land in the Mekong Delta. *Southeast Asian Studies* **12**, 135-142.

Thiramongkol, N. (ed.). (1989). "Proceedings of the Workshop on Correlation of Quaternary Successions in South, East and Southeast Asia," 1988, Department of Geology, Chulalongkorn University, Bangkok, Chulalongkorn University, Bangkok, 269pp.

Tong, T. D. and Vu, K. (2006). "Stratigraphic Units of Vietnam," Vietnam National University Publishing House, Hanoi, 526pp.

Törnqvist, T. E., van Ree, M. H. M., and Faessen, E. L. H. J. (1993). Longitudinal facies architectural changes of a middle Holocene anastomosing distributary system (Rhine-Meuse delta, central Netherlands) controlled by sea-level rise and subsoil erodibility. *Journal of Sedimentary Geology* **85**, 203-219.

Tran, B. V. (2000). Geology and minerals of mid-central Laos region Scale 1:200,000. Report of the Department of Geology and Mines, Lao People's Democratic Republic, 300pp.

Twidale, C. R. (2004). River patterns and their meaning. *Earth-Science Reviews* **67**, 159-218.

Voris, H. K. (2000). Maps of Pleistocene sea levels in Southeast Asia: Shorelines, river systems and time durations. *Journal of Biogeography* **27**, 153-167.

Wang, P. (2004). Cenozoic deformation of Asia and the history of sea-land interactions in Asia. *In* "Continental-Ocean Interactions in the East Asian Marginal Seas," Clift, P. D., Kuhnt, W., Wang, P., and Hayes, D. (eds.), pp 1-23. Geophysical Monograph Series, 149. AGU, Washington, DC.

Wang, P., Clemens, S., Beaufort, L., Braconnot, P., Ganssen, G., Jian, Z., Kershaw, P., and Sarnthein, M. (2005). Evolution and variability of the Asian monsoon system: State of the art and outstanding issues. *Quaternary Science Reviews* **24**, 595-629.

Wongsomsak, S. (1992). Preliminary investigation on Mekhong terraces in Nakhon Phanom Province: Distribution. Characteristic, age and imbrication. *In* "Proceedings of a National Conference on Geologic Resources of Thailand: Potential for Future Development," Department of Mineral Resources, Bangkok, pp 326-331.

Wood, B. G. M. (1985). The mechanics of progressive deformation in crustal plates—A working model for Southeast Asia. *Geological Society of Malaysia Bulletin* **18**, 55-99.

Workman, D. R. (1997a). Cambodia. *In* "Encyclopedia of European and Asian Regional Geology," Moores, E. M. and Fairbridge, R. W. (eds.), pp 122-127. Chapman & Hall, London, 804pp.

Workman, D. R. (1997b). Laos. *In* "Encyclopedia of European and Asian Regional Geology," Moores, E. M. and Fairbridge, R. W. (eds.), pp 493-497. Chapman & Hall, London, 804pp.

Workman, D. R. (1997c). Thailand. *In* "Encyclopedia of European and Asian Regional Geology," Moores, E. M. and Fairbridge, R. W. (eds.), pp 718-726. Chapman & Hall, London, 804pp.

Xang, L. Q. (1998). "Giao trinh gia chat hoc (Textbook of Geology)." Ministry of Education and Training, Cantho University, Cantho, Vietnam (in Vietnamese).

Zhisheng, A., Tunghseng, L., Yanchou, L., Porter, S. C., Kukla, G., Xihao, W., and Yingming, H. (1990). The long-term paleomonsoon variation recorded by the loess-paleosol sequence in Central China. *Quaternary International* **7-8**, 91-95.

CHAPTER

3

Geology and Landforms of the Mekong Basin

Avijit Gupta

Visiting Scientist, Centre for Remote Imaging, Sensing and Processing, National University of Singapore, Singapore 119260 & School of Geography, University of Leeds, Leeds LS2 9JT, UK

OUTLINE

1. INTRODUCTION

The Mekong flows in a pan-shaped basin shaped by regional geology (Fig. 3.1). The upper basin in China is a steep narrow valley, its geometry determined primarily by Himalayan orogeny. The drainage basin widens south of the Chinese border in Lao PDR and Thailand but remains mountainous. Long sections of the river and its tributaries are structure-guided and rock-entrenched in this part of the basin (Fig. 3.2). The valleyflat of the

Mekong widens after 4000 km from the source and it begins to wander across the wide alluvial lowland of Cambodia. Further downstream, the delta stretches into Viet Nam where the distributary channels are so linear that a structural control can be suspected. The basic physical features of the basin have been mapped (USGS, 1994) and described (Douglas, 2005; Gupta, 2005). Topographical maps and high-resolution satellite images are available for the basin but only partial details of its geology are in public domain, as in

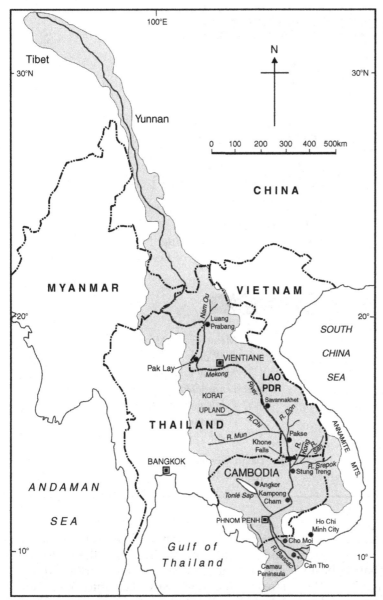

FIGURE 3.1 Location map: the Mekong Basin.

UN-ESCAP (1990, 1993). Available research publications in geology tend to concentrate on two areas: (a) the tectonics of the upper basin (Brookfield, 1998; Clark *et al.*, 2004; Clift *et al.*, 2004; 2006a,b; Fenton *et al.*, 2003), and (b) the evolution of the delta (Nguyen *et al.*, 2000; Ta *et al.*, 2001, 2002a,b, 2005). A few specul- ative accounts of an earlier Mekong are avail- able (Brookfield, 1998; Gupta, 2004, 2007; Hutchison, 1989), but such discussions primar- ily indicate the need for further research. The form and behavior of the river have been described recently (Gupta and Liew, 2007; Gupta *et al.*, 2002) but not the geomorphology

FIGURE 3.2 The Mekong downstream of Chiang Saen, Thailand. The river is in rock with very little accommodation space for sediment. Photograph: Avijit Gupta. (See Color Plate 4)

of the basin. Walling (2008) has reviewed the available literature, both published and unpublished, on the sediment load of the Mekong. The large lake of Tonlé Sap has attracted a number of researchers (e.g., Carbonnel and Guiscafré, 1965; Kummu and Sarkkula, 2008; Kummu et al., 2008; Penny, 2006). The recent papers not only deal with the physical environment of the lake but also attempt to determine its environmental future.

This account highlights the major characteristics of the geology and landforms of the Mekong Basin. The basin is part of the Eurasian Plate, an old continental plate regionally deformed by its Tertiary collision with the Indian Plate. This collision, which resulted in the formation of the Himalaya Mountains, modified the regional structure in Southeast Asia, redefining major river basins and river alignments. The Mekong, like the rest, was affected. The monsoon system of rainfall was also probably accentuated, and a number of climate proxies indicate that the monsoon of Early Holocene was stronger (Kale et al., 2003). The Mekong Delta started to form 6000-7000 years ago (Nguyen et al., 2000; Ta et al., 2001, 2002a,b). The present landforms and river channels in the Mekong Basin are outcomes of such a geological history.

2. GEOLOGY

2.1. General

The pre-Tertiary geology of Southeast Asia is complicated. A large part of the Mekong Basin comprises pre-Tertiary metamorphosed terrestrial and marine sedimentary rocks, rocks associated with suture zones, and intruded granites (Fig. 3.3 and Table 3.1). Metamorphosed rocks of the Kontum Plateau in Viet Nam (Fig. 3.4) have been dated to 2300 Ma (Hutchison, 2005). Cullen et al. (1994) reviewed the geology of northeast Thailand and south-central Lao PDR during the Jurassic and Cretaceous, when an en echelon system of half grabens was rapidly filled by fluvial sandstones and lacustrine mudstones. This was followed by extensive deposition of continental mudstone interrupted by several marine incursions. Indosinian orogeny in the Triassic appears to have modified the basement of Indochina. Tectonic activities later inverted the half grabens along preexisting extensional faults to large-scale anticlines, which were subsequently heavily eroded. Most of the basin in China, Lao PDR, and Thailand therefore is underlain by Palaeozoic and Mesozoic sedimentary rocks that influence the topography and river sediment.

Following the collision of India and Eurasia (about 50 Ma ago), the whole of Indochina

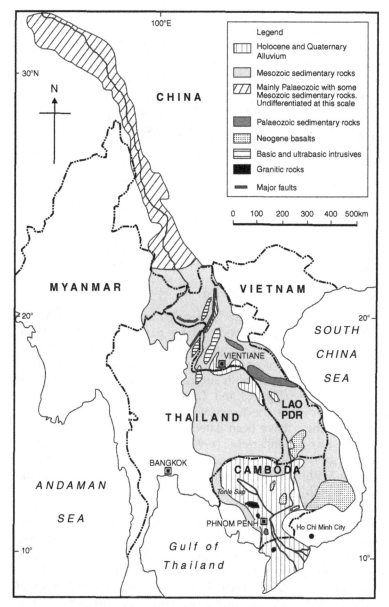

FIGURE 3.3 Generalized geology of the Mekong Basin.

appears to have moved toward the southeast in relation to the location of Central Asia (Clift *et al.*, 2006a). Large-scale strike-slip faulting opened up basins in Southeast Asia, the upper Mekong River came into existence, and the present drainage system of the Mekong started to evolve. Later, Neogene eruptive rocks and Quaternary alluvium covered various parts of the basement geology. The present delta was formed in the Holocene (Nguyen *et al.*, 2000; Ta *et al.*, 2001, 2002a,b, 2005).

TABLE 3.1 Generalized geology of the Mekong Basin

Age	Lithology	Distribution
Quaternary	Holocene fluvial alluvium	Mekong valley flat, along major tributaries, across central Cambodia, Tonlé Sap basin, hillslopes
	Holocene deltaic deposits	Mekong delta in Cambodia and Viet Nam
	Quaternary fluvial alluvium	Mekong valley (Vientiane-Savannakhet), intermontane valleys, Tonlé Sap, and central valley of Cambodia
	Quaternary Basalts (including alkaline varieties, often continuing from Neogene)	South Annamite Mountains, east of Pakse, small outcrops in Lao PDR and Thailand
Tertiory and Mesozoic	Neogene terrestrial deposits	Intermontane valleys in Lao PDR, overlying grabens in Cambodia
	Neogene basalts	
	Middle to late Mesozoic primarily terrestrial sediments, mainly sandstones with evaporites towards the top	Scattered outcrops
	Triassic granites	Mesozoic granites and metasediments occur in the basin panhandle in China, extensively in Lao PDR, in highlands of Cambodia, and probably the graben basements
	Early Mesozoic sediments, mainly marine (limestone)	
	Mainly acidic volcanic rocks	
Palaeozoic	Sedimentary rocks, predominantly marine (mainly limestone), metamorphosed to form phyllites, schists, marbles, and also blueschists from igneous rocks in the upper basin	China, Lao PDR (mainly north), and Thailand
	Widespread volcanic rocks	
	Ultrabasics and basics (gabbro) in narrow belts	
	Granites and associated rocks	
Proterozoic	Granites and schists	China and Lao PDR (scattered outcrops)

Source: Choubert and Faure-Muret (1976), UN-ESCAP (1990, 1993), Brookfield (1998), Clark *et al.* (2004), Ta *et al.* (2005), Heppe *et al.* (2007), Wood *et al.* (2008).

Geology of the basin has been summarized in Table 3.1. The basin panhandle is a narrow valley in China, 40–50 km across, eroded into granitic and sedimentary rocks of Palaeozoic and Mesozoic age. The geology of this region has been summarized by Wang *et al.* (1998), Clift *et al.* (2006a), and Heppe *et al.* (2007). In northern Lao PDR granitic rocks, folded Palaeozoic sedimentary and metamorphic rocks, and Mesozoic sedimentary deposits outcrop. Westward the Mekong drainage basin extends over the Korat Upland (mainly in Thailand, also known as Korat Plateau), where Mesozoic terrestrial rocks rest on an ancient basement, and Quaternary alluvium is found in the valleys of the Mun and Chi Rivers. Sandstones, slates, schists, cherts, limestones, and evaporites form the common lithology in this part of the basin in Lao PDR and Thailand. Volcanic exposures, granitic intrusions, and basic and ultrabasic rocks associated with an ancient suture zone in the Pak Lay region of northern Lao PDR also occur locally. Downbasin from Vientiane to Savannakhet, a shallow

FIGURE 3.4 Detailed location map of the Mekong Basin below the panhandle, showing place names used in the text.

Quaternary alluvium, mainly fine to coarse sand with some clay and gravel, covers Palaeozoic and Mesozoic rocks in the valley of the Mekong. Mesozoic sandstones and evaporites reappear in the Mekong valley further south beyond this Quaternary cover, locally overlain by shallow alluvium. Here, numerous rock ribs are exposed in the Mekong River and rock outcrops occur in the channels of its tributaries. Volcanic rocks are exposed again near the

Lao-Cambodia border (Choubert and Faure-Muret, 1976; UN-ESCAP, 1990). Neogene and Quaternary basalts, including an alkaline variety, form the extensive plateaus of Boloven and Kontum on the eastern divide and occur locally in northern Lao PDR and the southern Korat Upland (Fig. 3.3). With such geology, a coarse sediment load is carried by the Mekong.

In Cambodia, the basin widens to about 500 km with an alluvium of variable thickness as surface cover. Permian carbonates, Triassic sedimentary rocks, and Neogene basalts form the divides, and with granite emerge from underneath the alluvial cover elsewhere to form low isolated hills. A limited amount of information is in the public domain regarding the subsurface geology below the alluvial plains of Cambodia, but several grabens or half grabens are likely to exist below the alluvium (http://www.pecj.or.jp/japanese/division/division09/asia_symp-4th). These grabens are filled with sediment, Neogene to Quaternary in age. Recent surveys by the Cambodian National Petroleum Authority (CNPA) and Japan National Oil Corporation (JNOC) have identified two 10^3 m deep sedimentary basins (probably Tertiary grabens) to the north and west of Tonlé Sap. The lake is situated off-graben, but the course of the Mekong to the east of the lake appears to be aligned along a narrow structural depression.

The present Mekong River and its basin achieved their current appearance finally after the Neogene post-Himalayan realignment of the headwaters, integration of the drainage net, and formation of the Holocene delta.

2.2. Alignment of the Headwaters

The Mekong Basin in Tibet and Yunnan is a narrow strip (Fig. 3.1) where most of the surface originally had little relief but was later deeply dissected to form the gorge of the Mekong and the deep valleys of its short tributaries. Three parallel 30-50 km wide upper basins of the Yangtze, Mekong, and Salween occur next to each other. After flowing across the Tibetan Plateau over low gradients, these rivers turn south to descend through close, steep, and parallel gorges (the Three Rivers Region) before traveling in different directions (Fig. 3.5). The gorges of the Mekong and Yangtze are 2000–3000 m deep at the steepest section of the plateau margin. In contrast, part of the divide between the basins of the Mekong and the Yangtze has a gentle slope (Fig. 3.6).

Clark et al. (2004) have described the current interfluve surface as representing a remnant landscape not yet in equilibrium to the present-day environment.

Low-relief interfluve surfaces, deep gorges of the major rivers, strath terraces in the gorges, and accelerated erosion all have been attributed to the collision of India with Asia resulting in crustal shortening, differential shear, and clockwise rotation. The early stage of the collision was soft, but lithospheric shortening of about 1000 km between 50 and 25 Ma probably was responsible for eastward lateral extrusion of Southeast Asia as proposed by Tapponnier et al. (1986). Brookfield (1998) produced a diagram (his fig. 22) generalizing the possible changes from the precollision to the postcollision stage in the river systems of South and Southeast Asia (including the Mekong). The gradient of rivers in this area, such as of the Mekong, varies considerably due to cross-faults, although lithological changes could be locally important. According to Brookfield (1998), the cross-faults not only caused dislocations and river gradient changes but also allowed headward capture by tributary streams.

The explanations offered for the extraordinary change in these river systems from the precollision to the postcollision patterns range between (1) treating rivers as passive strain markers and attributing the metamorphosis of former drainage systems to collision-related crustal shortening and horizontal shear (Hallet and Molnar, 2001), and (2) disintegration of the precollision Sông Hóng (Red River) drainage net (which is believed to have included the former upper Yangtze, Mekong, Salween, and Tsanpo) due to progressive river capture following uplift of the Tibetan Plateau (Clark et al., 2004). A former drainage net disintegrated to form the headwaters of separate major rivers. Probably a combination of both hypotheses is needed in order to explain the present major drainage systems of this high area, including that of the Mekong.

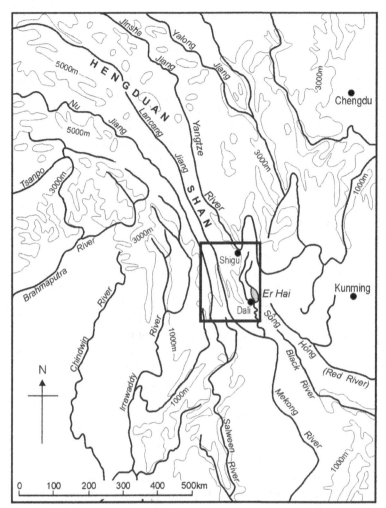

FIGURE 3.5 The convergence of major rivers in eastern Tibet and Yunnan. The rectangle indicates location of the low-relief interfluve between the Yangtze and the Mekong. It also marks the southern part of the Three Rivers Region where the Yangtze, Mekong, and Salween flow very close to each other. Hypothetically, a former drainage net was altered by tectonic movements and/or river capture to form the present pattern of headwaters of separate large rivers. Clark *et al.* (2004) suggested that the headwaters of the Tsanpo, Irrawaddy, Salween, Yangtze, Mekong, and Sông Hóng all were parts of the upper drainage net of the proto Sông Hóng (Red River), a drainage net later disrupted by progressive river capture.

The estimated increase in sediment volumes in the offshore basins at the other end of the major rivers (Métivier *et al.*, 1999) supports the hypothesis that landforms of this part of Asia underwent active erosion following Himalayan orogeny and a post-Neogene stronger monsoon. A slow accumulation rate from the Palaeocene to the beginning of the Oligocene has been computed for the offshore basins. Sedimentation increased in exponential fashion from the Oligocene onwards, peaking during the Quaternary. Other sediment budgets, reconstructed for sediment deposited from the coast to the abyssal plain (Clift *et al.*, 2003, 2004), similarly identified increased offshore sedimentation rates but later, during Early to Middle Miocene (20-15 Ma). Clift *et al.* (2004) conclude that the sharp increase in sedimentation rates in East Asia resulted from a combination of strong uplift of the Tibetan Plateau and intensification of the summer monsoon. The resulting strong erosion on the edge of the Tibetan Plateau cut the deep gorges of these large rivers releasing large volumes of

FIGURE 3.6   Upper Mekong Valley near Er Lake. Photograph: Avijit Gupta. (See Color Plate 5)

sediment. This area of active tectonic strain, river incision, and moderate precipitation still continues to be an important source of sediment for the Mekong (Clift *et al.*, 2006a). The large-scale transfer of the sediment derived by erosion of the upper Mekong to the South China Sea implies that by the end of the Himalayan orogeny the current Mekong drainage network was in operation, the summer monsoon operated actively, and volumes of sediment were pouring through the gorges. Clark *et al.* (2005) using (U-Th)/He thermochronology dated the incision of these gorges to start at 13-9 Ma, although other estimates put the date of the gorges earlier (Clift *et al.*, 2006b). In general, a sequence of lateral extrusion of rigid crustal block along the strike-slip faults, plateau uplift, monsoon activation, river capture, and gorge incision appears to have followed the Himalayan orogeny. A major part of the Mekong sediment is still derived from this region.

2.3. Mekong Delta

An early account of the geology of the Mekong Delta was provided by Gagliano and McIntire (1968). Recently, a series of boreholes

sunk into the deltaic sediments have expanded our understanding of the deltaic geology (Nguyen *et al.*, 2000; Ta *et al.*, 2001, 2002a,b, 2005). Undifferentiated Late Pleistocene deposits form the basement strata for the delta, and laterites and oxidization in these sediments indicate a period of subaerial exposure (Ta *et al.*, 2005). On the basis of the presence of foraminifers and molluscs and radiocarbon dating, these sediments have been related to the Last Glacial Maximum (LGM) and a low sea level (Ta *et al.*, 2005). Southeast Asia experienced a low sea level stand at −120 m during the LGM and a wide exposure of the shallow Sunda Shelf (Biswas, 1973; Gupta *et al.*, 1987 and references therein). The Mekong River of that time must have extended its course and eroded its valley. As sea level rose in the post-LGM time, this incised/eroded valley was filled with sediment. A sharp contrast in lithology and color occurs between the basement and this overlying sediment which starts with what Ta *et al.* (2005) have described as a transgressive incised valley fill. A rapid rise of the sea level deposited a relatively thin layer of estuarine to marine sediment over the interfluves but filled existing river valleys to greater thickness. Marshy and estuarine sediments from the lower part of this fill have been radio-carbon dated to 13-8 ka. The fill ends with a sandy lag deposit coinciding with the mid-Holocene sea-level highstand. The transgressive fill is overlain in turn by Holocene deltaic sediments associated with a slowly receding sea level that allowed an accommodation space for the early Mekong Delta. Ta *et al.* (2005) identified the maximum flooding surface as between 8 and 7 ka in age. Deltaic deposits are widely developed above this surface (Fig. 3.7).

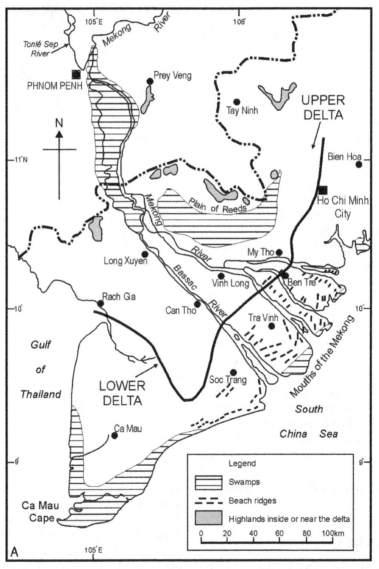

FIGURE 3.7 The Mekong Delta. (A) Plan of the delta indicating both the tide-dominated upper delta and the tide- and wave-dominated lower delta. The boundary adopted from Ta *et al.* (2005).

(Continued)

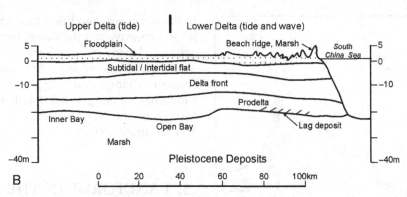

FIGURE 3.7—cont'd (B) Generalized north-south stratigraphic section through the Mekong Delta. Simplified from Ta *et al.* (2005).

The posthighstand slow recession of the sea provided the accommodation space for the large sediment load of the Mekong to construct a progressively outward-building delta in two stages The inner part was built as a tide-dominated delta, which about 3000 years ago was replaced by a tide- and wave-dominated one, as the delta extended beyond a headland east of Ho Chi Minh City and the delta face was exposed to strong wave action. Over the last 3000 years, the Mekong Delta was built at a nearly constant rate of 20 km per thousand years. In all, the delta shoreline extended for 200 km from a previous shoreline near the Cambodia-Viet Nam border to its present location (Nguyen *et al.*, 2000). The average annual rate of deltaic sedimentation has been determined as 144 ± 36 million t (Ta *et al.*, 2002a), practically matching the present rate of sediment discharge at the mouth of the Mekong of 150-170 million t (Meade, 1996, MRC, various years), and implying no significant anthropogenic increase in sediment production in the Mekong Basin, a characteristic not seen for any other major river in Southeast Asia (Ta *et al.*, 2001, 2002a,b).

2.4. Integration of the Drainage Net

The present drainage net of the Mekong is probably more than 20 Ma old, as shown by accelerated sediment accumulation in the South China Sea (Clift *et al.*, 2004; Métivier *et al.*, 1999). Any account of its formation need to explain a number of characteristics: the short dendritic net at the extreme head of the basin, the very deep gorge and strath terraces at the edge of the Tibetan Plateau, the right-angled turns of the tributaries, the sharp direction changes and U-bends cut into rock by the Mekong, the south to east direction change (discussed later) that prevents the river from flowing across Korat Upland into the central plains of Thailand, the change from a rock-cut channel to an alluvial one in Cambodia, and the formation of the current delta about 6000 years ago (Fig. 3.7). The proto-Mekong probably was aligned along a late Triassic suture southward into the Yom River in Thailand. At present, it turns sharply east from a southwest-flowing direction into central Lao PDR very close to the Thai border (Fig. 3.8). The morphology and behavior of the Mekong tend to change conspicuously several times along its course (Gupta and Liew, 2007). This suggests that parts of different preexisting drainage nets were integrated together to form the present drainage system as had happened for other large rivers of the world such as the Danube or the Zambezi (Lóczy, 2007; Moore *et al.*, 2007). The sharp direction changes and U-bends in rock have been attributed to

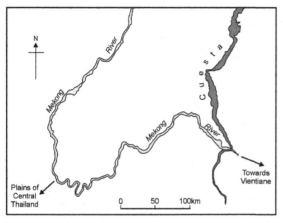

FIGURE 3.8 The Mekong River in northern Lao PDR. Note the southwest-flowing river suddenly changes direction to flow east first through a set of meanders in rock and then in straight reaches joined at right angles downstream. The river then flows through a cuesta in a steep canyon. See Gupta *et al.* (2002) for a detailed discussion. It has been suggested that an earlier Mekong used to flow into the plains of central Thailand and was diverted to the present course by river capture. No real evidence, so far, has been forwarded. The Mekong Lowland (Section 3.3) begins downstream of the canyon eroded through the cuesta.

various causes: selective alignments along regional and local faults (Fenton *et al.*, 2003; Wood *et al.*, 2008), reversal of movements along faults (Lacassin *et al.*, 1998), and river capture. A holistic explanation has yet to emerge.

The geology of the Mekong Basin explains the turbulent gorge and steep strath terraces of the Mekong in China and Lao PDR with long reaches of the river along structural lineations and short rougher reaches across it. In China, Lao PDR, and Thailand, the Mekong follows several large faults. In the lowlands of Cambodia and Viet Nam, the alluvial fill covers preexisting structures. It is likely that the Mekong had to traverse subsurface grabens and half grabens similar to those that have formed Tertiary basins in Thailand (Lorenzetti *et al.*, 1994). The middle part of the basin could be the oldest, being part of the Eurasian Plate. The panhandle

to the north was modified following the Himalayan orogeny, and the downstream part of the present basin in southern Cambodia and Viet Nam was added after the Quaternary sea-level changes. Although certain basic indicators have emerged, the detailed evolutionary history of the Mekong remains an unknown and fascinating story.

3. LANDFORMS IN THE MEKONG BASIN

The 4880 km long Mekong rises in a meadow at an elevation of about 5000 m near Rupsa Pass in Qinghai (China). The first 4000 km of the Mekong drains a mountainous basin that gradually loses elevation. Areas with low relief are limited. Plains are found only in the valley flat of the Mekong and extending a short distance along the major tributaries. In contrast, the lowland in Cambodia is extensive and highlands are restricted to the divides or isolated hills rising from middle of an alluvial plain. The transition between the Mekong of the hills and the Mekong of the plains is marked by a group of rapids and cataracts (the Khone Falls) that in combination carry the divided flow of the Mekong, the main channel flowing over the Phapheng Falls.

The basin can be divided into a number of physical units (Fig. 3.9):

1. The mountainous panhandle
2. Mountains of northern Lao PDR and Thailand
3. Mekong Lowland
4. Korat Upland (also known as Korat Plateau)
5. Cardamom and Elephant Hills
6. Annamite Range of mountains
7. The delta

The NW-SE flowing Mekong emerges out of the mountains to enter the lowland and ultimately the delta. The Korat Upland and Cardamom and Elephant Hills form the western divide of the lower basin; the Annamites form the more prominent eastern one.

3.1. The Mountainous Panhandle

The upper part of the panhandle lies across the Tibetan Plateau. It is a narrow (about 40 km wide) part of the old surface in eastern

FIGURE 3.9 The Mekong Basin: physical units.

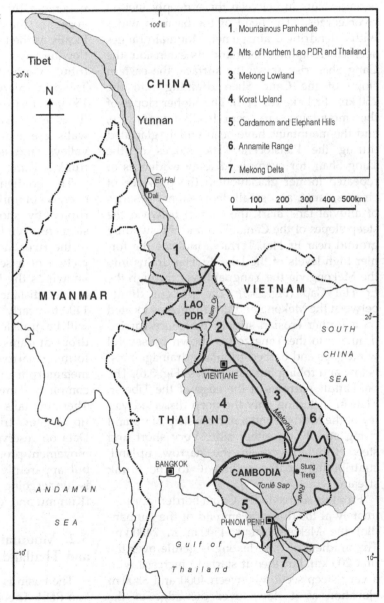

1. Mountainous Panhandle
2. Mts. of Northern Lao PDR and Thailand
3. Mekong Lowland
4. Korat Upland
5. Cardamom and Elephant Hills
6. Annamite Range
7. Mekong Delta

0 100 200 300 400 500km

Tibet that has been disrupted by tectonic movements and river incision. Clark *et al.* (2004) mentioned that flat Eocene sedimentary strata overlay the old undisturbed surface, which they described as a "regionally continuous relict, low-relief landscape preserved over large aerial extents in between these deeply incised river gorges." The head of the basin is wider with a dendritic drainage net. Mountain ranges such as Hengduan Shan and its extension, the Cang Shan rise above the surface, the eastern slope of the Cang Shan draining into the 250 km^2 Er Lake (Er Hai). The higher slopes of the mountains are periodically under ice, and the mountains have been widely glaciated during the Pleistocene. The slopes of the Cang Shan, for example, display evidences of repeated former glaciation in the presence of cirques, tarns, and small depressions. A series of alluvial fans mark the contact between the steep slopes of the Cang Shan and the swampy ground near Er Hai. Terraces indicate the former high levels of the lake. Er Hai drains into the Mekong via the Yangpai River through the Er Hai Gap (Fig. 3.5). The present divide between the Mekong and the Yangtze is located on a former erosion surface that once carried drainage to the Yangtze (as shown by several wind gaps and barbed tributary drainage nets) is low and rolling in many places (Fig. 3.6). The panhandle narrows at the edge of the Tibetan Plateau including only the deep dissected valley of the Mekong and a narrow strip of upland rising steeply on either side. Very short and steep tributaries drain the narrow upland, contributing coarse sediment to the trunk stream.

From its source to the China border (approximately at the downstream end of the panhandle) the Mekong drops 4500 m in 2400 km. The gradient of the Mekong is gentle over the first 200 km, but then it starts to increase with a very steep section between 4000 and 3500 m elevation as it flows over the edge of the Tibetan Plateau and across faults. Here, the river follows a winding course through deeply incised bedrock gorges (up to 2-3 km in depth) with narrow strath terraces suspended on gorge walls (Brookfield, 1998; Clark *et al.*, 2004). Satellite imagery indicates a number of landslides, caused by both tectonic and anthropogenic activities, on the steep slopes of the gorge of the Mekong. The Mekong (known locally as the Lancang) is a turbulent river here, flowing over rapids created by rock ribs, tributary-mouth fans, and landslide screes. The river emerges from the narrow gorge about 150 km upstream of the Chinese border, and flows through a wider valley whose convoluted walls are marked by steep spurs, tributary valleys, and isolated hills rising next to the trunk stream.

The gradient of the Upper Mekong, however, is not uniformly steep. It is repeatedly disrupted by short reaches where the gradient steepens even further. The nature and behavior of the river, however, would change on completion of a series of eight dams, collectively known as the Lancang Cascade, planned over a river distance of 800 km, from elevation of 1400 to nearly 500 m. The gradient of the river will be transformed to a staircase of vertical drops at dams separated by impounded pools forming narrow and winding reservoirs. This metamorphism has already happened for two complete dams (Manwan and Dachaoshan). There the tails of the reservoirs have extended up bigger tributaries for short distances. Data on reservoir sedimentation in this mass-movement-prone steep mountains are limited but apparently the reservoir behind Manwan is undergoing rapid sediment accumulation (Kummu and Varis, 2007).

3.2. Mountains of Northern Lao PDR and Thailand

The basin is considerably wider in northern Lao PDR. This is an area of alternate mountain ridges and steep valleys, the relief being progressively low downbasin and the valleys wider. Few mountain peaks rise above 2000 m although local relief between mountains and valley bottom can exceed 1000 m in 10 km. Northern Lao PDR is a country of steep-sided, sharp-crested, near-parallel, and northeast-southwest ridges. The 1100-1400 m high lime-stone plateau located north of Xieng Khunan, part of which is known as the Plain of Jars, is the only area without advanced dissection. The valleys are usually V-shaped, locally resembling slot canyons. The Mekong and its major tributary, the Nam Ou, flow through wider valleys that are bounded by steep ridges. Isolated hills rise from the valley bottom of both rivers. Sheer walls of limestone flank a num-ber of the eastern tributaries of the Mekong. Destruction of vegetation, high seasonal rainfall events, and tectonic movements resulted in a number of slope failures. Stored valley-floor sed-iment (Fig. 3.10) then travels episodically during the rainy season as demonstrated by the hydro-graphs of the tributaries of the Mekong display-ing short-term spikes (Gupta and Liew, 2007).

Rivers in the mountains of Lao PDR tend to flow in straight courses that change direction sharply, possibly due to both river capture and faulting. U-bends are common, often with a tributary joining the main stream at the top of the convex bend (Gupta and Liew, 2007). The possibility of river capture is high as demonstrated by sharp elbows of many streams and convergence of the headwaters on either side of a short low divide. The relief is lower and the valleys wider on the western (Thai) side of the Mekong. The morphology of the Mekong itself varies between canyon-like narrow valleys and wider reaches with narrow floodplains as at Sop Ruak near the China bor-der or about 4 km south of Chiang Saen where the Mekong flows through an S-loop (See Wood *et al.*, 2008, Fig. 2 for a detailed map and a Landsat image of the area). Such varia-tions have been attributed to lithology and tec-tonics. Faulting has determined the course of some of the tributaries, developed shutter ridges (fault-related ridges that intercept for-mer drainage lines), offset streams, widened valleys, and formed swampy depressions as near the confluence of the Kok and Kham riv-ers with the Mekong where the development of the swamp has been explained as a tectonic sag with active left-lateral shear along the Mae Chan Fault (Fenton *et al.*, 2003; Wood *et al.*, 2008).

FIGURE 3.10 Sediment stored in the channel and narrow valley of the Mekong. Photograph: Avijit Gupta. (See Color Plate 6)

About 200 km river distance upstream of Vientiane, the south-flowing Mekong sharply turns east, away from a possible entry into Thailand. As the Mekong turns east (Fig. 3.8), it flows through a set of incised meanders, a pattern that changes into a series of short straight reaches linked by sharp turns of the river (Gupta and Liew, 2007; Gupta et al., 2002). The valley flat is still narrow although a number of abandoned channels and short tributaries are linked with the river in this reach. The river crosses a cuesta in a small canyon and enters the next division, the Mekong Lowland.

3.3. Mekong Lowland

The Mekong Lowland stretches north-south along the river from the northern mountains to the delta. The lowlands widen from Vientiane as the river angles cross-basin toward the eastern divide marked by the northern Annamite Range. The Mekong in this reach receives more than 10% of its annual discharge from the short steep left-bank streams that drain the northern Lao mountains and the northern part of the Annamite Chain: Nam Ngum, Nam Nhiep, and Nam Theun (MRC, various years). To the west, the Nam Songkhram and Huai Luang drain part of Korat Upland that rises gently to form the western divide.

The Mekong turns south near the Annamite Mountains, and enters a narrow valley again, flowing on both rock and alluvium. Here, the lowland alternates between wider flat stretches where larger tributaries (such as the Bang Hiang) join the Mekong and narrow rocky valleys between two ranges of hills. The Mun River, one of the longest tributaries of the Mekong, joins the trunk stream from the west, draining the Korat Upland. The Mun is a low-gradient stream which has built a small alluvial fan at its confluence. From about 25 km upstream of Savannakhet to the Lao-Cambodia border, the Mekong is visibly in rock and the valley of the Mekong between the Korat Upland to

the west and the Annamite Chain to the east narrows to 20-125 km. The lowland widens downbasin near Pakse, and the island-studded river flows south to an anastomosing pattern of multiple islands and channels on Mesozoic basalt at 4000 Islands, immediately upstream of the Khone Falls. Below the falls, the alluvial plain widens rapidly and the Mekong changes from a set of parallel alluvial channels into a single wide thread with sharp bends. The river begins to wander freely only from Kampong Cham in central Cambodia.

The western extension of the alluvial plain in Cambodia is bounded to the north by the Dong Rek Escarpment of the Korat Upland. Although it rises precipitously from the flat plain of Cambodia, the top of the plateau only reaches 600-700 m. The Mekong Lowland is linked to plains in Thailand beyond the basin via a low saddle between the Korat Upland and the steep Cardamom Hills toward west. The drainage in this wide western embayment of the Mekong Lowland between the Dong Rek Range and the Cardamom Hills (Fig. 3.9) collects in the large lake of Tonlé Sap. The lake comprises a large basin to the northwest and a smaller basin to the southeast linked by a narrow strait. It is surrounded by a low swampland, but partly forested low steep hills occur locally. Tonlé Sap is connected to the Mekong by the 120 km long Tonlé Sap River that leaves the Mekong at Chatomuk, north of Phnom Penh, and enters the smaller of the two lacustrine basins to build a small underwater delta. The bottom of the lake is only 0.5-0.7 m above sea level (Kummu et al., 2008), indicating the low gradient and flat nature of the Mekong Lowland in Cambodia.

Sediment-laden outflow from the Mekong enters the lake via the link river during the wet season when the river stage is high. In contrast, clear water flows out of the lake to the Mekong during the dry season. The area of the lake increases from 2500 km^2 in the dry season to 15,000 km^2 during the wet with a corresponding rise of 5.4-9 m of its water level.

The volume of the lake changes approximately from 1.3 km^3 in the dry season up to 75 km^3 in the wet. About 57% of its annual inflow arrives from the Mekong via the Tonlé Sap River and the floodplain, 30% from local streams flowing into the lake from various directions, and the remaining 13% from direct precipitation. About 78% of total suspended sediment is derived from the Mekong, the rest from local tributaries. The computed average annual suspended sediment flux into the lake totals about 7 trillion kg, only about a fifth of which flows out through the link river during the dry season (Kummu et al., 2008). Sediment in the lake consists primarily of medium to very fine silt (Penny, 2006).

The sediment accumulates not directly in the lake but mostly around its shores: on the floodplain, in the swamps, and in flooded forests of the wet season dominated by *Barringtonia acutangula* and *Diospyros cambodiana*. This sediment is essential for the productivity of the lake region, which in turn supports two main ingredients of Cambodia's rural resources, rice and fish. A small part of the sediment is resuspended during the dry season but there is no marked shallowing of the lake by sediment accumulation as has been suggested occasionally. On the basis of a number of cores, Penny (2006) has demonstrated a very low rate of sedimentation in the lake from Mid-Holocene onwards. Similar results have been reached by modeling studies. Kummu et al. (2008) suggest that the misconception of a shallowing lake is partly due to misunderstanding of the early study of Carbonnel and Guiscafrè (1965) who also cored through the lake sediment.

Isolated sandstone hills rise above the alluvial plain between Dong Rek Escarpment and Tonlé Sap but the general landscape is flat with 10^2 km long meandering streams flowing to the Mekong on low gradient. The eastern lowland is different, brooded over by the steep southern Annamite Range which has been eroded back by three large tributaries of the Mekong: the

Srepok, San, and Kong. The headwaters of these rivers have built alluvial fans at the contact between the mountains and lowland, thereby locally softening the precipitous relief. South of Kampong Cham small hills rise through the valley alluvium, but the lowlands are that of a large meandering river: a mosaic of a levee-bounded large channel, abandoned former river courses, ox-bow lakes, and backswamps. Large floods episodically inundate the backswamps for weeks (Gupta, 2007).

3.4. Korat Upland

Most of this 150,000 km^2 upland ultimately drains eastward into the Mekong. It has a gently sloping, saucer-shaped top surface but the sides could be steep, especially toward the south where the Dong Rek Range separates it from the plains in Cambodia. Gently folded and uniclinally dipping Mesozoic and Tertiary strata are overlain by gravel, ferricrete, and aeolian and fluvial sands. On top, a few hills rise to 1000 m but the surface is only about 200 m near its northwest corner and as low as 50 m at its southeastern end. The northwest-southeast trending Phu Phan Hills separate the upland into the Korat basin to the south and Sakon Nakhon Basin to the north (Thiramongkol, 1983). A number of lakes, whose areas fluctuate seasonally, occur on top along with cuestas, hogbacks, and water and wind gaps eroded out of regional sedimentary rocks. Toward the north, where the Upland extends beyond the river into Lao PDR, it drains directly into the Mekong. Almost elsewhere, it is part of the Mun Basin, one of the major tributaries of the Mekong. The Mun and its main tributary, the Chi, exhibits a wide range of palaeomeanders that, along with the aeolian and fluvial sands, suggest climate change during the Quaternary as proposed by Nutalaya et al. (1989) and Udomchoke (1989). These rivers flow on a gentle gradient through steeply cut wide shallow valleys.

3.5. Cardamom and Elephant Hills

This small but steep hilly region, with the Cardamom Hills to the northwest and the Elephant Hills toward the southeast, separates the southwestern corner of the Mekong Lowland from the coastal plains of southeastern Thailand. The contact between the hills and the lowland is around 100 m. The lower slopes are gentle and actively eroded by small streams, especially toward the south. The higher slopes are steep, ending in a set of flat-topped hills crossing 1000 m. The two highest points are Phnum Samkos (1717 m) on the main range and the outlier of Phnum Aoral (1771 m). The northern hills slope gently toward Tonlé Sap, drained by small low-gradient streams to the lake. Toward the southeast, the hills reach almost to the coast.

3.6. Annamite Range

The Annamite Range is a collection of mountains and plateaux that forms the eastern divide of the Mekong Basin, running from the northern mountains of Lao PDR to the South China Sea. Peaks rise above 2000 m throughout its length, and the range can be crossed only through a limited number of passes. Several plateaux occur to the west of the crestline, of which three stand out. Khammuan is the northernmost of these plateaux which drops steeply to the Mekong River. It is on limestone with well developed karst features and deep river gorges. Boloven and Kontum in basalt are the two other large plateaux, located further to the south between rugged hills. The hills rise steeply from about 500 m, and the contact with the lowland is softened in many places by alluvial fans, isolated rocky hills, and rock outcrops. Toward the south in Cambodia, three large tributaries of the Mekong, the San, Kong, and Srepok, have eroded an embayment about 150 km deep into the hills. About 17% of the

discharge of the Mekong comes from these tributaries and probably also a considerable amount of sediment (MRC, various years). The slopes of the southern Annamite Range are exposed not only to the seasonal rainfall of the southwestern monsoon but also to the episodic tropical cyclones that reach the Mekong Basin.

3.7. The Delta

The Mekong has one of the largest deltas in the world although estimations of its size vary surprisingly between 62,520 and 93,781 km^2 (Ngyuen et al., 2000). The delta begins near Phnom Penh, where the Bassac, as the first distributary, leaves the main channel 330 km from the sea (Fig. 3.7A). The delta is perceived as the slightly curved triangular piece of flat land between Phnom Penh to the north, the mouth of the Saigon River to the east, and the southwestern cape of the Ca Mau Peninsula to the west (Ta et al., 2005). The Mekong and Bassac flow southeast in almost linear courses. The Mekong later divides into three channels, each of which split in two with a large island in between. The river, therefore, flows into the South China Sea through a number of mouths. The position of the main channels has not changed appreciably over the last 2000-3000 years (Ta et al., 2002a). Ta et al. (2005) are of the opinion that morphological differences in the Mekong delta are related to both past variations in the coastal environment and the rate of delta progradation.

Both sides of these channels are bounded by levees. Sand and silt bars occur midstream. Backswamps and depressions, where floodwater accumulates regularly during the wet southwestern monsoon, occur between the levees bordering the main channels and also between the channels and the eastern and western edge of the delta. The large swamp to the northeast is known as the Plain of Reeds. Not many interconnecting channels are found in the delta. The first connection between the Mekong and the Bassac, after the latter's separation, occurs at Cho Moi, about 200 km downstream. The general level of the swamps and floodplains is about 0.5-1.5 m above the mean sea level, the levees rise to another 2 m, and a 3-5 m high Pleistocene Terrace occurs to the north (Ta *et al.*, 2005).

The upper delta plain is dominated by fluvial processes as described earlier but the nature of landforms changes in the lower delta which is tide and wave dominated. Here divergent and bifurcating northeast-southwest beach ridges rise 3-10 m above the sea level, separated by interridge depressions 1.5 and 2.5 m above sea level (Ta *et al.*, 2005). This beach ridge and salt marsh topography dominates the eastern lower delta, having evolved through wave action. The major channels are toward the east, flowing out to the sea through beach ridges with bars and islands in midstream. The Ca Mau Peninsula to the west is different, a coastal plain with tidal channels and marshes and a 2200 km^2 mangrove forest. Its western edge is remarkably straight, probably due to structural control, although this characteristic has been attributed to waves and currents (Ta *et al.*, 2002a). The extreme outer edge of the delta is a coastal wetland, changing between mangrove swamps and tidal flats. The mean tide range is around 2.5 m although the tidal range at the mouth of the Mekong may exceed 3 m (Ta *et al.*, 2005).

Parts of the delta has been settled, modified, and utilized for about 2000 years. A network of canals distributes water of the Mekong across the delta. The land use pattern is a mosaic: settlements and roads on levees and canal embankments; rice in swampy depressions normally inundated by overbank flows in the wet season; depleted *melaleuca* forests in the Ca Mau Peninsula, Long Xuyen Quadrangle, and Plain of Reeds; mangroves near the coast partially replaced by aquaculture, casuarinas, oil palms; and settlements on the beach ridges near the southeast with mangroves and *Nypa fruticans* swamps in between ridges. A strongly anthropogenic component is visible across the Mekong Delta.

4. BASIN GEOMORPHIC PROCESSES AND THE RIVER

The upper Mekong River flows through steep-sided hills or mountains for most of its course and, in spite of this being one of the largest rivers in the world, through a valley flat only 1-10 km wide. The valley widens between Vientiane and Savannakhet, but further south steep hills restrict its width between the Annamite Mountains and the Korat Upland. Only inside Cambodia, the river begins to flow across a wide alluvial fill.

Before the construction of the dams of Lancang Cascade, about 18% of water and about half of the sediment (especially the coarser fraction) used to arrive from the panhandle in China (MRC, various years). The sediment was derived mostly from mass movements on slopes and debris flows along short steep tributaries that contributed directly to the main river. The sediment accommodation space is limited to the narrow valleyflat between steep hills, leading to the primarily coarse sediment being stored mostly inside the channel on rock. We do not know how much of the stored sediment is a relict of the Quaternary past when the river probably was more active. A substantial part of the present sediment is sand and fine pebbles, transferred as bed load (Plinston and He, 2000).

The rest of Mekong's annual water discharge (82%) mainly comes from four sources: (a) mountains of northern Lao PDR through a number of tributaries; (b) the southern Annamite

Mountains primarily via the San, Kong, and Srepok; (c) the Mun-Chi system draining a very large part of the Korat Upland; and (d) the drainage outflow from the Tonlé Sap during the dry season. The northern Lao Mountains and the Annamite Range are also major sources of sediment to the river. The average annual discharge is 15,000 m³/s at the mouth of the river and the annual sediment released to the South China Sea has been estimated to be between 150 and 170 million t (Meade, 1996; MRC, various years).

This is a seasonal river that rises during the southwestern monsoon (June-October) when most of the sediment transport also takes place. The discharge in the headwaters is augmented temporarily in May by water from Tibetan snowmelt. Sediment is transported along the tributaries during the southwestern monsoon, but episodically, with most of the sediment traveling only during the peak flows of the wet season or in floods. Sediment transport in the Mekong itself also is episodic and seasonal, and happens primarily in the high flows of the wet season when the water depth increases to 10-30 m (Gupta and Liew, 2007). Wood *et al.*

(2008) have described a clockwise hysteresis effect that shows higher sediment concentration at the beginning of the wet season when the river remobilizes sediment resting in the channel. Upstream of Cambodia, sediment is generally confined within the rock-cut channel of the Mekong where in-channel storages are determined primarily by channel relief rather than channel geometry. Sediment, therefore, piles up inside the channel as bank insets, against rock ribs, and as bars and islands around boulder piles and high spots of the irregular rocky bed of the channel. The larger islands are high and stable enough to support 30 m trees and paddyfields (Fig. 3.11). The finer fraction of the sediment that reaches the low plains of Cambodia spreads overbank into the flood basins. This pattern holds even for large floods as those of 1999, 2001, and 2002. Satellite images indicate that the floodwater remains confined within the channel almost up to Stung Treng before spreading overbank across the alluvial plains and into the Tonlé Sap Lake. The overbank floodwater, however, does not reach the delta face, probably due to the diversion into the large number of canals

FIGURE 3.11   Rock-cored islands showing sediment accumulation around a rocky barrier. Photograph: Avijit Gupta. (See Color Plate 7)

and the presence of beach ridges across the outer delta (Gupta, 2006; Gupta and Liew, 2007). Seasonal variation in the size of sediment plumes offshore, as indicated by MODIS imagery, also demonstrates the seasonal pattern of sediment storage and transfer. The images also indicate a westward passage of sediment offshore toward the Ca Mau Peninsula, especially during October and November (Gupta *et al.*, 2006).

Up to the end of the twentieth century the basin of the Mekong was little affected anthropogenically. Anthropogenic alterations of the hillslopes and the ongoing impoundments in the Lancang channel look likely to disrupt this pattern. This may lead to rather interesting modifications of the channel, the Tonlé Sap Lake, and the fertile delta of the river (Gupta and Liew, 2007; Gupta *et al.*, 2006). A detailed knowledge of the physical nature of its basin and the operating geomorphic processes, therefore, is needed for any future environmental planning or economic development of the Mekong basin.

On the basis of high-resolution SPOT and IKONOS imagery and field visits, Gupta and Liew (2007) divided the Mekong into eight river units for the 2000 km of its course between the China border and the South China Sea. Each unit is morphologically unique, storing and transferring sediment in a different fashion. Some of these differences may have been inherited from the geological history of the river that integrated parts of former drainage channels to constitute the present Mekong and its drainage net. The velocity of the river, storage and transfer of sediment, and channel morphology, all these can vary considerably depending on the location. It is, therefore, necessary to recognize the difference in morphology and behavior between the Mekong of the hills and the Mekong of the plains, and at a localized level, between the separate river units.

ACKNOWLEDGMENTS

I thank the Centre for Remote Imaging, Sensing and Processing (CRISP), National University of Singapore for the satellite images used for writing this article. The Mekong River Commission graciously allowed visits to their library and access to the hydrological yearbooks. The maps and line diagrams were prepared by Lee Li Kheng. The careful reviews of Ian Campbell and Spencer Wood are gratefully acknowledged.

References

Biswas, B. (1973). Quaternary changes in sea-level in the South China Sea. *Bulletin, Geological Society of Malaysia* **6**, 229-256.

Brookfiel, M. E. (1998). The evolution of the great river systems of southern Asia, during the Cenozoic India-Asia collision: Rivers draining southwards. *Geomorphology* **22**, 285-312.

Carbonnel, J. P. and Guiscafré, J. (1965). "Grand Lac du Camboge: Sedimentologie et Hydrologie 1962-63." Museum National d'Histoire Naturelle de Paris, Paris.

Choubert, G. and Faure-Muret, A. (1976). "General Co-Ordinators. Geological World Atlas (1/10 000 000)." UNESCO, Paris.

Clark, M. K., Schoenbohm, L. M., Royden, L. H., Whipple, K. X., Burchfiel, B. C., Zhang, X., Tang, W., Wang, E., and Chen, L. (2004). Surface uplift, tectonics, and erosion of eastern Tibet from large-scale drainage patterns. *Tectonics* **23**, TC1006, doi:10.1029/2002TC001402.

Clark, M., House, M. A., Royden, L. H., Whipple, K. X., Burchfiel, B. C., Zhang, X., and Tang, W. (2005). Late Cenozoic uplift of southeastern Tibet. *Geology* **23**, 525-528.

Clift, P. D., Clark, M. K., and Royden, L. H. (2003). An erosional record of Tibetan Plateau uplift and monsoon strengthening in the Asian marginal seas. *Geophysical Research Abstracts* **5**, 04300.

Clift, P. D., Layne, G. D., and Blusztajn, J. (2004). Marine sedimentary evidence for monsoon strengthening, Tibetan uplift and drainage evolution in East Asia. *In* "Continental-Ocean Interactions in the East Asian Marginal Seas," Clift, P. D., Wang, P., Hayes, D., and Kuhnt, W. (eds.). Monograph 149, pp 255-282. American Geophysical Union, Washington, DC.

Clift, P. D., Carter, A., Campbell, I. H., Pringle, M. S., Ngyuen, V. L., Allen, C. M., Hodges, K. V., and Mai Thanh Tan (2006a). Thermochronology of mineral grains in the Red and Mekong Rivers, Viet Nam: Provenance and exhumation implications for Southeast Asia. *Geochemistry, Geophysics, Geosystems* **7**, Q10005, doi:10.1029/2006GC001336.

Clift, P. D., Blusztajn, J., and Nguyen, A. D. (2006b). Large-scale drainage capture and surface uplift in eastern Tibet-SW China before 24 MA inferred from sediments of the Hanoi Basin, Viet Nam. *Geophysical Research Letters* **33**, L19403, doi:10.1029/2006GL027772.

Cullen, P. J., Wright, S. C., Kearney, C. J., and Pink, A. T. (1994). Exploration in the Savannakhet basin, Peoples Democratic Republic of Laos. *American Association of Petroleum Geologists* **78**, doi:10.1306/A25FE56D-171B-11D7-8645000102C1865D.

Douglas, I. (2005). The Mekong River basin. *In* "The Physical Geography of Southeast Asia," Gupta, A. (ed.), pp 193-218. Oxford University Press, Oxford.

Fenton, C. H., Charusiri, P., and Wood, S. H. (2003). Recent paleoseismic investigations in Northern and Western Thailand. *Annals of Geophysics* **46**, 957-981.

Gagliano, S. M., and McIntire, W. G. (1968). Reports on the Mekong River delta. Technical Report 57. Louisiana State University, Coastal Studies Institute.

Gupta, A. (2004). The Mekong River: Morphology, evolution and palaeoenvironment. *In* "Journal of Geological Society of India, Special Issue on Progress in Palaeohydrology: Focus on Monsoonal Areas," Kale, V. S., Gregory, K. J., and Joshi, V. U. (eds.) Geological Society of India, Bangalore, Volume 64, pp 525-533.

Gupta, A. (2005). Landforms of Southeast Asia. *In* "The Physical Geography of Southeast Asia," Gupta, A. (ed.), pp 38-64. Oxford University Press, Oxford.

Gupta, A. (2006). The changing physical environment of Southeast Asian deltas: A case study on the Mekong. *In* "Hubs, Harbours and Deltas in Southeast Asia: Multidisciplinary and Intercultural Perspectives," pp 455-469. Royal Academy of Overseas Sciences, Belgium, Brussels.

Gupta, A. (2007). The Mekong River: Morphology, evolution, management. *In* "Large Rivers: Geomorphology and Management," Gupta, A. (ed.), pp 435-455. Wiley, Chichester.

Gupta, A. and Liew, S. C. (2007). The Mekong from satellite imagery: A quick look at a large river. *Geomorphology* **85**, 259-274.

Gupta, A., Rahman, A., Wong, P. P., and Pitts, J. (1987). The Old Alluvium of Singapore and the extinct drainage system to the South China Sea. *Earth Surface Processes and Landforms* **12**, 259-275.

Gupta, A., Lim, H., Huang, X., and Chen, P. (2002). Evaluation of part of the Mekong River using satellite imagery. *Geomorphology* **44**, 221-239.

Gupta, A., Liew, S. C., and Heng, A. W. C. (2006). Sediment storage and transfer in the Mekong: generalizations on a large river. *In* "Sediment Dynamics and the Hydromorphology of Fluvial Systems," Rowan, J. S., Duck, R. W., and Werritty, A. (eds.), pp 450-459. International Association of Hydrological Sciences Publication 306, Wallingford, UK.

Hallet, B. and Molnar, P. (2001). Distorted drainage basins as markers of crustal strain east of the Himalaya. *Journal of Geophysical Research* **106**, 13697-13709.

Heppe, K., Helmcke, D., and Wemmer, K. (2007). The Lancang River Zone of southwestern Yunnan, China: A questionable location for the active continental margin of Paleotethys. *Journal of Asian Earth Sciences* **30**, 706-720.

Hutchison, C. S. (1989). "Geological Evolution of Southeast Asia." Clarendon Press, Oxford.

Hutchison, C. S. (2005). The geological framework. *In* "The Physical Geography of Southeast Asia," Gupta, A. (ed.), pp 3-23. Oxford University Press, Oxford.

Kale, V. S., Gupta, A., and Singhvi, A. K. (2003). Late Pleistocene-Holocene palaeohydrology of Monsoon Asia. *In* "Palaeohydrology: Understanding Global Change," Gregory, K. J. and Benito, G. (eds.), pp 213-232. Wiley, Chichester.

Kummu, M. and Sarkkula, J. (2008). Impact of the Mekong River flow alteration on the Tonle Sap flood pulse. *Ambio* **37**(3), 185-192.

Kummu, M. and Varis, O. (2007). Sediment-related impacts due to upstream reservoir trapping, the Lower Mekong River. *Geomorphology* **85**, 275-293.

Kummu, M., Penny, D., Sarkkula, J., and Koponen, J. (2008). Sediment: Curse or blessing for Tonle Sap Lake? *Ambio* **37**(3), 158-163.

Lacassin, R., Replumaz, A., and Leloup, P. H. (1998). Hairpin River loops and slip-sense inversion on Southeast Asian strike-slip faults. *Geology* **26**, 703-706.

Lóczy, D. (2007). The Danube: Morphology, evolution and environmental issues. *In* "Large Rivers: Geomorphology and Management," Gupta, A. (ed.), pp 235-260. Wiley, Chichester.

Lorenzetti, E. A., Brennan, P. A., and Hook, S. C. (1994). Structural styles in rift basins, interpretation methodology and examples from Southeast Asia. *Bulletin, American Association of Petroleum Geologists* **78**, 1152.

Meade, R. H. (1996). River-sediment inputs to major deltas. *In* "Sea-Level Rise and Coastal Subsidence: Causes, Consequences and Strategies," Milliman, J. D. and Haq, B. U. (eds.), pp 63-85. Kluwar, Dordrecht.

Mekong River Commission. (various years). "Lower Mekong Hydrologic Yearbook." Mekong River Commission, Bangkok, Phnom Penh.

Métivier, F., Gaudemer, Y., Tapponier, P., and Klein, M. (1999). Mass accumulation rates in Asia during the Cenozoic. *Geophysical Journal International* **137**, 280-318.

Moore, A. E., Cotterill, F. P. D., Main, M. P. L., and Williams, H. B. (2007). The Zambezi River. *In* "Large

Rivers: Geomorphology and Management," Gupta, A. (ed.), pp 311-332. Wiley, Chichester.

Nguyen, V. L., Ta, T. K. O., and Tatheishi, M. (2000). Late Holocene depositional environments and coastal evolution of the Mekong River delta, Southern Viet Nam. *Journal of Asian Earth Sciences* **18**, 427-439.

Nutalaya, P., Sophonsakulrat, W., Sonsuk, M., and Wattanachai, N. (1989). Catastrophic flooding—An agent for landform development of the Khorat Plaeau: A working hypothesis. *In* "Workshop on Correlation of Quaternary Successions in South, East and Southeast Asia, Proceedings," Thiramongkol, N. (ed.), pp 95-115. Chulalongkorn University, Bangkok.

Penny, D. (2006). The Holocene history and development of the Tonle Sap, Cambodia. *Quaternary Science Reviews* **25**, 310-322.

Plinston, D. and He, D. (2000). Water resources and hydropower in the Lancang River basin. *In* "Policies and Strategies for Sustainable Development in the Lancang River Basin," pp 234-266. Asian Development Bank, TA 3239, Manila.

Ta, T. K. O., Nguyen, V. L., Tateishi, M., Kobayashi, I., and Saito, Y. (2001). Sediment facies, diatom and foraminifera assemblages in a late Pleistocene-Holocene incised valley-sequence from the Mekong River Delta, Bentre Province, Southern Viet Nam: The BT2 core. *Journal of Asian Earth Sciences* **20**, 83-94.

Ta, T. K. O., Nguyen, V. L., Tateishi, M., Kobayashi, I., Tanabe, S., and Saito, Y. (2002a). Holocene delta evolution and sediment discharge of the Mekong River, southern Viet Nam. *Quaternary Science Reviews* **21**, 1807-1819.

Ta, T. K. O., Nguyen, V. L., Tateishi, M., Kobayashi, I., Saito, Y., and Nakamura, T. (2002b). Sediment facies and Late Holocene propagation of the Mekong River Delta in Bantre Province, southern Viet Nam: An example of evolution from a tide-dominated to a tide- and wave-dominated delta. *Sedimentary Geology* **152**, 313-325.

Ta, T. K. O., Nguyen, V. L., Tateishi, M., Kobayashi, I., and Saito, Y. (2005). Holocene delta evolution and depositional models of the Mekong River Delta, southern Viet Nam. *In* "River Deltas—Concepts, Models, and Examples," Giosan, L. and Bhattacharya, J. P. (eds.), pp 453-466. SEPM Special Publication 83, Tulsa, Oklahoma.

Tapponnier, P., Peltzer, G., and Armijo, R. (1986). On the mechanics of the collision between India and Asia. *In* "Collision Tectonics," Coward, M. P. and Ries, A. C. (eds.), pp 115-157. Geological Survey of London Special Publication 19, London.

Thiramongkol, N. (1983). Reviews of geomorphology of Thailand. *In* "Geomorphology and Quaternary Geology of Thailand," Thiramongkol, N. and Pisutha Arnond, V. (eds.), pp 6-23. Chulalongkorn University, Bangkok.

Udomchoke, V. (1989). Quaternary stratigraphy of the Khorat Plaeau area, Northeastern Thailand. *In* "Workshop on Correlation of Quaternary Successions in South, East and Southeast Asia, Proceedings," Thiramongkol, N. (ed.), pp 69-94. Chulalongkorn University, Bangkok.

United Nations Economic and Social Commission for Asia and the Pacific. (1990). "Atlas of Mineral Resources of the ESCAP Region Lao People's Democratic Republic: Explanatory Brochure," Volume 7. United Nations, New York.

United Nations Economic and Social Commission for Asia and the Pacific. (1993). "Atlas of Mineral Resources of the ESCAP Region Cambodia: Explanatory Brochure," Volume 10. United Nations, New York.

United States Geological Survey. (1994). Southeast Asia. *In* "East Asia Geographic Map Series." Sheet 3, Coordinating Committee for Offshore Prospecting, Technical Secretariat, Bangkok.

Walling, D. E. (2008). The changing sediment load of the Mekong River. *Ambio* **37**(3), 150-157.

Wang, E., Burchfiel, B. C., Royden, L. H., Chen, L., Chen, J., Li, W., and Chen, Z. (1998). Late Cenozoic Xianshuihe-Xiaojiang, Red River, and Dali fault systems of southwestern Sichuan and central Yunnan, China. *Geological Society of America Special Paper* **327**, 1-108.

Wood, S. H., Ziegler, A. D., and Tharaporn, B. (2008). Floodplain deposits, channel changes and riverbank stratigraphy of the Mekong river area at the 14th century city of Chiang Saen, Northern Thailand. *Geomorphology*, **101**, 510-523.

The Hydrology of the Mekong River

Peter T. Adamson,[1] Ian D. Rutherfurd,[2] Murray C. Peel,[3] and Iwona A. Conlan[3]

[1]Consultant to the Mekong River Commission, and Senior Visiting Research Fellow, School of Mathematics and Statistics, University of Adelaide, South Australia

[2]Department of Resource Management and Geography, The University of Melbourne, Victoria 3010, Australia

[3]Department of Civil and Environmental Engineering, The University of Melbourne, Victoria 3010, Australia

OUTLINE

1. INTRODUCTION

The Mekong River is one of the great tropical rivers of the world. The regular huge wet-season flow peak supports arguably the most productive river fishery in the world (MRC, 2003, Chapter 9). Unlike many major rivers in Asia, the channel of the Mekong, and its flood regime, remain relatively intact (Revenga *et al.*, 1998). Although there are 58 dams in the catchment of the Mekong, they are not large enough to appreciably alter the consistent flow regime. In this paper, we review patterns of flow along the river, and their general controls; compare the hydrology of the Mekong with other large rivers of the world; discuss the possible effects of historical land use changes on hydrology; and indicate likely effects of climate change on hydrology. We draw heavily on the excellent hydrology reports prepared by the Mekong River Commission, and come to the following key conclusions:

- The most basic feature of the hydrology of the Mekong River is the dominance of the single wet-season flow peak, which provides a 20-fold increase in discharge during August/September. The most remarkable feature of the river's hydrology is the consistency of the size and duration of this wet-season peak, and the regularity of its onset throughout this huge basin.
- The hydrology of the Mekong River is controlled by runoff from two distinct regions: snowmelt from the Tibetan plateau on the one hand, and the left-bank tropical tributaries of Laos, on the other. It is the Lao tributaries that provide most of the wet-season peak discharge, and most of the floods.
- Intense tropical storms, together with steep catchments, mean that the Mekong produces among the largest floods of any river in terms of runoff per unit catchment area.
- Apart from the intensity of its floods, and the low coefficient of variation of the annual flows, the basic hydrology of the Mekong is otherwise unremarkable among global rivers.
- There is scant evidence that deforestation, dams, or climate change are, as yet, having any measurable effect on the hydrology of the Mekong River. This is not surprising given the size and the large interdecadal fluctuations in the river's discharge.
- Climate change will affect the Mekong River by ultimately reducing the dry-season discharge from the Tibetan Plateau, and by

increasing the strength of the SW Monsoon. Both impacts will have their main effect on the size and duration of base flows, and on the timing of the onset, and the timing of the end, of the wet season. Given the fact that biological systems of the Mekong have developed in an environment of extreme predictability, even small shifts in low flows, and the timing of the wet season, could have deleterious effects on the ecology of this large river. There is no question that issues of climate change, and issues of water resource development, should be treated together in planning the future of the Mekong River.

2. THE GENERAL HYDROLOGY OF THE MEKONG BASIN

The Mekong rises on the Tibetan Plateau at an altitude of about 5200 m and flows 4800 km southeast to the South China Sea, through six developing countries: China, Myanmar, Laos, Thailand, Cambodia, and Viet Nam. The river drains a total catchment area of 795,000 km^2 with an annual runoff of over 475 billion cubic meters of water at its mouth in the South China Sea. This makes the Mekong the 10th largest river in terms of discharge (World Resources Institute, 2003). While the Mekong is considered one of the last great unregulated rivers of the world, this reputation is now only partially deserved. Fifty-eight small to medium dams have already been built on tributaries of the Mekong (WWF, 2007), with three medium-sized dams built on the Chinese portion of the Mekong river itself, large volumes of dry-season water are already removed for irrigation in the delta region (Haddeland et al., 2006). Furthermore, with 11 major dams (over 6 m high) planned in the catchment, the World Wildlife Fund have classified the Mekong as being among the top 10 threatened rivers of the world (Wong et al., 2007). Only eight major basins have more dams planned than the Mekong (WWF, 2007).

The Mekong catchment has an unusual shape. Most catchments tend to have a dendritic form, with the width of the catchment gradually decreasing downstream, producing a characteristic tear-drop shape. The Mekong catchment, by contrast, progressively widens down valley so that its widest point is immediately upstream of its delta (see Chapter 3). This unusual shape is a consequence of the upper reaches of the river being confined within the geological folds of the great structural complex at the edge of the Himalayas.

The shape of the Mekong Basin allows us to divide the catchment into three parts.

1. In the Upper Basin, in Tibet and China (where the river is called the *Lancang Jiang*), the river and its tributaries are confined by narrow, deep gorges. The tributary river systems in this part of the basin are short and small, with only 14 having catchment areas greater than 1000 km². The Upper Basin makes up 24% of the total basin area.
2. The portion of the basin from Yunnan in China, downstream to Cambodia (2200 river km), is known as the Lower Mekong Basin. At the Cambodian border, the flood-season river flows back into the Tonle Sap floodbasin.

The Tonle Sap Lake in Cambodia is the largest body of freshwater in southeast Asia, and forms a key part of the Mekong hydrological system. Its mean surface area changes from 3500 km² during the dry months (January to April-May) to a maximum of up to 14,500 km² during the wet season (MRC, 2005).

3. The third distinct section of the river is the Delta, which begins at Phnom Penh, and extends into Viet Nam. This article concentrates on the hydrology of the Lower Mekong below Yunnan.

The hydrology of the Mekong River is characterized by a huge mean annual discharge; concentrated in an extremely regular wet-season peak. The size of the wet-season peak, as well as its highly predictable timing, are the defining characteristics of large tropical monsoonal rivers. The discharge volume of the wet-season peak averages 65 km³ at the Chinese border, increasing to 350 km³ at the Cambodian floodplain.

Although the river is usually characterized as having a dry and a wet season, it is more useful to divide the year into four distinct seasons. The annual minimum daily discharge usually occurs in early April (point 1 on Fig. 4.1). The doubling of this discharge, generally in late

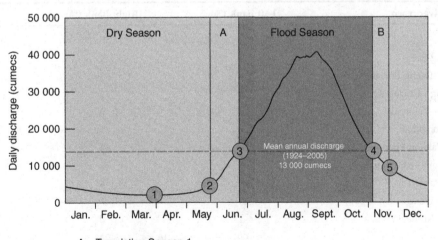

A = Transistion Season 1

B = Transistion Season 2

FIGURE 4.1 The four seasons of the Mekong River's hydrological year (Adamson, 2006).

May, defines the start of the first transition season (point 2). This ends when the flood season starts (point 3). Note that the onset of the flood season (the point at which the discharge exceeds the mean annual discharge for that station) occurs within a few days at the end of June. The second transition season defines the period between the end of the flood season (point 4) and the start of the dry (point 5), which occurs when rates of daily flow decrease become typical of "base-flow" recession. On average, the dry-season onset is in late November. The flood season lasts for just over 130 days (Table 4.1). The fact that the start and end of the annual flood can be guaranteed to occur within a period of just 2 weeks is a remarkable and defining characteristic of this huge river system.

The timing of the onset and the duration of the seasons is virtually identical at the upstream (Vientiane) and downstream (Kratie) ends of the Lower Mekong tract (Fig. 4.2) (*see Fig. 4.3 for the location of gauges along the river*). This is despite the fact that the hydrology of the former is dominated by flows originating on the Tibetan Plateau, while at Kratie the hydrological regime is largely dictated by flows entering the mainstream from the large left-bank tributaries in Lao PDR, downstream of Vientiane. The homogeneity of the temporal aspects of the system is remarkable. As will be discussed later, the regularity of these seasonal onsets and durations, means, first, that the biology of the river will be finely tuned to these regularities; and second, that the river system is likely to be very sensitive to any artificial changes in these characteristics.

In the upper part of the Lower Mekong system, at Vientiane, the flow originating from China and Burma, the so-called "Yunnan Component," not only provides most of the dry-season flows, but in addition, most of the floodwater during the majority of years. Average contributions range from over 75% during the low-flow months in April and May, to over 50% during the peak-flow months of July, August, and September. The year-to-year range of the contributions is, none the less, quite wide and indicates a complex and varying net contribution. Much further down the system at Kratie, the large left-bank tributaries in Laos provide

TABLE 4.1 Percentage flow contributions for mainstream reaches (from MRC, 2005, p. 27)

River reach	Left bank (%)		Right bank (%)	Total (%)
China		16		16
China—Chiang-Saen	1		4	5
Chiang Saen—Luang Prabang	6		3	9
Luang Prabang—Chiang Khan	1		2	3
Chiang Khan—Vientiane	0		0	0
Vientiane—Nongkhai	0		1	1
Nongkhai—Nakhon Phanom	19		4	23
Nakhon Phanom—Mukdahan	3		1	4
Mukdahan—Pakse	5		6	11
Pakse—Stung Treng	23		3	26
Stung Treng—Kratie	1		0	1
Total	60	16	24	100

Note the large contribution from the left-bank tributaries draining into the Nongkhai and Pakse reaches.
Note: percentages rounded to nearest whole number

FIGURE 4.2 Historical percentage frequency of the weeks during which the annual minimum discharge and flow season transitions occurred on the Mekong mainstream at Vientiane (1913-2005) and Kratie (1924-2005). The numbers in brackets are the standard deviation in weeks of each statistic. (from Adamson, 2006)

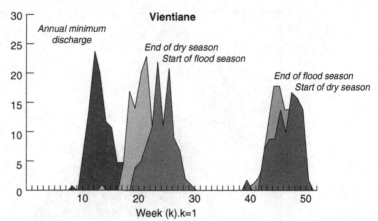

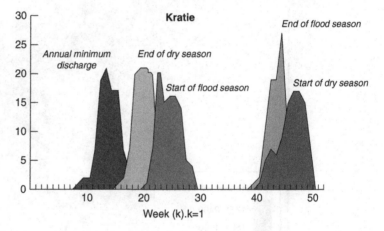

most of the flood-season flow on the mainstream such that the "Yunnan Component" is reduced to a modest 15–20%. However, it can provide over 40% of the dry-season flow in April (Adamson, 2006).

However, for the total discharge of the Lower Mekong, only about 16% comes from China and 2% from Myanmar (Table 4.1). The majority of flow volume in the Mekong originates from the tributaries below the Lao PDR border (Fig. 4.3). The left-bank tributaries drain the high-rainfall areas of Lao PDR downstream of Nongkhai (Table 4.1); and the right-bank tributaries (mainly the Mun and Chi rivers) drain the low relief, lower rainfall regions of northeast Thailand. It is the left-bank tributaries of Laos that contribute to the major wet-season

flows (Fig. 4.3). The Mun-Chi Basin that drains northern Thailand is the largest single tributary of the Mekong by basin area, but it contributes relatively little discharge because it drains a relatively dry, flat region. Shortly after the Mekong enters Cambodia, below Stung Treng, over 95% of the flows have already joined the river (MRC, 2005). From here on downstream, the terrain is flat and water levels rather than flow volumes determine the movement of water across the landscape.

2.1. Flooding in the Mekong River

Even though floods can cause major devastation along the Mekong River, the peak discharge of the largest floods tends to be only about double

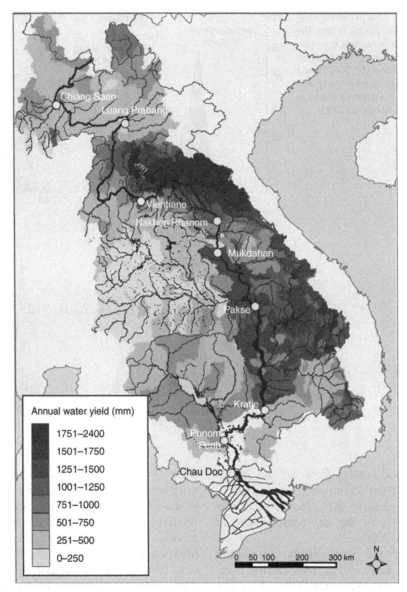

FIGURE 4.3   Annual water yield of the Lower Mekong Basin (from MRC, 2008). (See Color Plate 8)

the size of the bankfull discharge (Table 4.2, Fig. 4.4). The floodplain of the Mekong River is relatively narrow for most of its length. This means that, down to the Cambodian border, floods are restricted to several kilometers in width. In Cambodia, however, the flood spreads

out over hundreds of kilometers, and up to 5 m in depth (Fig. 4.4). There is no doubt that the flooding of the Tonle Sap basin is one of the world's great hydrological phenomena, and it is likely that the quantitative and temporal aspects of the regime have remained virtually

TABLE 4.2 Mekong at Vientiane

Recurrence interval (years)	2	5	20	100	200
Discharge (cumecs)	16,000	18,500	22,000	25,500	27,000

Estimated annual recurrence intervals of annual maximum discharge in cumecs (Generalized Extreme Value Model using the 1960-2003 data) (MRC, 2005).

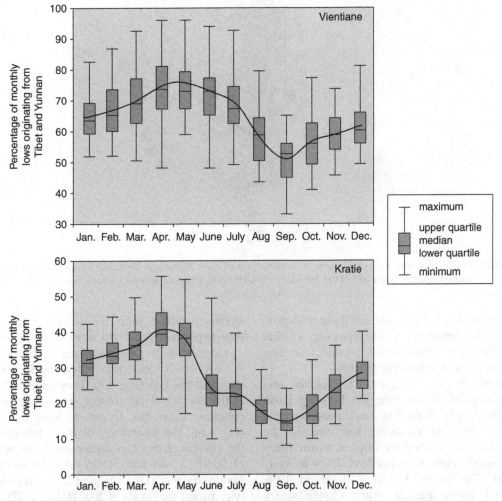

FIGURE 4.4 Historical contribution of flows from Tibet and Yunnan to monthly flows on the Mekong mainstream at Vientiane (1913-2005) and Kratie (1924-2005) (from Adamson, 2006).

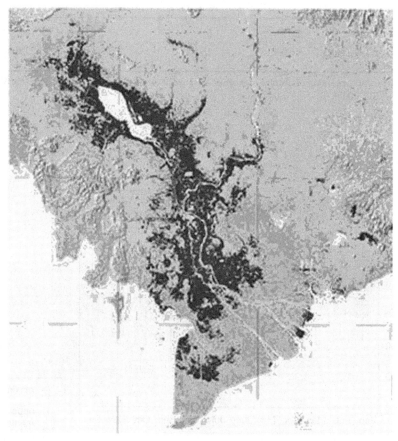

FIGURE 4.5 The blue area shows the extent of flooding in the Lower Mekong in 2006. The backflow into Tonle Sap is clear, as is the flooding at the upper end of the delta (Source: Dartmouth Flood Observatory image used in MRC, 2007). (See Color Plate 9)

unchanged over the last 5000-6000 years (Penny, 2006). It is interesting to note from Fig. 4.5 that the delta region tends to be flooded less than does the Tonle Sap basin upstream.

Historically, the incursion of cyclones and severe tropical storms over the Mekong Basin from the South China Sea, have generated the largest floods. At Vientiane, for example, the 2008 flood was caused by tropical storm Kammuri, which blew over northern Laos in early August. The largest historical flood, in 1966, occurred when tropical storm Phyllis struck China's southwestern Yunnan province. In the downstream end of the basin, these severe tropical storms combine with the southwestern Monsoon to produce floods in Cambodia and the delta.

Flooding in the Lower Mekong is dominated by the left-bank Laotian tributaries, the Se Kong, Se San, and Sre Pok (Fig. 4.3). They contribute 18% of the average flood volume, exceeding the contribution from the Upper Mekong in China (Fig. 4.6). The hydrology of these tributaries is the pivotal element in determining the severity of flooding and inundation on the Cambodian floodplain and in the Mekong Delta, as indeed it was during the floods of 2000 (MRC, 2007).

The timing of the Mekong flood is also fairly consistent. Over the 10-year period between

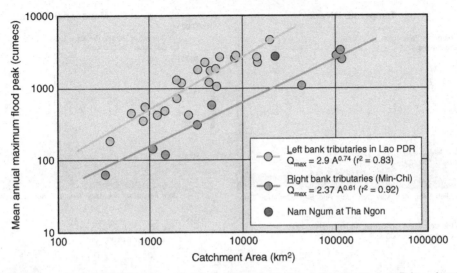

FIGURE 4.6   Contrast between flooding from the left- and right-bank tributaries, Vientiane to Pakse, showing the relationship between catchment area and mean annual flood within the left-bank tributaries in Lao PDR and the right-bank tributaries in Thailand.

1992 and 2003, the earliest that the flood peak occurred at Pakse was on 20 July and the latest was on 29 September. On average, it arrived on 1 September with a standard deviation of 23 days (Chapter 16, this volume).

3. THE GLOBAL CONTEXT OF MEKONG RIVER HYDROLOGY

The definitive characteristics of the Mekong River are its single, smooth, and regular flood peak and the consistent size and regularity of that peak. When compared with other large rivers on the globe, the most characteristic feature of the Mekong is its large runoff per unit area. In regard to other hydrological indices, neither the Mekong's annual flow characteristics, nor its flood characteristics, are remarkable.

As discussed earlier, the definitive feature of the Mekong's annual flow regime is the single annual flood hydrograph produced by the SW Monsoon. The vast spatial extent of the drainage systems means that the individual storm events caused by monsoonal depressions coalesce and accumulate into a single seasonal flood hydrograph. The highly seasonal and integrated nature of the flood hydrograph can be seen by comparing it with that of two temperate catchments, and the monsoonal flows of a much smaller river system in Pakistan (MRC, 2005; Fig. 4.7). The flood hydrology of the temperate zone rivers is nonseasonal, with seemingly random flood pulses throughout the year. This is the case even for extremely large river basins such as the Rio Uruguay in South America (Fig. 4.7). Smaller monsoonal catchments, such as the Chenab (a large tributary of the Indus), produce a clearly defined flood season that extends over 5–6 months. However, the smaller catchment leads to a great number of large, but short-term, fluctuations in discharge. With a catchment area that is 20 times larger than the Chenab, these fluctuations are smoothed out in the Mekong catchment. Such "monomodal" flood regimes are referred to as "flood pulsed" (Junk and Wantzen, 2004).

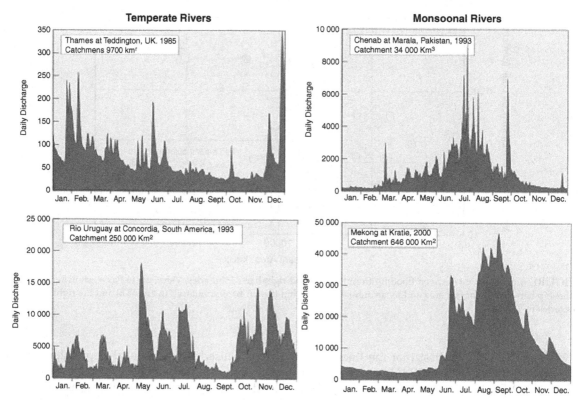

FIGURE 4.7　Annual hydrological regimes of temperate (left) and monsoonal (right) river systems (units are cumecs) (taken from MRC, 2005).

3.1. Average Annual Discharge on the Mekong: A Global Comparison

It is interesting to compare the hydrology of the Mekong River with the hydrology of other large rivers of the world, and more specifically, with other tropical rivers. We compare the annual flow statistics of three stream gauges from the main-stem of the Mekong (Table 4.3), and one from the Nam Mun tributary in Thailand, with comparable records from a global data set of 893 stations. We also compared the mean annual flow characteristics with just the tropical rivers in the data set.

Comparing the mean annual flow of the Mekong with other rivers demonstrates that its hydrology is not unusual for basic measures such as mean annual flood magnitude, mean annual runoff per unit area, coefficient of variation (Cv) of the annual discharge, and skewness. In general, there is more difference between the gauging stations on the Mekong than between the Mekong and other rivers.

One variable in which the Mekong does stand out as being unusual is the Cv of the annual discharge. The three main-stem gauges on the Mekong have an unusually low Cv for their catchment area (Fig. 4.8). This means that the volume of flow from year to year is very consistent. The same is not true for the gauge on the Thai tributary (Fig. 4.8), which has a coefficient that falls in the middle of the distribution for streams of comparable size. Although tropical rivers tend to have lower

TABLE 4.3 Four Mekong gauging stations used for comparison of mean annual discharge with a global data set of rivers

River	Station	Latitude	Longitude	Area (km^2)	Start	End
Mekong	Chiang Saen	20.27	100.1	189,000	1961	1986
Mekong	Mukdahan	16.53	104.73	391,000	1924	1986
Mekong	Nakhon Phanom	17.4	104.8	373,000	1962	1986
Nam Mun	Ubon	15.22	104.87	104,000	1956	1986

Note that the Nam Mun is a tributary.

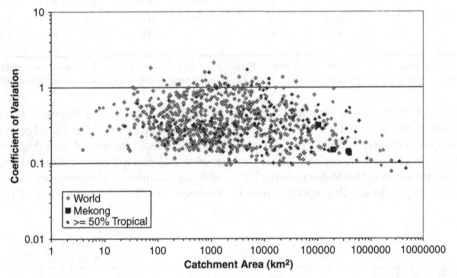

FIGURE 4.8 Coefficient of variation of the annual flow on the Mekong plotted against catchment area (Global data compiled by Murray Peel).

Cv (Fig. 4.8), the Mekong has unusually low Cv even for tropical streams. We now turn to the hydrological characteristics of flood flows.

3.1.1. Flooding on the Mekong: A Global Comparison

To summarize, when compared with other rivers, floods on the Mekong are relatively small in comparison to the size of the average annual flood, but they are relatively large in terms of the volume of runoff per unit catchment area. The size of the floods is also very predictable from year to year. Table 4.4 summarizes various statistics for the annual flood series for 10 hydrological stations on the Lower Mekong River. Some of these statistics are compared with a large data set of global flood records.

In common with many tropical rivers, the largest floods on the Mekong are not substantially larger than the average flood for any given year. For example, the 1939 flood was one of the largest recorded on the Mekong, but its maximum daily discharge was not even twice the daily peak of the average annual flood. The Fly River in Papua New Guinea,

TABLE 4.4 Various statistics for the annual flood series for 10 hydrological stations on the Lower Mekong River

Station name	Start year	End year	N (years)	Area (km^2)	MAF	SMAF	Iv	Cs	Lag-1
Chiang Saen	1960	2007	48	189,000	10,555	0.06	0.26	1.76	0.04
Luang Prabang	1939	2008	70	268,000	15,127	0.06	0.24	0.22	0.07
Vientiane	1913	2008	96	299,000	16,768	0.06	0.20	0.41	0.10
Chiang Khan	1967	2007	41	292,000	15,842	0.05	0.22	−0.10	0.12
Nong Khai	1969	2007	39	302,000	16,494	0.05	0.22	−0.24	0.18
Nakhon Phanom	1924	2007	84	373,000	25,979	0.07	0.18	−0.24	0.16
Mukdahan	1923	2007	85	391,000	28,617	0.07	0.17	−0.07	0.02
Pakse	1923	2007	85	545,000	37,609	0.07	0.15	0.15	−0.07
Stung Treng	1950	2007	58	635,000	51,286	0.08	0.18	0.01	0.05
Kratie	1924	2007	84	646,000	51,477	0.08	0.17	0.05	0.17

MAF, mean annual flood; SMAF, specific mean annual flood (Q per unit catchment area); Iv, index of variability (standard deviation of the natural logs of the annual floods); Cs, coefficient of skewness; Lag-1, Lag-1 autocorrelation of annual floods.

another tropical river, has a similar Q_{100}/Q_2 ratio. This compares with many Australian rivers, for example, that have 100-year flood peaks that are 10-20 times the annual peak.

For its catchment area, the Mekong River has an unusually large flood. The specific mean

annual floods for 10 gauging stations on the Lower Mekong plot at the top edge of the distribution for their catchment area (Fig. 4.9). Intense monsoonal precipitation, combined with reasonably mountainous catchments in Laos, generates extreme runoff. The coalescence of runoff from

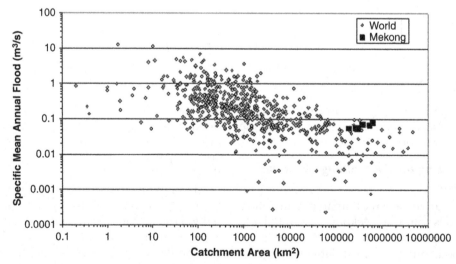

FIGURE 4.9 Comparison of the specific mean annual flood (runoff per unit catchment area) for 10 stations on the Lower Mekong River, compared with 624 global stations (data set compiled by Murray Peel).

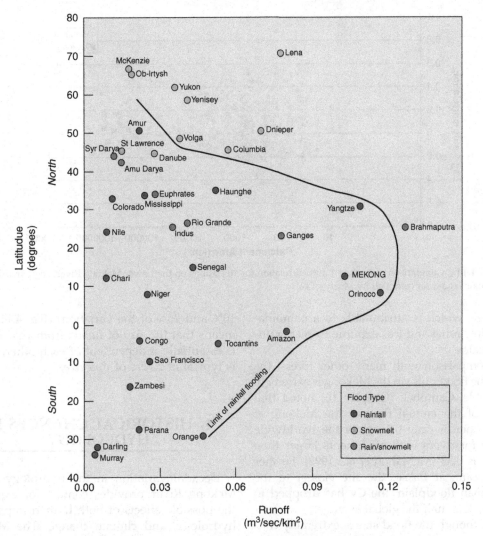

FIGURE 4.10 Comparing maximum floods on the Mekong with the largest meteorological floods "reliably" observed for global river basins exceeding 500,000 km² in area. The data are expressed as peak discharge per unit catchment area (cumecs km⁻²) (taken from MRC (2007), and based on O'Connor and Costa (2004) also IAHS (2003)).

monsoonal floods and tropical storms, into a single seasonal flood peak, means that the Mekong experiences among the largest meteorological floods of any river (Fig. 4.10).

There is a pronounced global pattern to the distribution of rivers that experience the extreme meteorological flood yields such as that experienced on the Mekong. Such rivers are generally confined to areas of the tropics between 10° and 30° N in Asia, and between 10° north and south of the equator in South America (Fig. 4.10). Such systems include the Brahmaputra, Ganges, Yangtze, Mekong, and Huang He (Yellow) River Basins. The large rivers of tropical Africa, such as the Congo, have relatively modest flood regimes in terms of unit

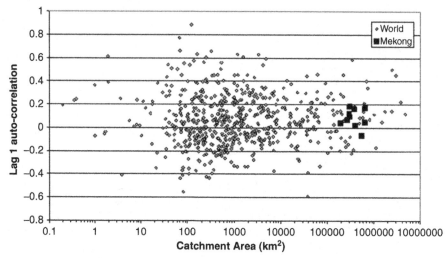

FIGURE 4.11 Comparison of the Lag-1 autocorrelation for 10 stations on the Lower Mekong River, compared with 624 global stations (data set compiled by Murray Peel).

discharge, which is attributable to a combination of low relief and less extreme tropical rainfall climates.

By comparison with many other rivers, the size of the flood pulse in the Mekong is extremely predictable. Campbell (Chapter 16) noted that the Cv of the annual flow of the Mekong at Chiang Saen is only 0.2, while the worldwide average for rivers with catchments larger than $10^5\,km^2$ is 0.33 (McMahon *et al.*, 1992). Further downstream, at Pakse, the site closest to the Cambodian floodplain, the Cv has dropped to 0.16, less than half the global average.

Even though the flood size is extremely consistent from year to year, this does not mean that the size of the flood in one year is a good guide to the size of the flood in the next year. The Lag-1 autocorrelation explores the relationship between floods in successive years. In catchments with high positive correlation, of 0.6 for example, this would mean that the magnitude of the flood in one year was quite a strong predictor of the size of the flood in the next year. In general, the floods on the Mekong are only weakly, positively correlated with the floods of the year before (explaining between

10% and 20% of the variation) (Fig. 4.11). This means that the size of floods from year to year is essentially independent of each other, which is typical of rivers of this size.

4. HISTORICAL CHANGES IN HYDROLOGY

Decadal variation in the hydrology of the Mekong River provides context for exploring the possible effects of both human impacts on hydrology, and climate change. The Mekong experiences quasi-periodic discharge fluctuations at an interdecadal scale. This fluctuation is visible in the annual Mekong dry-season flows at Vientiane (Fig. 4.12), particularly the substantial decrease in dry-season flows between the 1940s and 1950s.

Human impacts on hydrology can be classified as direct and indirect. The major direct impact on water volumes in the Mekong is diversion for irrigation. Indirect effects are caused by dams and changes in land use, in particular the conversion of forest to agriculture. These impacts

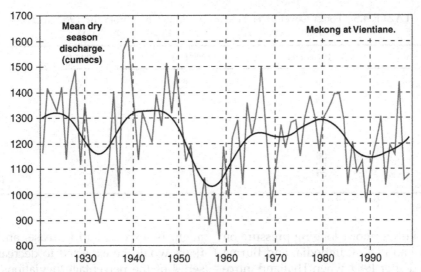

FIGURE 4.12   Quasi-periodicity of the annual low-flow hydrology at Vientiane. (the smooth function is a Gaussian time series filter with a wavelength of 20 years) (from Adamson, 2008).

can alter the gross volume of water in the river, as well as the timing and duration of flows. There has been much speculation about the effect of human impacts on flow regimes, but little investigation of the evidence, as described below.

4.1. Irrigation

Removal of water for irrigation is the largest direct hydrological impact on the Mekong River. Simulations of a 20-year flow period for the Mekong river basin indicates irrigation water requirements of 13.4 km^3 year, which corresponds to a 2.1% and 2.3% decrease in the mean annual streamflow at the outlet (Haddeland et al., 2006). Half of the diverted water is estimated to be lost via evapotranspiration, and half returned to the river (Jackson et al., 2001). While this is a substantial volume of water, when compared with irrigation water demand from other large rivers, this is a modest diversion. For example, 37% of the total volume of the Colorado River in North America is diverted for agriculture (Haddeland et al., 2006).

Although the volume of water diverted for irrigation is modest, it is important to note that this diversion occurs in the dry season, when the relative effect is greatest. For example, in the delta at Phnom Penh in February, March, and April, it is estimated that 60%, 45%, and 40% (respectively) of the flow is abstracted for irrigation (MRC, 2003). It is worth noting that the majority of dams planned for the Mekong Basin are designed for hydropower generation rather than for water extraction. The effect of these dams will be to increase dry-season flows (Podger et al., 2004) which could compensate for increases in dry-season irrigation extractions.

4.2. Effects of Deforestation

Forest degradation in the Mekong Basin has, according to Giril et al. (2001), been occurring at an unprecedented rate and scale, particularly from the 1960s onwards (Table 4.5). On the Korat Plateau in Thailand, which includes the Mun and Chi tributary systems, forest cover was reduced from 42% in 1961 to 13% in 1993

TABLE 4.5   Forest Cover Over Yunnan and Indo-China from the 1960s to 2000
(After Stibig et al., 2004, Compiled in Adamson, 2006)

Country	1960s–1970s (%)	1980 (%)	1990 (%)	2000 (%)
Cambodia	>70	>70	67	53
Lao PDR	60	–	47	41
Thailand	53	34	28	29
Viet Nam	42	–	28	30
Burma	58	–		52
Yunnan	55	–		33

(MRC, 2005). Furthermore, logging pressure on the forests of Lao PDR, Cambodia, and Burma was intensified after 1989, when Thailand introduced a logging ban within natural forests, and consequently sought increased imports from its neighbors.

Two potential hydrological impacts of deforestation might be distinguished:

1. *Total water yield* may be increased as annual evapotranspiration decreases, and
2. *Seasonal distribution of flows* may be modified as flood runoff increases and dry-season flow decreases.

The MRC (2005) proposed that land use changes in the catchment could be expected to reduce the storage of water resulting in less water flowing into the river during the dry season from December to April. Less catchment storage capacity would also tend to increase the proportion of runoff during the wet season, producing increased flood volumes. No one has yet found any conclusive evidence in the 90 years of historical data for any significant changes in rainfall-runoff relationships (MRC, 2005). In the following, we have summarized some of the evidences.

The high rates of deforestation might be expected to produce a long term, relatively smooth and systematic trend in aspects of the Mekong's flow regime. Annual flow volumes might be expected to increase, and dry-season flows would be expected to decrease. The time series of the percentage deviations (anomalies) above and below the long-term mean annual flows at Vientiane and Kratie over the 46 years between 1960 and 2005 have been plotted (Fig. 4.13), and reveal just how difficult it would be to confidently isolate any systematic pattern in the data that could be attributed to human activity. There is little evidence from the last 45 years of data of any systematic changes in the hydrological regime of the Mekong. Statistically, there is no significant upward or downward shift in the average magnitude of the flow in the years from 1960 to 2004 (Adamson, 2006).

The view that deforestation has increased the frequency and severity of flash floods is widely held, particularly in northern Thailand. Figure 4.14 shows the regional incidence of floods more than one standard deviation above the mean annual maximum flood peak, between 1970 and 2003. Even if the figures are corrected for the number of station years available in each decade, the increased proportion and incidence of such events from 1990 onwards remains apparent. Of the 25 events observed since 1970, 16 (or 65%) occurred after 1990 (MRC, 2008). Interestingly, none at all occurred during the 1980s. The regional median annual maximum event remained unchanged over the period as a whole.

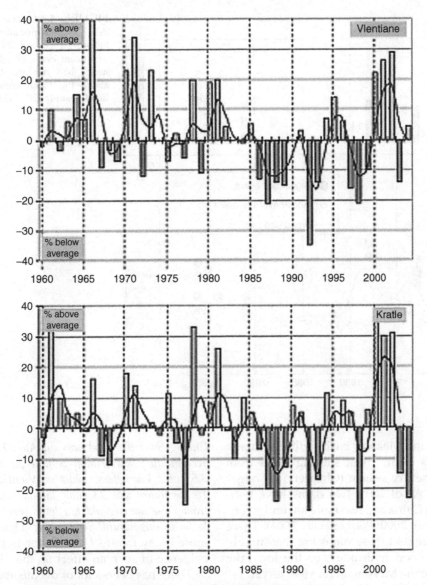

FIGURE 4.13 Mekong at Vientiane and Kratie: Percentage deviations of annual flows above and below the long-term mean, 1960-2005. The smooth lines are the 3-year moving averages (after Adamson, 2006).

4.3. Effects of Hydropower Dams

Over the next 20-30 years, the major area of water resource development is anticipated to be hydropower, the fundamental feature of which is the shift of water from the wet to the dry season via reservoir storage. Since hydropower schemes are in principle nonconsumptive, the "at site" mean annual flow remains the same except for evaporation. There is no doubt that the numerous hydropower dams

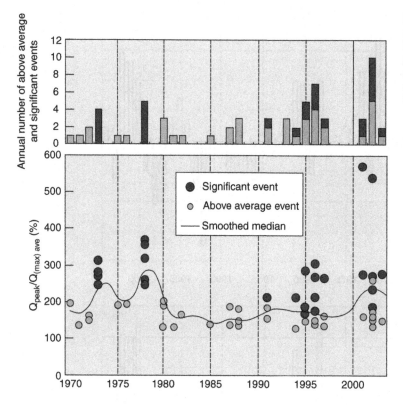

FIGURE 4.14 The regional incidence of "significant flood events" between 1970 and 2003 (defining a significant event as a flood that was more than one standard deviation above the mean annual maximum flood peak) (MRC, 2008).

planned for the Mekong basin will eventually change flows in the river, shifting flows from the wet to the dry season (Chapter 16; Podger et al., 2004). As of 2008, two dams have been completed in China and two more are under construction (Hori, 2000; Podger et al., 2004). There is no evidence that these dams are measurably altering the flood hydrology, or the low-flow hydrology, of the Mekong main-stem as yet.

The popular press has blamed the severity of recent floods on the Lower Mekong on the Chinese dams—Manwan, Dachaoshan, and Jinghong. However, the MRC released a statement after the August 2008 floods stating that the "combined active storage capacity of the Chinese dams, is less than 1 cubic kilometer, only a small part of which could be released within the period that the floodwater accumulated ... Given that at

Chiang Saen the flood peak on Aug. 12 showed an accumulated flood runoff volume for the month of 8.5 cubic kilometers, while at Vientiane on Aug. 15 the figure was 23 cubic kilometers, any release from these dams could not have been a significant factor in this natural flood event." Dams are much more likely to affect low flows, but there is no evidence of such an effect on the Mekong.

There has been a lot of debate about the dry-season hydrology of the mainstream Mekong and there is a widespread belief that there has been significant change due to upstream reservoir storage in China. Figure 4.15 shows the minimum daily discharge averaged over a sequence of 90 days in each year from 1960 to 2004 for Vientiane and Kratie (MRC, 2005). Such a "long duration" statistic can be regarded as an effective measure of dry-season flow conditions

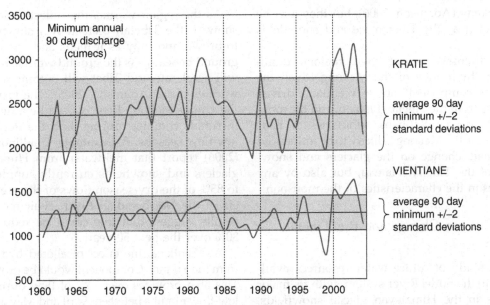

FIGURE 4.15   Mekong mainstream annual minimum 90-day mean discharges observed at Vientiane and Kratie between 1960 and 2004. The shaded bands indicate a range of ±2 standard deviations either side of the average 90-day low flow. Years when the minima fell outside this range would be considered exceptionally above or below normal dry-season conditions (MRC, 2005).

from year to year. There is no evidence of any systematic change in the low-flow hydrology, either in terms of a long-term increase or decrease in dry-season discharge. A study by Lu and Siew (2005) shows that the impact of the Manwan Dam (located on the Mekong River in China) on water discharge was largely restricted to the upper reaches of the river very close to the dam itself.

There was concern that the low-dry season flows in the 2003-2004 season was caused by the dams in China. However, the low flows in 2003-2004 were more intense downstream, in Cambodia, than closer to the dams in Laos and Thailand, indicating that they were caused by low rainfall in the lower basin, rather than any activities upstream (Campbell and Manusthiparom, 2004).

The conclusion that there has been no measurable change in the statistical properties of flows in the Mekong Basin over recent decades

is supported by Campbell (2007). Analysis of 43 years of flow records at Pakse, showed no significant trend in the maximum annual daily flow, the number of inundation days, or the number of flood years per decade.

5. POTENTIAL EFFECTS OF CLIMATE CHANGE ON THE HYDROLOGY OF THE MEKONG RIVER

Arora and Boer (2001) predict that anthropogenic climate change would result in lower mean annual flows and floods for the Mekong River, but that the seasonal distribution of water would remain the same. They predict that flood season volumes would decrease by 15%. However, as yet, there does not appear to be significant evidence that any such changes are manifesting themselves in the hydrology of the Mekong (Adamson, 2006), but they may be anticipated as the Tibetan ice and snowfields retreat.

As discussed above, two regions dominate the hydrology of the Mekong River: the "Yunnan component" above Vientiane, driven by flows from the Tibetan Plataeu, and the monsoonally driven left-bank tributaries of Laos PDR. Thus, the Mekong is likely to be influenced by climate change on the glaciers and snowfields of the Tibetan Plateau, but also by any changes in the characteristics of the monsoon.

5.1. Climate Change and the Tibetan Plateau

In a study of future water resources availability in the Sutlej River system, with its major sources in the Himalayan glacial snowfields, Singh and Bengtsson (2002) found that the main impact of climate change was on seasonal rather than annual water availability. Reduction of spring and summer meltwater would have severe implications for future regional water resources at times of the year when hydropower and irrigation demand are at their peak. The same would be true for the Mekong.

Evidence is accumulating of accelerating rates of glacial and snowfield recession. On the Qinghai-Tibetan Plateau, during the past 40 years or more, glacial extent has shrunk by some $6600 \, \text{km}^2$ out of a total of $110,000 \, \text{km}^2$ (WWF, 2005). Presently, 95% of global glacial systems are in retreat, such that the long-term implications for the freshwater resources of much of Asia are immense. The Tibetan Plateau may be particularly vulnerable to anthropogenic climate change. Permafrost on the Tibetan Plateau begins at an average altitude of about 5000 m, at the edge of the ecological limit for vegetation. The boundary between intermittent or seasonal permafrost areas on the Tibetan Plateau are likely to shift toward the center of the Plateau and the whole region will become warmer.

As the region warms, the reflective ice and snow of the Tibetan plateau will slowly turn to brown and gray as it melts and reveals the ground beneath. As the ground warms, melting will accelerate and Tibet will become a much warmer place (Adamson, 2008). The result for the Upper Mekong River would be a decrease in runoff volumes, during both the flood and low-flow seasons. For example, Challinor et al. (2006) report that meltwater from Himalayan glaciers and snowfields currently supplies up to 85% of the dry-season flows of the great rivers of the North Indian Plain. Their modeling results suggest that this could be reduced to 30% over the next 50 years.

At Vientiane, the effect of altered hydrology from the Yunnan Component would be profound. This area provides over 75% of flow during the low-flow months between April and May, to over 50% during the peak-flow months of July, August, and September. But even much further downstream, at Kratie, the Yunnan Component can contribute over 40% of flow during April, the lowest flow month. Thus, it is not the flood season hydrology of the Mekong that is potentially the most vulnerable to climate change impacts on the Tibetan Plateau, but dry-season flows.

5.2. Climate Change and the Monsoon

The Tibetan Plateau also plays a major role in the climate system of Asia and, in particular, upon the timing of the SW Monsoon system, through both thermal and mechanical (uplift) influences. Consequently, any change in the thermal regime of the plateau of Tibet, through snowmelt and glacier retreat, has the potential to disrupt the pattern and intensity of the monsoon itself (Wu and Zhang, 1998). Kripalani and Kulkarni (2001) examined seasonal summer monsoon (June-September) rainfall data from 120 East Asian stations for the period from 1881 to 1998. They concluded that the summer rainfall trends in East Asia do not support claims of intensified monsoonal conditions as a

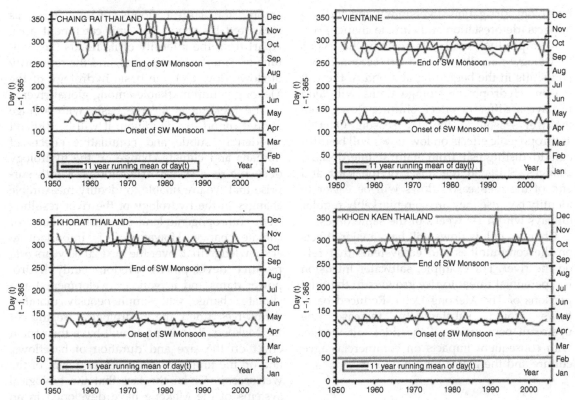

FIGURE 4.16 Historical onset and withdrawal dates of the SW Monsoon at selected sites in the Lower Mekong Basin (from Adamson, 2006), using the rainfall criteria described in Khademul *et al.* (2006).

result of CO_2-induced global warming. Adamson (2006) examined rainfall records at four locations in Laos and Thailand to see if there was any evidence of a shift in the date of monsoon onset and withdrawal over the flow record (Fig. 4.16). Two conclusions were drawn. First, the onset and end of the monsoon occur at very similar times across the Mekong Basin. Second, there is no evidence that, over the last 50+ years, the timing and duration of the SW Monsoon has changed significantly.

Thus, there is little evidence of any recent shifts in the monsoon, but the consensus is that under climate change, the SW Monsoon will both intensify, and become more variable between years. Turner *et al.* (2007) propose that the standard deviation of seasonal monsoonal

rainfall totals could increase by up to 14%, leading to larger floods, but with droughts just as likely as at present. One aspect that is generally agreed is that the incidence and severity of tropical cyclones will increase. Historically, the incursion of cyclones and severe tropical storms over the Mekong Basin from the South China Sea has played the major role in generating the largest observed floods.

5.3. Hydrological Change in the Mekong River: A Summary

When taken as a whole, there is scant evidence that humans have, as yet, altered the hydrology of the greater Mekong River. Nevertheless, both direct human impacts on the hydrology (irrigation diversions), and indirect impacts (deforestation and climate change) will tend to produce similar changes in the river's hydrology: namely, reduced dry-season flows, and shifts in the beginning and end of the wet season. Hydropower storage dams will tend to have a different effect, shifting flows from the wet season to the dry season. Overall, anthropogenic effects on low flows will be difficult to distinguish from natural interdecadal changes. Shifts in the onset of the start and end of the wet-season flows will be easier to identify because they are so remarkably regular (always within 1-2 weeks of the same dates).

Changes in low flows will have major effects on the communities and ecosystems that rely on the river. For example, saltwater intrusion is a perennial threat to rice growers in the tidal portions of The Mekong Delta. Reduced baseflows would substantially increase the upstream penetration of the tidal salt wedge, with consequent impacts on commercial agriculture and the global food economy.

6. CONCLUSIONS

The huge area of the Mekong basin means that the complex hydrological inputs to the river are homogenized into a single annual transition from low flow to high flow. The most remarkable feature of the river's hydrology is the consistency of the size and duration of this wet-season peak, and the regularity of its onset throughout this huge basin. The hydrology of the Mekong River is controlled by runoff from two distinct regions: snowmelt from the Tibetan plateau on the one hand, and runoff from the SW Monsoon entering from the left-bank tributaries of Laos, on the other. It is the Laos tributaries that provide most of the wet-season peak discharge, and most of the floods.

Intense tropical storms, coupled with steep catchments, mean that the Mekong produces among the largest floods of any river as measured by runoff per unit catchment area. Apart from the intensity of its floods (i.e., volume of runoff per unit area), and the regularity of flows (low Cv), the basic hydrology of the Mekong is unremarkable among global rivers.

The Mekong is a large basin with a huge annual runoff. By definition, it will be difficult to identify subtle and cumulative effects of humans and climate change on the hydrology of such a river system. Therefore, it is little surprise that we are unable to identify measurable changes in the hydrology of the river resulting from anthropogenic land use change, dams, or climate change. However, it is important to acknowledge that, over the next 20-50 years *both* resource development (predominantly hydropower dams and irrigation developments) and climate change will simultaneously manifest themselves upon the hydrological regime of the Mekong. Both impacts will have their main effect on the size and duration of baseflows, and on the timing of the onset, and end, of the wet season. Given the fact that the biological systems of the Mekong have developed in an environment of extreme predictability, we suggest that even small shifts in low flows, and the timing of the wet season, could have deleterious effects on the ecology of this large river. For example, there is good evidence that even small changes in dry-season lake levels produced by hydropower developments, will have deleterious effects on the ecology of the Tonle Sap system in Cambodia (Kummu and Sarkkula, 2008). There is no question that issues of climate change, and issues of water resource development, should be treated together in planning the future of the Mekong River.

References

Adamson, P. T. (2006). An evaluation of landuse and climate change on the recent historical regime of the Mekong. Final Report to the Mekong River Commission, Integrated Basin Flow Management Program, December 2006.

Adamson, P. T. (2008). Adaptive risk management in large tropical monsoonal rivers—Context, challenges and opportunities. *In* "International Association for Impact Assessment, Conference," Perth (WA), May 2008. Paper commissioned by the World Bank.

Arora, V. K. and Boer G. J. (2001). The effects of simulated climate change on the hydrology of large river basins. *Journal of Geophysical Research* **106**(D4), 3335-3348.

Campbell, I. C. (2007). Perceptions, data, and river management: Lessons from the Mekong River. *Water Resources Research* **43**, W02407.

Campbell, I. C. and Manusthiparom, C. (2004). "Technical Report on Rainfall and Discharge in the Lower Mekong Basin in 2003-2004," Mekong River Commission Secretariat, Vientiane, 7pp.

Challinor, A., Slingo, J., Turner, A., and Wheeler, T. (2006). "The Indian Monsoon: Contribution to the Stern Review on the Economics of Climate Change." HM Treasury, London.

Giril, C., Shrestha, S., and Levy, M. (2001). Assessment and monitoring of land use/land cover change in continental southeast Asia. *In* "Paper prepared for presentation at the Open Meeting of the Global Environmental Change." Research Community, Rio de Janeiro, 6-8 October.

Haddeland, I., Lettenmaier, D. P. and Skaugen, T. (2006). Effects of irrigation on the water and energy balances of the Colorado and Mekong river basins. *Journal of Hydrology* **324**, 210-223.

Hori, H. (2000). "The Mekong. Environment and Development," United Nations University Press, Tokyo, 398pp.

IAHS (2003). World catalogue of maximum observed floods, compiled by R. Herschy, Wallingford, International Association of Hydrological Sciences, Pub no. 284, pp. 285.

Jackson, R. B., Carpenter, S. R., Dahm, C. N., McKnight, D. M., Naiman, R. J., Postel, S. L., and Running, S. W. (2001). Water in a changing world. *Ecological Applications* **11**(4), 1027-1045.

Junk, W. J. and Wantzen, K. M. (2004). The flood pulse concept: New aspects, approaches and applications—An update. *In* "Proceedings of the Second International Symposium on the Management of Large Rivers for Fisheries," Volume II, Welcomme, R. L. and Petr, T. (eds.), pp 117-140. FAO, RAP Publication 2004/16, Bangkok.

Khademul, M., Molla, I., Rahman, S., Sumi, A., and Banik, P. (2006). Empirical Mode Decomposition Analysis of climate changes with Special Reference to Rainfall Data. *Discrete dynamics in Nature and Society* **2006**, 1-17.

Kripalani, R. H. and Kulkarni, A. (2001). Monsoon rainfall variations and teleconnections over south and east Asia. *International Journal of Climatology* **21**, 603-616.

Kummu, M. and Sarkkula, J. (2008). Impact of the Mekong River flow alteration on the Tonle Sap flood pulse. *Ambio* **37**(3), 185-192.

Lu, X. X. and Siew, R. Y. (2005). Water discharge and sediment flux changes in the Lower Mekong River. *Hydrological Earth System Science Discussion* **2**, 2287-2325.

McMahon, T. A., Finlayson, B. L., Haines, A. T., and Srikanthan, R. (1992). "Global Runoff. Continental Comparisons of Annual Flows and Peak Discharges," Catena Verlag, Cremlingen, Germany, 166pp.

MRC (2003). "State of the Basin Report," Mekong River Commission, Phnom Penh, 300pp.

MRC (2005). "Overview of the Hydrology of the Mekong River Basin," Mekong River Commission, Vientiane, 73pp.

MRC (2007). "Annual Mekong Flood Report 2006," Mekong River Commission, Vientiane, 76pp.

MRC (2008). "Annual Mekong Flood Report 2007," Mekong River Commission, Vientiane, 76pp.

O'Connor, J. E., and Costa, J. E. (2004). Spatial distribution of the largest rainfall-runoff floods from basins between 2.6 and 26,000km^2 in the United States and Puerto Rico, *Water Resour. Res.*, 40, W01107, doi:10.1029/2003WRO02247.

Penny, D. (2006). The Holocene history and development of the Tonle Sap, Cambodia. *Quaternary Science Reviews* **25**, 310-322.

Podger, G., Beecham, R., Blackmore, D., Perry, C., and Stein, R. (2004). "World Bank Mekong Regional Water Resources Assistance Strategy: Modeled Observations on Development Scenarios in the Lower Mekong Basin," World Bank, Vientiane, 122pp.

Revenga, C., Murray, S., Abramovitz, J., and Hammond, A. (1998). "Watersheds of the World: Ecological Value and Vulnerability." World Resources Institute, Washington, DC. http://pubs.wri.org/pubs_description.cfm?PubID =2900.

Singh, P. and Bengtsson, L. (2002). Hydrological sensitivity of a large Himalayan Basin to climate change. *Hydrological Processes* **18**(13), 2363-2385.

Stibig, H-J., Achard, F., and Fritz, S. (2004). A new forest cover map of continental Southeast Asia derived from satellite imagery of coarse spatial resolution. *Applied Vegetation Science* **7**, 153-162.

Turner, A. G., Inness, P. M., and Slingo, J. M. (2007). The effect of doubled CO_2 and model basic state biases on the monsoon-ENSO system. I: Mean response and interannual variability. *Quarterly Journal of the Royal Meteorological Society* **133**(626), 1143-1157.

Wong, C. M., Williams, C. E., Pittock, J., Collier, U., and Schelle, P. (2007). "World's Top 10 Rivers at Risk." World Wildlife Fund International. Gland, Switzerland.

World Resources Institute (2003). "Watersheds of the World_CD." The World Conservation Union (IUCN), the International Water Management Institute (IWMI), the Ramsar Convention Bureau, and the World Resources Institute (WRI), Washington, DC. http://multimedia.wri.org/watersheds_2003/index.html.

Wu, G. and Zhang, Y. (1998). Tibetan forcing and timing of the monsoon onset over South Asia and the South China Sea. *Monthly Weather Review* **126**(4), 913-927.

WWF (2005). "An Overview of Glaciers, Glacier Retreat, and Subsequent Impacts in Nepal, India and China,"

Himalayan Glacier and River Project, World Wildlife Fund, Nepal, Kathmandu, 70pp.

WWF (2007). "Rivers at Risk, Dams and the Future of Freshwater Ecosystems," World Wildlife Fund International, prepared with the World Resources Institute, Gland, Switzerland.

Geomorphology and Sedimentology of the Lower Mekong River

Paul A. Carling

School of Geography, Highfield, University of Southampton, Southampton SO17 1BJ

1. INTRODUCTION

This chapter provides an introduction to the geomorphology of the Lower Mekong River within a geological and physiographic setting.

Although the system is little researched, there is sufficient literature to provide a skeleton overview of the functionality of the River. The regional context of the Mekong River within Southeast Asia is provided by Gupta (2005a).

77

The Mekong River Basin extends from the Tibetan Plateau to the South China Sea, comprises around 795,000 km^2 and incorporates seven countries (MRC, 2003). The Mekong River is the 12th longest river in the world and has the 21st largest river basin (Douglas, 2005; Gupta, 2002a,b, 2005b). The length of the main river is around 4400 km with an average annual discharge in the lower course of 15,000 m^3 s^{-1} (ca. 470 km^3 a^{-1}; Berner and Berner, 1996, p. 175). The Basin can be divided into two units. The Upper Mekong Basin (or Lancang Jiang Basin) lies within China. The Lower Mekong Basin (LMB) lies to the south of the international border between China (Yunnan Province) and Laos (Lao PDR).

2. PHYSIOGRAPHY AND GEOLOGY

The primary controls on the river course, especially in the northern parts of the basin, are the tectonic forces that arise from the northeastern movement of the Indian continental plate against the Asian massif. Subaerial erosion processes and river migration are additional influences on the style of the river types especially in the more southern parts of the basin. However, and unusually for such a large river, tectonic influence and bedrock-confined river reaches are noted at many localities throughout the course of the Lower Mekong River. The regional geology is summarized by Workman (1972), Fontaine and Workman (1978), ESCAP (1990a,b), MRC (1997), Hori (2000), Rainboth (1996), Hutchinson (1989), Hutchinson (2005), and Hope (2005) and pertinent details are provided below and in Chapter 2.

Mainland Southeast Asia is characterized by extensions of the Himalayan mountain complex and the Tibetan Plateau which are manifested as a series of high mountain ridges and deep valleys extending roughly northeast to southwest (Fig. 5.1). In the lower part of the basin, the catchment boundaries are defined to the west by extensions of the high mountain ridges as a series of discontinuous ranges of hills, primarily the Phanh Hoei range in northeastern Thailand, south through the Cardamom mountains in western Cambodia. To the east, a more continuous range, collectively known as the eastern Highlands, extends southeasterly through the Laotian Highlands to the Annamite Cordillera, Bolovens Plateau, and the Kontum Massif in Vietnam. Thus, the course of the Mekong River cuts across six broad physiographic regions (MRC, 2003). These are the Lancang River Basin, the northern Highlands, the Khorat Plateau, the eastern Highlands, the Lowlands, and the Southern Uplands.

In China, the Mekong River is known as the Lancang Jiang ("the turbulent river"), the headwaters of which drain from an altitude of 4970 m on the Tibetan Plateau and flow for nearly 800 km in Tibet before entering Yunnan province in China, where it flows for a further 1200 km. The terrain is mountainous, the river, falling some 6.5 m km^{-1}, is largely confined in steep V-shaped mountain valleys except for some wider river valleys between 3000 and 1000 m. The river finally reaches an altitude of 310 m at the river port of Simao. The terrain and the nature of the river remain similar throughout the northern Highlands until just a few kilometers upstream of Vientiane. The total vertical drop in the river within China is about 4500 m. However, the river drops only about 500 m over the remaining 2600 km to the South China Sea, giving an average slope of about 0.0002 for the whole system (Fig. 5.2).

The mountains and hills of southern Yunnan, eastern Myanmar, northern Thailand, and Laos (Lao PDR) occupy an area between the eastern face of the Shan Plateau and the western part of the Indochinese peninsula. Elevations reach 2800 m and the mountains have a north-south alignment and, in the north, the valleys drain

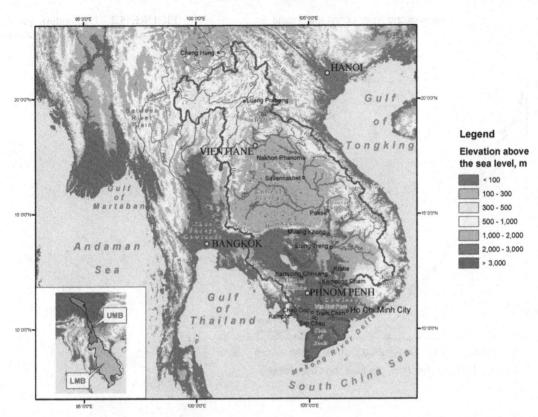

FIGURE 5.1 Mekong River Basin: topography and major towns and cities along the Mekong River. (See Color Plate 10)

into the Mekong whilst, in the south, four large rivers drain to the Chao Phrya River in Thailand. The geology of the northern Highlands is complex with extensive areas of Permian limestone with spectacular karst towers, but also with granites, sandstones, gneiss, and volcanic basalt overlays. The present course of the Mekong in northern Laos occupies the most easterly of a series of parallel valleys crossing several major Tertiary-Quaternary basins some of which have sediments up to 3 km thick. Considerable tilting occurred in this region during the Quaternary and the Mekong probably flowed through the most western of the aligned valleys in earlier times (Chapter 2). Thus, the course of the modern Mekong is

very recent when considering geological time-scales. In addition, between Chiang Saen and north of Vientiane, the river zig-zags considerably in response to fault movements in this region.

The Khorat Plateau at an average elevation of 150 m is composed of Triassic sandstones with scattered lava flows. It is fairly flat owing to extensive Pleistocene aeolian deposits blanketing complex bedrock topography but is traversed by the Mekong River in a deeply cut valley that often is gorge-like. A series of minor and larger tributaries have their sources within or traverse the plateau but owing to the relatively dry climate they deliver relatively little water to the Mekong.

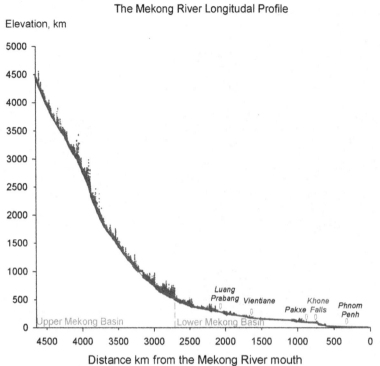

The Mekong River Longitudal Profile

Elevation, km

Distance km from the Mekong River mouth

FIGURE 5.2 Longitudinal profile of the Mekong River from source to delta.

Major tributaries, like the Mun and Chi rivers, at an elevation of 120 m, occupy deeply incised valleys now filled with around 150 m of Quaternary sediments and this deep incision and later fill probably is typical of other rivers crossing the Plateau. Deep incision may be explained by tectonic adjustments. However, the history of incision is complicated by, for example, the presence of lava flows (as young as 5720 years) across the river course near the mouth of the Mun River and at Khoné Falls on the Mekong River.

The eastern Highlands extend around 700 km from Laos through Viet Nam reaching altitudes of 2800 m and consist in part of Precambrian granites. A number of larger tributaries to the Mekong River rise in this region, notably the Se Kong, Se San, and Sre Pok rivers. The southerly limit of the Khorat Plateau is the prominent Dangrek Escarpment which roughly delimits the international borders between Thailand and Laos in the north and Cambodia in the south. The modern Mekong crosses this cuesta at the Khoné Falls and drops steeply onto the Cambodian Plain.

The Lowlands comprise chiefly the Cambodian Plain and the Mekong Delta and consist largely of Quaternary and Holocene river alluvium. During the Pleistocene, the course of the Mekong traversing the Plain south of Kratie varied and today this portion of the course remains naturally variable with evidence of course realignments during the Pleistocene and more recently. The area is essentially very flat with occasional bedrock outcrops projecting above the floodplain that locally influence the course of the river. During the wet season, high water levels in the Mekong causes reverse flow in a major right-bank tributary—the Tonlé Sap River—that links to the Tonlé Sap lake. This important relationship is described below.

The Southern Uplands are in southwestern Cambodia and consist of the Cardamon and Elephant Ranges. The Cambodian Plain lies between these uplands and the Kontum massif to the east in Viet Nam. Near Phnom Penh, the Mekong develops a major distributary, the Bassac River which flows parallel to the Mekong until further seaward where, within the delta, the Bassac and Mekong conjoin at the Van Nau pass before finally diverging into nine major distributaries (the nine dragons) and a further complex of minor distributaries and local water courses. This deltaic and Tonlé Sap hydrological system is very young. The Tonlé Sap lake formed due to subsidence of the Cambodian platform about 5720 years ago when lava flows occurred in the region at the same time. Prior to this date, the large river that flowed along the course of the modern smaller Stung Sen River and through the area that is now known as the Tonlé Sap may have drained directly into the Gulf of Thailand. More recently, probably during the "climatic optimum," the Mekong became connected to the Tonlé Sap system, in the vicinity of the present confluence, which today is known as Quatre Bras (or Chaktomuk Junction).

Parts of the delta area would have had an independent river system during the periods of sea retreat. A recent rise in sea level of around 120 m occurred in the last 17,000 years and is one of several that have influenced the lower course of the Mekong, including its unique relationship with the Tonlé Sap that is discussed later. The present mouths of the Mekong have formed during the Holocene (Ta *et al.*, 2002a,b). Presumably, as sea level fell, new channels were established much as seen today, whereas the previous mouth had been located near Kampot, south-southwest of Phnom Penh. The geology of the delta is summarized by Nguyen *et al.* (2000) and a geomorphological map provided by Haruyama and Shida (2008).

3. THE PROTO-MEKONG RIVER

Tertiary and Pleistocene tectonic adjustments and the development of an extended river basin across the Sunda Shelf, during the periods of Pleistocene lower sea levels, have influenced the modern course of the Mekong and the network of tributaries. A recent comprehensive assessment of the development of the "proto-Mekong system" has yet to be prepared (see Chapter 2), but a succinct summary of early ideas is presented by Rainboth (1996) who also summarizes the complex geology and, in particular, the Quaternary development of the Mekong River system (see also Hope, 2005; Hutchinson, 2005). Rüber et al. (2004) provide a brief synthesis of the early development of the regional drainage, including the history of the Mekong River (Clark *et al.*, 2004).

Before the Quaternary, the Mekong was not a major river but tectonic movements have led to extensive river captures, enlarging the Mekong system. What is now the Mekong River in southern Laos had its headwaters in the Khorat Plateau, whilst the upper portion of the proto-Mekong (China) drained first via the Red River in Viet Nam and latterly via the Chao Phraya in Thailand. During the Pleistocene, the Chao Phraya River lost its major headwaters owing to headward extension of the Lower Mekong system in the Khorat Plateau to give the Mekong River, much as evident today. At the same time as Pleistocene tectonic adjustments were causing changes in elevation and inclination of the Khorat Plateau, the Cambodian Plain was subject to strong movements accompanied by extensive lava flows, which also influenced the development of the Mekong drainage network (see Chapter 2).

4. GEOMORPHOLOGICAL ZONATION OF THE MEKONG RIVER

Adamson (2001), as reported in MRC (2001, p. 91) and MRC (2004, p. 8), recognized six distinct hydromorphological reaches along the Mekong River (Table 5.1). Gupta (2004) proposed an eight-division geomorphological categorization from which Carling (2006) derived a series of seven representative geomorphological reaches for the main stem Lower Mekong River plus the Tonlé Sap system. From this framework, the MRC recognize, for management purposes, six major geomorphological zones along the Mekong River that together with the riverine wetlands, especially the Tonlé Sap, can be regarded as constituting the key fluvial geomorphological attributes of the system (Table 5.1). Zone 1, from the source of the river in China to Chiang Saen in Laos, is a navigable waterway between Cheng Hung in China and Chiang Saen in Laos. Zone 2, from Chiang Saen to Vientiane, is bedrock confined and contains many rapids. Zone 3 extends from Vientiane to Pakse, wherein in the north the river is alluvial, largely single channel and sinuous, being described by Adamson as "broad and stately" but is increasingly bedrock confined towards the south. Zone 4, from Pakse to Kratie is a bedrock-confined multichannel complex with frequent rock shoals and islands and includes major rapids and water falls, notably the Khoné Falls on the Lao PDR-Cambodian border. Zone 5A is from Kratie to Phnom Penh wherein the river is again alluvial, meandering, and an anastomosed complex. Zone 5B is the Tonlé Sap system. Zone 6, from

Phnom Penh to the coast is flood-prone fluvial floodplains and the tidally influenced delta. Excluding Zone 1, the Lower Mekong River zones are considered below. At this point, the classification of channel geometry in plan requires some explanation.[1]

4.1. Zone 2: Bedrock Single-Thread Channel: Chiang Saen to Vientiane

From the China border to upstream of Vientiane, the Mekong River is a single-thread channel cut into bedrock and closely confined by mountain slopes. Locally, the river divides around in-channel bedrock outcrops and these are also often the nucleus for deposits of gravel and sand that form in-channel bars and small

---

[1]Generally the usage here follows convention with some significant variants that help define attributes of the Mekong system clearly. The channel may be relatively straight (sinuosity: 1-1.05), sinuous (sinuosity: 1.06-1.30, i.e., high radius bends), or meandering (sinuosity: 1.31-3.0, low radius bends). The river channel may exhibit a single channel or multiple interconnecting channels. In the case of interconnecting channels, the river might be described as braided, anabranching, or anastomosed. There is no convention on terminology for river reaches exhibiting multiple reaches, although several typologies have been proposed (Bridge, 1993; Nanson and Knighton, 1996). Here, braided river reaches are defined as channels with essentially a single channel at high flow (e.g., bankfull) but multiple channels at low flow. The channels are separated by banks of sand/gravel that are poorly fixed by vegetation, that migrate readily and are submerged by high flows. Divided (or wandering) reaches have up to three channels (Brierley and Fryirs, 2005, p. 119). Anastomosed reaches are defined here as multiple channels (>3) separated by large islands that sustain mature vegetation, that are fixed or migrate slowly by bank erosion and are rarely if ever inundated by high flows (e.g., bankfull). The islands are several hundreds of meters wide and possibly several kilometers long. Anabranch reaches are considered here to be where a channel (usually minor) deviates from the main stem and flows remote from the main river for a downstream distance of many kilometers before joining again with the main channel. Anabranch systems occur in Cambodia where the main channel is anastomosed. This latter planform is similar to that described by Bridge (1993, Fig. 5.4).

TABLE 5.1   Geomorphological and hydrological zonation of the Mekong River

Adamson	Gupta	Location	Carling (this chapter)	Representative reach characteristics
Zone 1	–	China	Zone 1: China	Not applicable
Zone 2	1a, 1b, 1c, 1d	1a: Chinese border to Nam Ou	Zone 2: Bedrock single-thread channel—Chiang Saen to Vientiane: deep pools, bedrock benches	1. Gradient: 0.0003
		1b: Nam Ou to 30 km upstream of Nam Loei		Channel width: 200-2000 m
		1c: 30 km reach upstream of Nam Loei		Reach length: 250 km
		1d: Nam Loei to 5 km upstream of Vientiane		Low flow depth: ca. 10 m
				Seasonal stage change: 20 m
Zone 3	2a, 2b, 3	2a: Vientiane to Pakranh	Zone 3: Alluvial single thread or divided channel—Vientiane to Pakse	2. Gradient: 0.0001
		2b: Pakranh to Mukdahan		Channel width: 800-1300 m
		3: Mukdahan to Mun confluence near Pakse		Reach length: 100 km
				Low flow depth: ca. 3 m
				Seasonal stage change: 13 m
				3. Gradient: 0.00006
				Channel width: ≤2000
				Reach length: 400 km
				Low flow depth: ≤5 m
				Seasonal stage change: 14 m
Zone 4	4, 5	4: Pakse to Muang Khong	Zone 4: Bedrock anastomosed channels: Pakse to Kratie, that is, Siphandone (4000 islands reach)	4. Gradient: 0.00006
		5: Muang Kong to Stung Treng		Channel width: 750-5000 m
				Reach length: 150 km
				Low flow depth: variable
				Seasonal stage change: 15 m
				5. Gradient: 0.0005
				Channel width: ≤15,000 m
				Reach length: 200 km
				Low flow depth: ca. 8 m
				Seasonal stage change: 9 m
Zone 5	6, 7	6: Stung Treng to Kampong Cham	Zone 5A: Alluvial meandering/anastomosed channels—Kratie to Phnom Penh: scroll bars, backwaters, overbank flooding, that is, upstream of confluence with Tonlé Sap River	6. Gradient: 0.000005
		7: Kampong Cham to Phnom Penh		Channel width: ≤4 km
				Floodplain width: 8-64 km
				Reach length: 50 km
				Low flow depth: ca. 5 m

Continued

TABLE 5.1   Geomorphological and hydrological zonation of the Mekong River—Cont'd

Adamson	Gupta	Location	Carling (this chapter)	Representative reach characteristics
			Zone 5B: Tonlé Sap Lake and River: Seasonally reversing flows	Seasonal stage change: 18 m
Zone 6	8	8: Phnom Penh to ocean	Zone 6: Alluvial Deltaic Channels—Phnom Penh to ocean: Distributaries, no marine influence in upper delta	7. Gradient: 0.000005
				Channel width: ≤3 km
				Delta inundation width: ca. 180 km
				Reach length: 330 km
				Low flow depth: 25 m
				Seasonal stage change: 15 m

vegetated islands that are inundated during the wet season (Fig. 5.3). Along the channel margins, there are extensive lateral silt "terraces" or benches (Wood *et al.*, 2008). Although there is some evidence of slight Holocene incision of the rockbed of the channel, the benches are not likely remnants of incised alluvial fill but rather are owing to deposition and preferential preservation of suspended silt along the slow moving and tree-lined channel margins during floods. The benches probably were built upward and outward during large prehistoric floods until the available space for deposition (the accommodation space) was fully filled. These benches are up to ca. 25 m above the modern river lowest bed level, are not inundated by the modern-flow regimen and are stable, supporting riparian bushes and trees, and are an important human habitation level. The outer edges are steep with vertical to 45° slopes. These high-level benches are composed of fine silt and clay and distinct bedding is usually absent, although small clay concretions are indicative of diagenetic processes. The outer edge of the benches are scoured by modern high flows to a level roughly 2 m below the bench surface, where tree roots are exposed and pop outs and other small mass movements

are common. Consequently, modern floods, including the historic high flow of 1966, do not inundate the bench levels. The complete absence of sand and gravel in the terrace suggests that this unit consists of ancient floodwater silts and clays deposited by floods that reached levels in the past that exceed the modern flood levels. Higher former flood levels may be related to the Quaternary and Holocene incision noted above or to larger discharge volumes during the Quaternary, in contrast to today. Although there is no information on the incision rates for the Mekong River in Zone 2, the degree of incision appears comparable with the neighboring Red River which has incised around 10 m in the last 200,000 years, with 3 m of this incision occurring in the last 66,000 years (Bacon *et al.*, 2008). While today fine sand is a major component of the Mekong sediment load, the dominance of silt in the benches might indicate deposition in a system that was essentially pristine with intact primary forest and little anthropogenic disturbance.

Banked or inset against the benches at lower levels, some 5-10 m above the river are one, two, or rarely three levels of sandy bars that are reworked annually during the flood season. These bars consist of light-colored fine-unconsolidated

FIGURE 5.3   View upstream from left bank near Luang Prabang. Width of channel is approximately 500 m.

well-bedded sand including cross bedding that indicates local flow in an upstream direction along the river banks. Thus, these bars are typical of "slack-water" deposits that form inflow separation and slow-flow areas close to the bank line and are best developed in reentrants to the bank line. The size of these sandy bars probably varies annually. These bars may result from anthropogenic disturbance within the system that has led to an increase in the flux of sand through the system in historic and modern times. From inspection of satellite images (Gupta and Liew, 2007), much of the course of the Mekong between Nakhon Phanom and Pakse is also a single-thread bedrock channel or, at least, has significant bedrock control alternating with alluvial reaches or exhibiting an alluvial overprint on a bedrock channel.

There is a gauging station at Luang Prabang within a bedrock section (Fig. 5.4). The lowest and the highest flow recorded are 485 m³ s⁻¹ and 25,200 m³ s⁻¹, respectively. At the gauge site, the water depth at which the terraces would flood is 30.2 m which would require an estimated discharge of 44,892 m³ s⁻¹. Thus, in the period of record (1960-2006), the benches have not been inundated and a degree of Quaternary and Holocene incision is inferred.

4.2. Zone 3: Alluvial Single-Thread or Divided Channel: Vientiane to Mun River Confluence

Between Vientiane and the Mun confluence, much of the river course is alluvial and consists of a fine sand bed with sand and silty sand river banks. However, downstream of Nakhon Phanom, and especially in the downstream reaches, bedrock increases in importance. In the alluvial reaches, the channel is not characterized by tight meanders but rather it is actively migrating laterally via long-radius bends. This process results in steep to vertical cut banks on one side and gentle alluviated banks on the opposing side. The key feature of the alluvial sections of this zone is a tendency for the river course to develop a divided channel (wandering) configuration. Large

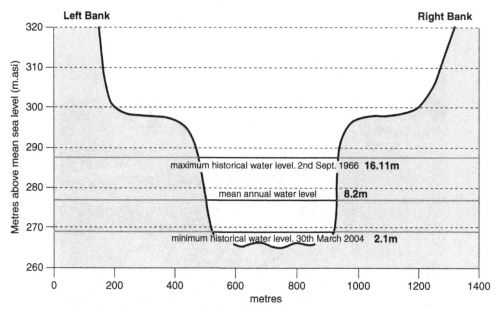

FIGURE 5.4   Channel cross section at Luang Prabang gauging station.

elongate or lozenge-shaped isolated islands individually known as "ban" occur that appear to be a relict portions of the floodplain rather than having developed by accretion of sand bars in the main channel. Evidence for this assertion is that (i) the top of the islands tend to be flat and at an elevation in accordance with that of the adjacent floodplain; (ii) the island stratigraphy and sedimentology are the same as seen in the floodplain cut banks; (iii) islands often support relict and degraded "primary" forest. A detailed 1:50,000 geomorphological map of the river showing one such island, palaeochannels, and other landscape features in the Vientiane reach was published as annex to books (Ōya, 1979, 1993) and was based on 1967 survey work.

Through time, islands may become connected to one or other river bank by progressive shoaling of one channel. Islands also extend downstream and upstream by progressive accretion at their extremities. In some examples, such as Ban Don Sang Khi, extension is augmented by aeolian dunes developing on

exposed sand flats during the low-water period. At Ban Don Sang Khi, there is a single-channel alluvial reach some 750 m wide and 7-8 m deep at low flows, immediately upstream of the confluence with the Pak Ngum River (Fig. 5.5).

Natural levées occur on both banks which suggest the river used to overtop them, although it does not do so under the modern-flow regimen except in the vicinity of tributary confluences. Beyond the levées are wetlands which have been artificially drained to varying degrees for agriculture. The river bank on the Laotian side is about 8-10 m high and is composed primarily of silt and clay. The bankline is in places vertical in the upper half to two-thirds with slumped material along the base or a cut surface inclined to about 45° extends from the low water up to the vertical sections. Most of the bank face is unvegetated but more gentle slopes are patchily covered by annuals, or cultivated for recession agriculture. Bank collapse is by block fall or locally rotational mass failure. Significant basal cutout is not evident and so high flows are responsible for

FIGURE 5.5   Oblique aerial view looking downstream during high-stage river discharge at Ban Don Sang Khi. Dashed line indicates gauged section. Width of channel at gauge is 800 m.

recession with negligible retreat during low flows. Although the modern bed is largely composed of fine sand, there are frequent outcrops (< 2 m high) of indurated fluvial pebble beds and black layers, ca. 30-50 cm thick, near the low water mark, which may be organic-rich or black inorganic silts. The outcrops of pebble beds indicate a degree of Holocene incision of the river. The black layers may represent reducing environments associated with back-swamp deposits being exposed as the river migrates laterally. The river is being actively cut back on the outside of long curving bends with recession rates of 0.1-0.5 m per year. Such recession is compensated by deposition on the other side of the river. Accreting banks are gently inclined and well vegetated with locally intensive recession agriculture.

The origin of the islands has yet to be confirmed. Some islands may develop by extension and vegetation of sandy bars. However, taking Ban Don Sang Khi as an example, the island appears to have formed by the development of a chute channel across the inside of a long-radius river bend cutting off a portion of floodplain. Such chutes form during very high flows when the river attempts to take a shortcut through the bend rather than follow the longer low-flow course. Ban Don Sang Khi is around 15 m high (Fig. 5.6), the same elevation as the floodplain, and has many large and mature diptocarp trees, which can take a century to grow. The sediments in the banks of the island are not typical of sandy bars but are similar to the floodplain sediments seen in the cut banks of the river. Thus, it is possible that avulsion occurred such that a chute isolated a remnant of floodplain. Further fluvial accretion is occurring at the upstream end of the island which is augmented by the growth of aeolian dunes as wind-blown sand, sourced from exposed parts of the river bed, is trapped by pioneer vegetation.

There is a gauging station at Vientiane and a rated section at Pak Ngum (dotted line on Figs. 5.5 and 5.7). The lowest recorded flow is $1250 \, m^3 \, s^{-1}$, the water depth at bankfull is 19 m, and the highest flow recorded is $25,900 \, m^3 \, s^{-1}$. At Pak Ngum, the river rarely has flowed overbank in the period of the gauge record ca. 1960-2000, although back-swamps can become flooded via Mekong water flooding out through tributary junctions.

FIGURE 5.6   View of Ban Don Sang Khi looking upstream.

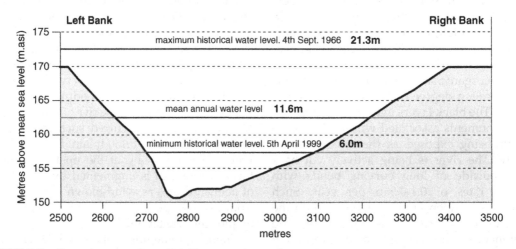

FIGURE 5.7   Channel cross section at Pak Ngum gauging station.

4.3. Zone 4: Bedrock Anastomosed Channels

Between the Mun River confluence and Stung Treng town, much of the channel is single thread but bedrock confined. However, close to the town of Muang Khong the channel is heavily anastomosed. At the large scale, this anastomosed pattern is alluvial (Fig. 5.8) but at the medium scale and small scale, channels are constrained by fault patterns and a myriad of bedrock islands (Fig. 5.9). This latter reach is called Siphandone, "four thousand islands," and terminates at the Khoné Falls on the Lao

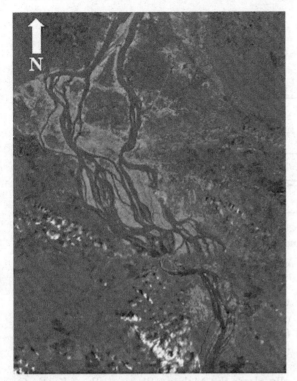

FIGURE 5.8 Anastomosed bedrock-controlled fluvial network—Siphandone. Landsat-7, horizontal field of view is approximately 40 km. (See Color Plate 11)

PDR-Cambodian border. The geology and physiography of this reach are described by Brambati and Carulli (2001). Anastomosed alluvial channels are often the result of the presence of a downstream constriction or base-level control that prevents adjustments in reach gradient. In the case of Siphandone, the presence of Holocene lava flows across the course of the river, most notably evident at Khoné Falls, may be the control on the large-scale alluvial "overprint," although this has yet to be investigated. Integrated into the "overprinted" alluvial pattern is an anastomosed bedrock-confined channel network that is structurally controlled by fault lines running NNW to SSE and a secondary set running E-W (Fig. 5.9). There is approximately twice as much discharge here than at Vientiane owing to major left-bank tributaries. The true right bank contributes little discharge. The major islands here lie around 5-8 m above the low water level and the biggest rise to ca. 10 m. There are gauging stations at Pakse and at Stung Treng. The lowest recorded flow at Pakse is 1750 $m^3 s^{-1}$, the water depth at bankfull is 17 m, and the highest flow recorded is 56,000 $m^3 s^{-1}$ (Fig. 5.10). The river has flowed

FIGURE 5.9 Oblique aerial view of 4000 islands reach of Mekong River (large island is approximately 900 m wide at widest point). (See Color Plate 12). (Photograph by Stuart Chape)

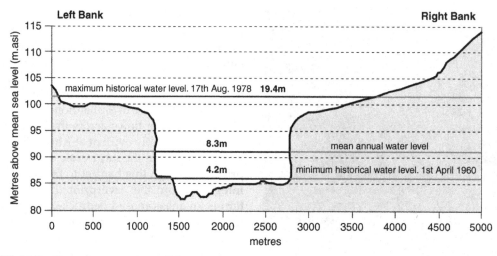

FIGURE 5.10   Channel cross section at Pakse gauging station.

overbank at Pakse and at Stung Treng in the period of the gauge record ca. 1960-2008: notably in 1978, 1979, and 2000.

4.4. Zone 5: Alluvial Meandering/ Anastomosed Channels

Downstream of Kratie, the river is a flood-plain meander complex with anabranch and anastomosed channels connecting in times of high water to Tonlé Sap lake (Fujiii *et al.*, 2003). The junction of the Mekong River and the Tonlé Sap River form the two northern arms of a complex channel junction known as Quatre Bras (or Chaktomuk Junction), that in 2002 was recorded as moving downstream by 10 m per year (Anonymous, 2002a). The southern two arms of the Quatre Bras consist of the Mekong River and a major distributary, the Bassac River, which divides from the Mekong River at point. As the monsoon rains commence, the Mekong River starts to rise and floods wetlands adjacent to the river. Local tributary inflow to the Tonlé Sap River and lake also commences. As the level of the Mekong at Phnom Penh continues to rise the flow of the Tonlé Sap reverses and Mekong River

water joins local runoff to fill the lake to a depth of around 10 m. As flow in the Mekong falls, the Tonlé Sap flow reverses and the lake is reduced to around 2 m deep.

The main flooding in this zone occurs annually along the Mekong River south of Kratie to the border with Viet Nam and also along the Tonlé Sap River. Much floodplain flow occurs parallel to the main rivers along land depressions and palaeochannels lateral to and between the main rivers. To the south of Kratie, much of the out-of-channel overbank flow is sustained by the guiding control of natural levées (Fig. 5.11) which are well developed, locally prevent return flows into the main rivers and also delay the same during flood recession. Channel migration in this region is extensive and natural but threatens developing infrastructure at some locations, such as at Kampong Cham (Uyen, 1989a,b) where flood revetment that protects large areas of the town from inundation during annual floods is at risk. The Quatre Bras (Chaktomuk) junction is morphologically complex and has been subject to considerable engineering study with a view to management and stabilization (Olesen, 2000). Near the town of Kampong Cham, this is a very

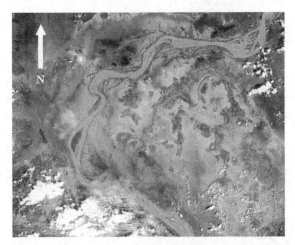

FIGURE 5.11 Satellite image of the Mekong River during flood season in vicinity of Kampong Cham. Landsat-7: field of view approximately 45 × 45 km. (See Color Plate 13)

complex and dynamic meandering system with anastomosed and anabranch channels. There is an actively meandering main channel developed in very fine sand and silt within an extensive floodplain. The latter is between 8 and 60 km wide. The insides of the meanders are accreting whilst the outsides are eroding rapidly. The sandy accretion areas inside the meanders are called point bars. Here, they are complex, consisting of a multitude of smaller elongate sedimentary bars separated by long wetted small channels that cut across the point bar complex. These small channels are called chute channels. Taken together, the small channel network forms a braided channel network on each point bar complex. The islands are sometimes very large with mature forest but more usually the islands are sandy with rank grasses and sedges. The islands erode readily and are thus individually has transitory features within a relatively stable but dynamic complex of islands and channels. The Tonlé Sap system is described elsewhere in this chapter, but it should be noted that little is known of the geomorphology of the northern Cambodian (Mekong) system and its relationship with the Tonlé Sap lake, Tonlé Sap River, and associated floodplains.

Basalt bedrock outcrops locally and is most evident on the west bank at the town of Kampong Cham. The bedrock outcrops control the overall gradient of the river and also cause "pinch-points" in the river network. It is noted that at Kampong Cham town, the river is reduced to a single channel as it passes the bedrock, but meanders freely and braids somewhat upstream and downstream of this point where it is not constrained. During and since the Quaternary, the river has meandered across much of the floodplain, shifting tens of kilometres. The evidence for this is shallow linear or curvilinear pools on the floodplain and long very low-amplitude ridges often occupied by tree lines or picked out by footpaths and field alignments. These ridges are called scroll bars which typically are curvilinear and subparallel with one another. Today, they are actively forming in the point bar complexes.

There is a gauging station at Kratie but the rating is very unstable, varying from year to year. A gauge at Kampong Cham (Fig. 5.12) is more reliable and records the lowest flow as 1880 m³ s⁻¹. The highest recorded flow is 69,000 m³ s⁻¹ and the channel depth at bankfull is 20 m. Overbank flows occur on an annual basis.

4.5. Zone 5: Tonlé Sap Lake and River System

The name Tonlé Sap is applied to the "great lake" that dominates central Cambodia and to the 147 km long river that connects the lake to the Mekong River. The lake has an area of some 3000 km² during the dry season increasing to 10,000-14,000 km² during the wet-season floods (MRCS/UNDP, 1998; MRCS/WUP-FIN, 2003; Tsukawaki et al., 1994) such that the surface area increases fivefold (Puy et al., 1999). Early investigations are reported by Carbonnel and Guiscarfé (1965) and Carbonnel (1972).

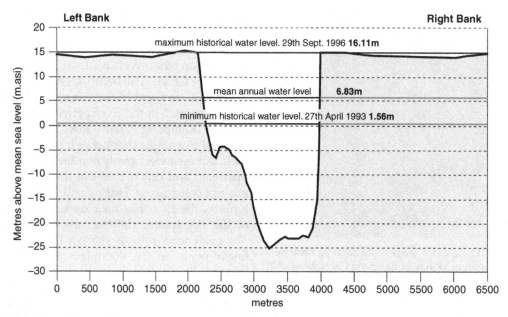

FIGURE 5.12   Channel cross section at Kampong Cham gauging station.

The elevation-volume relationship of the lake was estimated first by Carbonnel and Guiscarfé (1965) by means of a simple water balance. An alternative and more recent estimate of the capacity of the lake is 150 billion m^3, with a surface area of the lake 250,000-300,000 ha in the low season increasing to 1-1.4 million ha during the high-flow season (UN-ESCAP, 2000).

Each year between May and October, the southwest monsoon dominates the Lower Mekong system. As the level of the Mekong River rises, it causes water to back up in some tributary rivers. In the Tonlé Sap especially, the flow reverses its direction and the great lake is replenished by Mekong water (Anonymous, 2002a). Additional flood waters are contributed by direct rainfall into the lake but more especially by local tributary flow and some additional water may enter the lake by floodplain flooding to the south-west of Kampong Thom (Fig. 5.11). In total, it has been estimated that around 51,000 million m^3 of the Mekong's floodwaters are regulated in this manner

(Puy et al., 1999). The greater part of this water is later returned to the Mekong during the transitional and dry seasons when the Tonlé Sap River drains the lake waters toward the Mekong.

The lake is very shallow being around 2 m deep during the dry season (Fig. 5.13) and increases to about 10 m maximum depth during the wet season. Several publications have proposed that increase in sedimentation rates within the lake may be occurring owing to anthropogenic disturbance of the system. Unfortunately, this conclusion is largely speculative and is often a misrepresentation of early and limited studies of sedimentation in the Tonlé Sap lake. Detailed flow-balance studies have shown that most sediment entering the lake is effectively washload from the Mekong (Anonymous, 2002a). Penny et al. (2005) and Penny (2006) provide an apposite assessment of the current state of knowledge and the following is a summary of the situation as presented by Penny and colleagues.

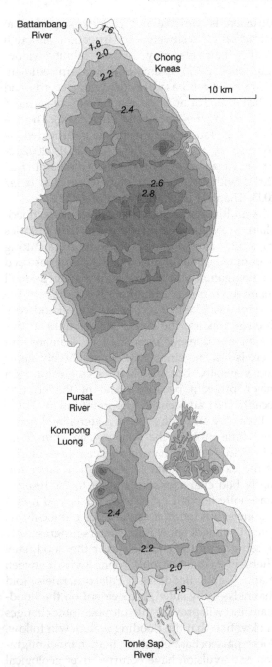

FIGURE 5.13 Bathymetry of the Tonlé Sap lake. Contours are in metres for dry-season water level. (See Color Plate 14)

The original estimates of sedimentation rates were developed by Carbonnel and Guiscarfé (1965) and Carbonnel (1972) from a single radio-carbon date of a single sediment sample obtained at a depth of 1.8 m below the lake bed. Given a date of 5720 ± 300 ^{14}C years BP an average sedimentation rate of 0.3 mm per year was determined. A modern sedimentation rate was also determined from estimates of the 1962-1963 sediment load. From this calculation, it was determined that the modern rate of sedimentation was 0.15 mm per year higher than the long-term rate determined from the lake core and consequently it was concluded that the lake was filling more rapidly with silt in modern times (Carbonnel and Guiscarfé, 1965). The suspended sediment data for the Tonlé Sap for the period 1950-1951 and 1955-1956 show around 4.5-6.0×10^6 tonnes per year entering the Tonlé Sap from the Mekong and around 3.0-6.7×10^6 tonnes per year passing into the Mekong from Tonlé Sap. The rate of sedimentation was thus estimated as less than 1 mm per year (Pantulu, 1986). More recently, though still limited, coring studies have determined that sedimentation may have declined through time, interpreted as owing to a lack of additional accommodation space for sedimentation in the northern part of the lake (Tsukawaki, 1997; Tsukawaki *et al.*, 2000a, 2000b; Tsukawaki *et al.*, 2005).

4.6. Zone 6: Alluvial Deltaic Channels

The Plain of Reeds is a trans-boundary ecosystem of 700,000 ha in Viet Nam and Cambodia lying to the northeast of the Mekong delta but hydrologically interrelated with the deltaic system. A large area of 368,000 ha in the Plain of Viet Nam (Dong Thap Muoi) is composed of acidic sulfate soils (SMEC, 1998). Except for areas of relatively high ground near the Cambodian border and along the river levees, the plain is low lying and is subject to seasonal flooding from the beginning of July until the end of January. Some 60% of flood waters

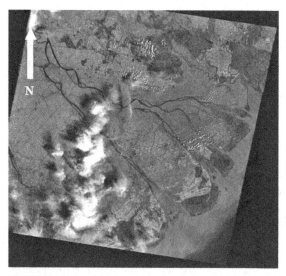

FIGURE 5.14 Satellite image of the Mekong River delta—Landsat-7: 185 × 185 km. (See Color Plate 15)

originate from outside of the immediate area. The maximum depth of flooding is around 4 m. In the dry season, the area dries out to leave only scattered ponds and swamps.

The Mekong delta (Fig. 5.14) can be considered to begin where the Mekong crosses into Viet Nam as two channels: the Bassac River to the west and the Mekong to the east. About 80% of the flow is in the Mekong and 20% in the Bassac. Around 50 km downstream of the Viet Nam border, a major connecting channel—the Van Nao pass—transfers around 40% of the Mekong flow across to the Bassac during high-flow conditions. Downstream of this point, the Bassac and Mekong have similar flow volumes during high flows. The delta is highly developed economically but the channels remain largely unengineered such that the various small distributaries still distribute flow from the main rivers through the natural silt levées to flood low-lying interdistributary areas, the latter lying only 2-3 m above sea level. Total delta area is 65,000 km^2. In the upper delta, this flooding may reach 10 m in depth but reduces closer to the coast where there are sandy chenier ridges. The degree of saline

intrusion is variable but is increasing due to extraction of freshwaters for irrigation; although locally, barrages may reduce salinity values (Campbell, 2007). Estimates of modern sedimentation rates are based on dredging records and indicate an annual accretion rate of around 1 mm (Vongvisessomjai and Phan, 2000). The history of Holocene sedimentation is provided by Nguyen *et al.* (2000), Ta *et al.* (2001, 2002a,b, 2005), Tanabe *et al.* (2003), and Murakami *et al.* (2004) within a regional framework (Sidi *et al.*, 2003; Thanh *et al.*, 2004; Woodroffe, 2000).

A multitude of minor channels drain the floodplain areas between the major distributaries (Fig. 5.15). In addition, there are cross-cutting man-made canals connecting distributaries used for navigation and a variety of managed small channels serving irrigation networks on the floodplains. The bed sediments are silt and very fine sand as are the banks. River banks in this region are densely populated by humans but there is little revetment. Banks are largely stable and vegetated but Viet Nam sees erosion as a major problem because some of the eroding localities are adjacent to important towns. There is little clay and fine organic material and hence very little mud in the system. The Tram Trim (Fig. 5.15) wetland nature reserve represents 1% of the extensive Plain of Reeds. The latter formerly had characteristics similar to the reserve, seasonally flooded grasslands, pools, and minor channels, but today is largely rice monoculture. Satellite images show flood waters progressively crossing from the channels over the floodplain. There is evident strong connectivity between channel and floodplain. Palaeochannels (old channels—fully silted) are evident on the floodplain and will provide microtopographic changes in elevation that the flooding waters will follow. These palaeochannels show that channel migration and avulsion have occurred over geological time but today the channels are relatively stable.

There are rated sections at Tan Chau (Fig. 5.16) on the Mekong River and at Chau Doc on the Bassac River; they are found to be tidal.

FIGURE 5.15 Example of distributary channel and floodwater flow over the Plain of Reeds; 20 July 2000—Horizontal field of view approximately 25 km. (See Color Plate 16)

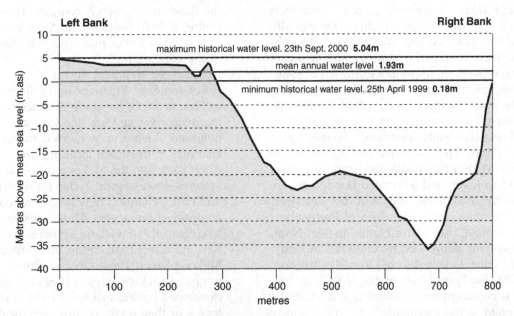

FIGURE 5.16 Channel cross section at Tan Chau gauging station.

4.7. Riverine Wetlands

Riverine wetlands are here defined as wetland areas adjacent to and directly connected with the Mekong River system either all-year round or seasonally and, as such, they rely on the connectivity and hydrological regimen of the Mekong system to a greater or lesser extent to maintain their internal hydrological processes and ecological functionality. Some of these wetlands can be regarded as permanently wet whilst others are seasonally dry. Information on Cambodia is contained within the study carried out be Van Oertzen (1999). A detailed inventory exists for Lao PDR for those wetlands that are within 50 km of the river (Claridge, 1996) and although the choice of 50 km is an arbitrary delineation of wetlands connected to the river system it is not unreasonable. The definition of a wetland varies between studies with dramatic differences in the consequent estimates of the natural floodplain wetland. Much former natural wetland is now highly managed floodplain rice paddy (and in Viet Nam—shrimp farms) which may still be classified as wetland areas. Claridge (1996) reports that estimates for Laos have ranged from 560 to 21,800 km² depending on definition employed. Much rice paddy functionally requires annual flooding from the river but otherwise lacks a natural hydrology and an intact natural ecosystem. Important examples of semi-natural wetlands in Laos include That Luang marsh near Vientiane and within Cambodia, the Stung Treng complex of islands and channels hosts seasonally inundated riverine forest habitat. Stung Treng was declared a Ramsar site in 1999. Within Cambodia, the Tonlé Sap lake, Tonlé Sap River, and associated floodplains are regionally significant and will be described below in more detail. The Plain of Reeds occupying around 13,000 km², mainly in Viet Nam, is a low-lying depression that floods seasonally, which is now largely devoted to rice cultivation. The 7588 ha Tram Chim National Park in Viet Nam is an exception, constituting a seminatural floodplain area surrounded by rice paddy and which represents around 1% of the Plain of Reeds. The 3280 ha Lang Sen, 23 km northeast of Tram Chim, is the only area within the Plain of Reeds where remnant natural *Melaleuca* forest is found adjacent to a river channel. Other major wetlands in Viet Nam are coastal and brackish, for example, the Ream National Park in Cambodia has extensive areas of mangroves and mud flats.

The MRC have mapped wetland distribution (Fig. 5.17). Wetlands are of several types depending on hydrological function:

• Some wetlands are associated with tributary rivers of the Mekong and as such changes in the tributary river flow regimen and not the changes in the regimen of the main river will moderate their hydrological function.
• Some important wetlands exist at the confluence of major tributaries with the Mekong and are inundated seasonally by the combined flood regimen of the Mekong and the individual tributary (Fig. 5.18). Thus, for these latter systems, the flooding regimen is complex as the timing of high flows within the tributary may be in or out of phase with the timing of the Mekong high flows.
• Other major wetlands, such as downstream of Vientiane and adjacent to Phnom Penh, are some distance from the main river and are flooded via overbank flows and by natural anabranch channels in upstream locations and drain back into the Mekong by similar systems downstream. Today these afflux and efflux channels may be controlled by engineered floodgates. Thus, substantial lengths of these wetland systems may have no direct local connection with the Mekong River. These large wetlands might be back-swamps that have developed behind extensive natural river levées or they might occupy tectonically controlled hollows in the landscape.
• Some smaller wetlands are back-swamps that are inundated directly by flows overtopping the river levées and/or by seepage through the levées.
• Major tracts of floodplain such as downstream of Kratie in Cambodia or in

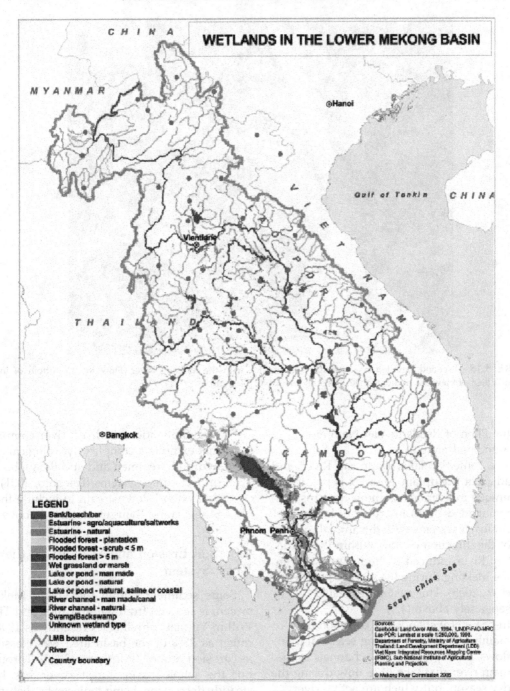

FIGURE 5.17 Distribution of wetlands (courtesy of MRC). (See Color Plate 17)

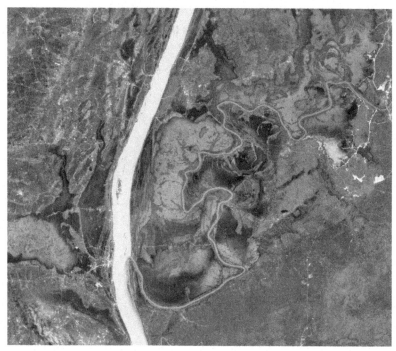

FIGURE 5.18 Wetland inundation at tributary confluence with the Mekong River. Flow top to bottom of image. Landsat-7: field of view approximately 30 × 30 km. (See Color Plate 18)

the Plain of Reeds in Viet Nam are inundated annually.

- The Tonlé Sap lake, Tonlé Sap River, and adjacent floodplain complex is possibly unique in as much as the wetland is maintained by an interplay of local drainage systems with the annual reversal of the direction of flow within the Tonlé Sap River induced by high river stage in the Mekong during the wet season (Penny *et al.*, 2005). This system is treated separately above (4.5).

- In-channel wetlands are common but mainly extant only during seasonal low flows. Examples are side channels to the main river that take little flow during the dry season, but which are active river channels during the wet season. A major and important exemplar is the Khoné

Falls-Siphandone bedrock river complex that exhibits a complex hydrological and hydraulic regimen and specific globally important ecosystems (Daconto, 2001) that extend downstream to include the Stung Treng Ramsar site noted above.

4.8. Soil Erosion and Delivery to the River System

Some sediment within the Lower Mekong system is sourced from within China and Tibet. Within Yunnan province as recently as 1998 as much as 28% of the basin area was classified as "erosion prone" (Puustjarvi, 2000). Throughout the whole basin, forest cover has been steadily decreasing, being replaced by disturbed secondary growth or agricultural systems prone to soil erosion. Vegetation changes in the LMB (and on similar topography in Thailand) and the impact on sediment erosion from slopes and delivery to the river have been considered using remote-sensing techniques (Chen *et al.*, 2000; Gupta, 1996, 1998; Gupta and Chen, 2001; Gupta and Krishnan, 1994; Gupta *et al.*, 2002a and b), which findings have been summarized by Gupta and Chen (2002) who observed, from satellite images, that bare ground on steep slopes was most prevalent in the first several weeks of the rainy season but soon reduced as vegetation became established. Direct measures of sediment transfer from the slopes to the rivers are lacking, and consequently from the earlier work cited above, Gupta and Chen developed an approximate rule that disturbed vegetated areas generated around 100 tonnes km^2 of soil per year whereas bare ground generated around 150 tonnes km^2 per year. Field visits demonstrated that this sediment entered first and second order streams via slope wash, gullies, and debris flows. The few available suspended sediment concentration data for the Mekong River (Anonymous, 2002a; Fu *et al.*, 2008; Kummu and Varis, 2006; MRC, unpublished data; Walling, 2005, 2008) tend to show higher concentrations before the end of December each year which tends to confirm that transfer to the main channel network occurs early in the wet season. However, conclusive basin-wide evidence of a significant change in suspended sediment loads and river hydrology owing to land use change is not available (Campbell, 2007; MRC, 2004). Gupta *et al.* (2006) argue that there is little opportunity for storage of sediments in the modern Mekong system for more than 4000 km of the course, with storage potential only available in the last 400 km.

4.9. Planform Adjustments and Bank Stability of the Mekong River

Although there is evidence for Mekong River channel adjustments in prehistoric and geological time-scales, only changes in modern time-scale are considered here. A series of basin-wide reconnaissance studies has been completed with a view to determining the location and means by which bank protection works (Bergado *et al.*, 1994) should be considered (e.g., Anonymous, 1988; de Vries and Brolsma, 1987a,b; Termes, 1987). These various reports inevitably contain information on channel planform and stability. For example, Rutherfurd and Bishop (1996) noted that the Mekong River planform can simply be classified as straight (rare), meandering, or braided; in addition, the categories of anastomosed and anabranching need to be added. Sitthisak (1989), Rutherfurd and Bishop (1996), and Rutherfurd *et al.* (undated) considered chiefly the reach of the river near Vientiane and noted that this reach lay in a transitional regimen between braided and meandering. Using hydrographic charts and cross sections surveyed at gauging stations, Rutherfurd and colleagues deduced approximate planform adjustments and adjustments in bed level through time and related these to a simple bank stability model. Channel stability has also been considered near Vientiane (Kummu *et al.*, 2007) and Kampong Cham (Uyen, 1989a,b) for various periods between 1961 and 2005 and for the Bassac River (Mansell, 2004). More recently, major studies have been undertaken with regard to the channel stability in the Vientiane (JICA, 2004) and the Phnom Penh areas where, for the latter area, bank recession rates are locally 10 m per year (Anonymous, 2002a).

Several reports contain information on near-bank velocity field, bed and bank sediment-grain size, stratigraphy, and bank-failure mechanisms. Bank recession through mass collapse occurs mainly during the annual flood recession (JICA, 2004). River banks throughout the system superficially often appear to be homogeneous in vertical section when considering grain size and stratigraphy, but subtle differences do occur and these translate into distinctive differences in bank profile and erosion mechanisms.

Bank erosion is locally a social, economic, and political issue along the border between Laos and Thailand (Rutherfurd *et al.*, 1996). Lateral movement of the river is a completely natural process, which can be understood and predicted to some degree by specialist studies. Erosion on one side of the river is balanced by deposition of sediments against the opposite bank such that the width of the river remains essentially the same although the lateral position of the main channel and the bank lines are changed. The zones of erosion or deposition tend to migrate upstream or downstream through time such that an eroding bank at one time may become a lateral accretion zone at another time and vice versa. Direct human intervention or unintentional activities can affect these processes. Examples are:

- Artificial infilling one side of the channel will accelerate erosion on the opposite side of the river.
- Construction of revetment along one bank alone can cause changes in the flow patterns and the patterns of erosion and deposition such that there are implications for the alignment of the opposing bank line.
- Extraction of aggregates can redirect the direction of the main flow causing changes in the patterns of erosion and deposition.

Figure 5.19A shows a short steep headwall that is largely inactive but subject to localize shear failures and occasional rotational slips. The base of the bank line consists of a complex of slumped units at an overall lower angle. Erosion of the toe of this complex by river

Sithantai near Vientiane

FIGURE 5.19   Example of bank profile near Sithantai within study reach 2. Redrawn from JICA (2004).

currents results in very slow mass movement away from the bankline toward the river. Much of the material is fine sand and silt, visually little differentiated in the vertical (Fig. 5.19A). However, close inspection reveals quasihorizontal lamination and variation in the character of the sediments (Fig. 5.19B). Often a basal gravel layer underlies sand or silt and sand or a basal sand layer is exposed close to the seasonal low water level. These differences result in composite bank profiles which are subject to different mass movement processes. The difference in failure mechanisms is also dependent on the ground-water conditions as percolation through the sediment mass and more especially along interfaces between sediment units is an important element in bank failure.

Figure 5.20 shows a near vertical cliff composed in the upper part of silty clay that is quite cohesive and, when dry, resistant to subaerial erosion. The unit is horizontally bedded and this is demonstrated by the slightly eroded horizon at about two-thirds height. The lower third of the cliff consists of stratified gravel close to the angle of repose (32°) but punctuated

by small steps indicating the stratigraphic horizons. The deep eroded alcoves are probably the result of seepage, enhanced by turbulence along the interface between the underlying gravel, and the superimposed silty clay.

4.10. Vertical Bed Level Adjustments and Deep Pools

Vertical scour and fill during the flood season and longer term changes in bed elevations have not been systematically studied, although some limited analysis of repeat cross-section surveys are contained within the literature (Bountieng, 2003). Throughout the Mekong system, both alluvial and bedrock reaches often contain very deep pools, reportedly up to 40-60 m deep (Anonymous, 2002b; Chan *et al.*, 2005; Conlan *et al.*, 2008; Viravong *et al.*, 2006) that seem to be temporally persistent within their spatial locations. These pools are important for fish conservation and as a fisheries resource throughout the system and locally are important for Irrawaddy dolphins (Poulsen *et al.*, 2002). However, although these pools

Sibounheuang-Muang Wa

FIGURE 5.20 Example of composite gravel and sand bank profile within study reach 2. Redrawn from JICA (2004).

may constitute important geomorphological features of the Mekong River, the mechanisms by which they are formed and maintained have not been studied within the Mekong. A cursory inspection of air photography, satellite images, and bathymetric maps indicates a possible connection with structurally induced flow constrictions in bedrock reaches and tributary junction scour in alluvial reaches, for example, but the issue is receiving further research.

4.11. Floodplain Flooding

Floodplains can be defined on a practical basis as those areas that are prone to flooding on a regular basis within modern times. The riparian levées may remain exposed whilst the floodplains are submerged (Fig. 5.21). Methods to determine floodplain areas are remote sensing and hydraulic modeling of flood inundation. An example of the former is provided by Ratanavong (1998) for Thabok town in Laos for the 1997 inundation and for the Vientiane plain by Gillespie and Inthiravongsy (1998) using RADARSAT (Hinel, 1998). Hydraulic modeling of flood inundation using the computational fluid dynamics package ISIS has been

FIGURE 5.21 Inundation of floodplain to left of riparian levée. (See Color Plate 19). (Photograph by Joe Garrison).

conducted by the MRC for the Lower Mekong downstream of Kratie (Fig. 5.22).

According to the MRC estimates, 80% of rural flood events and 20% of urban flood events are caused by tributary flows and not main river high flows. The four main flood-prone areas in Laos are situated along the Mekong near large tributaries: (i) Vientiane plain, (ii) Khammoune Province (Thakhek town), (iii) Savannakhet Province, and (iv) Champasak Province (Pakse town). In Thailand, the main flood-prone areas are limited riparian areas along the Mekong and along the tributaries in Nong Khai, Mukdahan, Nakhon Phanom, and Ubon Ratchatanee provinces.

The Mekong River annually floods the floodplain areas downstream of Kratie and the delta area, inundating approximately 1.2-1.8 million ha with flooding in some areas lasting between 2 and 6 months with depths of 0.5-5 m. During the 2000 record flood (1:50 year recurrence), approximately 38,900 km^2 were flooded throughout LMB. During 1966, the largest flood on record occurred for the upper and middle reaches of the basin. Impacts vary annually depending on factors such as flood duration, timing in relation to human activities, high water levels, and the significance of regionally or locally generated rainfall. In the case of the 2000 flood, monsoon flood waters were augmented by heavy rainfall especially over the Tonlé Sap and lower reaches of the Mekong. This was followed by intermittent rain showers over the same area. The Tonlé Sap lake was full by late July and thus August floodwaters in the Mekong could not be accommodated and these conjoined factors led to extensive flooding downstream. The flood depth in the delta reached 5.06 m (MRC, 2001).

4.12. Sediment Load and Sedimentation

Throughout the LMB, the sediment load is predominately fine sand with local gravel bars mainly evident upstream of Vientiane. There

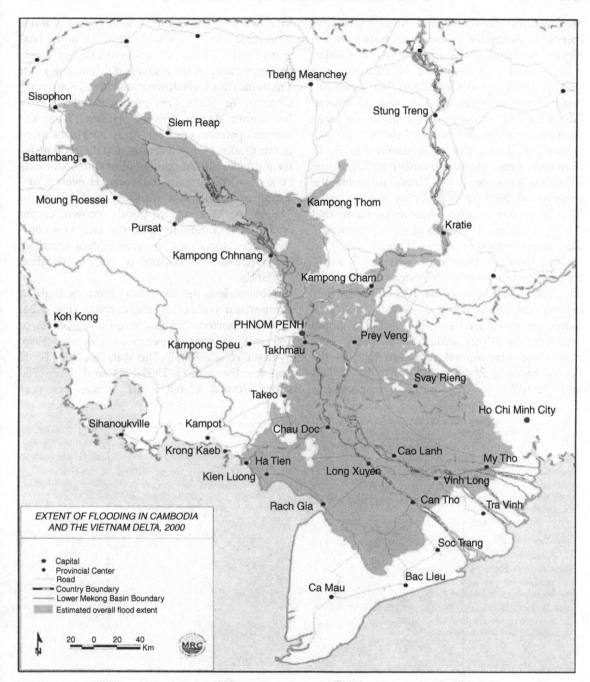

FIGURE 5.22 Extent of 2000 flooding effectively delimits the floodplain area in the lower Mekong River system.

are few grain size data and definitive measurements or estimates of bedload transport are highly localized (Anonymous, 2002a; Conlan et al., 2008), for example, Rutherfurd and Bishop (1996) report some 20 million tonnes of bedload per year for a reach near Vientiane. Given the high-stream powers in such a large river much of the fine bed material will go into suspension during the flood season (Fig. 5.23). However, a considerable quantity will continue to move as a bedload. Thus, although the majority of the load is probably a suspension load, it would be erroneous to presume that the bedload of the Mekong is insubstantial. The annual total suspended load (TSS) at Chiang Saen has been estimated as 67×10^6 tonnes per year, 109×10^6 tonnes at Vientiane and 132×10^6 tonnes at Khoné Falls. The organic content is about 6-8% (all these estimates are given by Pantulu (1986)). Milliman and Syvitski (1992) and Roberts (2001) estimated the total annual sediment load of the Lower Mekong River to be 160×10^6 tonnes and $150\text{-}170 \times 10^6$ tonnes, respectively.

Métivier and Gaudemer (1999) report a long-term (Quaternary) yield of 150×10^6 at the seaward limit. Notably, the concentration of TSS is higher upstream for a given discharge (e.g., at Vientiane) than it is downstream (e.g., Kampong Cham) (Fig. 5.23). This probably reflects both the higher stream powers resulting from the steeper upstream gradients in contrast to the more gentle gradients downstream, and the propensity for a greater storage potential of fine sediments at some downstream locations. However, the supply, storage, and transfer mechanisms of TSS in the Mekong are poorly known, Gupta et al. (2006) commenting on the lack of accommodation space for overbank deposition of fine sediment upstream of the Cambodian lowlands.

Nonetheless the total load remains high in comparison with other major rivers (Fig. 5.24) with a reported TSS discharge to the delta of 160 million tonnes per year (Meade, 1996; Syvitski et al., 2005). The data for the Tonlé Sap for the period 1950-1951 and 1955-1956 show around $4.5\text{-}6.0 \times 10^6$ tonnes per year

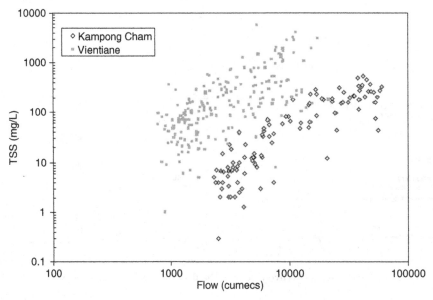

FIGURE 5.23   Relationship between TSS and discharge (courtesy of MRC).

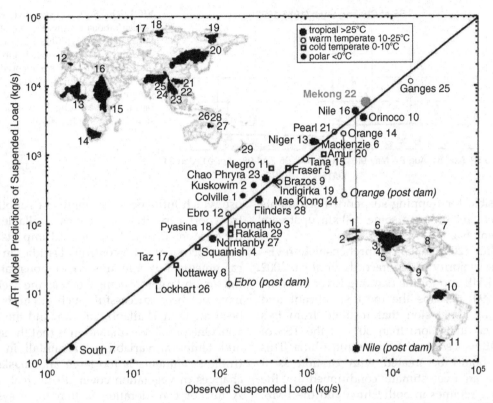

FIGURE 5.24 Comparison of TSS load of Mekong with other major rivers (redrawn from Syvitski et al., 2005).

entering the Tonlé Sap from the Mekong and around $3.0–6.7 \times 10^6$ tonnes per year passing into the Mekong from Tonlé Sap. The rate of sedimentation in the Tonlé Sap was thus estimated as less than 1 mm per year. This is similar to estimates in the delta based on dredging volumes of less than 1 mm per year. More recent water balance estimates have shown that the Tonlé Sap River tends to aggrade in the lower reach during the rising flow in the wet season and degrade during the falling stage (Anonymous, 2002a).

Sedimentation may be more significant at specific locations in the system, particularly near Snoc Trou where the Tonlé Sap joins the lake and around Phnom Penh (Anonymous,

2002a; SMEC, 1998). Most investigations have considered local sedimentation issues (Azam, 1975; Chinnarabri, 1990; Harden and Sundborg, 1992; San, 1999; Wolanski et al., 1998) rather than assessing the basin-wide situation.

The most evident change in TSS is a sudden reduction around January 1993 onward noted at Chiang Saen in a record from 1985 to 1992 (Fig. 5.25). Two possible factors may be considered for this: large-scale revegetation or dam installation (Kummu and Varis, 2006). If the former was the control then a slow adjustment might be anticipated. If dam closure is responsible then a sudden change is reasonable. The closure of the Manwan dam in China in 1992 (ICOLD, 2002) on the main river is probably

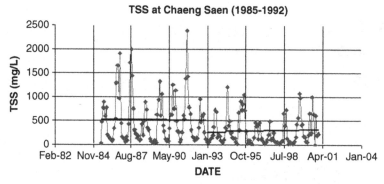

FIGURE 5.25 Suspended solids concentrations in the Mekong River at Chiang Saen, with linear regression lines fitted for the periods before and after Manwan Dam commenced filling in 1992 (source: Mekong River Commision, 2003; The State of the Basin, 2003. Mekong River Commission, Phnom Penh, 300pp).

responsible for trapping suspended sediments (Fu *et al.*, 2008), being some 350 km upstream of Chaeng Saen.

Walling (2005, 2008) comprehensively reviewed the majority of TSS records until ca. 2002 for the LMB and for the Lancang River in China until 1990. Perhaps the most significant and surprising conclusion that resulted from this work was that more than 50% of the TSS of the LMB was sourced from within China. This result should be treated with caution as it may be an overestimate conditioned by the sampling regimes in both China and the LMB. Nevertheless, it does indicate that a probable major source for TSS in the LMB is the Upper Mekong system. A significant consequence is that changes to the TSS flux should be detectable in Zone 2 during the next several years as river resource development continues within China.

A major environmental control on TSS is changes in basin-wide land use (Anonymous, 1998). An issue of concern in the LMB is the effect of forest clearance on hydrology and related effects such as flooding, soil erosion, and slope mass movements (Oughton, 1993). Much relevant literature was reviewed by Bruijnzeel (1990). With respect to TSS there are short-term consequences related to the release of solids during clearance and important long-term issues concerning sustained or reduced TSS loads following clearance. The impact will depend largely on the type of land use which influences the degree of runoff (i.e., transport capacity) as well as available soils for erosion. Typical hydrological impacts have been documented for controlled studies in small catchments but attempts to demonstrate the impact on larger systems (such as the Mekong) have not been successful. Such attempts have been made in Thailand, Taiwan, and the Amazon. One possible explanation is that the spatial and temporal variability of rainfall in large tropic catchments is too large and masks any changes in vegetation cover (Bruijnzeel, 1990). A further consideration is that large systems either store sediments "permanently" within the system such as on floodplains, or that there are substantial lag effects for coarser components moving down the system.

ACKNOWLEDGMENT

The Mekong River Commission is thanked for the opportunity to develop this review and for access to data and unpublished reports.

References

Adamson, P. (2001). Hydrological perspectives of the Lower Mekong. *International Water Power and Dam Construction*, 16-21.

Anonymous. (1988). Pilot project stage 1 of Mekong River bank protection (basin-wide). Interim Committee for Co-operation of Investigations of the Lower Mekong Basin, vol. 1, 12pp.

Anonymous. (1998). Soil erosion, sedimentation and flash flood hazards (basinwide). Review and Assessment Report Phase 1 1990-1996, MRC Report March 1998 MKG/R.98018.

Anonymous. (2002a). The Chaktomuk area, environment, hydraulics and morphology: A Comprehensive Study for the stabilization of the Chaktomuk area. Final Report July 2002, on CD available from MRC.

Anonymous. (2002b). Deep pools as dry season fish habitats in the Mekong River basin, MRC Technical Paper No. 4, 20pp.

Azam, M. A. (1975). Sediment movement of Bassac River. AIT Thesis No. 734.

Bacon, A. M., Demeter, F., Duringer, P., Helm, C., Bano, M., Vu, T. L., Nguyen, T. K. T., Antoine, P. O., Bui, T. M., Nguyen, T. M. H., Dodo, Y., Chablaux, F., and Rihs, S. (2008). The Late Pleistocene Duoi U'Oi cave in northern Vietnam: Palaeontology, sedimentology, taphonomy and palaeoenvironments. *Quaternary Science Reviews*, doi:10.1016/j.quascirirev.2008.04.017.

Bergado, D. T., Long, P. V., and Dezeure, J. (1994). *In* "Mekong River Bank Slope Protection of Thadeau and Mutangwa, Vientiane, Lao, Proceedings Landslides, Slope Stability and the Safety of Infrastructures Conference," Kuala Lumpur, Malaysia, 13-14 September 1994, 3pp.

Berner, E. K. and Berner, R. A. (1996). "Global Environment: Water, Air, Geochemical Cycles," Prentice-Hall, New Jersey, 376pp.

Bountieng, S. (2003). Cross-section profiles study Thanaleng-Nongkhai, Mekong River Commission, 14pp.

Brambati, A. and Carulli, G. B. (2001). Geology, geomorphology and hydrogeology of Siphandon wetlands. *In* "Siphandone Wetlands," Dacanto, G. (ed.), pp 24-48. +maps. CESVI, Vientiane, Lao PDR.

Bridge, J. S. (1993). The interaction between channel geometry, water flow, sediment transport and deposition in braided rivers. *In* "Braided Rivers," Best, J. L. and Bristow, C. S. (eds.), pp 13-71. Geological Society Special Publication 75, London.

Brierley, G. J. and Fryirs, K. A. (2005). "Geomorphology and River Management: Applications of the River Styles Framework," Blackwell, Malden, USA, 398pp.

Bruijnzeel, L. A. (1990). "Hydrology of Moist Tropical Forest and Effects of Conversion: A State of Knowledge Review," UNESCO International Hydrological Programme, Amsterdam, 224pp.

Campbell, I. C. (2007). Perceptions, data, and river management: Lessons from the Mekong River. *Water Resources Research* 43, doi:10.1029/2006WR005130.

Carbonnel, J. P. (1972). Le Quaternaire Cambodgien: Structure et stratigraphie, Paris, France. ORSTOM Memoire No. 60, 248pp.

Carbonnel, J. P. and Guiscarfé, J. (1965). "Grand Lac du Cambodge: Sedimentologie et Hydrologie 1962-1963," Museum National d'Histoire Naturelle de Paris, Paris, France, 403pp.

Carling, P. A. (2006). Geomorphology and Sedimentology: Integrated Basin Flow Management Specialist Report, Water Utilization Program/Environment Program, Mekong River Commission, 84pp.

Chan, S., Putrea, S., and Hortle, H. G. (2005). Using local knowledge to inventory deep pools, important fish habitats in Cambodia. *In* "Proceedings of the 6th Technical Symposium on Mekong Fisheries," Pakse, Lao PDR, 26-28 November 2003, Burnhill, T. J. and Hewitt, M. M. (eds.), pp 57-76. Mekong River Commission, Vientiane.

Chen, P., Lim, H., Huang, X., Gupta, A., and Liew, S. C. (2000). Environmental study of the middle Mekong basin using multi-spectral SPOT imagery. In "Proceedings of 2000 International Geoscience and Remote Sensing Symposium," Volume 7, pp 3237-3239. IEEE Publications, Piscataway, NJ.

Chinnarabri, C. (1990). Sediment transport and river bank protection of the Mekong River, Thailand and Lao, AIT thesis WA 90 7.

Claridge, G. (1996). An inventory of wetlands of the Lao PDR. The IUCN Wetlands Programme, The World Conservation Union, 287pp.

Clark, M. K., Shoenbohm, L. M., Royden, L. H., Whipple, K. X., Burchfiel, B. C., Zhang, X., Tang, W., Wang, E., and Chen, L. (2004). Surface uplift, tectonics, and erosion of eastern Tibet from large-scale drainage patterns. *Tectonics* 23, TC106, doi:10.1029/2002TC001402.

Conlan, I., Rutherfurd, I., Finlayson, B., and Western, A. (2008). The geomorphology of deep pools on the lower-Mekong River: Controls on pool spacing and dimensions, processes of pool maintenance and potential future changes to pool morphology. Final report submitted to the Mekong River Commission Secretariat, 98pp.

Daconto, G. (2001). "Siphandone Wetlands," CESVI, Cooperation and Development, Bergamo, Italy, 192pp.

de Vries, M. and Brolsma, A. A. (1987a). Mekong River basin wide bank protection project—Stage 1. Report on March-April mission, Delft Hydraulics, 37pp.

de Vries, M. and Brolsma, A. A. (1987b). Mekong River basin wide bank protection project—Stage 1. Report on August-September mission, Delft Hydraulics, 41pp.

Douglas, I. (2005). The Mekong river basin. *In* "The Physical Geography of Southeast Asia," Gupta, A. (ed.). Oxford University Press, Oxford, UK, 440pp.

ESCAP. (1990a). "Atlas of Mineral Resources of the ESCAP Region Series. (Separate Booklets Prepared by the Secretariat of the United Nations Economic and Social Commission for Asia and the Pacific in Cooperation with National Geological Agencies)," Volume 7. LAO People's Democratic Republic, United Nations publication, 1990. Explanatory Brochure in English, 19pp., with 2 atlas sheets in a back pocket, comprising the Geological and Mineral Resources

Map of the Lao People's Democratic Republic on the scale 1:1,500,000.

ESCAP. (1990b). "Atlas of Mineral Resources of the ESCAP Region Series. (Separate Booklets Prepared by the Secretariat of the United Nations Economic and Social Commission for Asia and the Pacific in Cooperation with National Geological Agencies)," Volume 10. United Nations Publication, Cambodia. Explanatory brochure in English, 87pp., with 2 atlas sheets in a back pocket, comprising the Geological and Mineral Resources Map of Cambodia on the scale 1:1,500,000.

Fontaine, H. and Workman, D. R. (1978). In "Review of the Geology and Mineral Resources of Kampuchea, Laos and Vietnam, Third Regional Conference on Geology and Mineral Resources of Southeast Asia," Bangkok, Thailand, 14-18 November 1978.

Fu, K. D., He, D. M., and Lu, X. X. (2008). Sedimentation in the Manwan reservoir in the Upper Mekong and its downstream impacts. *Quaternary International* **186**, 91-99.

Fujiii, H., Garsdal, H., Ward, P. B., Ishiii, M., Morishita, K., and Boivin, T. (2003). Hydrological roles of the Cambodian floodplain of the Mekong River. *International Journal of River Basin Management* **1**, 1-14.

Gillespie, V. and Inthiravongsy, S. (1998). Untitled. In "Flood Management and Mitigation in the Mekong River Basin," Proceedings of the Regional Workshop, Vientiane, Lao PDR, 19-21 March 1998, pp 65-74. FAO, Bangkok.

Gupta, A. (1996). Erosion and sediment yield in Southeast Asia: A regional perspective. In "Erosion and Sediment Yield: Global and Regional Perspectives," Walling, D. E. and Webb, B. W. (eds.), pp 215-222. Proceedings Exeter Symposium, July 1996, IAHS Publ. no. 236, Wallingford, UK.

Gupta, A. (1998). Rapid erosion risk evaluation in the middle Mekong basin by satellite imagery. In "Proceedings Euro-Asia Space Week," Singapore, pp 241-245. ESA SP-430. European Space Agency, Paris.

Gupta, A. (2004). The Mekong River: Morphology, evolution and palaeoenvironment. *Journal of Geological Society of India* **64**, 525-533.

Gupta, A. (2005a). "The Physical Geography of Southeast Asia," Oxford University Press, Oxford, 440pp.

Gupta, A. (2005b). Rivers of Southeast Asia. In "The Physical Geography of Southeast Asia," Gupta, A. (ed.), pp 65-79. Oxford University Press, Oxford, 440pp.

Gupta, A. and Chen, P. (2001). Remote sensing and environmental evaluation in the Mekong basin. In "Proceedings 22nd Asian Conference on Remote Sensing," pp 428-432. National University of Singapore, Singapore.

Gupta, A. and Chen, P. (2002). Sediment movement on steep slopes to the Mekong River: An application of remote sensing. In "The Structure, Function and Management of Fluvial Sedimentary Systems," Dyer, F. J.,

Thoms, M. C., and Olley, J. M. (eds.), pp 399-406. Proceedings of a Symposium, Alice Springs, Australia, September 2002, IAHS Publication No. 276, Wallingford.

Gupta, A. and Krishnan, P. (1994). Spatial distribution of sediment discharge to the coastal water of South and Southeast Asia. In "Variability in Stream Erosion and Sediment Transport," Olive, L. J., Loughran, R. J., and Kesby, J. A. (eds.), pp 457-463. Proceedings Camberra Symposium, December 1994, IAHS Publ. 224.

Gupta, A. and Liew, S. C. (2007). The Mekong from satellite imagery: A quick look at a large river. *Geomorphology* **85**, 259-274.

Gupta, A., Hock, L., Hunag, X. J., and Chen, P. (2002a). Evaluation of part of the Mekong using satellite imagery. *Geomorphology* **44**, 221-239.

Gupta, A., Lim, H., Huang, X., and Chen, P. (2002b). Evaluation of part of the Mekong River using satellite imagery. *Geomorphology* **44**, 221-239.

Gupta, A., Liew, S. C., and Heng, A. W. C. (2006). Sediment storage and transfer in the Mekong: Generalizations on a large river. In "Sediment Dynamics and the Hydromorphology of Fluvial Systems," Rowan, J. S., Duck, R. W., and Werritty, A. (eds.), pp 450-459. Proceedings of a Symposium, Dundee, UK, July 2006, IAHS Publ. 306, Wallingford, UK.

Harden, P. O. and Sundborg, A. (1992). "The Lower Mekong Suspended Sediment Transport and Sedimentation Problems," AB Hydroconsult, Uppsala, 71pp.

Haruyama, S. and Shida, H. (2008). Geomorphologic land classification map of the Mekong Delta utilizing JERS-1 SAR images. *Hydrological Processes*, doi:10.1002/hyp.6946.

Hinel, J. (1998). Untitled. In "Flood Management and Mitigation in the Mekong River Basin," Proceedings of the Regional Workshop, pp 89-96. Vientiane, Lao PDR, 19-21 March 1998, FAO of the UN, Bangkok.

Hope, G. (2005). The Quaternary in Southeast Asia. In "The Physical Geography of Southeast Asia," Gupta, A. (ed.), pp 24-39. Oxford University Press, 440pp.

Hori, H. (2000). "The Mekong, Environment and Development," United Nations University Press, Tokyo, 298pp.

Hutchinson, C. S. (1989). "Geological Evolution of Southeast Asia." Clarendon, Oxford.

Hutchinson, C. S. (2005). The geological framework. In "The Physical Geography of Southeast Asia," Gupta, A. (ed.), pp 3-23. Oxford University Press, Oxford, 440pp.

ICOLD. (2002). www.icold-cigb.org.cn/icold2000/st-a5-03.html.

JICA. (2004). The study on Mekong Riverbank protection around Vientiane municipality in the Lao People's Democratic Republic. Interim Report, February 2004. Japan International Cooperation Agency and the Government of Lao PDR.

Kummu, M. and Varis, O. (2006). Sediment-related impacts due to upstream reservoir trapping in the Lower Mekong River. *Geomorphology*, doi:10.1016/j.geomorph.2006.03.024.

Kummu, M., Lu, X. X., Rasphone, A., Sarkkula, J., and Koponen, J. (2007). Riverbank changes along the Mekong River: Remote sensing detection in the Vientiane-Nong Khai area. *Quaternary International*, doi:10.1016/j.quaint.2007.10.015.

Mansell, P. E. (2004). "Expert Report on RN21 Embankment Erosion," Ministry of Public Works and Transport, Kingdom of Cambodia, 51pp.

Meade, R. H. (1996). River-sediment inputs to major deltas. *In* "Sea-Level Rise and Coastal Subsidence," Milliman, J. D. and Haq, B. U. (eds.), pp 63-85. Kluwer Academic Publishers, Dordrecht, 369pp.

Mekong River Commission. (1997). "Mekong River Basin Diagnostic Study Final Reort," MRC, Bangkok, 249pp.

Mekong River Commission. (2001). Analytical review of the year 2000 flood damage in the Mekong River Basin. *In* "Consultation Workshop on Formulation of a Regional Strategy for Flood Management and Mitigation in the Mekong River Basin," Proceedings of a Consultation Workshop on Formulation of a Regional Strategy for Flood Management and Mitigation in the Mekong River Basin, Phnom Penh, Cambodia, pp 81-104. 13-14 February 2001.

Mekong River Commission. (2003). "State of the Basin Report, 2003." MRC, Phnom Penh, Cambodia.

Mekong River Commission. (2004). Overview of the hydrology of the Mekong basin, Water Utilization Program Start-up Project, Draft 24 November 2004, 78pp.

Métivier, F. and Gaudemer, Y. (1999). Stability of output fluxes of large rivers in South and East Asia during the last 2 million years: Implications on floodplain processes. *Basin Research* 11, 293-303.

Milliman, J. D. and Syvitski, J. P. M. (1992). Geomorphic/tectonic control of sediment discharge to the ocean: The importance of small mountainous rivers. *Journal of Geology* 100, 525-544.

MRCS/UNDP. (1998). "Natural Resources-Based Development Strategy for the Tonlé Sap Area, Cambodia," Cambodian National Mekong Committee, NEDCO, MIDAS, Phnom Penh, Report No. CMB/95/003, 64pp.

MRCS/WUP-FIN. (2003). "Modeling Tonlé Sap for Environmental Impact Assessment and Management Support: Water Utilization Program—Modeling of the Flow Regimen and Water Quality of the Tonlé Sap," Draft Final Report. Finnish Environmental Institute, Helsinki, Finland, 110pp.

Murakami, F., Saito, Y., Kinosita, Y., Tateishi, M., Nguyen, T. L., Luong, B. L., and Nguyen, T. T. (2004). High-resolution seismic reflection survey in the Mekong River delta, Viet Nam. *In* "Stratigraphy of Quaternary System

in Deltas of Viet Nam," Nguyen, T. V., Saito, Y., Nguyen, V. Q., and Ngo, Q. T. (eds.), pp 25-35. Department of Geology and Minerals of Viet Nam, Hanoi, Viet Nam.

Nanson, G. C. and Knighton, A. D. (1996). Anabranching rivers: Their cause, character and classification. *Earth Surface Processes and Landforms* 21, 217-239.

Nguyen, V. L., Ta, T. K. O., and Tateishi, M. (2000). Late Holocene depositional environments and coastal evolution of the Mekong River Delta, Southern Vietnam. *Journal of Asian Earth Sciences* 18, 427-439.

Olesen, K. W. (2000). Morphological modelling of the Chaktomuk junction. *In* "Proceedings of a Workshop on Hydrological and Environmental Modelling of the Mekong Basin," 11-12 September 2000, Al-Soufi, R. W. (ed.), pp 289-301. Mekong River Commission, Phnom Penh.

Oughton, G. A. (1993). Control of soil erosion, sedimentation and flash flood hazards. Government of Lao PDR report.

Ōya, M. (1979). "River and Development of the Basin." Taimeido, Tokyo (in Japanese).

Ōya, M. (1993). "The Development of River Basins and Alluvial Plains: Laying Stress on the Rivers in Southeast Asia." Taimeido, Tokyo (in Japanese).

Pantulu, V. R. (1986). The mekong river system. *In* "The Ecology of River Systems," Davies, B. R. and Walker, K. F. (eds.), pp 695-720. Dr. Junk Publishers, Dortrecht, The Netherlands.

Penny, D. (2006). The Holocene history and development of the Tonlé Sap, Cambodia. *Quaternary Science Reviews* 25, 310-322.

Penny, D., Cook, G., and Im, S. S. (2005). Long-term rates of sediment accumulation in the Tonlé Sap, Cambodia: A threat to ecosystem health. *Journal of Paleolimnology* 33, 95-103.

Poulsen, A. F., Ouch, P., Sintavong, V., Ubolratana, S., and Nguyen, T. T. (2002). Deep pools as dry season fish habitats in the Mekong Basin. MRC Technical Paper No. 4, Mekong River Commission.

Puustjarvi, E. (2000). Review of policies and institutions related to management of upper watershed catchments, Yunnan, PRC. Poverty and Environmental Management in the Remote GMS. Watersheds Project. Phase 1. ADB Regional TA. 5771.

Puy, L., Sovan, L., Touch, S. T., Mao, S. O., and Chhouk, B. (1999). Diversity and spatial distribution of freshwater fish in Great Lake and Tonlé Sap River (Cambodia, Southeast Asia). *Aquatic Living Resources* 12, 379-386.

Rainboth, W. J. (1996). "Fishes of the Cambodian Mekong," Food and Agricultural Organisation of the United nations, Rome, 265pp + 27 plates.

Ratanavong, N. (1998). Remote sensing for inundation mapping in the Lower Mekong Basin. *In* "Flood

Management and Mitigation in the Mekong River Basin." Proceedings of the Regional Workshop, Vientiane, Lao PDR, 19-21 March 1998. FAO, Bangkok.

Roberts, T. (2001). Downstream ecological implications of China's Lancang Hydropower and Mekong Navigation Project, International Rivers Network (IRN). http://www.irn.org.

Rüber, L., Britz, R., Kullander, S. O., and Zard, R. (2004). Evolutionary and biogeographical patterns of the Badidae (teleostei: Perciforms) inferred from mitochondrial and nuclear DNA sequence data. *Molecular Phylogenetics and Evolution* **32**, 1010-1022.

Rutherfurd, I. D. and Bishop, P. M. (1996). Untitled. *In* "Seminar on Mekong River Morphology," Vientiane, Lao PDR, 9–10 May 1996, 5pp.

Rutherfurd, I. D., Bishop, P. M., Walker, M. R., and Stensholt, B. (1996). Recent changes in the Mekong River near Vientiane: Implications for the border between Thailand and Lao PDR. *In* "Development Dilemmas in the Mekong Region," Stensholt, B. (ed.), pp 172-185. Monash Asia Institute, Clayton, Australia, 277pp.

San, D. C. (1999). Flood flow and morphology of the lower Mekong River in Viet Nam. Thesis from AIT, Bangkok. WM99-7.

Sidi, F. H., Nummendal, D., Imbert, P., Darman, H., and Posamentier, H. W. (2003). "Tropical Deltas of Southeast Asia: Sedimentology, Stratigraphy and Petroleum Geology," SEPM Publication, Tulso, OK, 76pp.

Sitthisak, S. (1989). Final report of the pilot project studies of Mekong bank protection in the Lao PDR, Mekong basin-wide bank protection, Stage II. *In* Basin-Wide Seminar on River Management, Nong Khai, 4-5 December 1989, 6pp.

SMEC International Pty Ltd. (1998). Water utilization program preparation project, final report. December 1998 36pp plus extensive appendices.

Syvitski, J. P. M., Vorösmarty, C. J., Kettner, A. J., and Green, P. (2005). Impact of humans on the flux of terrestrial sediment to the global coastal ocean. *Science* **308**, 376-380.

Ta, T. K. O., Nguyen, V. L., Tateishi, M., Kobayashi, I., and Saito, Y. (2001). Sediment facies and diatom and foraminifer assemblages of Late Pleistocene-Holocene incised-valley sequence from the Mekong River Delta, Bentre Province, Southern Viet Nam: The BT2 core. *Journal of Asian Earth Sciences* **20**, 83-94.

Ta, T. K. O., Nguyen, V. L., Tateishi, M., Kobayashi, I., Saito, Y., and Nakamura, T. (2002a). Sediment facies and Late Holocene progradation of the Mekong River Delta in Bentre province, southern Viet Nam: An example of evolution from tide-dominated to a tide-and wave-dominated delta. *Sedimentary Geology* **152**, 313-325.

Ta, T. K. O., Nguyen, V. L., Tateishi, M., Kobayashi, I., Tanabe, S., and Saito, Y. (2002b). Holocene delta evolution and sediment discharge of the Mekong River, southern Viet Nam. *Quaternary Science Reviews* **21**, 1807-1819.

Ta, T. K. O., Nguyen, V. L., Tateishi, M., Kobayashi, I., and Saito, Y. (2005). Holocene delta evolution and depositional models of the Mekong River delta, southern Viet Nam. *In* "River Deltas—Concepts, Models, and Examples," Giosan, L. and Bhattacharya, J. P. (eds.), pp 453-466. SEPM Spec. Publication 83, Tulso, OK.

Tanabe, S., Ta, T. K. O., Nguyen, V. L., Tateishi, M., Kobayashi, I., and Saito, Y. (2003). Delta evolution model inferred from the Holocene Mekong Delta, Southern Viet Nam. *In* "Tropical Deltas of Southeast Asia—Sedimentology, Stratigraphy, and Petroleum Geology," Sidi, F. H., Nummedal, D., Imbert, P., Darman, H., Posamentier, H. W. (eds.), pp 175-188. SEPM Special Publication 76.

Termes, A. P. P. (1987). Report on morphological computations for bank protection studies in the Vientiane/Nong Khai reach, Mekong basin-wide bank protection, Stage II. *In* Basin-Wide Seminar on River Management, Nong Khai, 4-5 December 1989, 45pp.

Thanh, T. D., Saito, Y., Huy, D. V., Nguyen, V. L., Ta, T. K. O., and Tateishi, M. (2004). Regimes of human and climate impacts on coastal changes in Viet Nam. *Regional Environmental Changes* **4**, 49-62.

Tsukawaki, S. (1997). Lithological features of cored sediments from the northern part of lake Tonlé Sap, Cambodia. *In* "The International Conference of Stratigraphy and Tectonic Evolution of Southeast Asia and the South Pacific," pp 232-239. Bangkok, Thailand.

Tsukawaki, S., Okawara, M., Lao, K. L., and Tada, M. (1994). Prelimary study in Lake Tonlé Sap, Cambodia. *Chigaku Zasshi (Journal of Geography)* **103**, 623-636.

Tsukawaki, S., Sieng, S., Im, S., Ben, B., Kamiya, T., Touch, S., Ozawa, H., and Kato, M. (2000a). Distribution and composition of surface sediments in lake Tonlé Sap, Cambodia. *In* "Guidebook, Optional Field Tour, Lake Tonlé Sap and Related Fluvial Systems, and Angkor Monument Complex in Cambodia," International Conference on DELTAS, Siem Reap, Cambodia, pp 15-16. 17-18 January 2005.

Tsukawaki, S., Mildenhall, D. C., Ben, B., Touch, S., and Oda, M. (2000b). Lithological features and radiocarbon ages of cored sediments from the northern part of lake Tonlé Sap, Cambodia. *In* "Guidebook, Optional Field Tour, Lake Tonlé Sap and Related Fluvial Systems, and Angkor Monument Complex in Cambodia," International Conference on DELTAS, Siem Reap, Cambodia, pp 18-19. 17-18 January 2000.

Tsukawaki, S., Sieng, S., Mildenhall, D. C., Okawara, M., Kamiya, T., Touch, S., Kato, M., and Akiba, F. (2005).

Environmental changes of lake Tonlé Sap and the lower course of the Mekong River system in Cambodia during the last 6,500 years—Results of Tonlé Sap 96 Project. *In* "Guidebook, Optional Field Tour, Lake Tonlé Sap and Related Fluvial Systems, and Angkor Monument Complex in Cambodia," International Conference on DELTAS, pp 12-13. Siem Reap, Cambodia, 17-18 January 2005.

UN-ESCAP. (2000). Press release REC/92. www.un.org/russin/dhl/resguide/rec-92.htm. Accessed on 23 December 2002.

Uyen, V. T. (1989a). Notes on bank erosion and general Mekong River morphology, Mekong basin-wide bank protection, Stage II. *In* Basin-Wide Seminar on River Management, Nong Khai, 4-5 December 1989, 6pp.

Uyen, V. T. (1989b). Guidelines on river bank protection in the Mekong basin for practicing engineers, Mekong basin-wide bank protection, Stage II. *In* Basin-Wide Seminar on River Management, Nong Khai, 4-5 December 1989, 33pp.

Van Oertzen, I. (1999). *In* "An Introduction to Wetlands, in Environmental Concepts and Issues: A Focus on Cambodia, UNDP/ETAP Reference Guide-Book," O'Brien, N. (ed.), pp 7-1 to 7-14. Ministry of Environment, Phnom Penh.

Viravong, S., Phounsavath, S., Photitay, C., Putrea, S., Chan, S., Kolding, J., Valbojørgensen, and Phoutavong, K. (2006). Hydroacoustic surveys of deep pools in Southern Lao PDR and Northern Cambodia. MRC Technical Paper No. 11, Mekong River Commission, Vientiane, 76pp. ISSN: 1683-1489.

Vongvisessomjai, S. and Phan, N. H. T. (2000). Morphological computation for estuaries. *In* "Proceedings of a workshop on Hydrological and Environmental Modelling of the Mekong Basin," 270-284, 11-12 September 2000, Al-Soufi, R. W. (ed.), pp 270-284. Mekong River Commission, Phnom Penh.

Walling, D. (2005). Evaluation and analysis of sediment data from the Lower Mekong River. Unpublished Report to the MRC, September 2005, 61pp.

Walling, D. E. (2008). The changing sediment load of the Mekong River. *Ambio* **37**, 150-157.

Wolanski, E., Nhanm, N. H., and Spagnol, S. (1998). Sediment dynamics during low flow conditions in the Mekong River estuary, Viet Nam. *Journal of Coastal Research* **14**, 472-482.

Wood, S. H., Ziegler, A. D., and Bundarnsin, T. (2008). Floodplain deposits, channel changes and riverbank stratigraphy of the Mekong River area at the 14th-Century city of Chiang Saen, Northern Thailand, *Geomorphology*. doi:10.1016/j.geomorph.2007.04.030.

Woodroffe, C. D. (2000). Deltaic and estuarine environments and their Late Quaternary dynamics on Sunda and Sahul shelves. *Journal of Asian Earth Sciences* **18**, 393-413.

Workman, D. R. (1972). "Mineral Resources of the Lower Mekong Basin and Adjacent Areas of Kmer Republic, Laos, Thailand and Republic of Viet-Nam." United Nations, New York.

The Sediment Load of the Mekong River

Des E. Walling

Department of Geography, University of Exeter, The Queens Drive, Exeter, Devon EX4 4QJ, UK

OUTLINE

1. INTRODUCTION AND CONTEXT

The sediment load of a river provides an important measure of its hydrology, morphodynamics, and the erosion and sediment delivery processes operating within its basin. The total sediment load can be divided into two components, namely the coarser bed load, which is transported in contact with the river bed, and the finer suspended load, which is transported in suspension. Because it is much easier to measure and generally dominates the load, suspended sediment loads are measured on many of the world's rivers. However, there are very few long-term measurements of bed load transport. Information on the sediment loads of the world's rivers is, therefore, generally restricted to the suspended sediment load, and it is frequently assumed that the bed load comprises about 10% of the total load (e.g., Gregory and Walling, 1973; Milliman and

Meade, 1983). The magnitude of the suspended sediment load transported by a river has important implications both for the natural functioning of the system, for example, through its influence on channel morphology, water quality and aquatic ecosystems and habitats supported by the river, and for human exploitation of the river system. In the latter case, the sediment load can exert an important control on the use of a river for water supply, transport, and related purposes. High sediment loads can, in particular, result in major problems for water resource development, through reservoir sedimentation, and the siltation of water diversion and irrigation schemes, as well increasing the cost of treating water abstracted from a river.

The suspended sediment load of a river is sensitive to both climate change and to a wide range of human activities within its drainage basin, which influence erosion and sediment mobilization and transfer. These include forest cutting and land clearance, expansion of agriculture, land-use practices, mineral extraction, urbanization and infrastructure development, sand mining, dam and reservoir construction, and soil conservation and sediment control programs (see Walling, 2006). Some of these activities, such as forest cutting and the expansion of agriculture, will result in increased suspended sediment loads, whereas others, including sand mining, dam construction, and soil conservation works, are likely lead to reduced sediment transport. Recent reports have, for example, highlighted how the annual suspended sediment load of the Lower Yellow River in China has progressively reduced over the past few decades from an average of ca. 1.1 Gt year^{-1} in the period extending from the 1950s to the 1970s, to a value less than 0.2 Gt year^{-1} in the early years of the twenty-first century, in response to lower rainfall, reservoir construction, and increased water use and extensive soil conservation programs (see Walling, 2006). Similarly, the present suspended sediment load of the lower River Indus in

Pakistan is currently only ca. 15% of that in the 1930s, primarily as a result of dam construction and water abstraction for irrigation (see Milliman *et al.*, 1984; Walling, 2007) and the suspended sediment load of the Chao Phraya River in Thailand has, in recent years, declined to about 20% of its former value as a result of the construction of two major dams in 1965 and 1972. In contrast, the annual sediment load of the Rio Magdalena in Columbia, South America, appears to have increased by ca. 40% between the 1970s and the late 1990s, in response to land clearance, land-use change, and mining activity (see Restrepo and Kjerfve, 2000; Walling, 2006). Although changes in the *quantity* or amount of sediment transported by a river are commonly the key factors, in many cases, changes in the *quality* of the sediment, and more particularly its grain size composition and geochemistry, including nutrient and contaminant content, may also be an important consideration.

Increased sediment loads can give rise to many problems, linked to accelerated loss of reservoir storage capacity through sedimentation, siltation of river channels, and water distribution systems with associated loss of conveyance capacity, and the increased turbidity of river water. These impacts can have several physical, ecological, and economic dimensions. Although decreasing sediment loads will frequently bring benefits in terms of reduced sedimentation and siltation, it is important to recognize that there can also be negative impacts, associated with reduced nutrient inputs to lake, floodplain, delta, and coastal ecosystems and with reduced sediment supply to deltas and coastal areas, which can result in delta recession and coastal erosion. In view of the potential impacts of changing sediment loads on river behavior, river use, and the ecology of the river system, consideration of current and potential future changes in the sediment load of a river should be seen as an important requirement for sound river basin management.

This chapter provides an overview of existing information and understanding related to, firstly, the magnitude of the suspended sediment load of the Mekong River and the key features of its sediment transport regime, and secondly, recent and potential future changes in the sediment load of the river. As a large river basin impacted by accelerated development in recent years, including population growth, land clearance, infrastructure development, and water resource and hydropower development, the sediment load of the Mekong River might be expected to have changed over the past few decades and the ongoing construction of a suite of large reservoirs on the headwaters of the river in China is likely to bring further changes in the future.

2. THE MEKONG RIVER IN A WIDER CONTEXT

The Mekong River (Fig. 6.1) is one of the major rivers of the world. It drains a catchment of ca. 795,000 km^2 and it has been variously ranked as the 12th longest river in the world and as the 8th largest in terms of water discharge (mean discharge $= 15,000 \, \mathrm{m}^3 \, \mathrm{s}^{-1}$) (Mekong River Commission, 2003). One of the key features of the river basin is its largely rural nature. The population density ranges from ca. 10 persons per km^2 in the hill regions to more than 500 persons per km^2 in the more densely populated delta. Because of the low population density over much of the basin, the lack of major extractive industries and industrial development, and the limited use of the river as a transport waterway, due to the many rapids, both the basin and the river are relatively unimpacted by human activity and Kummu and Varis (2007) describe the Mekong as one of the world's most pristine large rivers. As such, its suspended sediment load might be expected to be relatively low when compared, for example, with other large rivers of the region, with much higher population densities,

more intensive land use, and more ongoing development, all of which lead to increased disturbance of the land surface, increased erosion and sediment mobilization, and transport.

Existing estimates of the mean annual land-ocean suspended sediment flux from the Mekong basin reported in the literature (e.g. Milliman and Syvitski, 1992), place this at about 160 Mt year^{-1}. In view of the rapidly changing sediment loads of many rivers in Asia, such as the Yellow River, the Indus, and the Chao Phraya mentioned above, and the many uncertainties regarding the reliability of sediment load data, it is difficult to make comparisons with other large rivers in the region. However, Table 6.1 attempts to place this value into a broader context, by comparing it with estimates of the longer term mean annual sediment loads of other large rivers in the region, prior to the construction of major dams along their courses and based on the data collected largely during the middle and latter years of the twentieth century. Table 6.1 emphasizes that in absolute terms the suspended sediment load of the Mekong River cannot be seen as large, when compared with those of other large Asian rivers. The values of specific suspended sediment yield (i.e., sediment yield per unit area—t km^{-2} year^{-1}) given for the individual rivers, which arguably provide a better basis for comparisons between rivers, indicate that the specific suspended sediment yield for the Mekong River of ca. 200 t km^{-2} year^{-1} is considerably less than that of the Yellow River, the Red River, the Irrawaddy, the Ganges, and the Brahmapura, but is nevertheless of a similar order of magnitude to that of the Yangtze, the Pearl River, and the Indus. The values of specific sediment yield reported for the individual rivers in Table 6.1 cover a considerable range and the very high value associated with the Yellow River is a reflection of the highly erodible loess deposits that are found in the Middle Yellow River Basin. For the other rivers, the values broadly reflect the relative proportions

FIGURE 6.1   The Mekong basin.

of the basin occupied by steep mountainous headwaters, characterized by high sediment yields, and more lowland areas, which are characterized by lower sediment yields and frequently represent significant sediment sinks.

Like the Yangtze, the Pearl River, and the Indus, the Mekong has its headwaters in steep mountainous areas, but a considerable proportion of its basin is occupied by areas of intermediate and low relief, which contribute less

TABLE 6.1 A comparison of the mean annual suspended sediment flux of the Mekong with that of other major rivers in the region

River	Catchment area (km^2)	Mean annual suspended sediment flux (Mt year^{-1})	Mean annual specific suspended sediment yield (t km^{-2} year^{-1})
Yellow River	777,000	1100	1400
Brahmaputra	559,202	799	1429
Ganges	500,176	838	600
Yangtze	1,900,000	480	250
Irrawady	43,000	260	620
Indus	970,000	250	260
Mekong	790,000	160	200
Red	119,866	143	1190
Chao Phraya	160,000	11	68

Based on data reported in the FAO/AGL—Database of World Rivers and their Sediment Yields (FAO/AGL, 2005).

sediment and therefore reduce the overall sediment yield. In the case of the Chao Phraya, the lack of steep mountainous headwaters and the dominance of areas of low and intermediate relief result in a specific suspended sediment yield which is very substantially lower than those of the other rivers.

Studies of the sediment deposits in the Mekong delta reported by Ta *et al.* (2002) have suggested that its sediment load has remained relatively constant over the past 3000 years. Furthermore, there is currently no evidence of the major reduction in sediment load in recent years reported for some other large Asian rivers, such as the Yellow River, the Yangtze, the Chao Phraya, and the Indus. However, population growth, land clearance, land-use change, reservoir construction, and other infrastructure development can be expected to have caused some changes in the sediment load of the Mekong over the past 50 years and these will be considered further later in this chapter.

For some, if not many, major world rivers, the lack of longer term measurements of sediment transport precludes detailed analysis of their sediment loads and sediment transport regimes and of recent changes in their sediment loads. In the case of the Mekong, the available data have significant limitations, particularly in terms of the continuity and length of the records, but these data, nevertheless, afford a worthwhile basis for this analysis of the sediment load of the Mekong.

3. THE AVAILABILITY AND RELIABILITY OF SEDIMENT DATA FOR THE MEKONG

Any attempt to characterize the sediment load of a river system, and particularly to identify temporal trends in that load, is clearly heavily dependent upon the availability of sediment load data. This availability, in turn, reflects the number and location of the measuring stations, the length of record, and the reliability, and temporal resolution of the data. In many areas of the world, sediment load data are unavailable. Where sediment data are available, the record length clearly exerts an important constraint on the ability to establish a representative value for the mean annual sediment load and to identify trends. The reliability of the results obtained will depend heavily upon the nature of the sediment-sampling or -monitoring program and the accuracy of the resulting load estimates.

As indicated above, the availability of sediment data for the Mekong is limited and the available data possess a number of deficiencies that prevent a comprehensive analysis of recent trends in the annual sediment load. However, it is important to recognize that the situation for the Mekong is significantly better than that for many other major world rivers, in that some data are available and these data, although intermittent, relate to a period in excess of 40 years. This chapter focuses on those sediment-monitoring stations on the Middle and Lower Mekong where measurements have spanned significant periods. These comprise the stations at Chiang Saen (Thailand), Luang Prabang (Lao PDR), Nong Khai (Thailand), Mukdahan (Thailand), and Pakse (Lao PDR) (see Fig. 6.1). Some data for Jinghong on the Upper Mekong or Lancang River in China (see Fig. 6.1) are also considered. One key feature of the sediment data available for the Mekong is that at several of the above stations, measurements were initiated in the early 1960s and, although the subsequent records are discontinuous and frequently involve limited numbers of samples, these early measurements provide a useful baseline for assessing trends over the ensuing years.

Considering the available data as of 2005 in more detail, these involve information obtained from three different measurement programs. The first is the sediment-sampling program initiated on the Lower Mekong in 1960, within the framework of the Lower Mekong Project, funded by the US Agency for International Development and coordinated by the Harza Engineering Company (Harza Engineering Company, 1962). This has been continued intermittently by national agencies through to the present. This sampling program was based on existing US practice and used standard US designed isokinetic samplers and involved depth-integrated sampling in several verticals, in order to derive an estimate of the mean suspended sediment concentration in the cross section. Originally, this network involved the measuring stations at Chiang Saen, Luang Prabang, Mukdahan, and Pakse, and in 1972 the station at Nong Khai was also added, using the same basic procedures. The operation of these sampling stations appears to have been somewhat haphazard in terms of both the frequency of sampling and its continuity from year to year. Table 6.2 indicates the years during which sampling was undertaken and the number of samples collected during those years.

The second data source represents the sediment measurement program undertaken by the Chinese authorities on the Upper Mekong or Lancang River at Jinghong, China. Despite the international status of the Mekong River, access to these data is unfortunately restricted, particularly for recent years. However, annual load data for the years 1963, 1965, 1966, and 1967-1990 have been compiled by the author from secondary sources. Full details of the sampling regime at this site are unavailable, but existing information on sampling procedures in China suggests that sampling is likely to be frequent and probably daily.

The third data source is the water quality monitoring network established by the Mekong River Commission in 1985 and which includes three of the sites where sediment monitoring has been undertaken, namely Chiang Saen, Luang Prabang, and Pakse. This is primarily a water quality monitoring program, but the determinands include total suspended solids (TSS). The data from this source have two important potential deficiencies for sediment studies. Firstly, the sampling frequency is monthly and, secondly, the samples are collected near the surface of the river (0.3 m depth), using a bottle rather than a true sampler. As such they are not

TABLE 6.2 The coverage of the available sediment concentration data for the Lower Mekong at the five key sites

Year	Location				
	Chiang Saen	Luang Prabang	Nong Khai	Mukdahan	Pakse
1960	22	8		9	9
1961	20	105		60	109
1962	5	27		71	44
1963				32	
1964				42	
1965				38	
1966				35	
1967				42	
1968	38			45	
1969	73			66	
1970	83			73	
1971	71			58	
1972	65		58	72	
1973	33		89	74	
1974	33		87	71	
1975	9		33	36	
1976			27	16	
1977			46	26	
1978			47	26	
1979				27	
1980				25	
1981			21	22	
1982			16	19	
1983			18		
1984			16	20	
1985			4	1	
1986		22	18	18	
1987		43	15	4	
1988		41	14	6	
1989		44	20	11	
1990		37	22	14	
1991		18	14	19	
1992		37	23	21	
1993			19	19	

Continued

TABLE 6.2 The coverage of the available sediment concentration data for the Lower Mekong at the five key sites—Cont'd

Year	Location				
	Chiang Saen	Luang Prabang	Nong Khai	Mukdahan	Pakse
1994	48		24	22	
1995	45		15	18	
1996	32		20	19	
1997	39	12	25	11	11
1998	38	12	26	35	10
1999	40	7	29	43	12
2000	40	9	27	41	14
2001	38	11	30	42	13
2002	38	9	42	38	11
2003	36				

The years when sampling was undertaken at the individual sites and the number of samples collected in those years are indicated.

isokinetic and, since suspended sediment concentrations are known to increase with depth, they are likely to underestimate the true mean concentration in the cross section.

Since the emphasis of this chapter is on the longer term sediment load and sediment regime of the Mekong, and its variation along the course of the river, and on assessing recent trends in this load, attention focuses on the available information on annual sediment loads and the likely reliability of those data. In the case of the Lancang River at Jinghong, the existing records provide values of annual load, based on summation of the measured daily loads. Since these data are based on standard Chinese practice, involving frequent sampling using specialized sediment-sampling equipment, they are considered to be reliable.

The data available from the sediment measurement programs for the five main measuring stations on the Mekong at Chiang Saen, Luang Prabang, Nong Khai, Mukdahan, and Pakse, differ from those for the Lancang River,

in that they represent values of sediment concentration only for the occasions on which sampling was undertaken. Further processing is required to derive estimates of sediment load for individual years. Since trends are being assessed, it is important that the estimates of annual load should reflect any changes in sediment transport taking place within the basin. In this study, emphasis has therefore been placed on using the data available for individual years to obtain estimates of the annual load for those years, rather than on combining the data for several years to produce a sediment rating curve for that period and thereby producing an average sediment load for the period. In view of the variable sampling frequency from year to year, and the general lack of frequent sampling, it was deemed inappropriate to attempt to reconstruct the continuous record of sediment concentration from the infrequent samples or to use interpolation procedures to estimate the load. Emphasis has been placed on using rating curves established for individual years (see Walling, 2005). Detailed scrutiny of the available data and comparisons of annual load estimates derived using rating curves with those obtained by reconstructing the continuous record of suspended sediment concentration for the few years where larger numbers of samples were collected, demonstrated that a procedure based on a single rating curve of the form $C = aQ^b$ fitted to the sediment concentration (C) and daily mean water discharge (Q) data directly, using a nonlinear estimation (Solver) routine, rather than the more generally used standard log/log regression technique, provided the most reliable estimates of the annual sediment load. With standard log/log regression techniques, the logarithmic transformation of the data means that all points are given a similar weight by the least squares routine, when fitting the relationship. With a nonlinear estimation routine, the raw data are used and the least squares fitting routine gives much greater weight to the data points representing high values of concentration and load that account for a large proportion of the annual load. Furthermore, the use of untransformed data avoids

the bias introduced by using log-transformed data, which can result in underestimation of the annual sediment load (see Ferguson, 1986; Koch and Smillie, 1986).

Figure 6.2 presents an example of the sediment rating plot for Luang Prabang for 1961, a year in which a relatively large number of samples (105) was collected. Rating relationships of the form $C = aQ^b$ were fitted to these data using both log/log regression and the nonlinear estimation (Solver) routine. As for most datasets employed in the study, the rating relationship generated by the nonlinear estimation routine is characterized by a higher constant (a) and a lower exponent (b) than that fitted by log/log regression. This reflects the closer fit to the points representing high values of sediment concentration and discharge, which account for the majority of the load of a river. Although Fig. 6.2 indicates that the log/log regression produces a higher coefficient of determination (r^2) than that fitted by the nonlinear estimation routine, care is required in interpreting the significance of this measure of goodness of fit for the likely accuracy of the estimates of annual load obtained using the relationship. The r^2 value

FIGURE 6.2   A sediment rating plot for the Mekong River at Luang Prabang for 1961, based on the concentrations measured in the samples collected during this year, and showing the rating relationships established using linear regression on log-transformed concentration and water discharge data (designated log/log) and a nonlinear estimation procedure applied to the untransformed data (designated Solver).

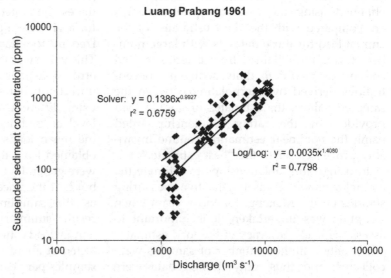

Luang Prabang 1961

Solver: y = 0.1386x$^{0.9927}$
         r^2 = 0.6759

Log/Log: y = 0.0035x$^{1.4080}$
          r^2 = 0.7798

Suspended sediment concentration (ppm)

Discharge (m³ s⁻¹)

for the log/log regression is based on the log-transformed data used to derive the relationship. Calculation of the r^2 value for the relationship between the measured sediment concentrations and those predicted from the discharge value via the log/log relationship, using the raw untransformed data, produces an r^2 value (0.65), which is lower than that for the relationship fitted using the nonlinear estimation routine (0.68). This situation was typical of most of the rating relationships fitted to the data from the five measuring stations. Other measures of goodness of fit further confirmed the advantages of the rating relationship fitted to the available data using the nonlinear estimation routine. For example, if the daily loads (t day^{-1}) (calculated as the product of the measured concentration and the daily mean flow) for the days on which the samples used to establish the rating relationship were collected, are summed and compared with the equivalent totals obtained using the concentration values estimated using the two alternative rating relationships, the value obtained using the rating relationship fitted using the nonlinear estimation technique commonly provided better agreement. Similarly, when the estimates of annual load obtained using the two rating relationships are compared with the "best estimate" of the annual load for those datasets with large number of samples, obtained from a reconstructed continuous record of daily sediment concentrations derived by interpolating between the sampled values, the estimates of annual load provided by the rating relationship fitted using the nonlinear estimation routine invariably proved closer to the "best estimate."

In using rating relationships to estimate the annual sediment loads for the five measuring stations on the Mekong, for those years when sampling was undertaken, it is important to assess the likely accuracy of the load estimates when only a limited number of samples were collected; and thus, only limited data were

available to establish the rating relationship. Similarly, there is a need to establish the minimum number of samples required to obtain a meaningful estimate of the sediment load for a given year. These considerations have been addressed by identifying nine station years of record, including at least 1 year from each of the five measuring stations, where the sampling frequency was relatively high and permitted reconstruction of the continuous record of suspended sediment concentration by interpolation and calculation of a "best estimate" of annual load, by combining the measured values of daily mean water discharge with the estimate of daily mean sediment concentration obtained from the reconstructed continuous record of suspended sediment concentration. The effects of sampling frequency on the reliability of the resulting load estimate was assessed by subsampling the reconstructed continuous record of sediment concentration, using sampling frequencies of 7, 14, and 28 days. In this way, seven replicate datasets were created by sampling on the same day each week throughout the year, and, using the same principle, 14 and 28 replicate datasets were created for sampling frequencies of 14 and 28 days, respectively. The variability of the load estimates generated from the replicate datasets for a given sampling frequency was characterized by the standard deviation of the values. This value was then used to calculate a standard error statistic indicating the likely reliability or uncertainty of the annual load estimates derived using different sampling frequencies, at a given level of confidence. The results indicated that the mean loads associated with the estimates obtained from the individual replicate datasets were generally close to the "best estimate" load but that the uncertainty increased considerably as the sampling frequency decreased. The results summarized in Table 6.3 indicate that annual load estimates derived from samples collected at about fortnightly intervals (i.e., ca. 25 samples per year) are likely to have an accuracy

TABLE 6.3 Estimates of uncertainty at the 95% level of confidence for annual load estimates derived using samples collected at different sampling frequencies, based on several representative datasets for sampling stations on the Mekong River

Station and year	Level of uncertainty in load estimates at the 95% level of confidence		
	7-day sampling (%)	14-day sampling (%)	28-day sampling (%)
Pakse, 1961	±2.5	±8.1	±22.0
Mukdahan, 1972	±2.7	±5.5	±25.5
Mukdahan, 1970	±0.5	±4.7	±11.1
Nong Khai, 1977	±9.4	±13.8	±36.8
Nong Khai, 1974	±1.7	±4.8	±19.8
Luang Prabang, 1961	±8.7	±13.7	±23.7
Chiang Saen, 1999	±2.5	±6.3	±18.9
Chiang Saen, 2001	±2.9	±4.6	±16.9
Chiang Saen, 2003	±2.4	±7.2	±20.9

of ca. ±10% at the 95% level of confidence. Although a higher level of accuracy might be desirable, this would only be achievable for a limited number of the available station years of record shown in Table 6.2. Equally, Table 6.3 shows that, if the sampling frequency is reduced to 28 days, the uncertainty associated with the resulting load estimates is likely to be of the order of ±20-25% or even greater, at the 95% level of confidence.

On the basis of the above analysis, load estimates were derived for all station years where the number of samples exceeded 20 and these were reasonably uniformly distributed throughout the year. These load estimates were judged to involve an uncertainty associated with the load estimation procedure of less than ±10-15% at the 95% level of confidence and it is important to recognize that this level of uncertainty is likely to be of a similar order of magnitude to that associated with the concentration and discharge data themselves. Where the number of samples was <20 but >10 per year (see Table 6.2) and these samples were suitably distributed throughout the year, load estimates were obtained by combining the data

for two adjacent years in order to establish the rating relationship, and in this case, the resulting load estimates were judged to have an equivalent uncertainty of ca. ±20%.

The data from the water quality network also comprise individual values of TSS for the days on which samples were collected and require further processing to derive estimates of the annual sediment load. However, based on the above analysis, the monthly sampling frequency was judged inadequate to permit the use of rating curves to provide reliable estimates of annual sediment load. The limited sampling frequency is clearly a major and serious limitation of these data, since sampling once a month is unlikely to provide representative information on sediment concentrations during the flood season. In addition, it is also necessary to consider the accuracy of the sediment concentration values, bearing in mind that the primary purpose of the sampling program was to characterize water quality and they were obtained from dip samples collected close to the surface of the river, rather than using specialist sediment-sampling equipment that can collect depth-integrated samples.

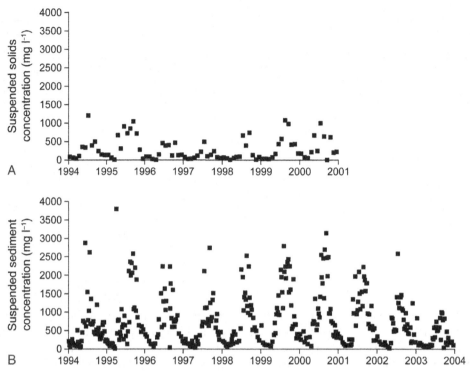

FIGURE 6.3 A comparison of the TSS concentrations reported by the water quality monitoring program for the sampling station at Chiang Saen (A) with the values of suspended sediment concentration obtained for the same site by the sediment-sampling program (B).

Since the finer fraction of the suspended sediment load is the most chemically active and exerts a key influence on water quality and on the sediment-associated transport of nutrients and contaminants, in a water quality sampling program there is less need to ensure that the coarser fractions of the sediment load are sampled and therefore to sample the complete depth profile and full cross section sampling is essential. Although it is not possible to make a direct comparison between the concentration values obtained using the two sampling methods, an indication of the potential errors associated with the surface dip samples is provided by a comparison of the magnitude of the concentration values reported for Chiang Saen for the years 1994-2001, for which both

datasets are available (see Fig. 6.3). During this period, the TSS concentrations reported for the water quality samples rarely exceeded 1000 mg l^{-1} and did not reach 1500 mg l^{-1} (see Fig. 6.3A). However, the data from the sediment-sampling program (see Fig. 6.3B) indicated that sediment concentrations exceeded 1000 mg l^{-1} for extended periods during the flood season and in many years individual samples exceeded 2500 mg l^{-1}. These findings cast serious doubt on the use of the TSS data provided by the water quality monitoring network for documenting the suspended sediment load of the Mekong and, although these data have been used by other workers (cf. Fu et al., 2006; Kummu and Varis, 2007; Lu and Siew, 2006), they have not been used in this study.

4. KEY FEATURES OF THE SEDIMENT REGIME OF THE MEKONG RIVER

As indicated above, the available sediment load data for the Mekong possess several limitations in terms of their lack of continuity and the limited sampling frequency for many years. However, they provide a reasonable basis for defining the key characteristics of the sediment regime of the river. Figure 6.4 presents the estimates of annual suspended sediment load for the five main sediment-sampling stations on the Mekong, derived for those years with sufficient samples, using the procedures outlined above. As indicated, the relatively small number of samples used to estimate the values of annual sediment load mean that these values involve uncertainties of the order of ±10-15% at the 95% level of confidence. Additional errors associated with the flow and concentration data could increase this uncertainty further. It is important that any attempt to compare values of annual sediment load between years or between stations should recognize these uncertainties. Equivalent data obtained from Chinese sources for the Lancang River or Upper Mekong at Jinghong are also presented in Fig. 6.4. Although the records for the six measuring stations span the period from the early 1960s to 2003, they are discontinuous. There are no years where estimates of annual suspended sediment load are available for all six stations and, for much of the period, estimates are only available for three or four of the stations. Any attempt to use the data presented in Fig. 6.4 to characterize the sediment load of the Mekong must also recognize that the records are unlikely to be stationary, due to changes in land use within the upstream catchments and the construction of dams on the headwaters of the river in China during the 1990s. Furthermore, the interannual variability of the annual load values makes it difficult to compare the values obtained for the individual rivers across different periods.

Although the discontinuous nature of the records of annual sediment load presented in Fig. 6.4 and the likely nonstationary nature of the records preclude detailed analysis of interannual variability, it is useful to consider the more continuous records available for the Mekong River at Nong Kai and Mukdahan. The coefficient of variation (CV) of the annual load values affords a useful measure of the interannual variability of the sediment record for a river and the CV values for these two rivers are 27% and 44%, respectively. Bearing in mind that both records are likely to reflect a degree of nonstationarity, these values suggest that the Mekong is characterized by a CV value of the order of ca. 25-35%. Considering the global assessment of CV values presented by Walling and Kleo (1979), a value in the range 25-35% must be seen as relatively low, in terms of both the overall global range and the range for river basins with an annual runoff similar to that of the Mekong. This finding indicates that the regular flood season associated with the well-defined seasonal monsoon climate serves to reduce the interannual variability of the sediment load of the Mekong. However, there is some evidence to suggest that the recent occurrence of dry years, characterized by both low annual runoff and low sediment yields, could reflect an increase in the variability of the climate.

4.1. Spatial Variability of the Sediment Load

Notwithstanding the limitations of the available data related to stationarity and interannual variability, highlighted above, Fig. 6.5 attempts to consider the downstream evolution of the suspended sediment load of the Mekong, by plotting the range of annual loads documented for each monitoring station versus the catchment area of the individual stations. It is

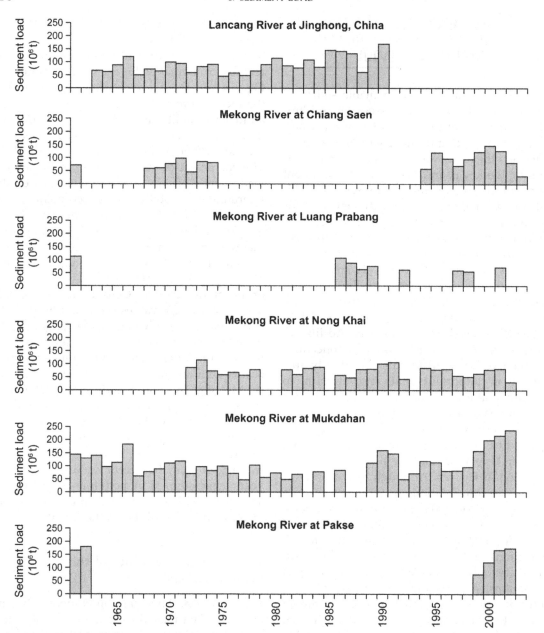

FIGURE 6.4 The available estimates of annual sediment load ($\times 10^6$ t) for the five designated sites on the Mekong River in Thailand and Lao PDR and equivalent data for the period 1983-1990 for the Upper Mekong or Lancang River at Jinghong, China.

FIGURE 6.5 Downstream trends in the annual sediment loads documented for the six measuring stations on the Mekong River, in relation to increasing catchment area.

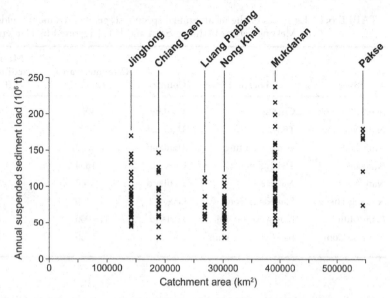

important to recognize that the most downstream station at Pakse still only represents about 70% of the total catchment area of the Mekong. However, there are no detailed sediment data available for stations further down the river. Probably, the most striking feature of the data presented in Fig. 6.5 is the absence of a well-defined increase in the sediment load, with increasing catchment area. The loads documented for Pakse are not substantially greater than those for Jinghong, despite the contributing catchment area being nearly four times greater at Pakse. This situation was highlighted by Roberts (2001), who indicated that about 50% of the suspended sediment load of the Mekong is contributed by the upper part of the basin in China, with that part of the basin accounting for only about 25% of the total area of the basin and about 18% of its total water discharge. Figure 6.5 also suggests that the annual sediment load may decrease between Jinghong and Nong Kai, before increasing further downstream. These two features of the sediment records could reflect two possible controls. Firstly, the sediment contribution from those parts of the Mekong basin below

Jinghong may be relatively low, resulting in only a limited downstream increase in the total sediment load through to the basin outlet. Secondly, some of the sediment moving through the main channel system may be deposited and stored within the system, rather than being transferred downstream, resulting in only a limited downstream increase in the sediment load, and even a decrease. These possibilities will be considered further below.

A marked reduction in the sediment input to the Mekong below Jinghong is consistent with the data for other large Asian rivers presented in Table 6.1, which, as discussed above, suggests that areas of intermediate and low relief are characterized by relatively low specific sediment yields, as compared to the steeper mountainous parts of their catchments. The specific sediment yield of the Chao Phraya, basin, which might, for example, be seen as more comparable with the lower parts of the Mekong basin, is only about 68 t km^{-2} $year^{-1}$, whereas that for the headwaters of the Mekong above Jinghong is an order of magnitude greater at around 700 t km^{-2} $year^{-1}$. Although information on the sediment loads of the Mekong tributaries is very

TABLE 6.4 Estimates of the mean annual specific suspended sediment yields of some of the tributaries entering the Mekong between Chiang Saen and Pakse reported by Harden and Sundborg (1992)

River	Location	Country	Catchment area (km²)	Mean annual sediment yield (t × 10³)	Mean annual specific sediment yield (t km⁻² year⁻¹)
Nam Mae Kok	Chiang Rai	Thailand	6060	799	132
Nam Mae Ing	Thoeng	Thailand	5700	300	53
Nam Loei	Wang Saphung	Thailand	1240	197	159
Nam Lik	Tha Ngone	Laos	1650	796	48
Nam Kam	Na Kae	Thailand	2360	96	41
Se Bang Hieng	Ban Keng Done	Laos	19,400	3200	165
Nam Mun	Kaeng Saphu Tai	Thaiand	116,000	4386	38
Nam Se Done	Ban Souvannakhili	Laos	5760	1321	229

limited, available estimates of the specific sediment yield of several of the tributaries joining the Mekong between Jinghong and Pakse reported by Harden and Sundborg (1992) are listed in Table 6.4. These data confirm the relatively low specific sediment yields of these tributary catchments and thus the limited sediment input from the part of the Mekong catchment below Jinghong. Although the information on annual suspended sediment loads for the measuring station at Pakse is very limited, the data presented in Fig. 6.5 suggest that the mean annual suspended sediment load is about 150 Mt. If this is compared with an equivalent load of ca. 90 Mt at Jinghong, the increase in sediment load of 60 Mt is equivalent to a specific suspended sediment yield from the intervening basin area of ca. 145 t km⁻² year⁻¹. This value is consistent with the data presented in Table 6.3.

The possibility that the limited downstream increase in sediment load and, perhaps more importantly, the possible reduction in sediment load between Jinghong and Nong Khai suggested by both Figs. 6.4 and 6.5 could reflect deposition and storage of sediment within the channel and floodplain system must also be considered. Existing information on the sediment load of the Yangtze River in China provides an indication of the potential for such conveyance losses. The mean annual load of the Yangtze River at Datong located near the basin outlet, with a drainage area of ca. 1,700,000 km², is generally slightly lower than that measured in the middle reaches of the river at Yichang, where the catchment area of ca. 1,000,000 km² represents only about 60% of the total basin area. Since there is a significant sediment input from the lower 40% of the basin, the reduced sediment load at the basin outlet, when compared with the middle reaches, must reflect depositional losses in the lower reaches of the basin, particularly in association with the large lakes connected to the river system. The more constrained channel of the Mekong between Jinghong and Nong Kai, the course of which is frequently incised and lacking in an extensive floodplain, affords much less opportunity for conveyance loss and sediment storage (see Gupta and Lieuw, 2007). Nevertheless, the data presented in Figs. 6.4 and 6.5 suggest that a significant proportion of the sediment load transported from upstream is sequestered with this reach.

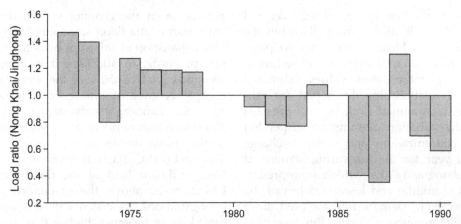

FIGURE 6.6 The ratio of the annual suspended sediment load recorded for the Mekong River at Nong Khai to that measured on the Lancang River at Jinghong and changes in this ratio over the period 1972-1990. No ratio value has been plotted for the years 1979, 1980, and 1985, due to the lack of a sediment load data for Nong Khai for these years.

Figure 6.6 provides further evidence of such storage by plotting the ratio of the annual sediment load at Nong Khai to that at Jinghong, for the period of which records are available for both stations for most years. Figure 6.6 indicates that during half of the years for which estimates of annual sediment load are available, the sediment load at the downstream site was significantly less than that at the upstream site, suggesting that a substantial proportion of the load was deposited within the intervening reach.

Taking account of the magnitude of the increase in drainage area between Jinghong and Nong Khai, and assuming an upstream sediment load of 100 Mt year^{-1}, and a likely sediment contribution of ca. 150 t km^{-2} year^{-1} from the intervening catchment area (see above and Table 6.2), a ratio of 1.0 would indicate deposition of ca. 25 Mt within the reach between Jinghong and Nong Khai and this value would increase to ca 50 Mt for a ratio of 0.7. Further work is clearly required to assess the field evidence for such deposition and to consider whether the apparent reduction in sediment load between Jinghong and Nong Kai might reflect measurement errors, rather than deposition

within the intervening reach. Interestingly, prior to 1980 the suspended sediment load at Nong Khai was generally around 20% or more greater than that upstream at Jinghong, suggesting that deposition was of limited importance. However, after 1980, the situation changes, with most of the years providing evidence of significant deposition. In a subsequent section, it is noted that the sediment load of the Lancang River at Jinghong evidenced a significant increase after 1980 and thus it might appear that the shift to significant deposition within the reach between Jinghong and Nong Kai after 1980 reflects a response to the increasing sediment loads entering the reach and thus a buffering of the increased load by increased deposition.

4.2. Intraannual Variability in Sediment Transport

Although the above discussion of the sediment load of the Mekong River has placed emphasis on the magnitude of the annual sediment loads and the associated values of specific suspended sediment yield, and their interannual variability, it is important to also consider the intraannual variation of sediment transport in response to the flow regime of the river and the seasonal distribution of rainfall within the catchment of the Mekong. In order to place emphasis on actual measurements of sediment concentration, rather than values estimated from rating relationships developed for estimating the total annual load, Fig. 6.7 presents plots of the variation of sampled suspended sediment concentrations and water discharge through a year for the measuring stations at Luang Prabang and Pakse, which are representative of the middle and lower reaches of the river, respectively. Data for 1961 are used, since the sampling frequency during this year was particularly high for both stations. A synthesis of the hydrology of the Mekong presented by the Mekong River Commission (MRC, 2005), indicates that the floods of 1961 were of average magnitude at Luang Prabang, whereas at Pakse they were more extreme, with the flood volume having a return period of ca. 5-10 years. The plots for both stations emphasize that most of the suspended sediment load of the Mekong is transported during the main flood season which extends from June to November.

A number of other important features of the plots presented in Fig. 6.7 can usefully be highlighted. First, the sediment concentrations recorded at Luang Prabang show clear evidence of hysteresis, with the highest concentrations occurring early in the flood season, in mid-July. After mid-August, concentrations begin to decline, well before the occurrence of the peak flood discharges in early September. Concentrations recorded after early September for a given discharge are substantially lower than those recorded during the early part of the flood season. This behavior must be seen as being similar to that of other large rivers with a distinct flood season and can be seen as reflecting an "exhaustion effect," wherein the first flood flows of the season are more heavily charged with sediment, than the later flows, due to the mobilization of readily available sediment that has accumulated on the catchment

surface or in the channel system during the early part of the flood season and the progressive exhaustion of this supply as the flood season proceeds. In addition, it is important to recognize that ca. 50% of the wet-season discharge recorded at Luang Prabang originates from the Lancang headwaters in China (the Yunnan component) and ca. 50% from areas further downstream in Myanmar, Laos, and Thailand (MRC, 2005). In view of the relatively high sediment load of the floodwaters from China, noted above, the suspended sediment concentrations associated with this component are likely to be much higher than those associated with the flood runoff from the areas further downstream. If the arrival of upstream floodwaters, which partly reflect snowmelt on the Tibetan Plateau, precedes that of the downstream areas, the early part of the flood season will be characterized by higher concentrations. In 1961, the highest discharges of the flood season were recorded in mid-August at Jinghong on the Lancang River, whereas further downstream at Luang Prabang the maximum discharge was recorded in mid-September. Although still evident, the hysteresis identified in the sediment record for Luang Prabang is less clear at Pakse further downstream. This situation can be related to the much larger size of the basin and the fact that the flood runoff from the headwaters in China represents only about 20% of the total flood season runoff and its high sediment concentrations will therefore be readily diluted and their effect masked, by the flood runoff from the downstream areas. Furthermore, variations in the timing of sediment contributions from different subbasins are likely to further blur any evidence of hysteresis.

A second important feature of the records of sediment concentration presented in Fig. 6.7 is the marked reduction in the magnitude of the suspended sediment concentrations between Luang Prabang and Pakse. Whereas concentrations at Luang Prabang exceed 1500 mg l^{-1} for several weeks of the flood season, and are in

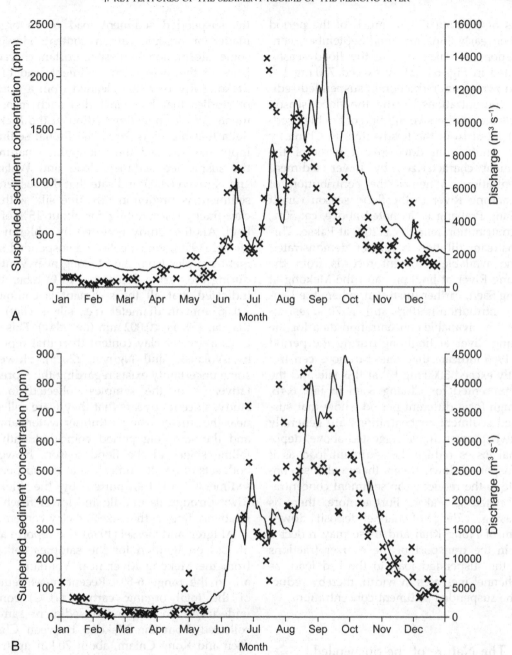

FIGURE 6.7   The record of water discharge and the measured suspended sediment concentrations for the Mekong River at Luang Prabang and Pakse, for 1961.

excess of 1000 mg l^{-1} for much of the period between early June and mid-September, concentrations at Pakse during the flood season recorded in Fig. 6.7 rarely exceed 750 mg l^{-1}. This downstream reduction in suspended sediment concentrations during the flood season reflects the progressive dilution of the high sediment input from the headwaters in China by flood runoff from downstream parts of the catchment characterized by lower sediment concentrations. Whereas the contribution of the Lancang River to the flood season runoff at Luang Prabang is, as noted above, ca. 50%, this contribution falls to ca. 20% at Pakse. This downstream dilution is further demonstrated by the available sediment records from the Lancang River at Jinghong and the Mekong at Chiang Saen, further upstream, where the Lancang contribution is 100% and ca. 70%, respectively. The available concentration data for the Lancang River at Jinghong during the period 1963-1966 indicate that concentrations can frequently exceed 3000 mg l^{-1} at this site and the data presented for Chiang Saen in Fig. 6.4B, although for a different period, show that suspended sediment concentrations are generally somewhat lower. If, as suggested above, depositional losses reduce the sediment load as it is transported downstream, these will also contribute to the reduction in sediment concentrations noted at Pakse. Furthermore, there is increasing evidence of channel deposits downstream of Nong Khai and these may reflect a shift in the transport of the coarser fractions from the suspended load to the bed load, as the channel increases in width, thereby reducing the suspended sediment concentrations.

4.3. The Nature of the Suspended Sediment Load

The existing sediment-monitoring program on the Mekong River does not include routine measurements of the grain size composition of the suspended sediment load or its organic matter or organic carbon content. However, some information on these important characteristics of the suspended sediment load of the Mekong River can be gleaned from a number of studies that have included such measurements. Work on sedimentation in the Mekong delta that has considered the fluvial sediment input has stressed the fine-grained nature of the suspended sediment load and Wolanski and Nguyen (2005) indicate that the suspended sediment is predominantly fine silt, with the clay fraction accounting for about 15% of the load. Another study reported by Ahlgren and Hessel (1996), which collected suspended sediment samples from the Mekong near Ventiane during the period September to October 1994, indicated that all the sediment is commonly <0.062 mm in diameter (i.e., silt + clay) and that ca. 45% is <0.002 mm (i.e., clay). This suggests a greater clay content than that reported by Wolanski and Nguyen (2005). However, some uncertainty exists regarding the representativeness of the samples collected in this study, since it appears that they were collected near the surface using a Ruttner water sampler and the sampling period coincided with the falling stages of the flood season. However, both sets of results indicate that the suspended sediment load transported by the Mekong River through its middle and lower reaches is relatively fine, with a sizeable clay content.

Ahlgren and Hessel (1996) also report values of loss on ignition for the samples collected from the Mekong River near Ventiane. These are in the range 6-8%. Recent measurements of the total organic carbon (TOC) content undertaken on suspended sediment samples collected from the Mekong between Chiang Saen and Kong Chiam, about 70 km upstream of Pakse, reported by Keenan et al. (2006) have indicated that organic carbon concentrations in sediment collected from several sampling points on the river are typically less than 5%. These would be equivalent to an organic matter content of <10%, which is consistent with the loss on ignition values reported by Ahlgren and Hessel (1996).

5. RECENT CHANGES IN THE SEDIMENT LOAD OF THE MEKONG RIVER

The available data on annual suspended sediment loads for the Mekong covering the period from the beginning of measurements in the early 1960s to 2003, assembled for this study and presented in Fig. 6.4 emphasize that any attempt to investigate recent trends in the sediment load of the Mekong River is likely to be compromised by the lack of long time series of annual loads, as well as the inherent interannual variability of the sediment record. For the six measuring stations represented in Fig. 6.4, only those at Jinghong, Nong Khai, and Mukdahan have records that span a significant proportion of the past 40 years. The lack of more recent data for the period post 1990 for the station at Jinghong precludes detailed analysis of the impact of the dams constructed on the Lancang River on its sediment load, and the value of the time series for Nong Khai, while continuing to the present, is limited by its more restricted length, having commenced only in 1972. However, notwithstanding these important limitations, the available data provide some basis for assessing and interpreting recent trends in the suspended sediment load of the Mekong River.

5.1. The Changing Sediment Load of the Lancang River

It is appropriate to commence with the evidence for changing suspended sediment loads within the Mekong basin provided by the Lancang River, since, as indicated above, previous work has suggested that the portion of the Mekong basin in China contributes approximately 50% of the downstream sediment load of the Mekong River. The data presented for the Lancang River at Jinghong in Figs. 6.4 and 6.8B provide clear evidence of a trend for annual sediment loads to have increased in recent years. A simple trend line fitted to the available annual load data in Fig. 6.8B provides evidence of a statistically significant (>99%) increase in sediment load over the period, with average loads increasing from ca. 60 Mt in the mid-1960s to about 115 Mt in the late 1980s. In contrast, the discharge record for this period (Fig. 6.8A) shows no evidence of a statistically significant trend, although the data arguably show some evidence of a minor decline in annual runoff, which could reflect increased water abstraction. The increasing sediment loads evidenced by the Lancang River can be linked to the marked expansion of population and associated intensification of land use in the middle and lower reaches of this basin in the period commencing in the 1970s. A cumulative double mass plot of the sediment load and discharge data (Fig. 6.8C) indicates that the impact of these changes on sediment loads was apparent from around 1980. A study reported by You (1999) suggests that most of this increase was generated within the middle Lancang Basin and that it was closely related to the rapidly increasing population and the resultant land clearance and intensification of agricultural production. Using multiple regression, he was able to demonstrate a significant improvement in the level of explanation of the interannual variation in sediment load over the period 1965-1987, provided by the magnitude of the annual runoff by including an estimate of the magnitude of the population for each year. Fu et al. (2006) suggest that the introduction of improved soil conservation measures into this area of China since 2000, through the "Grain for Green" program, has caused a reduction in erosion and sediment yield, but the lack of sediment load data for

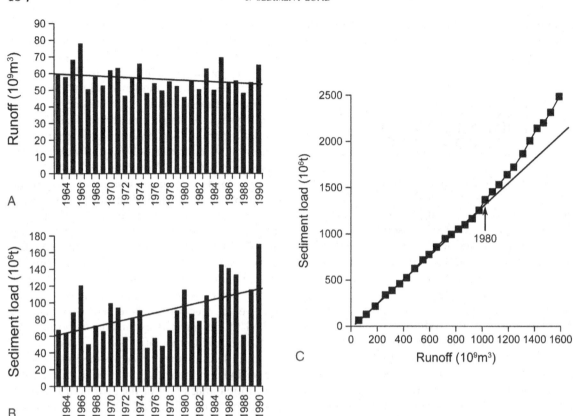

FIGURE 6.8 Variation in the annual runoff (A) and sediment load (B) of the Lancang River at Jinghong over the period 1963-1990 and a double mass plot (C) of annual sediment load versus annual runoff.

this period precludes further analysis of the impact of this program. Furthermore, it is likely to prove difficult to separate its effects from those of dam construction since the 1990s, which will also have reduced the sediment load of the Lancang River.

The construction of hydropower dams on the Lancang River in the 1990s can be expected to have caused a decrease in its suspended sediment load. The Manwan Dam was constructed during the early 1990s and was fully operational by 1996 and the Dachaoshan Dam was built in the late 1990s and completed in 2003 (see Figs. 6.1 and 6.9 for location). Although important, both these dams are relatively small by world standards, with total storage volumes

of 920 and 890 Mm^3 and active capacity/ annual inflow ratios of only ca. 0.007 and 0.009 for the Manwan and Dachaoshan dams, respectively (see Kummu and Varis, 2007). Kummu and Varis (2007) estimate the trap efficiencies of these dams to be 68% and 66%, respectively, but these values could overestimate the true situation. Furthermore, both dams are located towards the middle reaches of the Lancang basin, and the upstream area contributes only part of the overall sediment load generated within the Lancang basin. However, estimates of sedimentation rates based on the reservoir surveys undertaken by the Chinese authorities suggest that over the period from 1993 to 2005 the Manwan Dam trapped ca.

490 Mm^3, which is equivalent to an annual sedimentation rate of ca. 50 Mt year^{-1}. Since 2001 the Dachaoshan Dam, downstream from the Manwan Dam has also been trapping sediment and over the period November 2001 to April 2003 the total amount of sediment trapped was equivalent to ca. 30 Mt. Taken together, it would seem likely that the two dams are together currently trapping ca. 70-80 Mt year^{-1} and this could clearly be expected to cause a major reduction in the sediment load of the Lancang River. Restrictions on release of sediment load data for the period post 1990 imposed by the Chinese authorities unfortunately preclude detailed analysis of the impact of these changes on the sediment of the Lancang River itself, although any substantial reduction should be detectable further downstream.

Some information on changes in the suspended sediment concentrations measured in the Lancang River at Jinghong some 401 and 314 km downstream of the Manwan and Dachaoshan dams, respectively, have, however, recently been reported by Fu *et al.* (2006) and these are presented in Fig. 6.9A. These data suggest that the closure of the Manwan dam in the early 1990s resulted in a significant reduction in the suspended sediment concentrations measured at Jinghong and that the closure of the Dachaoshan Dam in 2001 caused a further decrease in concentrations. The particularly low concentrations that were reported for 2003

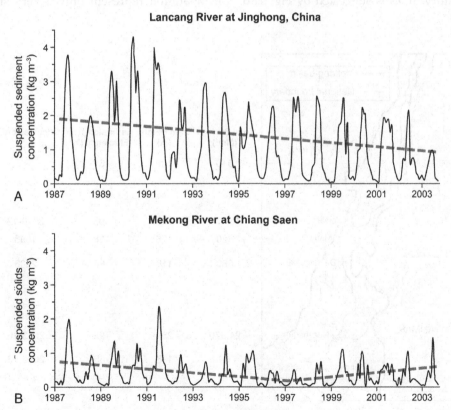

FIGURE 6.9 The suspended sediment concentration records for the period 1987-2003 for the Lancang River at Jinghong (A) and the Mekong River at Chiang Saen (B), as reported by Fu *et al.* (2006).

may, however, reflect the lower annual runoff associated with this year. If it is assumed that the closure of the two dams has caused little change in either the magnitude or the timing of the annual water discharge of the Lancang River, due to their relatively small storage volumes and active capacity to storage ratios (see above), the record of changing sediment concentrations presented in Fig. 6.9A for the Lancang River at Jinghong provides a surrogate for the record of annual sediment loads. Mean concentrations have fallen by around 50% since the late 1980s suggesting that the annual sediment loads transported by the Lancang River at Jinghong may have fallen to only about 50% of those recorded in the late 1980s and shown in Figs. 6.4 and 6.8B, by the early years of the current century. If, as is suggested by Fig. 6.8B,

the annual sediment load of the Lancang at Jinghong had increased to around 120 Mt year^{-1} by 1990, as a result of land clearance and intensification of land use, it would seem likely that the annual sediment load at Jinghong has subsequently decreased to around 60 Mt year^{-1}.

It is clear that future planned dam construction is likely to have an even more significant impact on the sediment load of the Lancang River, since some of the proposed dams will be larger and some will be located further downstream and therefore control a larger proportion of the basin. Further details on both the location and the key features of the proposed cascade of eight dams on the Lancang River are provided by Fig. 6.10. This information emphasizes that the two reservoirs currently in operation represent only a very small part

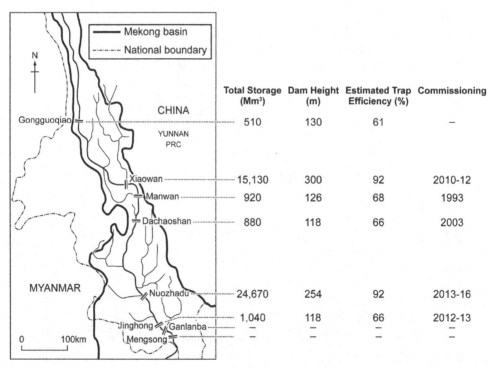

FIGURE 6.10 The cascade of dams planned for the Lancang River. (Based on the information reported by MRC (2003), Lu and Siew (2006), and Kummu and Varis (2007).)

of the overall scheme, which is planned to provide a total power generation capacity of more than 15,000 MW. The planned cascade of dams will result in string of reservoirs extending over ca. 750 km with a total fall of 800 m and a total storage of ca. 49 km^3 (see Plinston and He, 1999), that will trap most of the sediment transported by the river. Two of the dams, the Xiaowan dam and the Nuozhadu dam, are very large, with estimated trap efficiencies in excess of 90%, and the Xiaowan dam, which will have a height of ca. 300 m, will be one of the largest dams in the world. This is currently under construction and will have a total storage volume of >15,000 Mm^3, which is more than eight times that of the existing Manwan and Dachaoshan dams combined.

A recent paper by Guo *et al.* (2007) suggests that the dead storage associated with the Manwan reservoir will be filled by around the year 2008 and since then it will trap little sediment. In the absence of further dam construction, a similar situation is predicted for the Dachaoshan dam after about 30 years. However, the construction of the much larger Xiaowan dam upstream of the Manwan and Dachaoshan dams, which is due to be completed in 2010 will ensure that dams continue to exert a strong and indeed increasing influence on the sediment load of the Lancang River. Model predictions reported by Guo *et al.* (2007) suggest that the Xiaowan dam will have a long-term trap efficiency as high as ca. 96% and that, although the release of clear water into the reservoir behind the Manwan dam will remobilize some of the sediment stored in that reservoir, the amount of sediment passing the Manwan dam will gradually reduce from about 40% of the natural load immediately after the closure of the Xiaowan dam to ca. 10% about 30 years later. Recognizing that sedimentation behind the Dachaoshan dam will continue to reduce the sediment flux passing the Manwan dam for another ca. 50 years, as it gradually fills with sediment, it

can be predicted that after the closure of the Xiaowan dam in 2010 the sediment load passing the Dachaoshan dam is likely to be less than 10% of the natural load of the Lancang River. The sediment load of the Lancang River is likely to reduce still further in the future as further dams are constructed. In particular, the proposed construction of the large Nuozhadu dam, immediately below the Dachaoshan dam, which has a projected storage volume about 50% greater than the Xiaowan dam, will further reduce the sediment input into the lower Mekong system from the Lancang River to a value significantly below 10% of its natural contribution. In short, it must be recognized that the current impact of the Manwan and Dachaoshan dams shown by Fig. 6.9A is likely to increase very substantially in the future, when the Xiaowan dam is closed and further dams are constructed. It seems likely that the sediment input from the Lancang River to the Mekong River will progressively fall to about 5% of its natural value.

5.2. The Changing Sediment Load of the Lower Mekong River

Considering the sediment load data available for the stations further downstream in Thailand and Laos, there is less evidence of the clear increase in sediment load in the 1980s, shown for the Lancang River in Fig. 6.8B. However, a comparison of the annual sediment loads at Chiang Saen for the period 1968-1974 (mean annual load = 72.3 Mt) with those for period 1994-2000 which are likely to have already been impacted (reduced) by the dam construction (mean annual load = 101.4 Mt) arguably provides some evidence of this increase. Since Figs 6.4 and 6.8B and the above discussion suggest that the annual sediment load of the Lancang River at Jinghong progressively increased through to the early 1990s, after which the closure of the Marwan dam is likely to have caused a reduction by about 50 Mt year^{-1}, it can be suggested that the annual load at Chiang Saen might have been as high as 150 Mt year^{-1} in the late 1980s and early 1990s. However, the lack of reliable data for the period between 1980 and 1990 means that this can be only speculation.

The annual sediment loads for Nong Khai during the 1980s presented in Fig. 6.4 appear to be essentially stationary and show no clear evidence of the increase shown by the Lancang River at Jinghong over this period. The time series plot of the relative magnitude of the annual sediment loads at Nong Khai and Jinghong over this period presented in Fig. 6.6 and discussed previously shows that post 1980 the annual sediment load at Nong Khai was commonly less than that at Jinghong, suggesting that any increase recorded upstream was more than offset by increased deposition in the reach between Jinghong and Nong Khai. The essentially stationary response shown by the time series of annual sediment loads for the Mekong River at Nong Khai, over the period extending from the early 1970s to the late 1990s, appears to be mirrored by the record for Mukdahan for the 1970s and 1980s. At this station, however, the time series suggests that the loads have increased during the 1990s and early 2000s. Since there is no evidence of this increase in the records for Nong Khai, it is likely to reflect increased inputs from tributaries downstream of Nong Khai, increased bank erosion within the reach downstream of Nong Khai or a significant change (i.e., reduction) in conveyance losses. The first potential cause would seem to be the most likely and could reflect a similar impact of land clearance and land-use intensification within the Mekong tributaries in Lao, to that recorded in the Lancang basin in the 1980s.

There has been increasing concern for the potential downstream impact of dam construction on the Lancang River on the sediment load of the Middle and Lower Mekong River in recent years (see Lu and Siew, 2006; Kummu and Varis, 2007) and several reports of major changes in the downstream load of the river have appeared in the recent literature. For example, Saito et al. (2007) refer to major reductions in the sediment load of the Mekong and cite Kummu and Varis (2007) in indicating that the construction of the Manwan Dam in 1993 caused the annual sediment load at Chiang Saen to reduce from 71 to 31 Mt year^{-1} and that at Pakse to reduce from 133 to 106 Mt year^{-1}. Kummu and Varis (2007) also report similar reductions in the annual sediment load for Luang Prabang. Similarly, Lu and Siew (2006) report that the closure of the Manwan Dam in 1992 caused the annual sediment load at Chiang Saen to reduce by more than 50% from a predam (1962-1992) annual average of 74.1 Mt year^{-1} to a postdam (1993-2000) value of 34.5 Mt year^{-1}, but they also report that the equivalent values of mean annual load at Mukhadan increased from 97.5 to 131 Mt year^{-1}. It is, therefore, clearly important to use the available time series of sediment loads to assess the recent and current impact of dam construction. As indicated above, the lack of sediment load data for the Lancang River after 1990 unfortunately precludes analysis of changes in the sediment inputs from the Upper Mekong and emphasis must be placed on the evidence provided by the five designated stations further downstream. This is again limited by the absence of sediment load data for the station at Chiang Saen between 1975 and 1993, the sporadic nature of the data coverage for Luang Prabang in recent years and the absence of annual load data for Pakse between 1963 and 1998. The lack of sediment load data for Chiang Saen for the early 1990s is particularly unfortunate, since this is when the first impact of the construction of the Manwan dam might be evident.

Initial inspection of Fig. 6.4 provides no clear evidence of major changes in the annual sediment load of the Lower Mekong, such as those cited above. Furthermore, the data for Mukdahan show clear evidence of an upward trend since the early 1990s. Although it is tempting to link the low annual sediment load recorded at Chiang Saen in 2003 to the impact of dam construction, and more particularly the commissioning of the Dachaoshan dam, this apparent reduction is more likely to be a reflection of the low water discharge in the same year. It is important to recognize that the reports of a reduction in sediment load at Chiang Saen, Luang Prabang, and Pakse cited above relied heavily on the monthly water quality samples collected by the Mekong River Commission at these sites. Analysis of TSS data obtained from the water quality archive and described in the Mekong River Commission State of the Basin Report for 2003 (MRC, 2003) suggested that a marked reduction in suspended sediment concentrations could be identified in the data for Chiang Saen post 1992 (the year of dam closure). Monthly water quality data for this site extended back to 1985 and a clear distinction between the pre and post 1992 data was demonstrated and the data suggested that the mean concentration fell by about 50% during this period and this reduction was maintained till 2001. However, the limitations of the TSS values obtained from the water quality sampling program in representing the true suspended sediment concentrations, related to both the low sampling frequency and the representativeness of the resulting concentration values, have already been emphasized (see Fig. 6.3). This problem is further highlighted in Fig. 6.9, where the sediment concentration record for the period 1987-2003, for the Lancang River at Jinghong derived from the suspended sediment-sampling program at that site is compared with the concentration record for the Mekong at Chiang Saen, approximately 660 km downstream, obtained from the MRC water quality monitoring program. The magnitude of the concentration values provided by the water quality monitoring program for Chiang Saen would appear to be very low compared to those measured at

Jinghong upstream and to be unrepresentative of the river. This situation is clearly confirmed by the concentration data provided for Chiang Saen for the period 1994-2003 by the suspended sediment-sampling program presented in Fig. 6.3B, which closely match the magnitude of those recorded for the Lancang at Jinghong. Furthermore, the data presented in Fig. 6.9A for the Lancang River at Jinghong, while showing some evidence of declining concentrations, show no clear evidence of the major reduction in concentration commencing in 1993 suggested by the data presented in the MRC State of the Basin report for 2003 (MRC, 2003).

The overriding impression provided by the annual sediment load data presented in Fig. 6.4 is therefore one of relative stability of the sediment loads transported by the Mekong River over the past 40 years. The availability of sediment measurements for four of the measuring stations for 1961 (i.e., Chiang Saen, Luang Prabang, Mukdahan, and Pakse) affords a useful baseline against which to assess the magnitude of any subsequent changes. Any attempt to compare the suspended sediment loads of 1961 with those of more recent years must, however, take account of variations in water discharge between years and in Table 6.5 the loads for 1961 for individual stations are compared with those for a recent year with a similar water discharge. In the case of Luang Prabang and Pakse, there is little difference between the two load values, but in the case of Chiang Saen and Mukdahan there would appear to be evidence of increases of ca. 29% and 38%, respectively. Further evidence for increases at Chiang Saen and Mukdahan was presented previously. In both cases, however, the increase is not discernible at the next measuring station downstream (i.e., Luang Prabang and Pakse). The period of record covered by Fig. 6.4 will have coincided with significant land-use change and intensification leading to catchment disturbance, as well as the construction of dams in several tributary basins and in

TABLE 6.5 A comparison of the annual suspended sediment loads of the Mekong River at Chiang Saen, Luang Prabang, Mukdahan, and Pakse for 1961 with the loads for a recent year with a similar water discharge

Station	Sediment load 1961 ($\times 10^6$ t)	Water discharge 1961 ($\times 10^9$ m^3)	Recent sediment load ($\times 10^6$ t)	Recent water discharge ($\times 10^9$ m^3)
Chiang Saen	71.3	92.0	81.1 (2002)	89.2 (2002)
Luang Prabang	112.4	126.6	118.4 (1997)	118.4 (1997)
Mukdahan	144.5	283.3	199.1 (2000)	296.6 (2000)
Pakse	165.8	384.3	168.0 (2001)	388.0 (2001)

the headwaters on the Lancang River. The absence of major changes in sediment loads over the period of record may, therefore, reflect the balancing of increases caused by catchment disturbance by reductions associated with dam construction and associated sediment trapping within the tributary basins. In this context, it seems reasonable to suggest that the apparent lack of a major change in the sediment load of the Mekong River between 1961 and the early years of the present century, demonstrated by Table 6.5, despite the construction of a major dam on the Lancang River, could reflect a situation where sediment loads in the headwater regions increased during the 1980s and 1990s, but were balanced by the reductions caused by the sediment trapping associated with the Manwan dam in the period since the early 1990s. Equally, as demonstrated previously, the Mekong basin also appears to demonstrate at least some capacity to "buffer" changes in sediment load occurring in different parts of the basin. This characteristic was also highlighted by Ta et al. (2002) in their study of sediment supply to the Mekong delta over the past 3000 years, which demonstrated that the sediment input to the delta appears to have changed little over this period.

Although Table 6.5 suggests that the sediment load of the Mekong River has remained essentially stable since the 1960s, it would seem likely that the planned expansion of the cascade of dams on the Lancang River, and particularly the commissioning of the Dachaoshan dam in 2003 and the large Xiaowan and Nuozhadu dams in 2010-2012 and 2013-2016, respectively, will result in a further major reduction in the sediment input to the Mekong River from the Lancang headwaters, with this input possibly reducing to only ca. 10% of the "natural" value and that the impact of this reduction will be felt throughout the Lower Mekong. Recent proposals to construct up to six hydropower dams on the Lower Mekong, including a dam in the vicinity of Luang Prabang, could clearly introduce further changes to the sediment load of the Mekong River and the outcome of those proposals must be awaited with interest.

6. CONCLUSION

Although the sediment load data available for the Mekong River possess a number of important limitations, particularly in terms of their continuity, they, nevertheless, provide valuable information regarding the magnitude of the annual suspended sediment loads and their inter- and intraannual variability as well as recent trends in the sediment load of the river in response to human impact. These impacts include land cover and land-use change and land-use intensification and the construction of hydropower dams on the headwaters of the river in China. The problems associated with attempts to use TSS data derived from water quality sampling programs as a surrogate for traditional suspended sampling highlight the need for rigorous data collection and the revitalization and continuation of the existing sediment measurement program. It is hoped that the program of sediment monitoring will be expanded and intensified in the future, in order to provide more reliable data, particularly as the likelihood of changes in the sediment load of the river increase. In particular, a coordinated program of sediment sampling at all the main stations identified in this study would provide a good basis for further investigation of sediment sequestration within the main channel system. An expanded sediment-monitoring program could also usefully include a station closer to the outlet of the basin than those currently used and also further coordinated sampling stations on key tributaries, in order to establish a more detailed sediment budget for the main-stem system. A more detailed budget for the main-stem system would greatly facilitate prediction of the likely future impacts of reduced sediment inputs from the headwaters of the river in China due to dam construction. Use of modern technology, including ADCP (Acoustic Doppler Current Profiler) devices to document the variation of suspended sediment concentrations and flow velocity in the cross section and the continuous monitoring of surrogate variables, such as turbidity would clearly provide opportunities to both increase the reliability of the data collected and reduce the labor demands of traditional sampling activities. Inclusion of measurements of the grain size composition of suspended sediment, its organic matter content, and other physical and geochemical properties in future sampling programs would also provide valuable additional information.

Although the available data appear to emphasize the longer term stability of the sediment load of the Mekong River, it seems likely that the future will bring significant changes. The lack of data for the Lancang River for the period post 1990, when the impact of the hydropower dams constructed on the river is likely to be increasingly evident, inevitably hinders attempts to assess the current impact of the dams and to predict future changes in downstream sediment fluxes. However, the existing evidence suggests that completion of the cascade of dams on the Lancang River will greatly reduce the sediment input to the middle and lower reaches of the river and it seems inevitable that this will, in turn, cause significant changes to the sediment load passing through the middle and lower reaches of the river. A reduction in the downstream sediment flux could have important implications for the river, including reduced nutrient inputs to Tonle Sap lake and the delta and also for the longer term stability of the Mekong delta (cf. Campbell, 2007; Saito et al., 2007).

ACKNOWLEDGMENTS

The data and results presented in this paper draw heavily on a review of the available sediment load data for the Lower Mekong River undertaken for the Mekong River Commission by the author. The generous assistance of Ian Campbell in collating the original data and in securing additional information and the assistance of Robert Stroud with data processing and of Helen Jones in producing the figures are gratefully acknowledged.

References

Ahlgren, J. and Hessel, K. (1996). The suspended matter of Mekong. A study of the chemical constituents adsorbed to suspended matter of the Mekong River, Lao PDR. Working Paper 306, Swedish University of Agricultural Sciences, International Rural Development Centre.

Campbell, I. (2007). Perceptions, data, and river management: Lessons from the Mekong River. *Water Resources Research* **43**, W02407, doi:10.1029/2006WR005130.

FAO/AGL. (2005). "World River Sediment Yields Database." FAO, Rome.

Ferguson, R. I. (1986). River loads underestimated by rating curves. *Water Resources Research* **22**, 74-76.

Fu, K., He, D., and Li, S. (2006). Response of downstream sediment to water resource development in mainstream of the Lancang River. *Chinese Science Bulletin* **51**, 119-126.

Gregory, K. J. and Walling, D. E. (1973). "Drainage Basin Form and Process." Edward Arnold, London.

Guo, Q., Cao, W., Lu, C., and Lu, Q. (2007). Sedimentation of Dachaoshan Reservoir in cascade hydroprojects on Lancang River. *In* "Proceedings of the Tenth International Symposium on River Sedimentation," Volume VI, pp 51-59. Moscow State University, Moscow.

Gupta, A. and Lieuw, S. C. (2007). The Mekong from satellite imagery: A quick look at a large river. *Geomorphology* **85**, 259-274.

Harden, P. O. and Sundborg, A. (1992). "The Lower Mekong Basin Suspended Sediment Transport and Sedimentation Problems," Hydroconsult, Uppsala, Sweden, 71pp.

Harza Engineering Company. (1962). Final Report Lower Mekong River Project, 122pp.

Keenan, H. E., Dyer, M., Songsasen, A., Bangkedphol, S., and Homchan, U. (2006). Environmental monitoring of the sediment pollution along the Thai:Laos Mekong. *Journal of ASTM International* **3**, doi:10.1520/JAI13374.

Koch, R. W. and Smillie, G. M. (1986). Bias in hydrologic prediction using log-transformed regression models. *Water Resources Bulletin* **22**, 717-723.

Kummu, M. and Varis, O. (2007). Sediment-related impacts due to upstream reservoir trapping, the Lower Mekong River. *Geomorphology* **85**, 275-293.

Lu, X. X. and Siew, R. Y. (2006). Water discharge and sediment flux changes over the past decades in the Lower Mekong River: Possible impacts of the Chinese dams. *Hydrology and Earth Systems Science* **10**, 181-195.

Milliman, J. D. and Meade, R. H. (1983). World-wide delivery of river sediment to the oceans. *Journal of Geology* **91**, 1-21.

Milliman, J. H. and Syvitski, J. P. M. (1992). Geomorphic/ Tectonic control of sediment discharge to the ocean: The importance of small mountainous rivers. *Journal of Geology* **100**, 325-344.

Milliman, J. D., Quraishee, G. S., and Beg, M. A. A. (1984). Sediment discharge from the Indus River to the Ocean: Past, present and future. *In* "Marine Geology and Oceanography of Arabian Sea and Coastal Pakistan," Haq, B. U., and Milliman, J. D. (eds.), pp 65-70. Van Nostrand Rheinhold, New York.

MRC. (2003). "State of the Basin Report: 2003," Mekong River Commission, Phnom Penh, 300pp.

MRC. (2005). "Overview of the Hydrology of the Mekong Basin," Mekong River Commission, Ventiane, 71pp.

Plinston, D. and He, D. (1999). "Water Resources and Hydropower, Policies and Strategies for Sustainable Development of the Lancang River Basin," ADB TA-3139 PRC, Asian Development Bank, Manila, 27pp.

Restrepo, J. D. and Kjerfve, B. (2000). Magdalena River: Interannual variability (1975-1995) and revised water discharge and sediment load estimates. *Journal of Hydrology* **235**, 37-149.

Roberts, T. (2001). Downstream ecological implications of China's Lancang Hydropower and Mekong Navigational Project. International Rivers Network (IRN) Report.

Saito, Y., Chaimanee, N., Jarupongsakul, T., and Syvitski, P. M. (2007). Shrinking Megadeltas in Asia: Sea-level rise and sediment reduction impacts from case study of the Chao Phraya delta. *LOICZ Inprint* **2007**(2), 3-9.

Ta, K. T. O., Nguyen, V. L., Tateishio, M., Kobayashi, I., Tanabe, S., and Saito, Y. (2002). Holocene delta evolution and sediment discharge of the Mekong River, southern Vietnam. *Quaternary Science Reviews* **21**, 1807-1819.

Walling, D. E. (2005). Evaluation and analysis of sediment data from the Lower Mekong River. Report to the Mekong River Commission, 61pp.

Walling, D. E. (2006). Human impact on land-ocean sediment transfer by the world's rivers. *Geomorphology* **79**, 192-216.

Walling, D. E. (2007). Global change and the sediment loads of the world's rivers. *In* "Proceedings of the Tenth International Symposium on River Sedimentation," Volume I, pp112-130. Moscow State University, Moscow.

Walling, D. E. and Kleo, A. H. A. (1979). Sediment yields of rivers in areas of low precipitation: A global view. *In* "The Hydrology of Areas of Low Precipitation," 479-483. Proceedings of the Canberra Symposium. International Association of Hydrological Sciences Publication Publication No. 128, Canberra.

Wolanski, E. and Nguyen, H. N. (2005). Oceanography of the Mekong River estuary. *In* "Mega-Deltas of Asia— Geological Evolution and Human Impact," Chen, Z., Saito, Y., and Goodbred, S. I. (eds.), pp 113-115. China Ocean Press, Beijing.

You, L. (1999). A study on temporal changes of river sedimentation in Lancang River basin. *Acta Geographica Sinica* **54**, 93-100.

Vegetation in the Mekong Basin

Philip W. Rundel

Department of Ecology and Evolutionary Biology, University of California (UCLA),
Los Angeles CA 90095, USA

OUTLINE

1. REGIONAL FOREST FORMATIONS

There have been a variety of classification systems proposed over the past century to describe the major forest communities of the Mekong Basin as well as the broader area of mainland Southeast Asia. These classification systems have largely been based on geographic, climatic, and floristic traits that influence forest structure and composition (Rundel, 1999).

Climatic conditions associated with elevation and rainfall regimes provide some of the critical traits in determining the available species pool for forest systems. The elevation of occurrence, strongly associated with temperature conditions, separates montane uplands from lowland forests. Beyond this simple dichotomy, there have been many systems separating out more refined elevational patterns of forest community distribution (e.g., Santisuk, 1988). With higher latitudes, as in the upper Mekong watershed in

China, elevations are high enough to be separate distinctive upper montane, subalpine, or alpine where winter conditions promote a dominance of species with temperate forest affinities.

The topographic heterogeneity of the Mekong Basin produces highly variable mosaics of seasonal and total rainfall. Mean annual rainfall within the Mekong Basin varies from as little as 600 mm to perhaps 5000 mm or more in portions of the upper Annamite Range. Total rainfall generally shows a strong correlation with the seasonality of rainfall, with areas of lower rainfall exhibiting a longer dry season. Areas of high rainfall and a short dry season allow for the survival of a tropical rainforest flora with stronger links to the extensive Indo-Malaysian rainforest to the south. Drier conditions and a long dry season selects for a very different forest structure lower in stature and diversity, in many cases with affinities to the Indian monsoon forests to the west.

Ecological traits and floristic affinities are also used in delineating major forest associations. For example, biogeographers often recognize generalized categories of forest communities on the basis of the dominant leaf phenology, separating deciduous from evergreen forests. This approach has considerable traction in remote sensing of forest associations as this trait can be identified from satellite sensors. Floristic affinity as a tool in classification can also be seen in systems that utilize such qualifiers as tropical, subtropical, and temperate in describing forest associations. Finally, the presence of special elements in a community, as with a dominance codominance by conifers, often has lead to a separation of an association as distinct. Thus, many classification systems describing forest vegetation in the Mekong Basin refer to broad-leaved evergreen forests as distinct from conifer forests with a codominance of hardwoods and conifers.

Much of the Lower Mekong Basin has been included within terrestrial ecoregions identified by Wikramanayake et al. (2002). Ten of these ecoregions cover sections of the basin— the Luang Prabang montane forests, northern Annamite rain forests, Southeast Indo-China dry evergreen forests, Tonle Sap—Mekong peat swamp forests, Tonle Sap freshwater swamp forests, northern Thailand—Laos moist deciduous forests, northern Khorat Plateau moist deciduous forests, northern Triangle subtropical forests, northern Indo-China subtropical forests, and Indo-China mangroves.

With a long history of human habitation and accelerating rates of deforestation and land cover changes in recent decades, much of the Lower Mekong Basin is now covered by secondary forest landscapes. Although these secondary forests differ significantly from the original primary forest cover, they nevertheless play a highly significant role in maintaining basic ecosystem functions and biodiversity (Heinimann et al., 2007).

2. FOREST VEGETATION OF THE UPPER MEKONG BASIN

The upper watershed of the Mekong River drainage in northwestern Yunnan, Tibet, and Qinghai Provinces of China provides a transition from tropical and subtropical forests of Southeast Asia to temperate forests to the north (Li and Walke, 1986; Zhu et al., 2006). Lower valleys below 1000 m elevation with relatively mesic conditions characteristically support semievergreen forest. These habitats show floristic relationships to the seasonal forests of Burma and other parts of mainland Southeast Asia. The canopy exhibits a mixed dominance of both evergreen and deciduous broad-leaved species, with deciduous tree species attaining their major importance on drier valley slopes.

Subtropical broad-leaved evergreen forest once covered most of the lower elevation slopes of the upper Mekong watershed at elevations of 1000-2000 m, or 2500 m to the south (Zhang and Cao, 1995; Zhu *et al.*, 2006). There is a strong mixed dominance of tree species, with a canopy height of 20-30 m. This association has a strong floristic relationship to the montane evergreen forests of mainland Southeast Asia. The ecologically most important families in terms of dominance are the *Fagaceae*, *Theaceae*, *Lauraceae*, and *Magnoliaceae*, with important genera including *Castanopsis*, *Lithocarpus*, *Chrysobalanus*, *Magnolia*, *Machilea*, *Anneslea*, and *Schima*. A small but significant component of conifers is also present.

A large area of the upper Mekong drainage at intermediate elevations is characterized by a dissected cover of subtropical conifer forest with *Pinus yunnanensis* occurring over a broad range but commonly at 1500-3000 m, and *Pinus armandii* at 2500-3000 m. This association is a secondary community that forms following disturbance of the once widespread subtropical evergreen broad-leaved forest. Canopy height is typically 15-25 m. *Keteleeria evelyniana* is a common associate at middle elevations, along with species of *Cyclobalanus*, *Lithocarpus*, *Michelia*, *Alnus*, and *Rhododendron*. Reforestation programs have greatly increased in scale and scope in northwestern Yunnan Province during the 1990s. While well intentioned, these programs have not been successful in mitigating erosion and surface flow out of montane watersheds (Weyerhaeuser *et al.*, 2005).

Montane coniferous cover much of the watershed area above about 2700 m. The lower montane zone from 2700 to 3100 m is dominated by a hemlock forest with *Tsuga dumosa* in forest community mixed with broad-leaved evergreen trees and other conifers such as *Abies georgii*, *Picea likiangensis*, and *P. armandii*. Typical canopy heights are 30 m, with a relatively open and humid understory.

A montane pine forest dominated by *Pinus densata* is widespread in northwestern Yunnan from about 3000 to 3400 m on warmer dry slopes. This species, a stabilized natural hybrid that has become adapted to high mountain environments where neither of the parental species can normally grow (Wang *et al.*, 2001), often forms a monodominant over the core area of this association, with an open canopy 10-25 m in height. The understory exhibits 20-50% shrub cover, with species of *Lyonia* and *Rhododendron* as dominants.

Forest cover on more mesic slopes from 3100 to 3800 m consists of spruce forests with *Picea likiangensis* or *P. brachytyla*. These are humid forests reaching as much as 40 m in height. The fir species, *Abies forrestii* and its variety *A. f. georgei* may also be present. The forests exhibit a distinctive lower stratum, with an understory of sapling spruce, *Salix*, and *Betula* at 10-15 m, and a shrubby bamboo layer of *Sinarundinaria* 2-5 m in height that covers as much as 80% of the ground surface.

Then uppermost conifer zone is a fir forest at 3500-4300 m elevation, and occasionally reaching 4450 m. The forest is dominated by *A. forrestii* with *P. likiangensis* as an associated species. Below the canopy level at 17-30 m height is a shrubby understory including tall *Rhododendron* and lower bamboos.

Larch forests dominated by *Larix potaninii* or *Larix griffithii* are widespread at 2700-4000 m elevation. Although florisitically diverse, these larch forests are thought to be a secondary formation formed after the destruction of *Picea* or *Abies* forest (Li and Walke, 1986). These are likewise humid forests with a dense shrub understory and thick moss mat on the forest floor.

Subalpine shrublands and alpine meadows form mosaics above 4000 m (Deng and Zhou, 2004). Recent studies using archived climatedata and historic photographs indicate that treeline temperatures warmed steadily over the past two decades of the last century, causing glaciers to retreat and treeline to advance upward (Baker and Moseley, 2007).

3. LOWLAND FOREST VEGETATION IN THE LOWER MEKONG BASIN

3.1. Wet Evergreen Forest

Wet evergreen rainforest, typical over much of the Indo-Malaysian region to the south, presents classic tropical rainforest with multistoried canopies to heights of 40 m or more and abundant lianas. These forest formations occur in areas of humid climate with more than 2000 mm of rainfall annually. These are all areas where the topography captures large amounts of rainfall, fog, and mists from monsoon winds.

When defined on a floristic basis, the wet evergreen forests of mainland Southeast Asia are distinct from the better known wet evergreen forests of Malaysia and Indonesia, and generally exhibit at least a moderate dry season despite high annual total of rainfall. Strictly speaking, Indo-Malaysian forest formations do not extend north of the Kra Isthmus in southern Thailand at latitude of about 6°. Examples of wet evergreen forest can be found locally; however, as with Khao Yai National Park in Thailand (Smitinand, 1968).

Although structurally similar to Indo-Malaysian rainforests, these Indo-China communities are less floristically rich and distinctive in species composition. At the generic level, there is a strong floristic relationship, but at the species level there are obvious differences. While some trees in Indochinese wet evergreen forests represent the northern limit of distribution for Indo-Malaysian rainforest species, these forests are rich in endemics.

3.2. Semievergreen Forest

Semievergreen forests are widespread in a band across northern and central Thailand into Lao, Cambodia, and Viet Nam (Blasco *et al.*, 1996; Rundel and Boonpragob, 1995; Schmid,

1974; Vidal, 1979). These forests comprise one of the most significant forest associations with respect to the area of the Mekong drainage in northern Thailand and Lao. Semievergreen forest, adapted to conditions of relatively moderate amounts and highly seasonality distribution of rainfall are unique to mainland Southeast Asia. It is not surprising, therefore, that this association has evolved a large number of endemic species as well as species which extend to Malaysia only in areas with seasonal climatic regimes. However, regional species endemism within semievergreen forests is quite low. Thus, semievergreen forest floras in Lao, Cambodia, and Viet Nam are not strikingly different from those present in Thailand.

Semievergreen forest has been commonly designated as *dry evergreen forest* by many researchers, but this name is misleading because of the strong presence of deciduous tree species in the forest canopy of this community. The term *semievergreen* to describe this forest is much more accurate and will be used here. This forest association has also been termed *seasonal evergreen forest, tropical semievergreen forest*, or *semideciduous forest* (Champion and Set, 1968; Santisuk, 1988; Stamp, 1925), or *forêt dense* and *forêt hemiombrophile a dipterocarpacées* in the French literature (Blasco *et al.*, 1996).

Semievergreen forest generally occurs in humid and subhumid climatic regions where mean annual rainfall is generally between 1400 and 2600 mm with a significant dry period (generally 3-6 months) that occurs each year. These forests are often present in mosaics made up of stands of mixed deciduous and/or deciduous dipterocarp communities, with the semievergreen formations in areas where soil-moisture conditions are most favorable, as in northern Thailand (Elliot *et al.*, 1989; Santisuk, 1988).

While the overall structure of semievergreen forests is similar to that of wet evergreen forest in the presence of multilayered canopy, the canopy is less massive and continuous at about 30-40 m height, and more open in structure. Species richness of forest trees is lower in semievergreen forest as well. This reduction in richness can be particularly seen in the lower diversity of Dipterocarpaceae in individual forest stands compared to the richness of this family in wet evergreen forest communities.

Detailed studies of dry evergreen forest in Thailand outside of the Mekong Basin have shown that the structure and floristic composition of these forests arise from a disturbance history operating at multiple spatial and temporal scales and varying intensities. Tree ring studies suggest a combination of rare catastrophic disturbances as well as more frequent disturbances of low to moderate intensity have influenced patterns of stand development (Baker et al., 2005).

The distribution of semievergreen forest habitats across a landscape catena is largely a function of gradients of soil-moisture availability, with soil parent material relatively unimportant. Thus, it is common to see semievergreen forest on both calcareous and crystalline rock substrates, but with a relatively deep soil profile and good capacity for soil-moisture storage as a common characteristic.

In broad valley areas, semievergreen forest often occurs as a fringing gallery forest along streams, grading out into deciduous dipterocarp forest on drier sites with shallower soils. Massive trees of species such as *Dipterocarpus turbinatus*, *Dipterocarpus alatus*, *Dipterocarpus costatus*, and *Anisoptera costata* once formed dense stands in these forests, along with a diversity of other Dipterocarpaceae. Many semievergreen forest species begin to drop out at about 700-800 m elevation, where a distinctive assemblage of tree species dominates in broad ravines and moist slopes at elevations at the transition from semievergreen forest to montane evergreen forest in northern Thailand (Santisuk, 1988). Montane evergreen forests begin to dominate above this transition.

Although as many as half or more of the tree species in semievergreen forest may be deciduous, particularly those in young forest stands colonizing gaps, the presence and dominance of evergreen species means that this habitat never takes on a leafless appearance. The leafless period for deciduous species is relatively short, with about 60% of these deciduous species leafless for no more than 1 month (Vidal, 1956-1960). Leaf fall centers in February and March, but is poorly synchronous.

Semievergreen forests are generally relatively intolerant of fire. In comparison to species in adjacent deciduous dipterocarp communities, woody species in semievergreen forest resprout poorly after fire. Similarly, semievergreen forest species are relatively sensitive to drought, apparently due to less well-developed root systems. In regions subject to intense human occupation where wildfires are commonly used for land clearance, with the Khorat Plateau of Thailand as a primary example, areas of semievergreen forest have been converted to savanna or open deciduous dipterocarp woodland.

3.3. Mixed Deciduous Forest

Mixed deciduous forest, or *monsoon forest* as it is sometimes termed, forms an extensive forest cover in a broad belt extending from the Ganges Basin of India through Burma to northern Thailand and Lao, and thus covers extensive areas of the Mekong Basin. This is an area of strong seasonality in climate, with the dry season typically lasts 5-6 months. Because of this strong seasonality, frequent fire is a natural ecological factor. The canopy of mixed deciduous forest is typically closed and high, often reaching 30 m or more. Beneath this canopy, the understory is relatively open despite a diverse assemblage of small trees, shrubs, and bamboos forests. Unlike the evergreen-forest formations, lianas and vascular plant epiphytes are uncommon (Rundel, 1999; Rundel and Boonpragob, 1995).

A floristic characteristic of mixed deciduous forest is a strong importance of members of the Fabaceae, Lythraceae, and Rubiaceae, together with a relatively low occurrence or absence of Dipterocarpaceae. Dense local stands of bamboo are often present, particularly in areas with significant human impacts. Teak (*Tectona grandis*) was once the dominant tree through much of this formation, but this species has been intensely lumbered over the past century in Burma and Thailand. Other widespread and codominant tree species include members of the Fabaceae (*Xylia kerrii, Afzelia xylocarpa, Pterocarpus macrocarpus*, and *Dalbergia* spp.), Combretaceae (*Terminalia* spp.), and Lythraceae (*Lagerstroemia* spp.). It has been argued that these are the most species-rich tropical deciduous forests in the world (Elliot et al., 1989).

The phenology of canopy trees exhibits a complete dominance by a deciduous growth habit, with species almost equally split between those losing their leaves for less than 1 month and those with a longer leafless period. When the dry season commences in late November, as in northern Thailand, leaf fall typically begins 1-2 months later and continues until the forest becomes leafless by the end of March. This leafless period extends for 4-5 months. To the east in the Mekong basin of Lao, however, the dry season commences somewhat later and leaf fall is delayed proportionately.

The southern boundary of mixed deciduous forest across Thailand and Lao is difficult to plot precisely. Teak whose distribution characterizes much of the range of mixed deciduous forest is not present in southern Lao or northern Cambodia. Complicating the matter further is the fact that *Lagerstroemia* may be a natural associate in both mixed deciduous forest and semievergreen forest. Areas of semievergreen forest in southern Lao and northern Cambodia dominated by deciduous *Lagerstroemia* have often been classified as mixed deciduous forest rather than more appropriately as a deciduous forest form of semievergreen forest. In fact, these two forest associations grade together in this region in a manner that makes it difficult to clearly characterize the correct forest association. Heavy disturbance by logging of these forests has led to replacement by agriculture or secondary dipterocarp forests and woodlands and further complicated this situation.

Vidal (1956-1960) treated mixed deciduous forests individually in each of three climatic regions of Lao. These correspond geographically with the lower Mekong valley and its tributaries from Vientiane to southern Lao, the lower elevational areas of northern Lao, and the western slopes of the Annamite Range. He suggested that many of these forests represent degraded communities that were once dominated by dry evergreen forests. For the Mekong valley area, the floristic structure of mixed deciduous forests strongly supports an intermediate ecological condition of this habitat between more mesic semievergreen forests and more xeric deciduous dipterocarp forests. Mixed deciduous forests in northern Lao have a structure and high species diversity similar to that present in northern Thailand.

The historically most significant species in the mixed deciduous forest has been teak (*T. grandis*). The natural range of teak extends from northern India and Burma across northern Thailand to northern Lao. The best stands of teak were reported by Vidal (1956-1960) to occur in the Golden Triangle area, where the borders of Burma and Thailand join that of Lao, and further south in the Mekong valley around Pak Lay. Teak generally is not present on lateritic soils. The great majority of large teak has been cut out of these forests beginning in Burma in the mid-nineteenth century and extending into Thailand and Lao at the end of that century. Heavy and indiscriminate cutting of teak and other large forest trees began at this time as European and Burmese companies vied for leases over the most favorable areas. Logging practices in this era, described in wonderful detail by Campbell (1935), involved girdling standing trees to allow them to dry for 2 years before they were felled and hauled by elephant into major rivers for transport in log rafts.

Despite poor logging practices, teak is inherently a good colonizing species so long as fire is present to promote seed germination. Many areas that were once heavily logged for teak and other forest trees now support good stands of young teak. The strong seasonality of the mixed deciduous forest environment, and the consequent accumulation of leaf litter on the forest floor during the dry season, promotes flammability in these forests.

3.4. Deciduous Dipterocarp Woodland

Deciduous dipterocarp forest forms a relatively low and open forest or woodland community dominated by deciduous trees. Community structure may range from virtually closed canopy forest of low trees 5-8 m in height, with occasional emergents reaching 10-12 m, to an open woodland structure with 50-80% canopy cover. In all cases, there is an open understory dominated by grasses. This forest formation has often been termed *dry dipterocarp forest*, but this designation is unfortunate in that it has led to confusion with *dry evergreen forest*. The use of deciduous dipterocarp forest or woodland is the more appropriate designation used here. This community has also been termed *idaing* in Burma (Stamp, 1925) and *forêt claire a dipterocarpacées* in Lao (Vidal, 1956-1960; 1960).

Deciduous dipterocarp woodland covers more area in mainland Southeast Asia than any other forest type. It extends from northeastern India and Burma (Champion and Set, 1968) through Thailand (Rundel and Boonpragob, 1995) to the Mekong River region of Lao (Vidal, 1956-1960), Cambodia (Aubreville, 1957; Pfeffer, 1969; Rollet, 1953, 1972), and Viet Nam (Schmid, 1974). Over this range, it characteristically occurs in areas with 1000-1500 mm rainfall and 5-7 months of drought. Potential evapotranspiration may exceed rainfall for up to 9 months per year.

Deciduous species of Dipterocarpaceae form the dominant element of deciduous dipterocarp woodlands. Only six species of the approximately 550 dipterocarps in the world are deciduous and all of these occur in this formation. Four of these, *Shorea siamensis, Shorea obtusa, Dipterocarpus obtusifolius*, and *Dipterocarpus tuberculatus*, generally form the dominant biomass and cover. Also present is a reasonable diversity of other small trees, particularly legumes. *Pinus merkusii* may be a codominant. Overall, deciduous dipterocarp woodland exhibits relatively moderate species richness, and a similar floristic structure extends broadly across mainland Southeast Asia. While the woodland association itself has endemic species, there are relatively few local endemics.

Fire is a frequent event in most deciduous dipterocarp woodlands. Areas with regular human impact commonly have fire at intervals of 1-3 years through both deliberate and accidental ignitions. Most fires occur between December and early March in Thailand when woodland conditions are driest. Dominant tree species in this formation exhibit adaptations to fire in the form of thick corky bark to protect cambium tissues and root crowns which readily resprout (Stott, 1984, 1986, 1988a,b, 1990; Stott *et al.*, 1990; Sukwong *et al.*, 1975).

Deciduous dipterocarp woodland may occur in higher rainfall areas where local soil conditions due to laterites or shallow rocky soil produce functionally dry edaphic conditions. Such sites are present in northern Lao where annual rainfall may be between 1500 and 2000 mm, as around Vientiane, or even wetter conditions to the north of Pakse. Vidal (1956-1960) suggested that deciduous dipterocarp woodland over its entire range is really an edaphic subclimax on shallow, rocky, and nutrient-deficient soils, with dry evergreen woodland as a climatic climax community. Human activities over centuries have also acted to promote the expansion of deciduous dipterocarp woodlands at the expense of semievergreen forest. Fire has undoubtedly been the most significant single factor in promoting these changes.

The deciduous dipterocarp woodlands are strongly deciduous, with virtually the entire community leafless in at least some part of the dry season. Vidal (1956-1960; 1960) reported that 95% of tree species and 84% of shrub species in the stands that he studied were deciduous. Woody species began to lose their leaves in December and reach a peak of deciduousness in February (85% of species deciduous) and March (60% of species deciduous) before beginning to leaf out in April. Some individual species such as *Dipterocarpus intricatus* begin to lose their leaves in December, while other species such as *S. siamensis* do not typically begin to drop leaves until February. The timing of leaf fall is highly variable between sites and years, however, depending on soil-moisture availability.

Lowland pine forests dominated by *P. merkusii* are common on thin, but well-drained soils under semihumid rainfall regime (Rundel, 1999). *Dipterocarpus obtusifolius* is the most frequent associate of *P. merkusii*, particularly in areas with a shallow impermeable hardpan. These open forests have scattered shrubby associates and a grassy understory. In many respects, these pine forests represent an extension of deciduous dipterocarp forest, with the addition of the pine as a dominant or codominant species. In the French literature, this community has been lumped with deciduous dipterocarp forest as *forêt claire* (Aubreville, 1956), or separated and termed *pinede à P. merkusii* (Maurand, 1943), *forêt claire de gymnospermes* (Vidal, 1956-1960), or *pinede et forêt claire à P. merkusii* (Schmid, 1974).

Under heavy human impact, particularly from repeated fires, lowland areas of deciduous dipterocarp woodland may be converted to open savanna woodlands. The presence of shallow rocky soils promotes such conversions. Savanna habitats with a dominance of grasses and sedges and a scattered distribution of woody species are common in southern Lao with the degradation of dry dipterocarp woodland (Vidal, 1956-1960). These savanna woodlands may maintain a few of the hardier tree species from typical dry deciduous dipoterocarp woodlands, particularly fire resistant species such as *S. siamensis* and *P. merkusii*. Other fire-tolerant tree species include *Careya arborea, Mitragyna parvifolia, Acacia siamensis, A. catechu,* and *P. macrocarpus* (Smitinand, 1988).

4. MONTANE FOREST VEGETATION IN THE LOWER MEKONG BASIN

4.1. Lower Montane Forests

Montane forests, or hill evergreen forests as they are often termed in Thailand, grow under humid conditions with cooler temperatures and higher rainfall than that present in adjacent lowland areas. These have been termed forêt dense and forêt dense humide in the French literature (Vidal, 1956-1960). Rather than a single community type, these montane forests include a variety of forest associations under the general name of hill evergreen forest (Rundel, 1999). Also included in these montane habitats are conifer forests which are treated separately below.

In northern Thailand and Lao, the boundary separating semievergreen and/or deciduous forest communities from montane evergreen communities generally occurs at 800-1000 (rarely 1200) m elevation, but at lower elevations of 600-700 m in northern Viet Nam and Lao. Annual rainfall is typically 2000 to as high as 4000 mm or more, and the mean temperature of the coldest month is below 15°C. Frosts may be present at the higher and more northern parts of this formation.

The relatively abrupt lower boundary of montane forest is due more to temperature conditions than to moisture availability. This boundary represents a sharp transition not only in forest structure, but also in floristic composition. The Dipterocarpaceae and other lowland groups drop out rapidly, and are replaced by a rich community dominated by tree species of families such as the Fagacaeae, Magnoliaceae, Lauraceae, Theaceae, and Juglandaceae. Overall, these communities exhibit a strong floristic relationship with temperate mountain floras of southern China, as well as with montane forests of Malaysia and Indonesia. There is, however, significant endemism at the species level in the montane areas of mainland Southeast Asia. This endemism is particularly notable in the Annamite Province, and on the extensive limestone karst areas of Lao. The montane areas of Lao remain poorly investigated by quantitative ecological studies of plant communities. The recent discovery of the rare conifer *Glyptostrobus penilis* in Lao (Coffman *et al.*, unpublished data) suggests that much remains to be found in this area.

The structure of montane evergreen forests is distinctive with an open to semi-open canopy of relatively low and twisted tree forms. Epiphytes are typically present and often abundant and diverse in these forests, with the Orchidaceae, a notable example. Much of the remarkable diversity of orchids from Southeast Asia is centered in these evergreen montane forests.

In describing the floristic and vegetation patterns of northern Thailand, Santisuk (1988) subdivided the montane forests into first lower montane at elevations up to about 1800 m, and then upper montane forests at 1800-2000 m and above. This boundary corresponds to an elevation in which frequent cloud cover and mists occur, making them equivalent to cloud forests that have been described for tropical montane forests in many other parts of the world. Santisuk further divided lower montane forests into three types—lower montane rainforest, lower montane oak forest, and lower montane pine-oak forest. No equivalent attempt has been made to classify montane forests in Lao and Viet Nam.

Montane evergreen forests at these lower elevations have been widely cleared in the past by the swidden agriculture of hill tribes and exist in a successional stage with lower tree diversity and very strong dominance of Fagaceae. Such human-impacted forests are typically relatively low in stature with an open canopy structure as a result of frequent fires and cutting. A ground cover of grasses, rare in pristine montane evergreen forest, is common, as is a cover of bracken fern (*Pteridium aquilinum*).

The communities of Fagaceae and Lauraceae give way at about 2000 m elevation to a mixed hardwood-conifer forest populated almost entirely by evergreen tree species. Present as dominant trees are *Fokienia hodginsii* (Cupressaceae) and *Podocarpus neriifolius* (Podocarpaceae), along with *Castanopsis gamblei* (Fagaceae), *Manglietia duclouxii*, and *Michelia floribunda* (Magnoliaceae), *Acer heptaphlebum* (Aceraceae) and *Elaeocarpus integripetalus* (Elaeocarpaceae) (Kerr, 1932; Vidal, 1966). Under relatively undisturbed conditions, this humid community accumulates thick layer of moist humus over the ground surface. An abundance of cool-environment epiphytes indicates the presence of humid environment.

A long history of human impact in Lao has led to the degradation of extensive areas of what was once montane evergreen forest. Such forests have been replaced by thickets or savannas with diverse structure depending on the degree of degradation and the time for recovery since major disturbance.

4.2. Montane Conifer Forests

Conifers play an important role in montane forests in the Lower Mekong Basin. The most significant of these are the widespread *Pinus kesiya* forests, which extend throughout much of northern Thailand into the mountains of Lao at elevations of 800-1500 m on skeletal soils of schist or sandstone. This pine may become a dominant species on drier montane sites with less than 2000 mm annual rainfall. Under reasonable growing conditions, *P. kesiya* reaches 20-30 m in height. The most common tree associates of *P. kesiya* are *Keeteleria evelyniana*, *Pieris ovalifolia* (Ericaceae), *Schima wallichii* (Theaceae), and *Tristania merguensis* (Myrtaceae). Frequent fire seems to promote the establishment and maintenance of these pine forests at the expense of hardwoods (Rundel, 1999).

In Lao, these drier montane conifer forests often contain *K. evelyniana*, which is absent in Thailand. This species shows wide ecological amplitude, growing on sites with relatively dry conditions to areas with more than 3000 mm of mean annual rainfall. Like the pines, *K. evelyniana* appears to be much more edaphically controlled than moisture controlled in its distribution. It is more characteristic, however, of drier pine forests. Another drier site conifer, *Calocedrus macrolepis* (Cupressaceae) may also be present in these forests.

The Annamite Range contains unique conifer forests and a rich assemblage of conifer species. Beyond the drier *P. kesiya / Keeteleria* forests described above, there are wet montane forests with mixed dominance of hardwoods and conifers in areas with more than 2000 mm annual rainfall. The cool growing season temperatures and short drought period in these habitats promotes the growth of large trees on sites with a strong development of humic soils. The most abundant of these mesic forest conifers are species of Podocarpaceae which often form very large forest trees reaching 1-2 m in diameter and 40 m in height. One such community is dominated by *Dacrycarpus imbricatus* (Podocarpaceae) which may occur in relatively pure stands, but more commonly mixed with Fagaceae and Lauraceae. These stands are best represented on deep clay soils formed over crystalline parent rock in very humid areas of the Bolovens Plateau and the Annamite Range. Individual trees reach 30 m in height.

A second community forming dense conifer stands is dominated by *Fokienia hodginsii* in humid habitats at 1500-2400 m elevation in northern Lao. Vidal (1966) reported that these imposing trees reach as much as 40-50 m in height, although Hiep and Vidal (1996) suggest a height limit of 30-35 m. These humid forests receive more than 3000 mm rainfall annually, and support a lush understory layer of mosses and ferns growing over 50-60 cm of humus. Mists often cover these forests during the dry season, helping protect them from desiccation.

A third form of dense conifer forest in northern Lao is dominated by *Cunninghamia lanceolata* (Taxodiaceae). This tall tree, reaching 30-40 m (or even 50 m) in height, is more typical of temperate mountain regions of Japan and China, but extends its range southward into the Annamite Range of northern Lao and Viet Nam. Its habitat preference is for wet humid conditions with high rainfall and deep humus, as described above for *F. hodginsii*.

High-elevation montane areas exposed to heavy cloud cover and fog drip experience a cool climate and high rainfall with a relatively brief drought period. Areas with this climatic regime support a cloud forest community of low canopy height and abundant epiphytes. The growth form of trees in this community is usually highly branched with a twisted main trunk. Species of Ericaceae frequently form a major component of the flora of these forests.

5. WETLANDS AND SWAMP FORESTS

5.1. Lao Wetlands

On the basis particularly of vertebrate species diversity, Claridge (1996) identified wetland areas of international significance within Lao. These include the Mekong River wetlands, the Xe Champhon-Nong Louang wetlands, the Bung Nong Ngom-Xe Pian-Xe Khampho complex of wetlands, the Khone Falls-Seephandon cataracts, the Xe Kong Plains, the Soukhoum wetlands, and the Nam Theun wetlands (Nakai Plateau). The Nong Louang and Xe Champhon regions form a contiguous area of diverse wetlands located about 45 km southeast of Savannakhet. The former area is most notable for the presence of a number of permanent lakes extending over an area of about 90 km². The largest lake is approximately 4 km² in area, thus making it one of the largest permanent bodies of water in Lao. A recent mapping of wetland land use in a corridor within 50 km of the Mekong in Lao, made using 1:250,000 Landsat images but with little ground-truth verification, provides a preliminary quantitative estimate of the extent and types of wetlands along this corridor.

The Bung Nong Ngom-Xe Pian-Xe Khampho complex of wetlands represent a mosaic of forest-woodland-wetlands in southern Lao beginning 25 km southeast of Champasak. The wetlands are made up of scattered small ponds that include both seasonal and permanent water bodies, with much of this area lying on old volcanic flows. The woodland areas have not been described in detail but these are seasonally flooded communities surrounding wetland grasslands and ponds which support an important diversity of wildlife.

5.2. Tonle Sap Wetlands

Swamp forests typically occur in areas permanently inundated with shallow freshwater. The French literature has referred to the swamp forest association in Cambodia as *forêt inondee* (Légris and Blasco 1972; Rollet, 1972). Care must be taken, however, in separating the seasonal swamp forests that characterize extensive areas of the Tonle Sap basin and low-lying floodplains of major Cambodian rivers from these classic swamp forests of Southeast Asia with permanent flooding. Conditions of permanent flooding compared to flooding for 6-8 months produce differential selective factors and thus a distinctive floristic assemblage for each of the two forms of swamp forest.

Seasonal conditions of inundation produce several notable characteristics of these floodplain habitats. Native palms, often a characteristic component of typical swamp forests, are entirely absent from the Tonle Sap floodplain with the exception of the local occurrences of rattans in some gallery forests. Neither pneumatophores nor aerial roots are present among the swamp forest trees, and vascular epiphytes are lacking (Rollet, 1972). The swamp shrublands and forest of the floodplain of Tonle Sap covers about 3600 km² today, but is thought to have included more than three times of this area before extensive cutting began in the 1930s.

Several broad forest associations have been described for the extensive floodplain area of Tonle Sap (Anonymous, 2004; Blasco *et al.*, 1996; Campbell *et al.*, 2006). Similar swamp forests are also present along floodplains of major rivers, although an evergreen gallery forest with tropical rainforest affinities in structure is also locally present along limited river areas (MacDonald *et al.*, 1997).

A zone of stunted swamp forest, 7-15 m in height, originally dominated the dry-season shoreline of Tonle Sap, covering about 10% of the floodplain, and forms the second forest association. A similar community once occurred as a gallery forest along the seasonal flood plains of many major rivers in southern Cambodia, following channels or other waterways and occasionally occurring in isolated depressions that hold surface water through the dry season. This community around Tonle Sap is generally flooded by 4-6 m of water for up to 8 months each year, during which time the majority of species lose their leaves. Rather than forming a continuous forest, this community is broken into a mosaic of stands of large trees and open areas with floating aquatic herbs typical of the lake itself. Two tree species, *Barringtonia acutangula* (Lecythidaceae) and *Diospyros cambodiana* (Ebenaceae) are the primary dominates of this community, and generally occur with woody lianas such as *Combretum trifoliatum* (Combretaceae), *Breynia rhamnoides* (Euphorbiaceae), *Tetracera sarmentosa* (Dilleniaceae), and *Acacia thailandica* (Fabaceae).

The strong seasonal cycle of flooding around the floodplain of Tonle Sap has made the great majority of woody species deciduous. Rather than lose their leaves in the dry season, these species lose their leaves when the plants are submerged as the lake deepens. New leaves are produced rapidly when water recedes, growth and flowering follows several months later with a peak in July and August. Fruit and seeds are subsequently dispersed by floodwaters and herbivorous fish. There are several woody species, however, which remain evergreen despite submergence for 6-8 months each year, including *B. acutangula* and *C. trifoliatum* (Campbell *et al.*, 2006).

The short tree-shrubland association forms the dominant vegetation cover over approximately 80% of the Tonle Sap floodplain in those areas with nonsaturated soil during the dry season. In general, the dominant woody species form a semicontinuous canopy of deciduous species reaching no more than 2-4 m in height. The height reached by individual species appears to be related to soil-moisture conditions, with the tallest individuals occurring closer to the permanent lake basin and smaller individuals present at the periphery of the floodplain area.

The flora of these short tree-shrublands is dominated by species of Euphorbiaceae, Fabaceae, Elaeocarpaceae, and Combretaceae (MacDonald *et al.*, 1997). The most common species are *Barringtonia acutangula* and *Barringtonia micrantha* (Lecythidaceae), *Elaeocarpus griffithii* and *Elaeocarpus madropetalus* (Elaeocarpaceae), *Hydnocarpus authelminthica* (Flacourtiaceae), and *Malotus anisopodum* (Euphorbiaceae). Some of these species with shrubby growth forms in this community are capable of reaching tree size in swamp forest habitats. Several alien woody species have widely invaded this habitat and

are greatly expanding their cover and dominance. *Mimosa pigra* (Fabaceae) currently presents the most serious problems (Cambodian National Mekong Committee, 1998).

The shallow shoreline of Tonle Sap supports dense mats of herbaceous vegetation 1-3 m tall that may be emergent from shallow water but are more typically floating (MacDonald *et al.*, 1997). Large clonal mats of these species float freely over the lake, thereby colonizing large openings and gaps within the swamp forest. Notable among these species is the floating legume *Sesbania javanica* with huge mats of thickened rhizome bearing aerial stems reaching to 2 m above the lake surface. Given the heavy human impact around Tonle Sap for more than a millenium, it seems likely that the structure and diversity of the aquatic flora is greatly altered today from its original composition.

A notable floristic character of the aquatic flora of Tonle Sap is the rarity of many families of rooted monocots and dicots that are typically important in other Southeast Asian wetland habitats. These families include the Cyperaceae, Hydrocharitaceae, Marantaceae, Najadaceae, Nymphaeaceae, Pontederiaceae, and Potamogetonaceae (MacDonald *et al.*, 1997). The large seasonal change in water level of the lake and its strong turbidity are the likely cause of the absence of these groups. One rooted monocot, the grass *Phragmites karka*, is a local dominant at the mouth of Tonle Sap and at other river inlets and exits.

Freshwater marshland has been described near Prasat Tuyo (Bassac marshes) southeast of Phnom Penh between the Bassac and Mekong rivers exhibits swamp forests and wetlands similar to those of Tonle Sap. This marshland is inundated by up to 3 m of water from July to November, but forms a wetland surrounding a narrow body of open water during the dry season. The vegetation of this area consists of scattered individuals of *B. acutangula* in a wetland swamp matrix.

5.3. Gallery Forests and Riverine Wetlands

Riverine or evergreen gallery forests occur in a scattered distribution along the channel of the Mekong and other major rivers. These flooded forests, often forming open and discontinuous belts along riverbanks, are best known from scattered areas along the Mekong River in Stung Treng and Kratie Provinces (Bezuijen et al., 2008). Riverine gallery forests may vary greatly in structure and diversity. At one extreme are relatively large and diverse gallery forests growing on alluvial soils subject to seasonal flooding which grade into semievergreen forests. Dominant tree species in such habitats are typically semievergreen forest dipterocarps such as *Dipterocarpus alatus, Dipterocarpus dyeri*, and *Hopea odorata*, but obligately riverine tree species may also be present.

The floras of these riverine forests have distinctive affinities to wet evergreen forests of the Cardamom and Elephant Mountains in southern Cambodia, and thus may have served in the past as migration corridors between these ranges and the Annamite Range to the east. Thus, they have a major significance in protecting ecological processes of gene flow between "islands" of wet evergreen forest.

Low-diversity woodlands characteristically inhabit rocky and sandy islands in the midstream of major river courses. These forests are commonly dominated exclusively by a single species, *Anogeissus rivularis* (Combretaceae), which often exhibits a stunted growth form (Rollet, 1972). Other riverine forest communities have been described islands within the channel of the Mekong north of Steung Treng. These have been reported to include species of *Barringtonia* (Lecythidaceae), *Eugenia* (Myrtaceae), *Acacia* (Fabaceae-Mimosoideae), and *Ficus* (Moraceae) in the tree stratum, and *Morindopsis* (Rubiaceae) in a shrub layer. Many of the tree species have reduced narrow leaves and exhibit aerial root production up for 1 m

above their soil line. This habitat receives inundation and very strong river flow for part of the year, but the unconsolidated sandy soils suggest that drought may be present in the dry season. This community has been described as distinctive in floristic composition from surrounding dry upland areas, although few details are known of its species composition.

Areas of poor drainage and along the flood plain of the Mekong and its major tributaries often support a hydromorphic savanna dominated by graminoids. This community is called a *veal* in Khmer and the French literature has adopted this term. These soils are commonly saturated for at least 6 months of the year. Characteristic species would include *Saccharum* and a diverse group of Cyperaceae (Légris and Blasco, 1972).

5.4. Mekong Delta Region

The delta region of the Mekong includes three major landforms whose climate regime, hydrology, and soil condition determine the nature of plant communities based on a complex of climatic, hydrological, and soil processes. The floodplain area forms the largest landform type, covering much of the northern and central part of the delta. The high floodplain known as the Plain of Reeds forms the areas where flooding to a depth of up to 3 m occurs for much of the year, and drainage of floodwaters is slow. Another section exemplified by the Ha Tien Plain occurs where the land gently slopes towards the Gulf of Thailand and thus floods more briefly to only 1.5-2 m depth. In both areas, the flood conditions produce soils with high acid sulfate content (Husson *et al.*, 2000).

The tide-affected floodplain of the delta forms the second major landform. Inundation depths here are low, typically reaching only 0.5-1 m in depth in September and October. Although acid sulfate conditions occur here, their influence is less because the rivers wash away much of the sulfate content making this part of the delta highly suitable for agriculture. The coastal areas of the delta floodplain are influenced by both marine and river environments. Coastal flats lie between 1 and 1.5 m above sea level and are not directly inundated by saltwater, though seawater can enter the soils by capillary action. Sand ridges lying parallel to the coastline reach somewhat higher, while flats between the ridges are covered by saltwater during the dry season and by freshwater during the wet season. Mangrove swamps are found around the east-facing river mouths and around the Ca Mau peninsula.

The broad depression occupying a large area in the south of the delta forms the third landform. It is largely isolated from the river system, and as a result freshwater is scarce in the dry season. Much of the area is frequently inundated with saltwater during the dry season. Much of the remaining natural areas of the broad depression are peat swamps of the U Minh wetlands which can be inundated to 1-1.5 m during the wet season. The peat soils hold a large volume of water throughout the dry season, providing an important source of irrigation water for surrounding agricultural areas.

Only relatively small areas of natural *Melaleuca* swamp forest and grassland and sedgeland remain today in the Mekong Delta (Buckton *et al.*, 1999; Safford and Maltby, 1997; Tran Triet *et al.*, 2000). There is evidence in the form of tree stump remains, suggesting that extensive areas of the delta were once forested (Le Cong Kiet, 1993). However, the long human habitation in this area has meant that little is known of the original vegetation (Torell *et al.*, 2003). The remaining natural or relatively natural vegetation reflects the patterns of topography, soil, and hydrological conditions found in the delta. Freshwater communities include swamp forest vegetation, herbaceous grasslands and sedgeland, riverbank vegetation, and aquatic vegetation (Le Cong Kiet, 1994). Saline communities consist largely of mangrove forest.

Swamp forests are largely composed of pure stands of *Melaleuca cajuputi* (Craven and Barlow, 1997). While some apparently seminatural stands exist, the majority of these swamp forests today are plantations. Regenerating *Melaleuca* forest is largely found on acid sulfate soil and old alluvial sediments, and consists of trees 2-6 m tall, but locally reaching 10-12 m. The benefits of *Melaleuca* planting have been widely recognized, and the area of this forest type has increased in recent years. Although *Melaeuca* swamps are low in plant diversity, they have a great significance in maintaining natural ecosystem function. These swamps reduce water flow in the wet season and thus minimize flooding, store freshwater water, reduce soil acidification, promote biodiversity of many aquatic organisms, and provide a sustainable source of wood for construction and fuel.

Herbaceous vegetation includes extensive areas of seasonally inundated wetlands dominated by grasses and sedges. These have been subdivided into four main groups separated by the amount and duration of flooding during the wet season (Tran Triet, 1999). The first are wetlands on areas of deep and prolonged freshwater inundation on acid sulfate soils, and dominated by *Eleocharis dulcis*, *Oryza rufipogon*, and *Phragmites vallatoria*. A second wetland community inundated with freshwater to a moderate depth and for a shorter duration are dominated by *Eleocharis dulcis*, *Eleocharis ochrostachys*, *Ischaemum rugosum*, and *Lepironia articulata*. Grasslands on sandy and old alluvium soils inundated to only a shallow depth and for a short time are dominated by *Eragrostis atrovirens*, *Setaria viridis*, *Mnesithea laevis*, and *Panicum repens*. Finally, wetlands affected by brackish water that are dominated by *Paspalum vaginatum*, *Scirpus littoralis*, *Zoysia matrella*, *E. dulcis*, and *Eleocharis spiralis*.

Key conservation areas within the Mekong Delta, both protected and unprotected, have been described in detail (Buckton and Safford 2004; Buckton *et al.*, 1999). Tram Chim National Park, established in 1998, protects the last remnants of the wetland ecosystem of the Plain of Reeds, an extensive wetland area (Husson et al., 2000) that once covered more than 700,000 ha. The U Minh Thuong Nature Reserve protects nearly 23,000 ha of mature seminatural *Melaleuca* forest, seasonally inundated grassland, and open swamp (Safford et al., 1998). The Ha Tien plain is the last remaining extensive area of relatively intact seasonally inundated grassland in the Mekong Delta, as well as having stands of *Melaleuca* scrub and *Nypa fruticans* swamp. This area has no official conservation status and most of it is earmarked for resettlement (Buckton et al., 1999).

The aggressive invasive *M. pigra* has become widely established in wetlands of the Mekong Delta. The invasion of *Mimosa* is particularly troublesome in protected areas such as Tram Chim National Park as it threatens to reduce biological diversity (Tran Triet et al., 2004).

Mangroves once covered an area of about 4000 km² in Viet Nam, with more than half of this area within the Mekong Delta. The extensive military use of defoliants and napalm during the Vietnam War (1962-1972) destroyed more than 1000 km² of mangrove area in the delta area (Spaulding et al., 1997). Mangrove forests had been slowing recovering under active programs of reforestation, but much of this area has scrubby growth or open mangrove plantations rather than the closed canopy cover that once existed (Williamson, 1990). However, rapid expansion of shrimp farming has substantially reduced the area of mangroves (Thu and Populus, 2007).

Mangrove diversity in the Mekong Delta area is relatively high. Of the approximately 50 species of true mangroves which are distributed in South and Southeast Asia, including Indonesia, 29 species occur in Viet Nam. Mangrove forests typically exhibit strong patterns of zonation. The pioneer species along the open coastline is typically *Avicennia alba* (Avicenniaceae). Next along a gradient of decreasing relative exposure and submergence by sea water are *Rhizophora apiculata* and *Brugiera parviflora* (Rhizophoraceae)

which become established after 5-6 years and grow to replace *Avicennia* after about 20 years. Higher ground subject to conditions of brackish water is dominated by *Avicennia officinalis* (Avicenniaceae), *Sonneratia caseolaris* (Sonneratiaceae), *N. fruticans* (Arecaceae), and *Phoenix paludosa* (Arecaceae). The upper elevational distribution of mangrove forests grade into swamp forests with dominates that are not classic mangrove species. These include *M. cajuputi* (Myrtaceae), *Acronychia laurifolia* (Rutaceae), *Canthium didynum* (Rubiaceae), *Alstonia spathulata* (Apocynaceae), and the ferns *Stenochlaena palustris* (Blechnaceae) and *Polybotrya appendiculata* (Aspleniaceae). The environmental conditions and floristics of mangrove forests in the Cape Saint Jacques area of the Mekong Delta were described in considerable detail by Cuong (1964).

6. LAND-USE DYNAMICS IN THE MEKONG WATERSHED

Changes in land use and land cover have occurred rapidly across Southeast Asia over the last few decades, including the Mekong watershed. These alterations in land use and land cover have been associated with changes in rates of economic development and the interactions of population growth and poverty. Shifting cultivation and associated conversion of forest land to agricultural land, together with intensification of agricultural activities and increase monoculture cropping, are major reasons for these changes (Giri et al., 2003).

Satellite data using AVHRR, MODIS, and SPOT imagery have now provided regional maps of major forest cover and land-use categories (Giri et al., 2003; Stibig et al., 2004; Zhan et al., 2002). Nevertheless, problems exist in these maps related to separating ecological forest communities, assessing forest fragmentation and land cover mosaics, and transitions between primary and secondary forests.

Swidden agriculture has remained the dominant land-use practice in montane areas throughout the past half-century. This has taken place despite government policies throughout the region to limit this practice by such means as resettling people into lowland areas, declaring protected forest reserves, and even banning shifting cultivation (Fox and Vogler, 2005).

References

Anonymous. (2004). Cambodia independent forest sector review. http://www.cambodia-forest-sector.net/index.htm. Accessed 11 June 2008.

Aubreville, A. (1957). An pays des eaux et des forêts. Impressions du Cambodge forestier. *Bois et Forêts des Tropiques* **52**, 49-56.

Baker, B. B. and Moseley, R. K. (2007). Advancing treeline and retreating glaciers: Implications for conservation in Yunnan, P.R. China. *Arctic, Antarctic and Alpine Research* **39**, 200-209.

Baker, P. J., Bunyavejchewin, S., Oliver, C. D., and Ashton, P. S. (2005). Disturbance history and historical stand dynamics of a seasonal tropical forest in western Thailand. *Ecological Monographs* **75**, 317-343.

Bezuijen, M. R., Timmins, R., and Seng, T. editors. (2008). Biological surveys of the Mekong River between Kratie and Stung Treng Towns, northeast Cambodia, 2006-2007. WWF Greater Mekong – Cambodia Country Programme, Cambodia Fisheries Administration and Cambodia Forestry Administration, Phnom Penh.

Blasco, F., Bellan, M. F., and Lacaze, D. (1996). "Vegetation Map of Cambodia." Ecocart, Toulouse, France.

Buckton, S. T. and Safford, R. J. (2004). The avifauna of the Vietnamese Mekong Delta. *Bird Conservation International* **14**, 279-322.

Buckton, S. T., Cu, N., Quynh, H. Q., and Tu, N. D. (1999). "Conservation of Key Wetland Sites in the Mekong Delta," Conservation Report No. 12. Institute of Ecology and Biological Resources. Bird Life International Vietnam Programme, Hanoi, 114pp.

Cambodian National Mekong Committee. (1998). "Natural Resources-Based Development Strategy for the Tonle Sap area, Cambodia." Volume 2: Sectoral Studies. 1. Environment in the Tonle Sap Area. Mekong River Commission Secretariat, UNDP, Phnom Penh.

Campbell, R. (1935). "Teak-Wallah." Oxford University Press, Oxford.

Campbell, I. C., Poole, C., Giesen, W., and Valbo-Jorgensen, J. (2006). Species diversity and ecology of the Tonle Sap Great Lake, Cambodia. *Aquatic Sciences* **69**, 355-373.

Champion, H. G. and Set, S. K. (1968). "The Forest Types of India: A Revised Survey." Manager of Publications, New Delhi.

Claridge, G. (1996). "An Inventory of Wetlands of the Lao PDR," IUCN, Bangkok, 287pp.

Craven, L. A. and Barlow, B. A. (1997). New taxa and new combinations in *Melaleuca* (Myrtaceae). *Novon* **7**, 113-119.

Cuong, V. V. H. (1964). Flore et vegetation de la mangrove de la region de Saigon-Cap Saint Jacques, Sur Viet Nam. Dissertation. Universite de Paris, Paris, France.

Deng, M. and Zhou, Z. K. (2004). Seed plant diversity of scree from northwest Yunnan. *Acta Botanica Yunnanica* **26**, 23-24.

Elliot, S., Maxwell, J. F., and Beaver, O. P. (1989). A transect survey of monsoon forest in Doi Suthep-Pui National Park. *Natural History Bulletin of the Siam Society* **37**, 137-141.

Fox, J. and Vogler, J. B. (2005). Land-use and land-cover change in montane mainland Southeast Asia. *Environmental Management* **36**, 394-403.

Giri, C., Defourny, D., and Shrestha, S. (2003). Land cover characterization and mapping of continental Southeast Asia using multi-resolution satellite data. *International Journal of Remote Sensing* **24**, 4181-4196.

Heinimann, A., Messerli, P., Schmidt-Vogl, D., and Wiesmann, U. (2007). The dynamics of secondary forest landscapes in the lower Mekong Basin. *Mountain Research and Development* **27**, 232-241.

Hiep, N. T. and Vidal, J. E. (1996). Gymnospermae. *Flore du Cambodge, du Laos et du Viêtnam* **28**, 1-166.

Husson, O., Verburg, P. H., Phung, M. T., and Van Mensvoort, M. E. F. (2000). Spatial variability of acid sulphate soils in the Plain of Reeds, Mekong delta, Vietnam. *Geoderma* **97**, 1-19.

Kerr, A. F. G. (1932). A trip to Pu Bia in French Laos. *Journal of the Siam Society of Natural History* **9**, 193-223.

Le Cong Kiet. (1993). Dong Thap Muoi—Restoring the mystery forest of the Plain of Reeds. *Restoration and Management Notes* **11**, 102-105.

Le Cong Kiet. (1994). Native freshwater vegetation communities in the Mekong Delta. *International Journal of Ecology and Environmental Sciences* **20**, 55-71.

Légris, P. and Blasco, F. (1972). Notice de la carte de végétation du Cambodge au 1/1,000,000. *Travaux de la Section Scientifique et Technique de l'Institut francais de Pondichéry* **11**, 1-238.

Li, X. and Walke, D. (1986). The plant geography of Yunnan Province, Southwest China. *Journal of Biogeography* **13**, 367-397.

MacDonald, J. A., Pech, B., Phauk, V., and Leeu, B. (1997). Plant communities of the Tonle Sap floodplain. Final Report in contribution to the nomination of Tonle Sap as a UNESCO Biosphere Reserve, UNESCO, IUCN,

Wetlands International and SPEC (European Commission), Phnom Penh, 30pp.

Maurand, P. (1943). "L'Indochine Forestière," Imprimerie d'Extreme Orient, Hanoi, 252pp.

Pfeffer, P. (1969). Considerations sur l'ecologie des forêts claire du Cambodge orientale. *Terre et Vie* **23**, 3-24.

Rollet, B. (1953). Notes sur les forêts claires du sud de l'Indochine. *Bois et Forêts des Tropiques* **31**, 3-13.

Rollet, B. (1972). La végétation du Cambodge. *Bois et Forêts des Tropiques* **144**, 3-15; **145**, 24-38; **146**, 4–20.

Rundel, P. W. (1999). "Forest Habitats and Floristics of Indochina: Lao PDR, Cambodia and Vietnam," World Wide Fund for Nature (WWF), Hanoi, 194pp.

Rundel, P. W. and Boonpragob, K. (1995). Dry forest ecosystems of Thailand. *In* "Tropical Deciduous Forests," Bullock S., Medina, E. and Mooney, H. A., (eds.), pp 93-123. Cambridge University Press, Cambridge.

Safford, R. J. and Maltby, E. (1997). The overlooked value of Melaleuca wetlands. *In* "Towards Sustainable Management of Tram Chim National Reserve, Vietnam," Stafford, R. J., Ni, D. V., Maltby, E., and Xuan, V.-T. (eds.), pp 109-112. Proceedings of a Workshop on Balancing Economic Development with Environmental Conservation. Royal Holloway Institute for Environmental Research, London.

Safford, R. J., Tran Triet, Maltby, E., and Duong Van Ni (1998). Status, biodiversity and Management of U Minh wetlands. *Tropical Biodiversity* **5**, 217-244.

Santisuk, T. (1988). "An Account of the Vegetation of Northern Thailand," Franz Steiner Verlag, Weisbaden, 103pp.

Schmid, M. (1974). Végétation du Viet-Nam: Le massif sud-annamitique et les régions limitrophes. *Mémoire ORSTOM* **74**, 1-243.

Smitinand, T. (1968). Vegetation of Khao Yai National Park. *Natural History Bulletin of the Siam Society* **22**, 289-297.

Smitinand, T. (1989). Thailand. *In* "Floristic Inventory of Tropical Countries," Campbell, D. G. and Hammond, H. D. (eds), pp 63-82. New York Botanical Garden, New York.

Spaulding, M., Blasco, F., and Field, C. (1997). "World Mangrove Atlas," The International Society for Mangrove Ecosystems, Okinawa, 178pp.

Stamp, L. D. (1925). "The Vegetation of Burma." Thacker, Spink and Co. Calcutta, India.

Stibig, H. J., Achard, F., and Fritz, S. (2004). A new forest cover map of continental southeast Asia derived from SPOT-VEGETATION satellite imagery. *Applied Vegetation Science* **7**, 153-162.

Stott, P. (1984). The savanna forests of mainland Southeast Asia: An ecological survey. *Progress in Physical Geography* **8**, 315-335.

Stott, P. (1986). The spatial pattern of dry season fires in the savanna forests of Thailand. *Journal of Biogeography* **13**, 345-358.

Stott, P. (1988a). The forest as Phoenix: Towards a biogeography of fire in mainland southeast Asia. *Geographical Journal* **154**, 337-350.

Stott, P. (1988b). Savanna forest and seasonal fire in South East Asia. *Plants Today* **1**, 196-200.

Stott, P. (1990). Stability and stress in the savanna forests of Thailand. *Journal of Biogeography* **17**, 373-383.

Stott, P., Goldhammer, J. G., and Werner, W. L. (1990). The role of fire in the tropical lowland deciduous forests of Asia. *In* "Fire in the Tropical Biota: Ecosystem Processes and Global Challenges," Goldhammer, J. G. (ed.), pp 32-44. Springer-Verlag, Berlin.

Sukwong, S., Dhanmanonda, P., and Pongumphai, P. S. (1975). Phenology and seasonal growth of dry dipterocarp forest species. *Kasetsart Journal* **9**, 105-113.

Thu, P. M. and Populus, J. (2007). Status and changes of mangrove forest in Mekong Delta: Case study in Tra Vinh, Vietnam. *Estuarine Coastal and Shelf Science* **71**, 98-109.

Torell, M., Salamanca, A. M., and Ratner, B. D. (eds.). (2003). "Wetlands Management in Vietnam: Issues and Perspectives," WorldFish Center, Penang, Malaysia, 89pp.

Triet, T. (1999). Freshwater wetland vegetation of the Mekong Delta: A quantitative study of the relationship between plant species distribution and the wetland environment. Ph.D. dissertation, University of Wisconsin, Madison.

Triet, T., Safford, R. J., Tran Duy Phat, Duong Van Ni, and Maltby, E. (2000). Wetland biodiversity overlooked and threatened in the Mekong Delta, Vietnam: Grassland ecosystems in the Ha Tien Plain. *Tropical Biodiversity* **7**, 1-24.

Triet, T., Kiet, L. C., Thi, N. T. L., and Dan, P. Q. (2004). The invasion by *Mimosa pigra* of wetlands of the Mekong Delta, Vietnam. *In* "Research and Management of *Mimosa pigra*," Julien, M., Flanagan, G., Heard, T., Hennecke, B., Paynter, Q., and Wilson, C. (eds.), pp 45-51. CSIRO Division of Entomology, Canberra.

Vidal, J. E. (1956-1960). La végétation du Laos, I. Conditions ecologiques. II. Groupements végétaux et flore. *Travaux Laboratoire Forestier de Toulouse* **5**, 1-582.

Vidal, J. E. (1960). Les forets du Laos. *Bois et Forêts des Tropiques* **70**, 5-21.

Vidal, J. E. (1966). Endémisme végétal et systématique en Indochine. *Compte Rendu Sommaire des Seances. Societe de Biogeographie, Paris* **41**, 153-159.

Vidal, J. E. 1979. Outline of the ecology and vegetation of the Indochinese Peninsula. *In* "Tropical Botany," Holm-Hensen, O. (ed.), pp 109-123. Academic Press, New York.

Wang, X. R., Szmidt, A. E., and Savolainen, O. (2001). Genetic composition and diploid hybrid speciation of a high mountain pine, *Pinus densata*, native to the Tibetan plateau. *Genetics* **159**, 337-346.

Wege, D. C., Long, A. J., Vinh, M. K., Dung, V. V., and Eames, J. C. (1999). "Expanding the Protected Areas Network in Vietnam for the 21st Century: An Analysis of the Current System with Recommendations for Equitable Expansion," Bird Life International, Hanoi, 74pp.

Weyerhaeuser, H., Wilkes, A., and Kahrl, F. (2005). Local impacts and responses to regional forest conservation and rehabilitation programs in China's northwest Yunnan province. *Agricultural Systems* **85**, 234-253.

Wikramanayake, E., Dinerstein, E., Loucks, C. J., Olsen, D. M., Morrison, J., Lamoreaux, J., and Pimm, S. L. (eds.). (2002). "Terrestrial Ecoregions of the Indo-Pacific: A Conservation Perspective," Island Press, Washington, DC, 643pp.

Williamson, M. (1990). War and forests: South Vietnam. *New Zealand Forestry* **34**, 18-21.

Zhan, X., Sohlberg, R. A., Townshend, J. R. G., DiMiceli, C., Carroll, M. L., Eastman, J. C., Hansen, M. C., and DeFries, R. S. (2002). Detection of land cover changes using MODIS 250 m data. *Remote Sensing of Environment* **83**, 336-350.

Zhang, J. and Cao, M. (1995). Tropical forest vegetation of Xishuangbanna, SW China, and its secondary changes, with special reference to problems in local nature conservation. *Biological Conservation* **73**, 229-238.

Zhu, H., Cao, M., and Hu, H. B. (2006). Geological history, flora, and vegetation of Xishuangbanna, southern Yunnan. *China Biotropica* **38**, 310-317.

CHAPTER

8

Fish Diversity in the Mekong River Basin

John Valbo-Jørgensen,[1] David Coates,[2] and Kent Hortle[3]

[1]Fisheries Department, Food and Agriculture Organisation of the United Nations, Rome, Italy
[2]Programme Officer, Secretariat, Convention on Biological Diversity, 413 Rue St Jacques Suite 800, Montreal H2Y 1N9, Canada
[3]Fisheries and Environment Consultant, 23A Waters Grove, Heathmont 3135, Australia

OUTLINE

1. INTRODUCTION

1.1. Ecological Classification of Fishes

Fish that are found in inland waters are usually categorized at family level based on their evolutionary history as "primary" freshwater fishes (a long evolutionary history in freshwaters), "secondary" (some salt tolerance), and "peripheral" (species from marine families that live in freshwaters for part or all or their lives) (Berra, 2001). The distribution of primary division fish reflects past and present river connections, whereas the distribution of the secondary and peripheral division fishes is also influenced by marine dispersal. From an ecological perspective, the "evolutionary" classification of fishes has limitations, because some peripheral division fishes have fully adapted to freshwaters; the ancestors of two primary division catfish families (Ariidae and Plotosidae) reinvaded the sea, with some species subsequently reinvading freshwaters; and some primary division fishes are found in brackish-waters, or even in full seawater as, for example, the catfish, *Pangasius krempfi* from the Mekong.

Known occurrences of fishes in fresh, brackish, and marine environments are documented on the website FishBase (Froese and Pauly, 2008). In the Mekong system (and in the Oriental region generally) the breeding requirements of peripheral division fishes are not well documented. Juveniles or adults may be commonly found in fresh, brackish, or salt water, but most peripheral division fishes that occur in freshwater probably require brackish or saline water

for development of their eggs or larvae. Many of these could be regarded as "facultatively catadromous," that is, adults or juveniles usually move into freshwaters, but require a connection with the sea to complete their life cycles. Well-known examples include the barramundi (*Lates calcarifer*), mangrove jack (*Lutjanus argentimaculatus*), and some mullets (Mugilidae), typical "freshwater" fishes that breed in coastal areas. Hence, the number of principal freshwater fishes within the peripheral division is not known with certainty for the Mekong. Lagler (1976) attempted to subdivide peripheral species in the Mekong into additional categories based on the degree or duration of their usual penetration into inland waters.

1.2. Fish Diversity in Inland Waters in the Oriental Region

Relative to area, inland waters support a disproportionate number of species of fish and many new species are being described each year. FishBase, in 2005, listed 28,900 species of which 13,000 (45%) were primary of secondary freshwater species (Froese and Pauly, 2008; Lévêque et al., 2008).

In 2005, there were about 4400 fish species, or about 15% of the world's total fish fauna (Lévêque et al., 2008), recorded from freshwaters in the Oriental region. This region, which contains the Mekong, extends from Pakistan through India, IndoChina, the Indonesian archipelago, and southern China. The regional freshwater fish fauna is dominated by the otophysan fishes, which include the cypriniformes (carps, barbs, minnows, and loaches) (about 1380 spp.) and siluriformes (freshwater catfish) (about 535 spp.) and this dominance is also apparent in the Mekong; of about 730 species that occur in freshwaters, about 500 or about 68% are otophysan fishes (cyprinids or catfishes) (Table 8.1).

Otophysan fishes possess characteristics that have allowed them to adapt to rivers and streams, environments which are frequently turbid, turbulent, with fluctuating chemistry and temperature (Moyle and Cech, 1988). These characteristics include a Weberian apparatus, a chain of bones that connects the swim bladder to the inner ear, which gives them an acute sense of hearing that is particularly useful while in turbid water or at night. These fishes are the dominant group on all continents except Australia and Antarctica (Briggs, 2005). By contrast, coastal and marine waters are dominated by nonotophysan groups of fishes, such as perchlike fishes (perciformes), which typically rely on sight and visual cues for feeding, intraspecific interactions, and predator avoidance.

Within the Oriental region, the fauna of rivers tends to be more similar in adjacent drainages which have been recently connected. On the basis of cyprinid species distribution, Yap (2002) found the Mekong fauna most similar to that of the Chao Phraya, and also found the fauna of the Mekong, mid-Mekong, Lower Mekong, and Chao Phraya are equally similar to each other, reflecting recent or continuing connections. This contrasts with earlier findings of Taki (1978), based on more limited surveys and species records, who considered that within these four subareas, the Lower Mekong primary fish fauna was most similar to the Lower Chao Phraya and the Middle/Upper Mekong fauna were more similar to the Upper Chao Phraya, a pattern that resulted from former connection of these rivers and an ongoing barrier effect for some species of the Khone Falls on the Great Fault Line. The Mekong and Chao Phraya group at a higher level

with the Mae Klong and the rivers that drain into the Sunda Shelf, reflecting past connections (Yap, 2002). Work on the genetics of individual species also shows that present distributions largely reflect faunal exchanges across the Sunda Shelf early in the Pleistocene and that regional populations have subsequently diverged in isolation (Dodson *et al.*, 1995; McConnell, 2004).

While geological history has influenced aquatic biodiversity throughout Southeast Asia where extraordinary numbers of species have accumulated in several river systems, river dynamics and hydrology play an important role in maintaining ecosystem diversity. The naturally fluctuating environmental conditions over space and time, both within and between years, drive this ecological diversity. Likewise, reducing this hydrological and ecological diversity is a driver of biodiversity loss. Variations in the hydrological cycle also counteract dominance by certain species because no species will have optimal conditions for long time. The result is that different species will be at an advantage at different times. The chances that certain species will be outcompeted to extinction are therefore reduced (Ward and Stanford, 1983). This is reflected, for example, in discrete changes in relative species compositions of catches between years for the same fishing gears used in the same habitats.

2. FISH SPECIES DIVERSITY AND ENDEMISM

The Mekong Fish Database (MFD) developed by the Mekong River Commission (MRC, 2003) provides the most comprehensive data source on Mekong fishes, although it does not include any information on species from areas upstream of Yunnan and in Myanmar. Froese and Pauly (2008) and other references have been used to supplement the MFD where necessary.

The MFD lists 924 named species (898 indigenous) most of which have been recorded

TABLE 8.1 Fish taxa recorded from MRC (2003) except that MRC (2003) classifies the noodlefish (Sundasalangidae) as part of Osmeriformes; however, there is now general agreement that the family should be classified under Clupeiformes. Habitat occurrence data from Fishbase. Numbers in parenthesis refer to number of taxa excluding introduced species.

Order	Common names for the main families	Families	Genera	Number of species	Number of species which occur in habitats classed by salinity					
					Freshwater only	Fresh and brackish	Fresh, brackish and marine	Brackish only	Brackish and marine	Marine only
Orectolobiformes	Bamboo sharks	1	1	3			1		1	1
Carcharhiniformes	Requiem sharks, catsharks, hammerheads	3	5	8			2		4	2
Rajiformes	Rays, sawfish	4	6	13	2	3	3		3	2
Osteoglossiformes	Bony tongues, featherbacks	2	3	5	4	1				
Elopiformes	Tarpons, tenpounders	2	2	2			2			
Anguilliformes	Eels	4	8	13			7		4	2
Clupeiformes	Herrings, shads, anchovies	3	17	33	4	5	9		13	2
Gonorhychiformes	Milkfish	1	1	1			1			
Cypriniformes	Carps, barbs, minnows, loaches	4	105	387 (371)	377 (364)	10 (7)				
Characiformes	Characins	1 (0)	1 (0)	1 (0)	1 (0)					
Siluriformes	Catfish, sheatfish	13 (12)	39 (37)	127 (124)	92 (89)	13	12		10	
Aulopiformes	Lizard fishes, Bombay ducks	2	3	7					3	4

Order	Common name									
Gadiformes	Codlets	1	1	1				1	1	
Batrachoidiformes	Toadfishes	1	2	2				2	2	
Lophiiformes	Anglerfishes	1	2	2				1	1	
Atheriniformes	Silversides, priaprium fishes	2	5	7				3	3	
Beloniformes	Needlefishes, halfbeaks, ricefishes, rivulines	5	12	29	6	5	5	8	4	1
Cyprinodontiformes	Live bearers	1	2	2 (0)					2 (0)	
Gasterosteiformes	Pipefishes, armoured sticklebacks	3	8	13		3	5	3	2	
Synbranchiformes	Swamp eels, spiny eels	3	6	13	9		4			
Scorpaeniformes	Scorpionfishes	2	7	7			3	4	3	
Perciformes	Perches, threadfins, croakers, archerfishes, gobies, sleepers, gouramies, snakeheads	29 (28)	105 (103)	207 (203)	38	37 (33)	62	18	50	2
Pleuronectiformes	Soles, tonguefishes	2	7	19	6	2	3	1	6	1
Tetraodontiformes	Puffers	1	9	22	13	1	3		5	
Total		91 (87)	357 (341)	924 (898)	554 (537)	89 (80)	115	47	115	4

upstream from the mouth of the river (MRC, 2003). Some coastal species that have not actually been recorded from the river have been included in the database because they are highly likely to be found within the river system at some time. About 60% of the listed endemic species are primary freshwater fishes while secondary freshwater fishes derived largely from marine families, and estuarine comprise about 40%.

With 24 orders and 87 families of indigenous fish listed in the MFD, higher level taxonomic diversity in the Mekong may exceed that in any other river in the world, even when it is considered that four orders are only represented by marine visitors (Table 8.1).

The Mekong shares most of its indigenous species (about 60%) with the other large Southeast Asian rivers. The nonostariophysan fishes of the Mekong (species that are not cypriniforms or catfishes) are widespread throughout Southeast Asia (Taki, 1975, 1978). Kottelat (1989) found that the Mekong and Chao Phraya had more than 50% of their fish fauna in common (Table 8.2). On the basis of fish distributions, Taki (1978), Kottelat (1989), and Rainboth (1991, 1996) reached the conclusion

that the Upper Mekong formed part of the Chao Phraya Basin in the past.

In contrast, the fish fauna of the Vietnamese rivers on the eastern slope of the Annamite chain is more similar to the fauna of the East Asian rivers and shares only a few species with the Mekong. It is rather species poor in comparison (Banarescu, 1972; Rainboth, 1991).

The high incidence of species of marine origin may be explained by the extensive estuarine zone and the high diversity of fishes associated with the large shallow water area known as the Sunda Shelf in the South China Sea. A number of large rivers feed into this area which has allowed many species to adapt into the estuarine environment. Some of these species have subsequently colonized habitats further upstream in the freshwater zone. In the Mekong, colonization has been made easier because there are no natural barriers along the 700 km from the Khone Falls to the mouth.

According to the MFD, the Mekong has 219 endemic species; all are "freshwater" fishes (freshwater or fresh-brackishwater occurrence), so about 35% of this group and 24% of all species occur only in the Mekong (Table 8.2).

TABLE 8.2 Percentage similarity (determined using the Jaccard Index) of the fish faunas of Southeast Asian rivers (based on Kottelat, 1989; adapted from Visser et al., 2003)

	Mekong	Chao Phraya	Salween	Malay Peninsula	Mae Khlong	SE Thailand + SW Cambodia	Annam
Mekong	*	56	11	30	31	21	7
Chao Phraya		*	11	34	35	22	7
Salween			*	11	14	11	6
Malay Peninsula				*	28	22	6
Mae Khlong					*	36	12
SE Thailand + SW Cambodia						*	18
Annam							*

By contrast, the coastal and marine fishes listed in the MFD all have broad distributions, with the range of many species extending throughout the tropical areas of the Indo-Pacific. About 76% (166) of the Mekong's endemic species are cypriniformes (cyprinids, loaches, and algae eaters) and 12% are catfishes. About 54% of the endemic species (118 species) occur either only in Laos or Yunnan or both Laos and Yunnan. Further, the proportion of the Mekong endemics in terms of the fauna is also much higher in Laos (41%) and in Yunnan (40%) than in Thailand (24%), Cambodia (15%), or Viet Nam (7%), where the Mekong catchment is mostly lowlands that are inhabited by widespread lowland or coastal fish species. Hence endemism in the Mekong is largely a result of speciation by representatives of several families of primary division fishes which are specialized for, and tend to become isolated within, upland tributaries. In the Mekong basin, the specialized species with restricted distributions in these upland habitats are at particular risk of extinction when habitats are drastically altered; for example, as a result of dam construction. Other land-use changes are also of concern; for example, of the 28 endemic species from Yunnan, 6 species are found only in Lake Erhai and these have been affected by eutrophication and the introduction of exotics (see Hortle, Chapter 9).

3. HABITAT DIVERSITY AND SPECIES ASSEMBLAGES

3.1. Mekong Mainstream and Major Tributaries

A very large number of fish species are found in the mainstream (Fig. 8.1). However, few if any species are confined to it, even the largest species such as the giant Mekong catfish (*Pangasianodon gigas*), and *Pangasius sanitwongsei* have been recorded from some of the larger Mekong tributaries.

Current speed and channel morphology change as the river flows downstream affect the species that favor specific parts of the river, and the upper and lower reaches of the Mekong harbor different fish faunas. Because of the heavy sediment load, the Mekong River mainstream and the lowland sections of tributaries are highly turbid during the flood season. Catfishes are well adapted to this environment with their sensitive barbels that allow them to locate food without the help of their eyes. The threadfins (Polynemidae) use their filamentous pectoral fins in a similar way and both groups are well adapted to an environment with almost no light penetration (Moyle and Cech, 1988). The golden spotted grenadier anchovy (*Coilia dussumieri*) possesses a series of light organs (photophores) along the flanks and belly and on the lower jaw (Whitehead *et al.*,

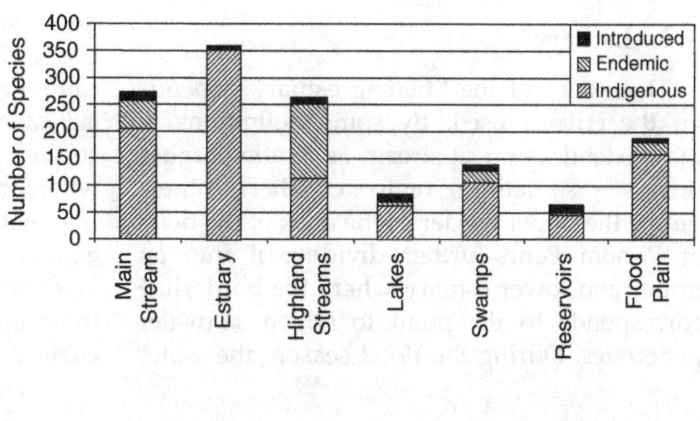

FIGURE 8.1   The number of species (species richness) in various aquatic habitats in the Mekong Basin (modified from Valbo-Jørgensen and Visser, 2003; data from MRC, 2003).

1988); however, their purpose is unknown in the murky waters of the Mekong.

Examples of species that show preference for certain parts of the mainstream are the endemic *Mekongina erythrospila* and *Labeo dyocheilus* that have a preference for fast-flowing water and are rare or absent downstream of Kratie where there are no rapids (Roberts and Warren, 1994). Although a number of species are capable of ascending the rapids at Khone Falls this still constitutes a barrier to marine visitors, and several secondary freshwater species, such as the anchovies (Engraulidae), threadfins (Polynemidae), and sea catfishes (Ariidae), do not occur upstream of the falls (Roberts and Baird, 1995).

A large proportion of the mainstream species including all the pangasiid catfishes are migratory some moving over long distances. *P. krempfi,* for example, migrates more than 700 km from the estuary to beyond Khone Falls to spawn (Hogan *et al.,* 2007) and the Mekong Giant Catfish moves from lower to upper reaches to reach its spawning grounds in the golden triangle or beyond. Other species migrate laterally into tributaries and floodplains to feed and spawn (Chan *et al.,* 2001). For many species, the lowland mainstream channels serve as migration corridors connecting habitats which serve different purposes in the fish's life cycle. Poulsen and Valbo-Jørgensen (2000) identified three major migration systems governed by the position of key habitats in the basin.

3.2. Estuary

The extent of the Mekong estuary depends on the criteria used. By some definitions, it may extend as far upstream as Kratie because tides are sometimes registered there, others define the upper border as the apex of the delta at Phnom Penh further dividing it into an upper and lower estuary where the borderline corresponds to the point to which saltwater penetrates. During the flood season, the water is fresh throughout the estuary; but during the dry season salt water intrudes to Can Tho and My Thuan in the Bassac and Mekong Rivers, respectively (MRC, 1997). However, the Mekong River provides a significant input of silt and nutrients to the South China Sea and this influence extends hundreds of kilometers beyond the river mouth (Tang *et al.,* 2004). This nutrient input is a major determinant of aquatic productivity in this area. In addition, the soil in the delta region contains high levels of sulfate, and during the dry season when the water exchange is low, the water in rice fields and canals becomes acidic creating an inhospitable environment for fish, allowing only hardy species to survive.

The estuary is the most species-rich part of the basin because of the mixture of marine, brackish-water, and freshwater species (Fig. 8.1). The mangroves serve as important nursery areas for many fish and invertebrate species living in both the river and along the coast. Marine and freshwater fishes move up or down the river according to the prevailing salinity and thus may occur at the same location in different seasons. Marine species mainly ascend the river during the dry season, the distance depending on their level of tolerance to freshwater. The opposite is true for freshwater species that descend further into the estuary during the flood season.

The estuarine species, the last group, are permanent residents and are able to tolerate variations in salinity that happens over the year. Gobies that are among the most prominent fish families in the estuary are, for example, renowned for their tolerance to variable salt concentrations. The mudskippers (*Periophthalmus* spp. and *Periophthalmodon* spp.) are so well adapted to the life in the tidal littoral where they have become semiterrestrial leaving the water to feed on the mudflats that are exposed at low tide. These fish are able to survive out of the water for very long time because they breathe air using their modified gill cavities.

Most estuarine species stay in the downstream reaches in Viet Nam. However, some may stray further upstream into Cambodia. The sawfish *Pristis microdon* has, for example, been recorded from the Tonle Sap and in the past it is known to have gone up the Mekong mainstream to Khone Falls on the border between Laos and Cambodia. However, none of these are permanent residents in the freshwater reaches of the river and most of them would not be able to complete their lifecyle without returning to the marine environment. These species are nevertheless an important part of the Mekong's ichthyofauna and depend upon the river's inputs of nutrients and organic material. They illustrate the complexity of the area whereby different components of the fish assemblage occur at different times of the year—thereby adding a temporal component to the heterogeneity of the environment.

3.3. Individual Tributary Basins

The faunasw of the large tributaries are similar to that of the Mekong River at least in the lower reaches. However, it is important to note that while the Mun and the Songkhram Rivers are lowland rivers with enormous floodplains, most of the rivers joining the Mekong from the east have their origin in the Annamite mountain range and part of their course is mountainous and their floodplains are small. The fish fauna of the Mun and the Songkhram, therefore, resembles the mainstream fauna more than the other river basins. The Se San, the largest of the eastern tributaries, is the result of the merge of three major rivers Sekong, Srepok, and Se San and it drains northeastern Cambodia, southern Laos, and the Central Highlands in Viet Nam meeting the Mekong River mainstream at Stung Treng. Many of the smaller streams in the Se San catchment flow through natural forest ecosystems and have high transparency due to low-silt load (Rainboth, 1996). Data on species diversity in individual Mekong tributaries is extremely scarce (Table 8.3), and the number of species cataloged from each subbasin is to a large extent a result of the intensity with which they have been surveyed. The Tonle Sap is, for

TABLE 8.3 The number of species recorded from Major Mekong tributaries and the Bassac River which is a distributary (and thus better described as part of the main stream)

Basin	Number of species	Dominating fish families
Nam Ou (Lao)	72	Cyprinidae (50%) Balitoridae (19%) Sisoridae (7%) Others (24%)
Nam Ngum (Lao)	122	Cyprinidae (47%) Balitoridae (12%) Bagridae (7%) Siluridae (6%) Cobitidae (5%) Others (24%)
Nam Mang (Lao)	57	Cyprinidae (47%) Balitoridae (9%) Bagridae (7%) Cobitidae (7%) Others (30%)

Continued

TABLE 8.3 The number of species recorded from Major Mekong tributaries and the Bassac River which is a distributary (and thus better described as part of the main stream)—Cont'd

Basin	Number of species	Dominating fish families
Nam Kading (Lao)	98	Cyprinidae (49%) Balitoridae (14%) Sisoridae (8%) Others (29%)
Nam Songkhram (Thailand)	181	Cyprinidae (45%) Siluridae (7%) Cobitidae (6%) Bagridae (5%) Pangasiidae (5%) Others (32%)
Xe Bang Fai (Lao)	157	Cyprinidae (48%) Cobitidae (8%) Balitoridae (8%) Others (36%)
Xe Bang Hiang (Lao)	160	Cyprinidae (42%) Cobitidae (9%) Balitoridae (8%) Siluridae (6%) Others (36%)
Nam Mun (Thailand)	176	Cyprinidae (46%) Bagridae (6%) Siluridae (6%) Pangasiidae (6%) Others (37%)
Se San (Sesan, Sekong, Srepok) (Cambodia and Viet Nam)	149	Cyprinidae (42%) Balitoridae (17%) Cobitidae (11%) Others (30%)
Great Lake (Cambodia)	167	Cyprinidae (37%) Bagridae (8%) Siluridae (7%) Pangasiidae (5%) Others (44%)
Tonle Sap (Cambodia)	228	Cyprinidae (38%) Bagridae (7%) Siluridae (5%) Pangasiidae (5%) Others (44%)
Bassac (Cambodia and Viet Nam)	155	Cyprinidae (23%) Gobiidae (8%) Bagridae (6%) Cobitidae (5%) Clupeidae (5%) Engraulidae (5%) Siluridae (5%) Cynoglossidae (5%) Others (39%)

example, relatively easy to access because of its vicinity to Phnom Penh, the river is fished intensively and has been subject to scientific studies since colonial times. It is, therefore, not surprising that the highest number of species (296 according to Baran *et al.*, 2007) have been recorded here. Also the Delta and the Thai Basins are relatively easy to access, while the upper reaches of rivers in Laos, and Myanmar are difficult to get to and therefore much less studied and the number of species here are likely much higher than has been recorded, so far.

Eleven fish families occur that each account for more than 5% of the species in a subbasin of the Mekong. The dominating family in all parts of the basin is Cyprinidae with between 23% and 50% of all species. All species of Cyprinidae have low tolerance to salt water and the family, therefore, has the lowest diversity in the Bassac. Hillstream loaches (Balitoridae) are important but only in the Laotian rivers and in the Se San system, accounting for 8-19% of the species. Other widespread families include loaches (Cobitidae) and the catfish families Bagridae and Pangasiidae. It should be noted that the group "others" (i.e., families with less than 5% of the species) is important everywhere and accounts for 24-44% of the species (Table 8.3).

3.4. Deep Pools

Chan *et al.* (2005) defined a deep pool as … *"significantly deeper than surrounding areas and holds water in the dry season, during which it may become disconnected from the main river. A deep pool is also defined ecologically as being of significance for the conservation of a number of fish species."* Viravong *et al.* (2004) reported that, based on a hydroacoustic survey, most of the deep pools in southern Laos and northeastern Cambodia are best described as canyons, fissures, or cracks in the bottom. They also stated

that deep pool morphology and depth are very variable and seem to be important factors in determining the number of fish seeking shelter in the pools. The fish showed preference for deep pools with serrated rocks with steep, almost vertical, sides. The contour of deep pools may, thus also, be important in determining which and how many species are present.

During a recent survey in Cambodia, 97 deep pools were identified by interviewing local fisherfolk (Chan *et al.*, 2005). Viravong *et al.* (2004) surveyed several of these pools using hydroacoustic equipment and found one of them to be 79 m deep. In southern Laos, fisheries officers report around 70 pools (Poulsen *et al.*, 2002). Deep pools are also present further upstream along the Thai-Lao border but have not been systematically surveyed.

Poulsen *et al.* (2002) listed 53 species that fishers have stated reside in deep pools during at least part of the year. Viravong *et al.* (2004) and Baran *et al.* (2005), from catch data, identified a wide range of catfishes (Pangasiidae, Siluridae, Bagridae, Bagriichthidae, and Ariidae), an algae-eater (Gyrhinocheilidae), and a number of cyprinids being caught in deep pools. Deep pools thus seem to serve as a dry-season refuge to a large number of species, and the absence or presence of these is one of the most important factors in determining species composition in a particular reach of the Mekong River mainstream (Poulsen and Valbo-Jørgensen, 2000). However, only the small-scale croaker (*Boesemania microlepis*) seems to be a definite deep pool species (Baird *et al.*, 2001). It is interesting to observe that the sheatfishes (Siluridae) that are among the most abundant deep pool species are morphologically similar to the electric knifefishes (Gymnotidae) that are abundant in deep pools in some South American rivers (Lundberg *et al.*, 1987). However, no Mekong species are known to possess electric organs similar to the ones found in the South American family.

3.5. Floodplains and Other Wetlands

The total area of wetlands (including all seasonally inundated land) in the Lower Mekong Basin is estimated at 185,000 km² (Hortle, Chapter 9), with additional areas inundated in some years when the Mekong flood extends into the Saigon River catchment. Aquatic habitats are extremely diverse throughout the basin and include rice fields (the most extensive habitats), rivers and their floodplains, natural swamps, lakes and reservoirs, and in the delta, canals, intertidal mudflats, and mangroves.

As a result of growing populations, more natural floodplains are being converted into rice paddies. In addition, rain-fed rice cultivation is extending the effective upper boundaries of floodplains as land has been cleared and leveled on terraces to create rice fields.

Floodplains are not evenly distributed in the basin. Along the Upper Mekong, floodplains are few and small. Plains are found only on the Plain of Jars in Laos and in association with the large tributaries Mae Kok and Mae Ing in Thailand. Along the Middle Mekong, flooding is much more intense, but floodplains along most of the mainstream are still small. The plains along the large western tributaries (Nam Songkhram and Nam Mun on the Khorat Plateau) are extensive, while floodplains on rivers with origin in the Annamites are much smaller. Most of the floodplains are found from Kratie to the South China Sea, particularly around the plain of reeds on the border between Cambodia and Viet Nam, and along the Tonle Sap and Great Lake enormous areas are deeply flooded every year. In June or July, the Tonle Sap reverses its flow and the Great Lake starts filling up, and its surface expands up to six fold each year from about 2500 km² in the dry season to 15,000 km² during the flood (Campbell, Chapter 10; van Zalinge et al., 2004). The duration of the flood varies

considerably among different parts of the basin. In the tributaries of the Middle Mekong, the floodplains are only inundated for 1-2 months, while in the Lower Mekong and in the Delta the floodplains may be under water for 6 months or more.

Along the Mekong and Tonle Sap Rivers, multiple channels and man-made canals cut through the levees and allow the water to enter the floodplains before the banks are overflowed. As water levels peak, the flow in the canals slows and eventually reverses as river level drops. These places provide important entry points for fish migrating onto the floodplain for spawning or feeding and for drifting eggs and larvae to become dispersed there.

Most medium to large Mekong fish species exhibit some degree of migratory behavior and utilize the floodplain during at least part of their life cycle typically for feeding and as nursery grounds for their juveniles. The large number of species found in floodplains is thus basically a subset of the species found in the main river channel. A range of the larger species takes advantage of access to direct inputs from terrestrial food sources such as insects, seeds, and fruits. At the same time, there is a boom in primary and secondary productivity in the warm shallow water which has abundant nutrients from the flood waters as well as from decomposing terrestrial plants in the newly flooded areas generating plenty of food for juveniles and small fish species, which in turn serve as food for piscivores (Poulsen and Valbo-Jørgensen, 2000). Production on floodplains is the major contributor to overall river productivity including being the main biological production source for species migrating through river channels. The flood cycle generates a dynamic environment where availability of, and access to, food and shelter continuously changes, favoring different sets of species at different times; preventing any one species becoming permanently dominant.

Small permanent floodplain water bodies such as marshlands and swamps are dry-season refuges and important for the recolonization of large areas during the flood. These water bodies are home to numerous small species, including *Boraras micros, Indostoma* spp., dwarf pipefish, *Nandus* spp., small anabantids, *Betta* spp., *Trichopsis* spp., and small gobies (Welcomme and Vidthayanon, 2003). Although fishes remain isolated during the dry season, many water bodies become connected during the flood which limits speciation between them.

Rice fields are habitats created by humans but rice is naturally a plant growing on river floodplains and in swamps, and the plant has been cultured in Asia for thousands of years. However, as a result of growing populations, more and more natural floodplains are converted into rice paddies and they now constitute one of the most widespread aquatic habitats in the Mekong with an estimated 10 million ha of wet-season rice (Dao Trong Tu *et al.*, 2004). Low-intensity cultivation with minor inputs of fertilizer and pesticides still allows fish to utilize rice fields the same way as they use the natural floodplains. However, fewer niches are probably available in the rice and there is less shelter to seek protection from predators. Nevertheless, rice fields flooded by the river usually still have a large number of species present. More than 80 species were recorded in surveys of Cambodia and Laos (Halwart and Bartley, 2005). Some of the most common species are walking catfish *Clarias batrachus*, snakeheads (*Channa* spp.), swamp eels (*Monopterus albus*), labyrinth fishes (Anabantidae, Osphronemidae, and Belontiidae), and a variety of small cyprinids such as *Rasbora* spp. However, juveniles of larger growing species also use the rice fields as nursery areas. Some large species have even been reported to spawn in the rice fields, including *Wallago attu* and *W. leeri* (Poulsen and Valbo-Jørgensen, 2000; Valbo-Jørgensen and Poulsen, 2000).

3.6. Highland Streams

In the numerous Mekong tributaries in Yunnan and in the upper reaches of tributary basins in Laos with origins in the Annamites, many species evolved in isolation. These species are adapted to high altitudes and are sometimes confined to single streams or caves. As a result, the mountain regions contain many endemic species. Water here is clear, cold, and oxygen rich, and the streams tend to be rocky with many rapids, and the current strong and turbulent as a result of the high gradient. The fish present mostly feed on algal films on stones and rocks and invertebrates.

The most species-rich families in highland streams are rheophilic primary freshwater species of the families Cyprinidae, Balitoridae, Cobitidae, and Sisoridae. A number of species possess special adaptations to survive in the torrential streams. Such species include the Balitorid genera *Homaloptera* with enlarged pectoral and pelvic fins, and *Sewellia* where the enlarged ventral fins together with the flattened belly form a ventral sucking disc. Such fish cling to rocks and stones, thereby allowing them to live in faster flowing water. The algae eaters (Gyrhinocheilidae) attach themselves to solid objects by using the mouth as a sucker, thereby preventing normal respiration through the mouth. Instead, they possess a small hole above the gill opening, leading into the gill chamber, through which they can inhale water for respiration.

Many species of sisorid and torrent catfishes (Amblycipitidae) can only be found in rapids here. Fish in the sisorid genus *Glyptothorax* possess small grooves under the head that constitute an adhesive apparatus (Thomson and Page, 2006) that enables the fish to stick to stones, etc., preventing them from drifting with the current. Freshwater species of gobies (Gobiidae) are often found in sections with very strong current, where they cling to stones and rocks with a sucker formed by the fused pelvic fins.

Given that many of these areas are difficult to access and some have never been surveyed, many additional and new species are likely to be found here.

3.7. Caves

There are a number of limestone caves in Laos and they probably harbor several species of fish that are unknown to science. Cave fishes tend to have a very limited distribution as they normally do not venture out of their habitat and they are sometimes restricted to a single cave system (Kottelat and Whitten, 1996). Unfortunately, this fauna is almost unstudied in the Mekong Basin, but at least three species of cave fishes *Poropuntius speleops*, *Troglocyclocheilus khammouanensis* (Cyprinidae), and *Schistura kaysonei* (Balitoridae) have been found. They are all either blind or have much reduced eyes showing that they evolved in the caves.

3.8. Lakes

True lakes are found in the upper parts of the basin including Yunnan where six lakes with a total surface area of 273 km² are located (Chen Kelin and Li Chun, 1999; Zhou Bo, 1999). These lakes are relatively species poor, Cibuhu Lake and Jianhu Lake both have seven species and the largest, Erhai Lake (250 km²), has 18 species (Yang, 1996).

Many small water bodies are found in the mountains between Laos and Viet Nam but nothing is known about their fish faunas. In the Central Highlands region of Viet Nam, the largest natural lake is Lak Lake, which is a shallow water body with an area of 600 ha. Some 49 species of fish live in it (Viet Nam Environment Protection Agency, 2005), but several of them are probably exotic because culture-based fisheries enhancements is the most popular way to manage fisheries in that region.

The Great Lake of the Tonle Sap is the largest water body in the system with a dry-season area of 2500 km² and it is the largest lake in SE Asia. The fish stocks in the lake are replenished in each flood season from three different sources: (1) the resident fauna that survives in the lake and in water bodies on the floodplain and spawn during the flood season; (2) migratory fishes (including some marine visitors) that enter the lake at the onset of the flood to feed and spawn along the shores and in the floodplains; and (3) eggs and fry from fish that spawn in the Tonle Sap River and upstream in the Mekong River that are carried into the lake by the incoming flow. Because of the connection with the Tonle Sap River, and thus the Mekong River, most of the species occurring there will likely eventually be recorded from the lake as well (Campbell *et al.*, 2006). More species can be anticipated to occur near the Tonle Sap River than in the northern part of the lake where no large rivers are present.

Of the confirmed records of 167 species of fish from the Great Lake (Table 8.3), only a little dragonet species, *Tonlesapia tsukawakii*, which was described recently, is endemic to the lake (Motomura and Mukai, 2006). Few fish species, all of marine origin, seem to be permanently restricted to the open lake including clupeids, croakers (Scianidae), and tongue fishes (Soleidae) (Lamberts, 2001). All the other species are mainly found in the flooded forest.

3.9. Reservoirs

About 25,000 small reservoirs have been created as water storages for irrigation purposes in the Mekong region (Sverdrup-Jensen, 2002). A number of large hydropower reservoirs have been formed on several large tributaries, including the Nam Ngum, the Mun, and the Se San Rivers and several are at various stages of planning and construction. For Laos alone, more than 60 dams (including five on the

Mekong mainstream) are at various stages of construction or planning (Powering Progress, 2007). Until now, only two mainstream dams have been built in China but several more are planned for in that part of the basin. Reservoirs have thus become important fish habitats in all the basin countries. However, as only few Mekong species are adapted to lacustrine conditions this environment is not optimal for them unless they are able to access flowing water periodically. It is especially the lack of suitable spawning grounds that prevent many species to establish themselves in the reservoirs, although this is partly mitigated where tributaries to the reservoir allow them to move to suitable spawning grounds upstream.

There are important differences between shallow small reservoirs and deep hydropower reservoirs. In deep reservoirs, most of the water mass may be completely devoid of fishes. The water is normally stratified and the bottom layers are hypoxic or anoxic and no fishes are able to live there, and few species in the river system are adapted to a pelagic life. Exceptions to the latter are the little clupeids, the Thai river sprat (*Clupeichthys aesarnensis*), Sumatran river sprat (*Clupeichthys goniognathus*), and the glass perch *Parambassis siamensis* (Chandidae) that thrive in the open water of many reservoirs, although none of these seem to be as abundant in the river. In some cases, the predatory cyprinids of the genus *Hampala* also become pelagic in reservoirs feeding on the clupeids (Welcomme and Vidthayanon, 2003). Most of the diversity in reservoirs is found along the shores and among old tree trunks where these have not been removed before reservoir filling. However, oscillations in water level resulting from the drawdown make the shoreline a dangerous place where the fish may be trapped and their eggs or juveniles destroyed.

Reservoirs are thus relatively species poor, and because stocking with alien species better adapted to this environment is the preferred way of compensating for the species lost

and enhancing fisheries throughout the basin, introduced species such as tilapia (*Oreochromis niloticus*), common carp (*Cyprinus carpio*), and Chinese and Indian major carps constitute a disproportionately large part of the fish species in reservoirs (Welcomme and Vidthayanon, 2003).

4. LIFE HISTORIES

4.1. Reproduction and Reproductive Migration

In tropical lowland areas, fishes are usually divided into guilds based on their migratory patterns: black fish, white fish, and sometimes an intermediate group of gray fish. A few species remain within the river channel and rarely move on to the floodplain (Taki, 1978). These include some large predators such as the drumfish, *B. microlepis* and the large predatory cyprinid *Aaptosyax grypus*. Other fish found in the lowlands include estuarine fishes as well as hill-stream fishes that are forced downstream during dry periods when flows are low.

According to Baran *et al.* (2007), of 296 fish species known from the Tonle Sap system the migration guilds of only about one-third are known. However, about 200 species are caught in the dai fisheries of the Tonle Sap and a similar river channel fishery in Tonle Touch in southern Cambodia (MRC data; Ngor *et al.*, 2005). It seems likely that most of these fishes are migrating downstream so would be categorized as white fish. Of the river fish recorded by Taki (1978) in the Middle Mekong based on repeated sampling, about half were considered migratory and half sedentary.

4.1.1. White Fish

Because of the large fluctuations in water level in the Mekong Basin, many Mekong fish species exploit different habitats in different seasons, and are thus migratory. Those that do so over longer distances, for example, from floodplains into and along main river channels, are termed "white fishes." These can be grouped into three main categories based on their migratory patterns: anadromous, catadromous, and potamodromous. Although the migrations appear to be obligatory in some species, in some they may be facultative with the fish simply responding to changes in living space and food caused by the annual flood patterns.

Anadromous fishes live most of their adult life in the sea, but must enter freshwater to spawn. Examples from the Mekong included the pangasiid catfishes *P. krempfi* (Hogan *et al.*, 2007) and *Pangasius mekongensis* (Gustiano *et al.*, 2002).

Catadromous fishes spawn in the brackish or saline water in the estuary or the sea; fry or juveniles enter freshwater where they grow until they are ready to return to the sea. Examples from the Mekong include the giant mottled eel (*Anguilla marmorata*, Anguillidae) and sea-bass or barramundi (*L. calcarifer*, Centropomidae).

Potamodromous fishes are the most important group in the Mekong, these fishes live their entire life in fresh water but migrate, often for long distances, within the river system in order to spawn, feed, or seek refuge. Potamodromous migrations can be further divided into longitudinal and lateral movements. The longitudinal migrations are along the main river channels, while the lateral migrations are from the main rivers into floodplain areas. Some species migrate both longitudinally and laterally (e.g., a longitudinal migration to spawning grounds followed by a lateral migration into feeding areas).

In a comparative study of the larval drift in the Mekong and Bassac branches, 127 species of fish belonging to 28 families were identified, of which most species were Cyprinidae, Siluridae, Gobiidae, Pangasiidae, and Bagridae (Nguyen *et al.*, 2001). The most common and abundant species were the cyprinids *Henicorhynchus* spp. and *Paralaubuca riveroi*, and the pangasiid catfishes *Pangasianodon hypophthalmus* and *Pangasius macronema*.

Most of the Mekong species spawn in a restricted period in the beginning of the flood, known exceptions to this rule are the *Probarbus* species, *Hypsibarbus malcolmi*, and *B. microlepis* that all spawn during the dry season; the former species in shallow water in the mainstream (Roberts and Warren, 1994), the latter two in deep pools (Baird and Phylavanh, 1999).

At the onset of the flood, most migratory species move upstream to spawn (Bouakhamvongsa and Poulsen, 2001) which takes place while the water level is still increasing ensuring that eggs and larvae are carried by the water into nursery areas on the floodplain. Observations on phenomena that correlate with fish migration were reviewed by Baran (2006) who referred to them as "triggers", a term not universally accepted. He found that one or several such phenomena had been identified for 30 species (Table 8.4). It is probable that the actual triggers are chemical cues, because water quality changes in a predictable way at the onset and during the annual cycle of flooding, and river fish have well-developed senses of smell and taste (Lucas *et al.*, 2001).

The actual spawning grounds for most Mekong fish species have still not been identified. Some species, such as the pangasiids, are believed to spawn in rapids in the main stream. The large quantities of ripe fish, of a variety of species, that are moving into many of the tributaries in

TABLE 8.4 Factors correlating with the timing of fish migrations, that have been identified for those Mekong fish species reviewed by Baran (2006)

Stimulus	Number of fish species responding
Water level and current	26
First rains	9
Changes in turbidity	9
Appearance of insects	3

Lao PDR, Thailand, and Northern Cambodia probably are spawning there. From the spawning grounds, eggs, larvae, and fry drift downstream with the current until they reach the floodplain areas that constitute the main feeding grounds.

After spawning, the spent adult fish also move into the flooded areas. During the flood season, the fish are feeding intensively in the flood zone, growing and building up fat layers for the following dry season, which is a time of starvation for most fish. When the water level falls and the floodplain dries up most fish leave the floodplain, and seek refuge in permanent water bodies. Of particular importance as dry-season refuges are deep pools (Poulsen and Valbo-Jørgensen, 2000; Poulsen et al., 2002). The fish following this generalized pattern thus utilize three distinct habitats: spawning and feeding grounds, and dry-season refuges. The existence of these three key habitats plays a decisive role in identifying the type of species and its span in a particular reach of the river (Poulsen and Valbo-Jørgensen, 2000).

In the Mekong, there may be hundreds of species of white fishes. They include many species of cyprinids, pangasiids and large river loaches which are the targets of river fishers, but although they are important to the fishery only a few species dominate the catch at any given locality.

4.1.2. Black Fish

Fish that reside permanently in swamps, marshes, canals, ponds, and similar environments, or use them for dry-season refuge, are known as black fish. In such places, high temperatures and low-oxygen levels prevail. This prevents more sensitive species from surviving. Many black fish possess accessory breathing organs, such as the labyrinth organ in labyrinth fishes (Anabantoidei), which allows these fish to breathe atmospheric air. The swamp eels are similarly capable of supplementing their oxygen intake by breathing air through an accessory respiratory organ, formed by two lung-like sacs that originate from the gill chamber. The featherbacks are capable of breathing air by inflating the swimbladder, and they come to the surface to breathe from time to time. Snakeheads and swampeels are among the hardiest species, and may survive deeply buried in the mud if their water body dries up. They may also crawl over land in search of a more suitable habitat as does the climbing perch (Anabas testudineus) and the walking catfish (C. batrachus) (Smith, 1945).

Black fishes are opportunistic breeders, and may have an extended spawning season. Some are able to reproduce all the year, but peak-spawning activity is normally during the flood season. The brood size is normally small but survival chances increased through various types of parental care that have developed in several unrelated fish groups: the swamp eel M. albus and several labyrinth fishes build a froth-nest; snakeheads construct a nest of vegetation; the male clown featherback (Chitala ornata) aerates the eggs, and keeps them free of sediments by fanning water over them with his tail (Smith, 1945). Clown featherback and snakeheads also defend eggs and juveniles against potential predators (Smith, 1945).

There is a relatively small number of species in the Mekong that are classified as black fishes, but they are disproportionately important to the fishery on the floodplain and in rice fields. They are valued by people there because they can live for days in little water, so they are easy to transport and require no preservation, as a result they are common in fish markets.

4.1.3. Gray Fish

Gray fish constitute an intermediate group that migrate into the floodplain for breeding and feeding at high water, but mainly seek shelter in the adjacent main river channel during the dry season. However, in some of the species, part of the population may stay on the floodplain while part moves to the river

(Chan *et al.*, 2001). They are less capable of surviving under extremely low-oxygen levels than the black fish, but do not have elaborate repeat spawning reproductive behaviors. They include some cyprinids and catfishes as well as some featherbacks, glassfish (Ambassidae), and needle fish (Belonidae).

4.2. Feeding Guilds

Out of the 391 species for which details on feeding habits are available, about half are carnivorous (Fig. 8.2). Insects, both aquatic and terrestrial, are clearly of great importance as they are reported to be eaten by the greatest number of fish species. To some extent, this apparent importance may be a result of dietary studies often being based on larger fishes. Larval and juvenile fishes typically feed on plankton, particularly in flooded areas, switching to larger items (including insects) as they grow. Fish is also an important food item, and several species feed on fins and scales and may be considered as parasites (Table 8.5). Among the more curious examples is the cyprinid *Luciocyprinus striolatus* that has been reported to feed on monkeys (Roberts, 2004). About 37% of the species are feeding on both animal and plant matter, while only 8% are strictly herbivorous.

TABLE 8.5   Food items recorded in the diets of Mekong fishes (from Valbo-Jørgensen, 2003)

Food item	Number of species
Insects	201
Fish	149
Algae	90
Insect larvae	80
Zooplankton	75
Detritus	69
Worms	67
Shrimps/prawns	57
Phytoplankton	48
Periphyton	44
Fruits	31
Crabs	25
Snails/gastropods	20
Terrestrial plants	12
Scales	11
Aquatic macrophytes	11
Fins	10
Scavenge	9
Flowers	6
Frogs	5
Bivalves	5

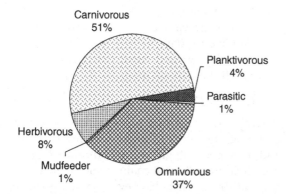

FIGURE 8.2   Proportion of Mekong Basin fish species with various diets.

The large numbers of species that feed on algae, phytoplankton, and periphyton probably reflect that these are the only food items available during the dry season. During the dry season when water levels drop, the water becomes less turbid and dense growths of filamentous algae develop in some parts of the river (Viravong *et al.*, 2004). The Mekong Giant catfish lacks teeth in adults (Roberts and Vidthayanon, 1991) and is thought to be herbivorous feeding on such algae. During the flood, a number of large species enter the forest to feed on fruits including several *Pangasius* species, the elephant ear gourami (*Osphronemus exodon*), and cyprinids like *Tor* spp., and mad barb (*Leptobarbus hoeveni*).

5. SIZE DIVERSITY

The fish species of the Mekong range from minute species such as *Sundasalanx mekongensis* (Sundasalangidae), *Oryzias* spp. (Adrianichthyidae), and various gobies and cyprinids, all growing to only a couple of centimeters and some become sexually mature at a length of less than 15 mm. The smallest species tend to be of marine origin as was also observed by Roberts (1972) in the Amazon.

On the other hand, some of the largest freshwater fishes in the world are also present including the 3-m-long Mekong giant catfish (*P. gigas*), the enormous stingray (*Himantura chaophrya*), which reportedly can attain a weight of 600 kg, and the giant barb (*Catlocarpio siamensis*) which is the largest cyprinid species in the world growing to a size of close to 3 m. Some of the marine visitors grow even larger (Fig. 8.3). The fish fauna thus spans over three orders of magnitude in length—as is normally seen in large rivers (Welcomme, 1985).

6. GENETIC DIVERSITY

The few Mekong fish species that have been studied show a high level of genetic diversity, which suggests the presence of distinct populations or stocks (Hara *et al.*, 1998; Na-Nakorn *et al.*, 2006; Sekino and Hara, 2000; So, 2006; So *et al.*, 2006; Takagi *et al.*, 2006). High intra-specific diversity is due to limited gene flow between populations, which is common amongst riverine faunas where populations in individual sub-catchments, or areas, become relatively isolated from each other due to a mix of behavioural and ecological factors. This phenomenon is enhanced in the Mekong where several sub-catchments were independent basins until relatively recent geological history (see Carling, Chapter 2). High intra-specific diversity means that management of the fish must consider genetic diversity. For capture fisheries each stock should be managed separately, as the likelihood of population crashes is higher as a result of smaller effective population sizes. High intra-specific genetic diversity is an advantage in aquaculture, because wild heterogeneous populations are the main sources of genetic material for improving farmed breeds. But the culture of

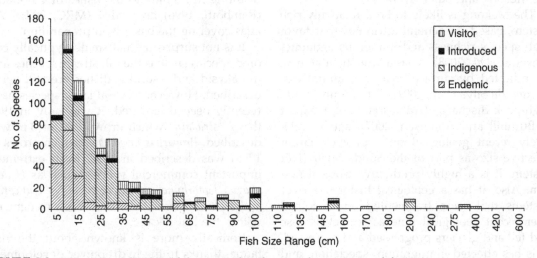

FIGURE 8.3   The number of Mekong fish species plotted against the maximum adult size in cm.

native species must consider the impact of releases or escapes on wild genetic diversity. The introduction of alien genes through aquaculture is a major cause of genetic erosion. Unless adequate information and management is in place, it should not be assumed that the culture of natives is less risk prone than that of alien species.

7. DETERMINANTS OF DIVERSITY

The number of fish species in the Mekong system appears to be high after size (as indicated by catchment area) is taken into account (Welcomme, 1985, p. 94). Kottelat and Whitten (1996) suggested that the Mekong system is a biodiversity "hotspot" and Groombridge and Jenkins (2002) identified it as globally important. However, comparisons between river systems are somewhat tenuous because recorded counts of species from different systems are not of equivalent accuracy nor do they cover the same ecological groupings, and different authors treat marine vagrants differently. Moreover, new species are being described and additional species recorded regularly from the Mekong and other rivers.

The Mekong is likely to be a relatively rich system, based on general attributes that favor high species richness and numerous endemics (Lévêque et al., 2008). It has a large habitat area, as indicated by surface area and annual discharge (de Silva et al., 2007), with an unusually high-peak discharge (Adamson et al., Chapter 4; Burnhill and Adamson, 2007); and in relatively recent geological time an even larger effective size as part of the Sunda Shelf river system. It is a highly productive tropical system. Also, it has a geological history of river systems that were fragmented and rejoined over recent geological time as sea levels rose and fell and glaciers progressed and regressed. This has affected immigration, speciation, and

extinction rates and allowed the Mekong species to evolve in a wide range of habitats while variation in hydrology and the variety of habitats allowed the persistence of many species which required different conditions (Rainboth, 1996). Finally, the orientation of the river system, running north-south, allows the resident fauna to move latitudinally in response to glacial progression and regression (see Oberdorff et al., 1997).

Probably, only the Amazon River system with more than 3000 fish species estimated (Dr Sven Kullander, personal communication) is richer in species than the Mekong. On a per unit area basis, the Mekong is in fact richer because the Amazon is more than eight times larger.

8. RESEARCH

During colonial times, French researchers studied the Tonle Sap and the Great Lake intensively. Prior to the American war in Viet Nam, substantial American sponsored research also took place especially in Laos. In the 1990s, the MRC and FAO began supporting fish research resulting in a guide to the fishes of Cambodia (Rainboth, 1996) and MFD (MRC, 2003). An atlas covering the basin is in preparation.

It is not surprising that small cryptically colored species such as the hill stream loaches and the akysid and sisorid catfishes are still being described. However, several large species have recently been discovered. It was only in 1986 that L. striolatus which grows to 2 or 3 m was described, Probarbus labeamajor which grows to 1.5 m was described in 1992. Even extremely important commercial species, such as Helicophagus leptorhynchus, Belodontichthys truncatus, and P. mekongensis have only been recognized within the last decade.

Ironically more is known about the fish fauna, thanks to the hydropower development which has taken place in the region. Taxonomic surveys in remote areas that were previously unstudied have been undertaken as part of hydropower environmental impact assessments (EIA). As a result Kottelat, for example, discovered 64 new species in Laos (Kottelat, 2000). However, as many of the species discovered in this way are endemic to the impacted basins, they are threatened. Worse, many rare species may not be collected at all during a short sampling effort in connection with an EIA. Hydropower projects are likely to continue to be the main supporters of taxonomic surveys in the years to come as most of the countries do not have the necessary human resources and their financial situation does not allow them to support this type of research which is mostly considered "redundant."

9. SOCIOECONOMIC VALUE OF FISH BIODIVERSITY

The riparian countries in the Mekong Region are often pictured as rice-farming economies, but the population of the Mekong Basin is culturally as diverse as its geography. There are more than 100 different ethnic groups present, all of them with ancient traditions that influence the way they use natural resources. Rural communities are highly diversified, incorporating a wide range of resource uses, and demonstrating high levels of seasonal and annual adaptation to their environments. The ubiquity and high diversity of aquatic organisms combined with the low cost of fishing gears made from local materials opens fishing as a part- or full-time occupation for large numbers of people. The capture and collection of a variety of aquatic animals and plants for a range of domestic and small-scale commercial purposes is among the most important ways by which poor people make use of common pool resources. Communities living

on the riverbanks and floodplains, and in floating homes, as well as the rice farmers are all heavily dependant on these resources. Here, fishing becomes an integrated part of the livelihood for entire communities. For most fishers, fishing is not the only occupation, and for many it is not the main business. But practically, the entire rural population of the Mekong is dependent in one way or another on living aquatic resources caught from the wild. For the poorest, such as landless people, fishing may be the only option during a large part of the year. The Mekong freshwater fishery underpins food security in the region.

The diversity of fish species, habitats, and seasonal influences leads to a highly diverse fishery where the diversity of methods of exploitation rivals that of the species themselves. This level of diversity promotes wide participation in the fishery across all age, gender, and ethnic groups. Thus, the high diversity of fishes in the Mekong contributes considerable socioeconomic benefits which are spread reasonably equitably among the population and, in particular, allow livelihood opportunities for underprivileged groups.

Fisheries take place in all water bodies: rivers and streams of all sizes and their associated wetlands many of which are seasonal. In densely populated areas, the waters of even the smallest ponds may be fished several times every day. As we have seen above, each type of habitat has its own community of aquatic organisms, which results in unique patterns of resource use by local people (Meusch et al., 2003). Fishing takes place when the opportunity arises (time or availability of fish). Women and children, for example, fish and collect other aquatic animals near their homes or set small traps while harvesting rice or other aquatic plants. Rice paddies constitute a multipurpose resource that are utilized for both rice production and capture fisheries. Seasonal integration of fishing with other activities constitutes the main livelihoods for millions of people throughout the Mekong Basin.

The agricultural calendar and the seasonality in abundance of fish are among the most important features governing local lifestyles. The variety of species ensures a spread of the risk involved with the exploitation of natural resources; in countries with weak social-security systems, the open access fishery provides a safety net that allows many people to survive during times of unemployment.

In the Mekong Basin, people use an extremely diverse collection of modern and traditional artisanal fishing gears such as traps, spears, and bows and arrows. In Cambodia, 150 gear types have been documented (Deap et al., 2003), and a similar diversity can be found in the other riparian countries. Traditional gears are often highly selective in terms of species and some of them have been developed over many generations to match the behavior of the fish. They may require considerable local knowledge to operate, as they are often closely adapted to the local environment. Sometimes, they can only be used under particular hydrologic conditions. Boats, large expensive gears, and more intensive methods may be necessary for fishing in larger rivers, but habitats like flooded forests, swamps, rice fields, and streams can be fished using low-intensity smaller scale gears.

Meusch et al. (2003) recorded almost 200 species of aquatic plants and animals that are frequently used by villagers. Apart from fin-fish, other aquatic animals being used include crabs, shrimps, frogs, shellfish, turtles, and insects—in some cases, these animals are as important for household consumption as fish (Meusch et al., 2003). The catch is often separated as the valuable species for sale and the remaining for own consumption. This often manifests itself in the large individuals selected for commercialization and small growing species for food. Under high fishing pressure, the fishing down process often means that only small opportunistic and prolific species are available for local consumption.

Fish abundance fluctuates as a result of reproduction and migration cycles and although fish are available all year round, their catchability varies. Fish are most numerous during the high water season, but they are dispersed in a large volume of water and are therefore more difficult to catch. The fish become much more vulnerable during migration periods especially when they have to pass narrow or shallow points, and at this time many people fish, including occasional fishers who may only fish at this time of the year. The fishers target the channels connecting the floodplain with the main stream, tributaries to reservoirs and lakes and calmer zones below waterfalls or rapids where fish tend to rest. Individual success depends on the experience and knowledge of the fisher. Some fishers follow the fish for hundreds of kilometers. Along mountain streams and river reaches without floodplains, fishing may be confined to the period when fish are moving through the area. Communities living here thus rely on habitats and ecosystems located elsewhere in the system (Mollot et al., 2003).

During the dry season the fish seek shelter in a few refuges which are vulnerable to fishing. In depressions on the floodplain, fish and other aquatic animals become increasingly concentrated in a decreasing volume of water as the dry season advances. The fish hiding here are thus relatively easy to catch and capture may be assisted by draining or pumping. Fish and other aquatic animals burying in the mud may be dug out. In deep pools in the main stream, it may require specialist knowledge and specially designed fishing gears to catch the fish (Viravong et al., 2004)—although the illegal use of explosives may kill large amounts of fish indiscriminately.

In the species-rich Mekong, the majority of the biomass of the total fish catch in any particular locality usually comprises 10 or fewer species (Hortle, Chapter 9). The role of diversity in sustaining total production is not well understood. Since the loss of energy at each step in a natural food chain is usually about 80-90% (Odum, 1971), "fishing down the food chain" by increasing fishing pressure can lead to increased total yields in a fishery, but reduced catches of larger more valuable species and decreased catch per fisher result in significant changes to the economic benefits of the fishery (Welcomme, 2001). Yields can also be increased in some systems by increasing nutrient loading and stocking with species that feed on algae. The highest yielding systems in inland waters are small eutrophic lakes stocked with a few species of plankton-feeding fishes. However, such fish are often of relatively low value per unit weight and the distribution of benefits to part-time fishers is frequently reduced as the fishery becomes commercialized. In enriched systems with fewer species, there are also greater risks of diseases, and water quality problems that may affect other uses.

All of the Mekong countries have policies to increase food production. This necessarily involves competition with other uses of water including maintenance of aquatic ecosystems and their fisheries. The conservation and sustainable use of biodiversity has also been identified as a priority by all lower Mekong governments. However, maintaining biodiversity while simultaneously increasing food production to support expanding populations presents a major technical challenge. Better integrated natural resources management is required in order to achieve these multiple goals. The Mekong River ecosystem provides a multitude of services that benefit its people. Many of these are undervalued and not included in formal economic analyses and subsequent decision making. In addition to fisheries, these include the role of the ecosystem in the regulation of climate, recycling of nutrients and providing sanitation, and the maintenance of the hydrological cycle, including drought and flood mitigation and the provision of drinking water. The best scenario for sustainable fisheries is not through simplistic trade-offs between fisheries and other food production practices, but through more holistic approaches which consider balancing the full suite of benefits that the maintenance of ecosystem integrity and function will bring to the region.

10. MAJOR DRIVERS OF BIODIVERSITY LOSS

Globally, the rate of biodiversity loss from freshwater ecosystems is the fastest of all of the world's major biomes (Secretariat of the Convention on Biological Diversity, 2006). For example, based on published data from around the world, the Living Planet Index (Hails, 2006) aggregates trends of some 3000 wild populations of species and shows a consistent decline in species abundance between 1970 and 2000 of about 50% for inland water species compared to 30% for marine and terrestrial species. This is hardly surprising considering the demands and pressures placed upon freshwater and the extent of conversion of freshwater and coastal aquatic habitat to alternative uses.

Human impacts on aquatic ecosystems in the Mekong Region have been at a low level until recently. However, expanding industrial and agricultural development is gaining momentum and the pressure on aquatic habitats is intensifying with negative consequences for aquatic species and therefore also for fisheries, food security, and human well-being. This development simply mirrors what has happened in countries throughout the world (FAO, 2007). Ricciardi and Rasmussen (1999), for example, found that the loss of freshwater species in North America is comparable to or higher than that of tropical forests species, and that contemporary extinction rates are 1000 times higher than natural levels. But due to the unparalleled diversity of organisms in the Mekong Basin, human population density and growth, the scale of the potential problem is far greater than in most other parts of the world and the consequences of biodiversity loss for human livelihoods severe.

10.1. Hydropower Development

Especially in the Upper and Middle part of the basin, countries have a high-hydropower potential on the mainstream and on major tributaries. However, extensive reservoir and dam constructions will have serious impacts on the aquatic fauna well beyond the actual location of the dam. Dams and weirs fragment the ecosystem by preventing the fish from reaching habitats that are crucial for them to complete their life cycles, and the transformation of the riverine habitat into a lake eliminates part of the fauna. Dams also hinder fish from moving downstream. Allowing fish to travel upstream through fish passes, etc., therefore is unlikely to solve the problem of enabling fish to complete their life cycle unhindered.

Dams on lower order mountain streams may well threaten a number of endemic species with very limited distribution including cave species that may become extinct and species adapted to life in strong current, which in most cases will not adapt to life in a reservoir. Dams on lowland high-order tributaries are more likely to affect species that are widespread in the system. However, the cumulative effect of a series of dams on tributaries may have similar or greater impact.

In the Nam Ngum Basin, the seven-line barb (*Probarbus jullieni*) (a species which is red listed by IUCN and for which trade is regulated by CITES) previously spawned in the Nam Ngum River. Since the Nam Ngum dam was built almost 40 years ago, the species can no longer reach its former spawning grounds, but it continues to migrate up the tributary Nam Lik that joins the Nam Ngum below the dam and is thus maintained within the Nam Ngum watershed. In contrast, the Mun River was dammed near the mouth, since then the migratory species coming from the Mekong were no longer able to ascend the river or any of its subbasins resulting in a considerable loss of biodiversity and fish production. However, after the gates were opened during the spawning season to mitigate the impact many of the species returned (Jutagate *et al.*, 2005).

Apart from physically blocking up and downstream movements of aquatic organisms, dams also cause perhaps more serious ecological disturbances by suppressing or displacing the flood. Lack of flooding prevents the fish and their eggs and larvae in reaching growth areas on the floodplains while excessive flows at the wrong time of the year can wash away young fish or wash drifting fry past the target floodplains resulting in the loss of most individuals. The physiological stimuli that trigger fish to migrate and spawn are also disturbed resulting in the wrong timing of movements and reproduction.

10.2. Navigation

Because of the poor development of rail and road infrastructure in the Mekong Basin, the Mekong mainstream and major tributaries are important arteries that connect distant cities, provinces, and countries. Both people and tradable goods are shipped for great distances up and down the river. Ocean going ships are only able to ascend the Mekong to Phnom Penh, but since French colonial times considerable efforts have been invested in developing navigation further.

Navigation in itself may have little impact on the fish fauna except in cases such as accidental spills of toxic substances such as oil. Some spills would have the potential to terminate all aquatic life over maybe hundreds of kilometers downstream. Depending on the season, the most sensitive areas would be vegetated areas with slow current such as floodplains where sensitive juveniles could be killed in large numbers. In the dry season, there would be a much smaller volume of water to dilute the spill, and the fish stocks are at their lowest, impacts may therefore be greatest at this time.

However, the engineering works involved with improving the water ways may have considerable both short-term and long-term impacts. The removal of rocks and rapids with the use of explosives may kill large quantities of fish and it will destroy spawning grounds and important habitats for specialized fish fauna. If explosives are used at the wrong time—entire populations of fish can be killed because some congregate seasonally in very confined areas.

The removal of rocks and shallows, which slow the current and provide shelters for fish, has negative implications for both the resident fauna and for migratory fish that normally rest behind such rocks. Deepening of the river by dredging the bottom may increase the silt load and lead to the silting up of spawning grounds downstream. If carried out at a sensitive place, dredging may severely affect biodiversity; one example is the Quatre Bras where even small changes in the flow patterns may result in eggs and fry no longer being carried up the Tonle Sap leading to a reduction in the number of species and potentially a substantial decrease in fish production in both the river and the Great Lake (Ngor 1999).

10.3. Land-Use Changes and Deforestation

Shrubs, forests, and mangroves are cleared for many reasons. People cut down the trees for fuel wood, for building houses and boats, and fishers build brushparks. Ground is also cleared for vegetation to allow for planting crops. In the Delta, the mangrove that was already severely reduced during the war is cleared for prawn aquaculture production on a large scale (Binh *et al.*, 1997). Large companies cut not only valuable trees but also destroy large areas of forest in the process. At the same time, they open up the forest for people to settle there. Many companies also violate the provisions stated on their concessions, and

there is little or no control with how these companies act.

The direct consequences of this development are that important allochtonous inputs (insects, fruits, seeds, leaves, etc.) to the aquatic ecosystem disappear or are reduced. Special habitats such as the flooded forest where the fish seasonally gorge on food of terrestrial origin disappear. Not only are such habitats naturally very diverse, but they also harbor some of the highest levels of biodiversity. The loss of mangroves will reduce biodiversity and production in the Delta and beyond as this is an important nursery for the species that enter the estuary and venture further upstream for feeding.

The clearance of vegetation also lead to increased erosion and thus in silt load. As a response, fish communities switch from visually sensing toward species with other means of orientation (such as tactile). The large amounts of sediment can also destroy spawning grounds and deep pools may fill up. Production of phytoplankton and algae that form the basis of the food chain and serve directly as food for many species will also be negatively affected due to reduced light penetration in the water. This impact will manifest itself the most on floodplains, including the Great Lake.

10.4. Agricultural Development

Agricultural development is the mantra of all the riparian countries. However, the loss and degradation of habitat for fish, such as seasonally flooded forests and wetlands, for conversion into rice paddy, may undermine food security if the important contribution of fish and other aquatic animals to local diets and livelihoods continues to be eroded. In particular, the considerable resources of food provided by healthy floodplain ecosystems should not be undermined (Mollot *et al.*, 2003).

The trend in agriculture has been to move away from diversity towards more efficient and intensive monocultures. Intensification of farming requires a controlled environment and high-yield crop varieties that are more vulnerable to pests and thus require more pesticides that are hostile to fishes and other animals living in the fields. Pesticides and fertilizers are currently mainly used in the most affluent areas of the basin and where crop production is most intense. The capacity among farmers to administer the use of these chemicals is generally low. This leads to fish kills and also makes fish toxic to eat.

With intensification of production, wetlands continue to be converted to other uses, even where it is not economically viable. Not only does this result in reduced overall production from fisheries but the fauna that characterizes these habitats are being lost and dry-season refuges for other species disappear as well.

Dikes are constructed along the rivers to prevent flooding of crops and settlements but these also prevent larvae and adult fish from accessing their nursery and feeding habitats on the floodplains, including rice fields. This considerably reduces habitat availability and complexity. In the past, people always considered the seasonal flooding as something positive generating production and prosperity (Petillot, 1911) and rural people still recognize that the seasonal fluctuations are crucial in maintaining their fisheries and livelihoods (Mollot *et al.*, 2003). The traditional varieties of rice that are well adapted to growing on the floodplains can therefore be promoted rather than replaced by modern varieties that do not tolerate deep flooding.

Enormous amounts of water are needed for irrigation. This water is abstracted from the river directly or from irrigation reservoirs. This leads to reduced water availability during the dry season with severe consequences for the fish already struggling to survive at this time of the year. It is of particular concern that less water reaches the Mekong Delta because it increases saltwater intrusion and

acidification of the water there. Water needs to be used more efficiently including through crop diversification toward less water consuming crops. The recently completed Comprehensive Assessment of Water Management in Agriculture (Molden, 2007) draws attention to the serious problems the world is facing resulting from the use of water by agriculture. It notes that a shift in thinking is required to a perspective of managing the ecosystem services that aquatic systems provide. Although agriculture is an important service, it is one of the many that needs to be considered.

10.5. Mining

Some of the mountainous regions in the basin are rich in minerals, and there is little control over the companies that receive mining concessions. Mining may lead to increased siltation and leachates from the mines may contain toxic minerals and residues of chemicals used during the processing of ore. Accidental spills of chemicals may also be lethal to aquatic organisms. Remote areas are home to many of the endemic species that may be eradicated by mining activities.

10.6. Industry

The absence of major cities in the Mekong Basin has until now spared the region from large-scale development of polluting industries—although this may change in the future. Pollution from industries includes the continuous discharge of effluents, the impact of which will depend on the degree to which the water is treated, diluted, or recycled, and accidental spills. Effluents will gradually diminish biodiversity because the fish that cannot survive in the polluted water will move away; especially sensitive stages such as eggs and larvae may be particularly vulnerable to pollutants. Polluted water may also deter migrating fish from entering certain areas and thus act as a barrier in the same way as a dam. Poisonous spills from factories can kill fish over large stretches of river, as it was seen with the 1992 dry-season spill from a sugar factory, when a 9 km expanse of molasses drifted down the Phong, Chi, and Mun Rivers to eventually reach the Mekong, leaving behind 500 tonnes of dead fish (Roberts, 1993; Sneddon, 2002).

10.7. Aquaculture and Culture-Based Fisheries Enhancements

Aquaculture is rapidly expanding throughout the Mekong Region following the world trend. In the upper countries, most production is based on the pond culture of alien species, especially tilapia. In Cambodia, and especially Viet Nam, production is dominated by cage culture of giant snakehead (*Channa micropeltes*) and *P. hypophthalmus*. About a dozen alien species are regularly bred in the Basin for food purposes (Welcomme and Vidthayanon, 2003) and more are likely to be introduced in the future. Two percent of the Mekong fish species are alien species that have become established, many of them through escapees from aquaculture installations like the African catfish (*Clarias gariepinus*), Tilapia, common carp, and silver carp (*Hypophthalmichthys molitrix*) (Welcomme and Vidthayanon, 2003). However, the riparian countries through the MRC are now exploring aquaculture of indigenous species (Phillips, 2002). There are also a significant number of alien aquarium fishes in the basin and some alien ornamentals are reared commercially in captivity. These activities are completely unregulated and several ornamental species (including *Gambusia affinis*, *Poecilia reticulate*, and *Hypostomus* spp.) are now widely established throughout the basin. Unless strict management is undertaken by the riparian countries, establishment of many more species in the future seems inevitable (Welcomme and Vidthayanon, 2003).

Various problems have materialized in connection with the use of alien fish in aquaculture and for stocking. (Welcomme and Vidthayanon, 2003): Introduced species can disturb habitats and in so doing, alter ecosystem characteristics to such a degree that native species are threatened; common carp, for example, searches for food in muddy bottoms of lakes and rivers and thereby remobilizes sediment and increases biochemical oxygen demand (BOD), this can lead to turbid conditions that reduce light penetration and plankton production (Welcomme and Vidthayanon, 2000); grass carp (*Ctenopharyngodon idella*) feeds on aquatic macrophytes and may change the composition of the aquatic vegetation and therefore also the species composition in water bodies it colonizes (Kottelat, 2001), it may also inflict damage on the rice if it is allowed to enter the paddies. Competition between introduced and native species is another cause of potential difficulty. Competition may be for breeding sites especially in nest-building species, or it may be for food. Another major cause of negative impacts has been the explosive expansion of populations of alien species with rapid growth and short life cycles such as Tilapia (*Oreochromis* spp.) and several cyprinid species. This is often accompanied by stunting, leading to dense populations of small individuals, which compete with, and reduce the numbers of, more valuable species (Welcomme and Vidthayanon, 2000).

The most extreme genetic impact is hybridization. The exotic African catfish (*C. gariepinus*) is capable of hybridizing with the indigenous *Clarias macrocephalus*. Also the two indigenous species *P. hypophthalmus* and the giant Mekong catfish (*P. gigas*) can hybridize in captivity, but they are not known to do that in the wild. Interbreeding between species or hybrids in the natural environment can pose risks, especially if the offspring are not sterile, because valuable adaptive characteristics, such as timing of migration and the ability to locate natal streams may be lost. Alternatively, the hybrid can prove more successful and vigorous than the parents, in which case the latter may become out-competed. Hybridization thus reduces natural genetic diversity. Concerns over genetic disruption are not limited to alien species since moving native strains and varieties of native species around the basin can cause similar effects. Most large rivers have a high degree of genetic diversity within many species groups and this is also apparent in limited data available from the Mekong. Where stocking material, including escapes from aquaculture, is drawn from a wide area and interbasin transfers of varieties occur, the risks of such effects are particularly high. For this reason, broodstock for the production of stocking material must be carefully selected. Material to be stocked should be derived only from parents drawn from the receiving subbasin (Welcomme and Vidthayanon, 2003). Too little diversity resulting from stocking with material derived from too few breeders is another risk which can result in a narrow genetic base leading to rapid degradation of the material used for stocking, which in turn will lead to poor growth and reproductive potential of the species concerned (Welcomme and Vidthayanon, 2003). This risk exists also in the case of stocking undertaken to support wild populations (such as the giant Mekong catfish).

For many species, culture is still based on wild seed. For example, snakeheads have a relatively low fecundity and due to their habit of guarding the young, the fry normally swarm around the adult in the surface rendering them very vulnerable to fishing by fry collectors. In smaller water bodies, snakehead may disappear due to this overfishing. *P. hypophthalmus* is a much more fecund species which is an adaptation to the high mortality of its eggs and young caused by spawning in the main river. The dai fishery for *P. hypophthalmus* fry

in the late 1990s caught about a billion fish a year, with many more fish killed as bycatch (Hortle and Lieng, 2005). Although it was never proved that the fry fishery was responsible for the declining catches of this species, the fishery was banned in both Cambodia and Viet Nam by 2000 and now virtually all aquaculture fish are a result of hatchery production (Poulsen *et al.* 2008). A *Pangasius* hook fishery continues on a small scale in Cambodia to supply fish from local pond culture, with about 3 million fry caught annually (mostly *Pangasius bocourti*) (Hortle and Lieng, 2005).

A large proportion of the aquaculture production in the Lower Mekong, particularly for snakehead and catfish, depends on low-value "trash fish" for feeding. Depending upon the season and locality, a significant proportion of that trash fish can come from the freshwater fishery. The "trash fish" which is used as feed for carnivorous species are caught using small meshed nets and is made up by a mixture of small species from where the more valuable elements have been picked out. Locally, this indiscriminate fishery may put an enormous pressure on fish resources and biodiversity. The tendency to overfeed the farmed fish may contribute to the creation of anoxic conditions at the local level where current is slow, and eutrophication of the river system. Alternative sources of feed are being developed and further improvements on the effectiveness of the utilization of feed need to be made in order to reduce pressure on wild resources.

10.8. Fishing

It is widely but incorrectly assumed that overfishing is the major cause of decline of the Mekong fishery. There is, in fact, no evidence of a decline in total production, but it is clear that populations of certain species especially the larger slower growing high-value species are in decline. For the Mekong species listed in the IUCN Red List, none are listed as in decline due to fishing alone. For most, the major factor determining their status is environmental change including loss and degradation of habitat. Neither is there evidence in the Mekong, nor in fact anywhere else, of fisheries being the primary driver of the extinction of fish species (Allan *et al.*, 2005; Bayley, 1995). Certainly, the fishery needs improved management, but the priority requirement is improved environmental management in order to sustain the fishery and the biodiversity that underpins it.

Fishing intensity is certainly very high even in some of the relatively remote parts of the basin and increased numbers of people fishing leads to catch per unit effort declining leading to further intensification. The fishery also uses unsustainable fishing methods such as explosives, poison, and electrofishing that kill large amounts of fish indiscriminately. Poison can empty small streams and water bodies completely from fish. Explosives are used to access fish in places where they would otherwise be impossible or difficult to reach with other methods such as deep pools where many species of fish, including the broodstock that will spawn the following flood season, are concentrated during the dry season and many of the individuals killed by this method are not even recovered. Another destructive method is the dry pumping of small natural water bodies which allows the capture of all the fish trapped in such ponds toward the end of the dry season thereby killing the fish that would have spawned and thus restocked the flooded areas next season.

The top-down approach to fisheries management currently in use in most parts of the basin is not efficient because of the vast size of the basin compared to the number of staff employed by the fisheries departments. The preferred solution is delegating responsibility for resource management to the communities through comanagement arrangements. However, to be effective this needs to be accompanied by greater influence over the management of the

environment upon which the fishery is based. The current institutional/government approach focuses on overfishing, in particular the failure to control access, which can undermine local management regimes. These regimes invariably, if allowed to flourish, have been quite supportive of biodiversity (Winnett, 1999). Biodiversity conserves the adaptive capacity of the ecosystem giving it an ability to buffer or absorb perturbations including exploitation by fishery dependent communities. Local communities understand such issues well and are likely to use their local knowledge to manage for diversity and sustainability, if they are given the opportunity.

10.9. Climate Change

Globally, climate change is emerging as the greatest driver of biodiversity loss. Certainly, climate change will influence the Mekong in significant ways, both directly, for example, in changes in rainfall patterns and rising sea levels, and indirectly, through shifts in demand and trade of commodities. The exact nature of these changes cannot be easily established for the Mekong and in any event there will likely be a wide margin of variability in predictions.

The life cycles of the Mekong fish species are closely adapted to the rhythmic rise and fall of the water level and changes to this pattern may disrupt many species. Changes in flooding patterns may lead fish to spawn at the wrong time of the year resulting in the loss of eggs and fry. Increasing flash floods may wash juvenile fish and eggs out of their normal habitats thereby increasing chances that they will die from starvation or predation. Prolonged periods of drought will reduce available habitat to the fish especially during the dry season.

What is very clear is that by maintaining the highest possible levels of biodiversity the Mekong River ecosystem stands the best chance of being able to adapt on its own to the changes that are already happening. Climate change is increasing the focus on the crucial role of the services that wetlands provide, for example, in the sustained delivery of freshwater, nutrient recycling, and the mitigation of extreme rainfall events (both droughts and floods), and the role of healthy coastal wetlands in mitigating the damage caused by extreme storms. Using nature's ability to cope with change is a sensible and cost effective response option to climate change and in this process, considerable benefits will also accrue to biodiversity and the fisheries reliant upon it.

While many environmental changes are unavoidable, the Mekong countries have options for how to respond. Maintaining diversity is the key. Appropriate direct responses to climate change are also critical. For example, in areas becoming drier, countries will likely secure their access to water by storing more of it which will escalate impacts on aquatic ecosystems. Climate change considerably increases the urgency and importance of implementing better basinwide environmental management plans which fully promote the optimal use of the full suite of services that the ecosystem provides to people and development.

11. BASIN PLANNING TO SUSTAIN ECOSYSTEM SERVICES

The Mekong River ecosystem provides an enormous range of services which support human well-being. Many of these services, which are underpinned by biodiversity, are greatly undervalued. In many respects, sustaining biodiversity is equivalent to sustaining ecosystem services and therefore sustaining human well-being. The "sector"-based approach to planning and development in the Mekong has resulted in increased service delivery in some areas, for example, food provided through agriculture, or timber through forestry. But this has resulted in other services being negatively impacted, for example, as outlined above, agricultural development can reduce the provisioning of fish through fisheries. To manage the

Mekong River environment better and to achieve more balanced and sustainable development, "ecosystem services" based approaches to policy and decision making need to be adopted instead of sector-based approaches which tend to lead to disparities in service delivery and inequities in benefits across the population. Further details of such reasoning and approaches are provided in the Millennium Ecosystem Assessment (2005a,b). This chapter has highlighted the important role of fish biodiversity to human welfare in the region but the service provided by this component of biodiversity is only one of many that should be managed collectively if the basin is to be sustainably developed.

Historically, national policies in most cases have favored the more visible economic sectors such as agriculture, and other sectors with political leverage. Donors also often preferred to invest in monumental projects with tangible and easily measurable outputs such as dams for flood protection, irrigation, or hydropower. The trouble is that it is difficult to balance the economic interests involved with power generation, navigation, agriculture, and industry because it is not easy to provide solid figures that demonstrate the true economic value of the intact aquatic habitat and its associated fish populations. However, as FAO stresses: "*While the precautionary approach should be applied to fisheries… there is an equal need to apply the approach to non-fisheries sectors whose capacity to damage the ecosystem is usually much greater than that of the fisheries themselves*" (FAO, 1997).

As a part of this process, better valuations of the fishery and greater awareness of the role of biodiversity, coupled with more transparent, informed, and impartial decision-making processes, are required in order to improve planning and management. Informal activities, such as subsistence fishing, that are undertaken outside Government control are poorly documented, and knowledge on their sizes, values, and nature are very fragmentary. For these activities to be taken into consideration when planning, crucial data and information are required. However, it is very difficult to find information about these types of activities through official channels. Most often data are not available or are produced by, for example, NGOs or academic institutions, and there is resistance at Government level to use such sources. Governments may also disregard such data because political priorities prevail.

Rural people, who depend directly on biodiversity resources, have significant knowledge about them and their value, and a greater interest in sustaining them. Thus, they need to be better involved in the decision-making process. However, to achieve this they need to be empowered. Biodiversity loss caused by hydropower, irrigation, and navigation development, for example, have seriously inequitable outcomes for development (Mollot *et al.*, 2003). Only by developing and implementing holistic land-use policies that emphasize user participation and an ecosystem/biodiversity-based approach to management will be possible to succeed in turning around the histories of non-sustainability of aquatic resource use (Parveen and Faisa, 2003; Pullin, 1999).

There are a number of international frameworks that can guide governments toward improving governance of natural resources and in all of these the focus is on sustaining benefits to people. These include the Ramsar Convention on Wetlands, the Convention of Biological Diversity, the Convention on Migratory Species, the Convention on Trade in Endangered Species, and the World Heritage Convention.

12. PROTECTING ENDANGERED SPECIES

A large proportion of the many endemic fish species in the Mekong Basin are endangered, and many of them increasingly so. To protect them from extinction, species-specific conservation measures are needed. While it may be necessary to protect some habitats that harbor many specialized endemics with limited distribution, the historical "protected area" approach to species conservation is not appropriate for many species of fish. Unless they are very carefully planned, single local initiatives will, for example, do little to benefit migratory species that depend on habitats that may be widely dispersed (Valbo-Jørgensen *et al.*, in press). Particularly for river systems, establishing protected areas to limit the impacts of specific local drivers of loss (e.g., overexploitation) will fail if they do not limit other threats (e.g., pollution or over extraction of water). Management interventions for isolated parts of such a complex system are likely to have negative impacts on other parts (Meusch *et al.*, 2003). Maintaining individual fish populations should be subordinate to the goal of sustaining the ecosystem that supports multiple species. As long as all rehabilitation actions are consistent with the overriding goal of restoring ecosystem processes and functions, habitats will be restored for multiple species (FAO, 2007). Key habitats such as spawning grounds, nursery areas, etc., for species with strict requirements for such, for example, the Giant Mekong catfish, must of course be protected. But closing part of a watershed for fishing may instead serve as a pretext for doing nothing else and result in habitat fragmentation (Valbo-Jørgensen *et al.*, in press). Degraded or destroyed habitats should be rehabilitated. A key system-wide requirement is to maintain ecosystem connectivity throughout the basin.

It is impossible to formulate management plans for each and every species. Better progress may be made through the identification of critical habitat types where management efforts can be focused, and by using species with stringent habitat requirements as indicators. At the same time, these can be promoted as "flagship species" and it may be easier to raise awareness about the need for implementing appropriate measures and to attract national as well as international funding for management programs. Large migratory species, for example, have strong requirements for habitat quality and ecosystem integrity, and it will not be possible to address the factors that make them vulnerable in isolation from the rest of the ecosystem (Poulsen, 2003).

References

Allan, J. D., Abell, R., Hogan, Z., Revenga, C., Taylor, B., Welcomme, R., and Winemiller, K. O. (2005). Overfishing of inland waters. *BioScience* **55**, 1041-1051.

Baird, I. G. and Phylavanh, B. (1999). "Observations of the Vocalisations of Spawning Mekong River Goldfin Tinfoil Barb *Hypsibarbus malcolmi* (Smith 1945) in Southern Lao PDR Below the Khone Falls," Technical Report Prepared for the Environmental Protection and Community Development in Siphandone Wetland Project. CESVI, Pakse, Lao PDR, 9pp.

Baird, I. G., Phylavanh, B., Vongsenesouk, B., and Xaiyamanivong, K. (2001). The ecology and conservation of the smallscale croaker *Boesemania microlepis* (Bleeker 1858–59) in the mainstream Mekong River, Southern Laos. *Natural History Bulletin of the Siam Society* **49**, 161-176.

Banarescu, P. (1972). The zoogeographical position of the East Asian freshwater fish fauna. *Review on Roumanian Biology, Zoological Section* **17**, 315-323.

Baran, E. (2006). "Fish Migration Triggers and Cues in the Lower Mekong Basin and Other Freshwater Tropical Systems," MRC Technical Paper No. 14. Mekong River Commission, Vientiane, 56pp.

Baran, E., Baird, I. G., and Cans, G. (2005). "Fisheries Bioecology at the Khone Falls (Mekong River, Southern Laos)," WorldFish Center, Penang, Malaysia, 84pp.

Baran, E., So, N., Leng, S. V., Arthur, R., and Kura, Y. (2007). Relationships between bioecology and hydrology among Tonle Sap fish species. In "Appendix 3: Study of the Influence of Built Structures on the Fisheries of the Tonle. Project Documents and Results." WorldFish Center, Phnom Penh, Cambodia.

Bayley, P. B. (1995). Sustainability in tropical inland fisheries: The manager's dilemma and a proposed solution. In "Defining and Measuring Sustainability: The Biophysical Foundations," pp 321–328, Munasinghe, M. and Shearer, W. (eds.) The United Nations University and The World Bank, Washington, DC.

Berra, T. M. (2001). "Freshwater Fish Distribution," Academic Press, San Diego, USA, 604pp.

Binh, C. T., Phillips, M. J., and Demaine, H. (1997). Integrated shrimp-mangrove farming systems in the Mekong delta of Vietnam. *Aquaculture Research* **28**(8), 599-610.

Bouakhamvongsa, K. and Poulsen, A. F. (2001). Fish migrations and hydrology in the Mekong river. In "Proceedings of the Third MRC Technical Symposium," Phnom Penh, December 2000, Mekong Conference Series No. 1, pp 114-121. Mekong River Commission, Phnom Penh.

Briggs, J. C. (2005). The biogeography of otophysan fishes (Ostariophysi: Otophysi): A new appraisal. *Journal of Biogeography* **32**, 287-294.

Burnhill, T. and Adamson, P. (2007). Floods and the Mekong River system. *Catch and Culture* **13**, 8-11.

Campbell, I. C., Poole, C., Giesen, W., and Valbo-Jørgensen, J. (2006). Species diversity and ecology of Tonle Sap Great Lake, Cambodia. *Aquatic Sciences* **68**(3), 355-373.

Chan, S., Chhuon, K. C., and Valbo-Jørgensen, J. (2001). Lateral fish migrations between the Tonle sap river and its flood plain. In "Proceedings of the 3rd Technical Symposium on Mekong Fisheries," Phnom Penh, 8–9 December 2000, pp 102-114. Mekong River Commission, Phnom Penh.

Chan, S., Putrea, S., Sean, K., and Hortle, K. G. (2005). Using local knowledge to inventory deep pools, important fish habitats in Cambodia. In "Proceedings of the 6th Technical Symposium on Mekong Fisheries," MRC Conference Series No. 5, pp 57-66. Mekong River Commission, Vientiane.

Chen, K. and Li, C. (1999). The wetlands of Mekong Basin in China. In "Proceedings of the Workshop: Wetlands, Awareness, Local People and the Ramsar Convention in the Mekong Basin. Can Local Communities Play a Role in the Wise Use of Wetlands?" Phnom Penh, Cambodia, 12-14 September 1999, pp 19-22. Mekong River Commission, Phnom Penh.

Dao Trong T., Geheb, K., Susumu, U., and Vitoon. (2004). Rice is the life and culture of the people of the Lower Mekong Basin Region. In "Mekong Rice Conference," Ho Chi Minh City, 15-17 October 2004. http://www.mrcmekong.org/download/programmes/AIFP/Mekong_rice_paper.pdf. Accessed 29 May 2008.

Deap, L., Degen, P., and van Zalinge, N. (2003). "Fishing Gears of the Cambodian Mekong," Cambodia Fisheries Technical Paper Series IV. Inland Fisheries Research and Development Institute of Cambodia (IFReDI), Phnom Penh, 269pp.

de Silva, S. S., Abery, N. W., and Thuy, N. T. T. (2007). Endemic freshwater finfish of Asia: Distribution and conservation status. *Diversity and Distributions* **13**, 172-184.

Dodson, J. J., Colombanis, F., and Ng, P. K. L. (1995). Phylogeographic structure in the mitochondrial DNA of a south-east Asian freshwater fish, *Hemibagrus nemurus* (Siluroidei; Bagridae) and Pleistocene sea-level changes on the Sunda shelf. *Molecular Ecology* **4**, 331-346.

FAO. (1997). "FAO Technical Guidelines for Responsible Fisheries No. 6: Inland Fisheries," FAO, Rome, 36pp.

FAO. (2007). "The State of World Fisheries and Aquaculture 2006," FAO, Rome, 162pp. ftp://ftp.fao.org/docrep/fao/009/a0699e/a0699e.pdf. Accessed 29 May 2008.

Froese, R. and Pauly, D. (eds.). (2008). "FishBase." World Wide Web Electronic Publication, www.fishbase.org, version (04/2008).

Groombridge, B. and Jenkins, M. D. (2002). "World Atlas of Biodiversity," UNEP and WCMC. University of California Press, Berkeley, USA, 340pp.

Gustiano, R., Teugels, G. G., and Pouyaud, L. (2002). Revision of the *Pangasius kunyit* catfish complex, with description of two new species from south-east Asia (Siluriformes; Pangasiidae). *Journal of Natural History* **37**(3), 357-376.

Hails, C. (ed.). (2006). "Living Planet Report," 44pp. http://assets.panda.org/downloads/living_planet_report.pdf. Accessed 29 May 2008.

Halwart, M. and Bartley, D. (eds.). (2005). "Aquatic Biodiversity in Rice-Based Ecosystems. Studies and Reports from Cambodia, China, Lao PDR and Viet Nam," [CD-ROM]. FAO, Rome.

Hara, M., Sekino, M., and Nankorn, U. (1998). Genetic differentiation of natural populations of the snake-head fish, *Channa striatus*, in Thailand. *Fisheries Science* **64**, 882-885.

Hogan, Z., Baird, I. G., Radtkek, R., and Van der Zanden, M. J. (2007). Long distance migration and marine habitation in the tropical Asian catfish, *Pangasius krempfi*. *Journal of Fish Biology* **71**, 818-832.

Hortle, K. G. and Lieng, S. (2005). Falling Vietnamese demand eases pressure on wild stock of catfish fry. *Catch and Culture* **11**, 7-9.

Jutagate, T., Krudpan, C., Ngamsnae, P., Lamkom, T., and Payooha, K. (2005). Changes in the fish catches during a trial opening of sluice gates on a run-of-the river reservoir in Thailand. *Fisheries Management and Ecology* **12**(1), 57-62.

Kottelat, M. (1989). Zoogeography of the fishes from Indochinese inland waters with an annotated check-list. *Bulletin Zoölogisch Museum Universiteit van Amsterdam* **12**(1), 1-55.

Kottelat, M. (2000). Diagnoses of a new genus and 64 new species of fishes from Laos (Teleostei: Cyprinidae, Balitoridae, Bagridae, Syngnathidae, Chaudhuriidae and Tertraodontidae). *Journal of South Asian Natural History* **5**(1), 37-82; 73 figs.

Kottelat, M. (2001). "Fishes of Laos," Wildlife Heritage Trust, Colombo, 198pp.

Kottelat, M. and Whitten, T. (1996). "Freshwater Biodiversity in Asia, with Special Reference to Fish," World Bank Technical Paper No. 343. World Bank, Washington, 59pp.

Lagler, K. F. (1976). "Fisheries and Integrated Mekong River Basin Development," Terminal Report of the Mekong Basinwide Fishery Studies. Executive Volume.

University of Michigan School of Natural Resources, Michigan, USA, 367pp.

Lamberts, D. (2001). "Tonle Sap Fisheries: A Case Study on Floodplain Gillnet Fisheries," AsiaPacific Fishery Commission, FAO Regional Office for Asia and the Pacific, Bangkok, Thailand, 133pp. RAP Publication No. 2001/11.

Lévêque, C., Oberdorff, T., Paugy, D., Stiassny, M. L. J., and Tedesco, P. A. (2008). Global diversity of fish (Pisces) in freshwater. *Hydrobiologia* **595**, 545-567.

Lucas, M. C., Baras, E., Thom, T. J., Duncan, A., and Slavik, O. (2001) "Migration of Freshwater Fishes," Blackwell Science, London, 420pp.

Lundberg, J. G., Lewis, W. M., Jr., Saunders, J. F., III, and Mago-Leccia, F. (1987). A major food web component in the Orinoco River channel: Evidence from planktivorous electric fishes. *Science* **237**, 81-83.

McConnell, S. K. J. (2004). Mapping aquatic faunal exchanges across the Sunda shelf, south-east Asia, using distributional and genetic data sets from the cyprinid fish *Barbodes gonionotus* (Bleeker, 1850). *Journal of Natural History* **38**, 651-670.

Meusch, E., Yhoung-Aree, J., Friend, R., and Funge-Smith, S. J. (2003). *The role and nutritional value of acquatic resources in the livelihoods of rural people - a participatory assessement in Attapeu Province, Lao PDR*. FAO Regional Office Asia and the Pacific, Bangkok, Thailand, Publication No. 2003/11, pp 34.

Millennium Ecosystem Assessment. (2005a). "Ecosystems and Human Well-Being: Biodiversity Synthesis," Volumes i-vi, World Resources Institute, Washington, DC, 86pp.

Millennium Ecosystem Assessment. (2005b). "Ecosystems and Human Well-Being: Wetlands and Water Synthesis," Volumes i-vi, World Resources Institute, Washington, DC, 68pp.

Molden D. (ed.). (2007). "Water for Food, Water for Life: A Comprehensive Assessment of Water Management in Agriculture," International Water Management Institute, London: Earthscan, and Colombo, 645pp.

Mollot, R., Phothitay, C., and Kosy, S. (2003). "Seasonally Flooded Habitat and Non-Timber Forest Products: Supporting Biodiversity and Local Livelihoods in Southern Lao PDR," WWF Lao PDR Program. Living Aquatic Resources Research Centre (LAReC), Livestock and Fishery Section of Savannakhet Province, Lao PDR, 27pp.

Motomura, H. and Mukai, T. (2006). *Tonlesapia tsukawakii*, a new genus and species of freshwater dragonet (Perciformes: Callionymidae) from Lake Tonle Sap, Cambodia. *Ichthyological Exploration of Freshwaters* **17**(1), 43-52.

Moyle, P. B. and Cech, J. J., Jr. (1988). "Fishes: An Introduction to Ichthyology," 2nd edition, Prentice-Hall, Englewood Cliffs, NJ, 559pp.

MRC. (1997). "Mekong River Basin Diagnostic Study," Final Report. Mekong River Commission, Bangkok, 196pp.

MRC. (2003). "Mekong Fish Database. A Taxonomic Fish Database for the Mekong Basin," CD-ROM. Mekong River Commission, Phnom Penh.

Na-Nakorn, U., Sukmanomon, S., Nakajima, M., Taniguchi, N., Kamonrat, W., Poompuang, S., and Nguyen, T. T. T. (2006). mtDNA diversity of the critically endangered Mekong giant catfish (*Pangasianodon gigas* Chevey, 1913) and closely related species: Implications for conservation. *Animal Conservation* **9**, 483-494.

Ngor, P. B. (1999). Catfish fry collection in the Mekong River of Kandal and Phnom Penh. *In* "Present Status of Cambodia's Freshwater Capture Fisheries and Management Implications," Van Zalinge, N. P., Nao, T., and Deap, L. (eds.), pp 124-134. Nine presentations given at the Annual Meeting of the Department of Fisheries, Phnom Penh, 19-21 January 1999. Mekong River Commission Secretariat and Department of Fisheries, Phnom Penh.

Ngor, P. B., Aun, S., Deap, L., and Hortle, K. G. (2005). The dai trey linh fishery on the Tonle Touch (Touch River), southeast Cambodia. *MRC Conference Series* **5**, 35-56.

Nguyen, T. T., Truong, T. T., Tran, Q. B., Doan, V. T., and Valbo-Jørgensen, J. (2001). Larvae drift in the Delta: Mekong versus Bassac. *In* "Proceedings of the Third Technical Symposium on Mekong Fisheries," Phnom Penh, 8-9 December 2000, pp 73-101.

Oberdorff, T., Hugueny, B., and Guegan, J. F. (1997). Is there an influence of historical events on contemporary fish species in rivers? Comparisons between Western Europe and North America. *Journal of Biogeography* **24**, 461-467.

Odum, E. P. (1971). "Fundamentals of Ecology," 3rd edition, W.B. Saunders Company, London, 574pp.

Parveen, S. and Faisa, I. M. (2003). Open-water fisheries in Bangladesh: A critical review. *In* Paper submitted to "Second International Symposium on the Management of Large Rivers for Fisheries: Sustaining Livelihoods and Biodiversity in the New Millennium."

Petillot, L. (1911). "La Peche et Les Poissons, une richesse du Cambodge," Librairie Maritime et Coloniale, Paris, 169pp.

Phillips, M. J. (2002). "Freshwater aquaculture in the Lower Mekong Basin." MRC Technical Paper No. 7. Mekong River Commission, Phnom Penh, 62pp.

Poulsen, A. F. (2003). Fish movements and their implication for River Basin Management in the Mekong River Basin. *In* Paper submitted to "Second International Symposium on the Management of Large Rivers for Fisheries," Phnom Penh, Cambodia. http://www.lars2.org/unedited_papers/unedited_paper/Poulsen%20migration.pdf.

Poulsen, A. F. and Valbo-Jørgensen, J. (eds.). (2000). "Fish Migrations and Spawning Habits in the Mekong Mainstream—A Survey Using Local Knowledge," AMFC Technical Report. Mekong River Commission, Phnom Penh, 149pp.

Poulsen, A., Poeu, O., Viravong, S., Suntornratana, U., and Nguyen, T. T. (2002). "Deep Pools as Dry Season Habitat in the Mekong River Basin," MRC Technical Paper No. 4. Mekong River Commission, Phnom Penh, 24pp.

Poulsen, A., Griffiths, D., So, N., and Nguyen, T. T. (2008). Capture-based aquaculture of Pangasiid catfishes and snakeheads in the Mekong river basin. *In* "Capture Based Aquaculture," Lovatelli, A. and Holthus, P. (eds.), pp 83-105. FAO Fisheries Technical Paper. No. 508. FAO, Rome.

Powering Progress. (2007). "Power Development Plan in Laos." http://www.poweringprogress.org/energy_sector/current_plants.htm. Accessed 29 May 2008.

Pullin, R. S. V. (1999). Developing policies for aquatic biodiversity. On the CD-ROM, distributed in: "Proceedings of the Conference on Sustainable Use of Aquatic Biodiversity: Data, Tools and Cooperation," Lisbon, Portugal, 3-5 September 1998. "ACP-EU Fisheries Research Initiative," Pullin, R. S. V., Froese, R., and Casal, C. M. V. (eds.). *ACP-EU Fisheries Research Report* **6**, 1-3.

Rainboth, W. J. (1991). Cyprinids of South East Asia. *In* "Cyprinid Fishes: Systematics, Biology, and Exploitation," Winfield, I. J. and Nelson, J. S. (eds.), pp 156-210. Fish and Fisheries Series 3. Chapman and Hall, London.

Rainboth, W. J. (1996). "FAO Species Identification Field Guide for Fishery Purposes. Fishes of the Cambodian Mekong," FAO, Rome, 265pp. i-xi, plus i-xxvii.

Ricciardi, A. and Rasmussen, J. B. (1999). Extinction rates of North American freshwater fauna. *Conservation Biology* **13**, 1220-1222.

Roberts, T. R. (1972). Ecology of fishes in the Amazon and Congo Basins. *Buletin of the Museum of Comparative Zoology* **143**(2), 117-147.

Roberts, T. R. (1993). Just another dammed river? Negative impacts of Pak Mun Dam on fishes of the Mekong basin. *Natural History Bulletin of the Siam Society* **41**, 105-133.

Roberts, T. R. (2004). Fluvicide: An independent environmental assessment of Nam Theun 2 hydropower project in Laos, with particular reference to aquatic biology and fishes. Reissue of 1996 Consultancy Report Submitted to the World Bank, 46pp. http://internationalrivers.org/en/southeast-asia/laos/nam-theun-2/fluvicide-independent-environmental-assessment-nam-theun-2-hydropowe. Accessed 29 May 2008.

Roberts, T. R. and Baird, I. G. (1995). Traditional fisheries and fish ecology on the Mekong River at Khone

Waterfalls in southern Laos. *Natural History Bulletin of the Siam Society* **43**, 219-262.

Roberts, T. R. and Vidthayanon, C. (1991). Systematic revision of the Asian catfish family Pangasiidae, with biological observations and descriptions of three new species. *Proceedings of the Academy of Natural Sciences, Philadelphia* **143**, 97-144.

Roberts, T. R. and Warren, T. J. (1994). Observations of fishes and fisheries in southern Laos and north-eastern Cambodia, October 1993-Febuary 1994. *Natural History Bulletin of the Siam Society* **42**, 87-115.

Secretariat of the Convention on Biological Diversity. (2006). "Global Biodiversity Outlook 2," Montreal, Canada, Secretariat of the Convention on Biological Diversity, 81 + viipp.

Sekino, M. and Hara, M. (2000). Genetic characteristics and relationships of climbing perch *Anabas testudineus* populations in Thailand. *Fisheries Science* **66**, 840-845.

Smith, H. M. (1945). "The Fresh-Water Fishes of Siam, or Thailand," United States National Museum, Bulletin 188. Smithsonian Institution, Washington, 622pp.

Sneddon, C. (2002). Water conflicts and river basins: The contradictions of co-management and scale in northeast Thailand. *Society and Natural Resources* **15**, 725-741.

So, N. (2006). High genetic diversity in cryptic populations of the migratory sutchi catfish *Pangasianodon hypopthalmus* in the Mekong River. *Heredity* **96**, 166-174.

So, N., van Houdt, J. K., and Volckaert, F. A. (2006). Genetic diversity and population history of the migratory catfishes *Pangasianodon hypophthalmus* and *Pangasius bocourti* in the Cambodian Mekong River. *Fisheries Science* **72**, 469-476.

Sverdrup-Jensen, S. (2002). "Fisheries in the Lower Mekong Basin: Status and Perspectives," MRC Technical Paper No. 6. Mekong River Commission, Vientiane, 95pp.

Takagi, A. P., Ishikawa, S., Nao, T., Hort, S., Nakatani, M., Nishida, M., and Kurokura, H. (2006). Genetic differentiation of the bronze featherback *Notopterus notopterus* between Mekong River and Tonle Sap Lake populations by mitochondrial DNA analysis. *Fisheries Science* **72**, 750-754.

Taki, Y. (1975). Geographic distribution of primary freshwater fishes in four principal areas of Southeast Asia. *Southeast Asian Studies (Kyoto University)* **13**(2), 200-214.

Taki, Y. (1978). "An Analytical Study of the Fish Fauna of the Mekong Basin as a Biological Production System in Nature," Research Institute of Evolutionary Biology, Tokyo, 77pp. Special Publication Number 1.

Tang, D. L., Kawamura, H., Van Dien, T., and Lee, M. (2004). Offshore phytoplankton biomass increase and its oceanographic causes in the South China Sea. *Marine Ecology Progress Series* **268**, 31-41.

Thomson, A. W. and Page, L. M. (2006). "Genera of the Asian Catfish Families Sisoridae and Erethistidae (Teleostei: Siluriformes)," Zootaxa 1345. Magnolia Press, Auckland, 96pp. http://www.mapress.com/zootaxa/2006f/zt01345p096.pdf.

Valbo-Jørgensen, J. (2003). Unchartered waters. *Catch and Culture* **9**(1): 12.

Valbo-Jørgensen, J. and Poulsen, A. F. (2000). Using local knowledge as a research tool in the study of river fish biology: Experiences from the Mekong. *Environment, Development and Sustainability* **2**(3-4), 253-276.

Valbo-Jørgensen, J. and Visser, T. (2003). The MRC Mekong Fish Database: An Information Base on Fish of a Major International River Basin. MRC Conference Series **4**, 55–69.

Valbo-Jørgensen, J., Marmulla, G., and Welcomme, R. L. (2008). Migratory fish stocks in transboundary basins—Implications for governance, management, and research. In: Rescue of sturgeon species in the Ural river basin; Proceedings; NATO Science for Peace and Security Series. C: Environmental Security (NATO), (no. 2008); NATO Advanced Research Workshop on Rescue on Sturgeon Species by Means of Transboundary Integrated Water Management of the Ural River Basin, Orenburg (Russian Federation), 13-16 Jun 2007 Lagutov, V. (ed.)/Dordrecht (Netherlands), Springer/NATO Public Diplomacy Div., 2008, pp 61-83 "Rescue of Sturgeon Species by Means of Transboundary Integrated Water Management in the Ural River Basin," Lagutov, M. V. (ed.). Springer-Verlag, Berlin.

Van Zalinge, N., Nuov, S., Choulamany, X., Degen, P., Pongsri, C., Jensen, J., and Nguyen V. H. (2004). The Mekong river system. *In* "Proceedings of the Second International Symposium on the Management of Large Rivers for Fisheries," Volume 1, Welcomme, R. L. and Petr, T. (eds.), pp 335-357. Food and Agriculture Organization & Mekong River Commission, FAO Regional Office for Asia and the Pacific, Bangkok. RAP Publication 2004/16.

Viet Nam Environment Protection Agency. (2005). "Overview of Wetlands Status in Viet Nam Following 15 Years of Ramsar Convention Implementation," Viet Nam Environment Protection Agency, IUCN Viet Nam, Mekong Wetlands Biodiversity Conservation and Sustainable Use Programme, Hanoi, Viet Nam, 72pp. http://www.mekongwetlands.org/Common/download/Vietnam_wetlands_status_English.pdf.

Viravong, S., Phounsavath, S., Photitay, C., Putrea, S., Chan, S., Kolding, J., Valbo-Jørgensen, J., and Phoutavongs, K. (2004). Deep Pools Survey 2003-2004 Final Report, Living Aquatic Resources Research Centre, Vientiane, 43pp.

Visser, T., Valbo-Jørgensen, J., and Ratanachookmanee, T. (2003). "Mekong Fish Database 2002. Introduction and Data Sources." Mekong River Commission, Phnom Penh.

Ward, J. V. and Stanford, J. A. (1983). The intermediate-disturbance hypothesis: An explanation for biotic diversity patterns in lotic ecosystems. *In* "Dynamics of Lotic Ecosystems," Fontaine, T. D. and Bartell, S. M. (eds.), pp 347-356. Ann Arbor Science, Michigan, USA.

Welcomme, R. L. (1985). River fisheries. FAO Fisheries Technical Paper Number 262, 230pp.

Welcomme, R. L. (2001). "Inland Fisheries. Ecology and Management," Fishing News Books, Blackwells, Oxford, UK, 358pp.

Welcomme, R. L. and Vidthayanon, C. (2003). "The Impacts of Introductions and Stocking of Exotic Species in the Mekong Basin and Policies for Their Control," MRC Technical Paper No. 9. Mekong River Commission, Phnom Penh, 65pp.

Whitehead, P. J. P., Nelson, G. J., and Wongratana, T. (1988). FAO species catalogue. "Clupeoid Fishes of the World (Suborder Clupeoidei). An Annotated and Illustrated Catalogue of the Herrings, Sardines, Pilchards, Sprats, Shads, Anchovies and Wolf-Herrings. Part 2—Engraulididae," Volume 7, Part 2, 579pp. FAO Fisheries Synopsis Number 125, FAO, Rome.

Winnett, A. (1999). The economics of aquatic biodiversity. *In* "ACP-EU Fisheries Research Initiative," Proceedings of the Conference on Sustainable Use of Aquatic Biodiversity: Data, Tools and Cooperation," Lisbon, Portugal, 3-5 September 1998. *Fisheries Research Report* **6**, 60-65. ftp://ftp.cordis.europa.eu/pub/inco2/docs/acp_6_proceedings_en.pdf. Accessed 30 May 2008.

Yang, J. (1996). The alien and indigenous fishes of Yunnan: A study on impact ways, degrees and relevant issues. *In* "Conserving China's Biodiversity (II)," Schei, P. J., Wang, S., and Xie, Y. (eds.), pp 157-168. China Environmental Science Press, Beijing. http://www.chinabiodiversity.com/shwdyx/technical-report-e/x-3e.htm. Accessed 30 May 2008.

Yap, S. Y. (2002). On the distributional patterns of Southeast-East Asian freshwater fish and their history. *Journal of Biogeography* **29**, 1187-1199.

Zhou, B. (1999). Use and management of wetlands by local people in Lancangjiang-Mekong River Basin in PR China. *In* "Proceeding of the Workshop: Wetlands, Awareness, Local People and the Ramsar Convention in the Mekong Basin. Can Local Communities Play a Role in the Wise Use of Wetlands?" Phnom Penh, Cambodia, 12-14 September 1999, O'Callaghan, B. (ed.), pp 23-26. Mekong River Commission, Phnom Penh.

Fisheries of the Mekong River Basin

Kent G. Hortle

Fisheries and Environment Consultant, 23A Waters Grove, Heathmont 3135, Australia

1. INTRODUCTION

The Mekong is one of the world's largest rivers (Chapter 4) with a basin that supports a population of about 70 million people, for most of whom the staple diet is rice, fish and other aquatic animals. On the basis of altitude, hydrology, and landforms, the basin is usually divided into the Upper Mekong (in China and Myanmar), comprising about 20% of the catchment, and the Lower Mekong (LMB) (in Laos, Thailand, Cambodia, and the Viet Nam delta and central highlands).

For about two-thirds of its length (from lower Yunnan to the sea), the Mekong is a lowland system, less than about 370 m ASL, and about 60% of the LMB comprises tropical lowlands where most people live, and where extensive and diverse aquatic habitats support a wide range of species which can be caught or cultured. Rainfall patterns and the flow of the Mekong and its tributaries are highly predictable from year to year, and the rural economies depend upon the annual alternation of dry and wet seasons, which in particular support "rain-fed" rice agriculture, fishing, and collection of other wild products. Aquatic organisms are well-adapted to the seasonal extremes, breeding and feeding in the vast area of seasonal wetlands where large quantities are caught each year, particularly as the waters recede.

Most people in the LMB are rural farmers with fisheries as a secondary occupation. "Fisheries" as a term covers the catching, collecting, or aquaculture of fish, as well other aquatic animals (OAAs) including shrimps, crabs, molluscs, insects, reptiles, and amphibians. Fisheries also include the processing, transporting, and marketing of products, and support many associated industries such as boatbuilding, the making of fishing gears, and the provision of ice and salt.

Fisheries products are predominantly used as food (the focus of this review) but also as fish- or stock-feed, as fishing bait, and even as fertilizer. Many Mekong system fish are important in the aquarium trade, and recreational or sport fishing is becoming increasingly popular, especially in more affluent areas, but there is very little published information on these two kinds of fisheries.

This review introduces the LMB fishery from a historical perspective, and then discusses some general features of the fisheries of the

Mekong system, including estimates of their size and value. Section 4 describes the main habitats of the LMB, their fisheries, and estimates of areal yield. Aquaculture is reviewed in Section 5, and finally Section 6 briefly covers some aspects of management of the LMB fisheries.

2. HISTORICAL PERSPECTIVE

Until about 5000 years ago, the LMB was inhabited by small groups of hunter gatherers who ate many kinds of wild foods, including fish, crustaceans, and molluscs (Gorman, 1971). Neolithic farming based on flooded rice developed as people settled on the most fertile areas, including the Tonle Sap-Great Lake system and other floodplains including along the Mun and Chi Rivers in Thailand and the Vientiane plain in Laos, as well as the highland plateau in Yunnan around Lake Erhai (see Chapter 3). Rice has long been the energy staple, with protein and micronutrients provided by fish and OAAs as is evident from archaeology. For example, excavations of old settlements at Angkor Borei in southeast Cambodia dated at 1600-2400 years ago included the bones of hundreds of species in at least 17 families of fish (Voeun, 2001). A "rice-fish" diet has been the norm for at least 2000 years in most of the Mekong Basin (MB). Dense settlements based on recession rice cultivation in fertile floodplains first developed in the delta region, during the Funan period, first to seventh century AD (Chandler, 2003; Fox and Ledgerwood, 1999) and then spread to other areas supporting the Angkor Empire (ninth to fifteenth century AD) which was centered on the Great Lake-Tonle Sap system. Later evidence for the importance of fisheries includes the many bas-reliefs of Angkor Wat, the Bayon, and other temples depicting fish, other aquatic animals, and fisheries (Roberts, 2002), as well as the observations of Zhou Daguan, a Chinese visitor to Cambodia in 1296-1297 AD, who commented on the richness of aquatic resources of the Great Lake, and the abundance of many kinds of fish as well as frogs, tortoises and turtles, lizards, large prawns, crocodiles, and molluscs (Zhou, 2001). French explorers during the late nineteenth century also noted the abundance of fish and the ease of catching very large fish, or large quantities of fish along the Mekong and its tributaries, and how, for example, the Vietnamese exploited the fruitful fisheries of the Tonle Sap (Garnier, 1996).

Environmental changes and fishing pressure became significant as the LMB population expanded during the twentieth century from about 5 million in 1900 to about 65 million now; Cambodia's population alone grew from about 1.1 million to about 14 million in 2008. The direct impacts of population growth include increasing catches for local consumption, competition for water by other industries, particularly agriculture, modification of hydrology and water quality by dams, and clearing of forests. The development of commercial fisheries for export from the Tonle Sap and northeast Thailand also increased fishing pressure, and led to conflicts with the growing rural subsistence populations (Bush, 2008). The "frontier" for underexploited fisheries is moving rapidly into the last inaccessible parts of Laos and northern Cambodia along new roads. Even in the 1990s in some relatively isolated and uninhabited areas with extensive natural aquatic habitats, fish were still so large and abundant that local people placed little economic value on them (e.g., Baird and Phylavanh, 1998). New gears and methods have also expanded from more populated areas along a "technological frontier"; efficient and cheap mass-produced gears such as nylon gill nets and cast nets as well as illegal gears such as fine-mesh nylon fyke nets and electrofishers, first introduced in the 1970s, have only recently reached some remote places.

Throughout the basin, the perception has become widespread that fisheries are experiencing "fishing down" or "ecosystem fishing." This is a process in which increasing effort leads to larger total catches but smaller catches per unit effort (CPUE); catches of larger fish and larger species—particularly carnivores—decline, while catches of smaller species—particularly herbivores or omnivores—increase (Allan et al., 2005; Baran and Myschowoda, 2008; Pauly et al., 1998).

The "Mekong Project," a United Nations project conceived in the early 1960s, envisaged basinwide economic development based on tributary and mainstream dams to provide hydroelectricity, to support irrigation and other industries, to control flooding and to improve navigation. Plans for coordinated research and development of fisheries throughout the LMB (rather than just by individual countries) were partly motivated by the increasing recognition that large dams would cause significant international transboundary impacts, as well as a realization that an unknown but probably significant number of species were transboundary migrants. As part of the Mekong project, the Mekong Basinwide Fisheries Studies of the 1970s (Lagler, 1976b; Pantulu, 1986a,b; Taki, 1978) included surveys of fish distribution, standing crop, fish migration, and general observations on fishing activities and aquaculture at locations throughout much of the basin and in the sea off the mouth of the Mekong. The fishery studies produced a general description of the main elements of the fishery and the distribution and migration patterns of many species, but suffered from limited coverage in some areas and a lack of any accurate data on the size of small-scale artisanal catches. This probably led to an underestimate of the size of the fishery, and an underappreciation of the importance of the flood pulse to production, a concept which was formalized later (Junk et al., 1989). "Development" at the time was typically framed in rather simple top-down "technical" terms of increasing production of goods and services, with impacts on capture fisheries to be compensated for by stocking of new reservoirs or by aquaculture. Relatively little consideration was given to the role of fisheries as a part of complex livelihoods, the differential impact of gains and losses between social groups, and effective approaches to implementing fisheries management (or any kind of development) in an equitable and sustainable manner.

The second Indo-China War (1964-1973), the Khmer Rouge regime (1975-1979), and subsequent conflicts impeded economic development in Viet Nam, Cambodia, and Laos, so that relatively few of the dams (and none of the mainstream dams) envisaged in the Mekong project were built. Warfare was very environmentally destructive, particularly as a result of the widespread use of defoliants and carpet bombing in Mekong watersheds and in the delta, where virtually all mangroves were destroyed by defoliants (Westing, 1976). Most fisheries development also stalled during this period as a result of disruption of social and government structures and traditional livelihoods, internal postwar migration, and subsequent rapid population growth. However, in northeast Thailand in the 1980s and 1990s, clearing of forests, intensification of agriculture, development of agricultural processing industries, and construction of many dams had a range of impacts; river-floodplain fisheries have no doubt declined (Roberts, 1993), but reservoir and rice field fisheries have increased in importance. In Laos and Cambodia, capture fisheries have been maintained through much of the mainstream Mekong and in many large tributaries where floodplains remain intact, as well as in rice fields which continue to expand in area. Development in the Viet Nam delta accelerated after 1975, with canalization and settlement of the extensive swampy areas of the plain of reeds and Long Xuyen quadrangle (Nguyen and Wyatt, 2006); internal migration added to population growth in newly settled areas, and large areas of regrowth mangroves were cleared along the coast for intensive shrimp culture (de Graaf and Xuan, 1998). Internal migration to the Viet Nam highlands after 1975 was supported by forest clearance for highland crops, such as coffee, and construction of many small irrigation reservoirs which support stocked fisheries.

A revived Mekong Committee in the early 1990s commissioned a review of LMB fisheries (Anonymous, 1992a), and its recommendations on research and development formed a basis for aid-funded programs in the 1990s. This review pointed out the continuing weak information base, the unreliability of official statistics based on estimates of commercial catches, and the lack of data on artisanal fisheries, which might be contributing an unmeasured 80-95% of total catches.

Partly as a response to the 1992 review, fishery issues were targeted in several aid-funded programs; these included support for several household surveys that did much to upgrade estimates of the likely size and value of the fishery (e.g., Ahmed et al., 1998; Funge-Smith, 1999a; Sjorslev, 2000 and others summarized in Hortle, 2007). National fishery agencies received direct support from many donors and a range of "basinwide" projects were funded through the MRC, with various other agencies such as the FAO, the Asian Institute of Technology, and the WorldFish Center also working on projects throughout the basin. Sverdrup-Jensen (2002) prepared a recent "sector review" of the basin's fisheries which updated yield estimates, taking into account preliminary data on subsistence catches and some new estimates of yield per unit area. The review also noted that the MRC's fisheries programs had done much to raise the profile of fisheries and the awareness of threats to capture fisheries, as well as encouraging dialog between LMB countries, including through the establishment of an international Technical Advisory Body (TAB). The 2002 review also reiterated many of the issues that were raised in the 1992 review, and suggested some general responses to ongoing problems in the sector. MRC (2003) and van Zalinge et al. (2004) provided similar more recent descriptions of the LMB fisheries.

The LMB fisheries reviews tended to identify technical issues and also to reflect the perspective of national governments, which is a consequence of review teams working mainly with staff of national fisheries agencies. The later reviews (Sverdrup-Jensen, 2002; van Zalinge et al., 2004) also reflect recent changes in emphasis of national policies by including recommendations for a more decentralized approach to management of local fisheries issues, for example, via comanagement.

With recent political stability, the growth in regional economies, and continuing pressure to raise living standards, the Mekong system's environment is entering a new phase during which some of the grand ideas envisaged in the Mekong project may be implemented: a massive increase in electricity production through hydropower, regulation of the Mekong mainstream, road building, conversion of forests to plantations, increased agricultural output through irrigation, and development of secondary industries. Much of the LMB lowland landscape has been altered, but in this new phase of intensive damming it is the hydrological regime—based on high and seasonal rainfalls in extensive mountain ranges—that will be brought under increasing control, causing a wide range of direct and secondary impacts to aquatic environments. There is a long-standing need to identify clearly the costs and benefits of developments, to support efforts to mitigate and manage impacts on fisheries and other existing uses, and to provide clear advice on which schemes are so harmful that they should be rejected outright.

A major and perhaps insurmountable challenge will be to develop an acceptable framework within which countries voluntarily modify their management of water and natural resources for the benefit of their neighbors. Dams impact downstream users, but it is less appreciated that impacts on fisheries—based on migratory resources—can be manifested up- and downstream, as well as across the Mekong itself, which is an international boundary between Laos and Thailand and Laos and Myanmar. While many agencies and NGOs play important roles in the development process, the MRC is the only international organization staffed and funded by national governments that is mandated to bring fisheries (and other sidelined issues) into considerations of water resources development from a transboundary perspective. Its Fisheries Program continues to support basinwide fisheries ecology and valuation studies, fisheries management, and aquaculture of indigenous species, as well as publishing and disseminating a range of fisheries information and supporting basinwide workshops and coordination meetings.

3. GENERAL FEATURES OF MB FISHERIES

3.1. Diversity and Fishing Pressure

The MB's fisheries are characterized by a diversity of species and habitats, as well as a diversity of gears and fishing activities, as are inland fisheries generally (Welcomme, 1985, 2001). About 900 species are recorded from or are likely to occur in the basin, and among them about 560 are purely freshwater fishes (Chapter 8). Catches at a locality may include more than 200 species, but typically 10 species would make up about 60-70% of the total catch by weight (e.g., Baran et al., 2005; Ngor et al., 2005, 2006; van Zalinge and Nao, 1999). The least diverse fisheries are in rice fields, where a few species make up most of the weight, whereas diversity is highest closer to the estuary where marine and estuarine species (marine vagrants) are caught with freshwater fishes.

Most of the Mekong system's fish species are small, growing to maturity in their first or second year and many are extremely fecund (e.g., Suwannapeng, 2002) or are repeat spawners adapted to take advantage of the great seasonal change in extent of available habitats. Other aquatic animals (OAAs), including shrimps, crabs, molluscs, insects, reptiles, frogs and toads, probably form about one-quarter of catches on average (Hortle, 2007), with higher proportions sometimes reported (e.g., de Graaf and Chinh, 2000). The extreme contraction of available habitat in the dry season causes stress and mortality of aquatic animals through desiccation, predation, competition, and disease, so the surplus production which can be harvested sustainably each year in inland waters (the catch or yield) in the monsoonal tropics is generally a very large proportion of the peak standing crop or biomass (Hoggarth et al., 1999a).

As fishing pressure intensifies, tropical inland fisheries generally exhibit a long-plateau phase of declining CPUE even as total catches continue to increase (Welcomme, 2001). Declining CPUE is widely reported by fishers throughout the LMB (e.g., Ahmed et al., 1998; Hortle and Suntornratana, 2008; Phanh et al., 2002; Sjorslev, 2002) and is usually attributed primarily to "fishing down" (variously reported as "too many fishers" or "too few fish per fisher") as well as environmental changes. Fishing down is consistent with an apparent dominance in catches in some areas by a few herbivorous or omnivorous species, as is evident for summary data on Cambodian commercial catches (van Zalinge and Nao, 1999) or by dominance of small planktivores in reservoir catches (e.g., Mattson et al., 2001). Declines in the catches of some giant species

have also been reported (Mattson et al., 2002) but in general it appears that the Mekong system is not heavily fished down, because medium-sized or large fish are still commonly caught and traded throughout the basin (e.g., Baran et al., 2005; Bouakhamvongsa et al., 2006) and predatory species are quite abundant in catches in some areas (e.g., Ngor et al., 2006). It is also worth noting that—despite widespread reports to the contrary—the only monitoring data that have been systematically collected during the last decade do not indicate a declining trend in CPUE, but do show that high-flow years have a large effect on catches in the same or subsequent years (Halls et al., 2008; Soukhaseum et al., 2007). Therefore, although fishing pressure is affecting some species, in general the capture fishery appears to be extremely resilient to fishing pressure, but is likely to be sensitive to hydrological and other environmental changes, which should therefore receive much more attention in fisheries management.

3.2. Migrations

Because of the large seasonal variations in water availability many fish exploit different habitats in different seasons, often moving in large schools which are the target of intense fishing pressure, either at particularly favorable localities or by fishers who move with the schools. Freshwater fishes may be broadly categorized as "white fish" (migrating between rivers and floodplains) (see Chapter 8), "black fish" (resident on floodplains and also common in rice fields and standing waters), "gray fish" (moving locally between floodplains and dry-season refuges), and "residents" of river channels. The maintenance of habitats (dry-season refuges, flood-season feeding and rearing habitats and spawning habitats) as well as maintaining connectivity and the basic pattern of hydrology is key element in environmental management to conserve fisheries (Poulsen *et al.*, 2002a). Coastal and estuarine fish are also important in fisheries, and many species move considerable distances into the river system. The migration patterns of other aquatic animals are not well-described, but the catadromous giant river prawn (*Macrobrachium rosenbergii*) is heavily fished and is one of the most valuable fishery species (Ngor *et al.*, 2006).

3.3. Participation

About 80% of the basin's population is rural and the economy is largely based on farming as the primary occupation and fishing or aquaculture as the secondary occupation and consequently overlooked in most censuses. Most people live along or near natural rivers, streams, or annually flooded areas in which they can fish, and many villages are close to permanent standing waters. As is generally the case in tropical inland fisheries, the Mekong system is dominated by small-scale or artisanal fishing on a part-time or occasional basis, with fishery products often sold to supplement a household's main income. There is generally a strong gender aspect to capture fisheries with most fishing done by men, whereas women are disproportionately involved in fish processing, marketing, and gear-making. Small-scale aquaculture—often promoted as a supplement or replacement for capture fisheries—tends to shift the burden of work onto women as it tends to be a "household" activity (Hatha *et al.*, 1995; So *et al.*, 1998). In a range of LMB studies reviewed by Hortle (2007), more than 80% of rural households in Thailand, Laos, and Cambodia and 60-95% of households in the Viet Nam delta were involved in capture fisheries. The slightly lower proportion of households involved in capture fisheries in some parts of the delta is offset by a very high involvement in aquaculture, with 15-90% of households culturing fish or OAAs, compared with up to about 15% in most other parts of the basin (Hortle, 2007).

Although most households sell or barter some fish, commercial, full-time, or professional fishing is typically practiced by fewer than 10% of households. In larger water bodies, commercial fishing may be more common, for example, 40% of households in fishing communes around the Tonle Sap Great Lake (Ahmed *et al.*, 1998), and 60% around Nam Ngum Reservoir (Mattson *et al.*, 2000).

Small-scale fishers typically fish in habitats that are accessible by foot or nonmotorized boats, such as rice fields and associated habitats (canals, small streams, and swamps), (Pham and Guttman, 1999) as well as in rivers and streams (Hortle and Suntornratana, 2008). Commercial fishers tend to use larger gear and motorized craft to access open-waters, or obtain licenses for fixed gears that filter large volumes of water. Although small-scale fishing contributes most to total catches, in the most productive areas, commercial fishers take a disproportionate share (Sjorslev, 2002) which may be most of the total catch (Phanh *et al.*, 2002). Where commercial aquaculture is significant, as in the Viet Nam delta, a small percentage of households that make a living purely from aquaculture may contribute most of the total yield (Sjorslev, 2002).

Typically, 1-2 members of the average 5-6-person household go fishing or collecting aquatic animals at some time, so about one-third of the basin's population—about 20 million people—are "fishers." The total annual catch of about 2.1 million tonnes (Mt) (Section 4.1) equates to about 100 kg fisher^{-1} year^{-1}, so if fishers are active for about 2-4 months per year, their daily catch could average about 1-2 kg person^{-1} day^{-1}, but would typically be less because the distribution of catches is usually highly skewed by commercial catches. The benefits of spreading the catch between many households (equity) should be judged against the relatively low return on effort for most fishers (efficiency) as well as the effects on professional fishers who must compete with the numerous subsistence fishers. Across the entire "wetland" area (Table 9.4), average fisher density is about 100 people per km^2, a high figure in comparison with inland fisheries generally (Halls et al., 2006) and consistent with the impression of generally high fishing pressure.

3.4. Fishing Gears and Methods

At least 100 types of gears can be distinguished, classified by principle of capture into 16 main groups (Claridge et al., 1997; Deap et al., 2003; Ko-Anandakul, 2004; Nguyen et al., 2006a) as follows:

1. Hand capture
2. Scooping devices such as hand nets
3. Wounding gears, for example, gaffs, forks, rakes, spears, bows, and guns
4. Hooks and lines, including pole and line, and longlines
5. Traps include many types and sizes, often in combination with fences
6. Gill nets and entangling nets, often set in multiple layers of differing meshes
7. Surrounding or seine nets
8. Dragged gears, including scrapers, trawl nets, and dragged fences
9. Push nets, hand or boat-mounted
10. Lift nets, hand or boat-mounted
11. Covering gears, including basket traps and cast nets
12. Bag nets, including fyke nets and stationary trawls or dais
13. Illegal methods
14. Anesthetic methods, including explosives, poisons, and electricity
15. Draining/emptying of ponds or watercourses by pumping
16. Attracting devices, including mobile and stationary brush parks
17. Scaring devices, including chains and marble boxes to create noise, and chemicals such as calcium carbide to scare fish into nets

Gears tend to be similar wherever habitats are similar, with local variants designed for certain conditions or target species. Gear usage also varies by time of year and water level. Traditional traps are the most diverse group, often specialized for certain habitats and taxa. Large-scale commercial gears include river barrages, stationary bagnets (dais), trawls, and seines, with a sequence of gear typically used through the fishing season. Traditional gears, particularly traps, are still common and widespread in and around rice fields and smaller water bodies, but in any open water bodies small-scale or family fishers throughout the basin tend to use gill nets, cast nets, and hook and line, now ubiquitous gears which tend to catch a range of species across habitats. Such modern gears have tended to supplant traditional methods because of their efficiency, widespread availability, low cost, and durability.

In all LMB countries, fisheries regulations prohibit the use of gear/methods 13–16 above, but these are still common anywhere conditions are suitable. The regulations are unenforceable in practice; moreover, the basis for each prohibition is not clear so little support can be expected for implementation. For example, brush parks (essentially large manmade snags) are used throughout the LMB to aggregate fish and OAAs, and they form the basis of a significant industry that supplies materials for their construction. Brush parks allow a planned harvest with a fairly predictable timing and yield; yields from brush parks can be high, between 1.2 and 15.5 t ha^{-1} year^{-1} (Welcomme, 2001). The negative effects of brush parks (destruction of forest and overharvesting) have never been properly assessed against the likely benefits to the system from increased production from the artificial habitat which forms shelter and a substrate for periphyton and food organisms, as well as being a source of fry from fish that spawn within them. The floating macrophyte *Eichhornia crassipes* (water hyacinth) is usually associated with brush parks, and is considered highly valuable to fisheries, providing habitat for a rich community of fish food organisms—invertebrates—that feed on periphyton that covers its roots, as well as habitat for fish and shelter for fry, as has been described in detail by Green *et al.* (1976). Brush parks (*kata*) in Bangladesh have been well-described as a successful way of intensifying fisheries production on floodplains, where the food chain to high-value fishes is based on the brush (Islam *et al.*, 2006). Rather than prohibiting brush parks, fisheries agencies might study their actual effects and seek to licence their use.

Apart from prohibition on certain types of gears, minimum mesh sizes are set for most gears in most countries; for example, in Cambodia nets, fences and traps must have a mesh aperture greater than 15 mm. In practice, many gears have smaller apertures—fences, traps, and fyke nets are now often constructed incorporating nylon mosquito netting—2 mm mesh—to retain virtually all small fish and shrimps. Fine-mesh gears, in particular traps set with long fences across floodplains or fish migration routes, are of particular concern and have been a focus of enforcement efforts in Cambodia in recent years.

3.5. Processing, Marketing, and Consumption

The seasonal excess of fish caught during the flood recession has led to the development of many methods of preservation based on salting, drying, and fermentation. Fish sauce and various kinds of fermented fish or OAAs are very characteristic features of cuisines in Asia; Phithakpol *et al.* (1995) describe procedures for 29 different commercial products made from aquatic animals. Fish sauce and fermented fish are eaten by most rural households throughout the dry season, and

salted dried fish from the Mekong system is sold throughout the region. Much of the seasonal excess of fish is fed to higher value species, particularly snakeheads (*Channa* spp.) as well as river catfishes (*Pangasius* spp.), which can be sold later to smooth out oversupply.

Most of the fish and OAAs caught or raised within the LMB are probably eaten directly by households, and the proportions that are sold in markets are not known. Markets are relatively more important for the supply of fishery products in Thailand and Viet Nam because of their well-established infrastructure and more diversified economies. Wholesale markets operate at landing sites wherever there are significant commercial fisheries (e.g., in the Tonle Sap-Great Lake system and at various places along the Mekong and in large reservoirs), and wholesalers may make various private arrangements to collect products from fishers or growers, so the marketing chain may be rather opaque. Fishery products are often traded through family-based systems, for example, the catch of many fishers is sold each day by their wives; typically over 90% of sellers are women (Khay and Hortle, 2005). Traders generally sell one main category of product (e.g., inland fresh fish or inland preserved fish). Variable numbers of part-time and occasional sellers complicate the task of assessing the size of a market and quantities of product sold, and sellers may employ variable numbers of people to process fish.

Markets throughout the basin are generally owned by local governments and run privately by lessees. LMB markets operate at three levels, province (capital), district, and village. The number of markets is not known for the basin, but in the entire Viet Nam delta in 2004 there were officially 1181 markets, of which about 20 were large wholesale/retail provincial markets, about 130-200 were district markets (including suburbs), and about 1000 were village markets (Phan Thanh Lam, personal communication, 2008). As the delta has about one-third of the LMB's population, the total number of markets is likely to be 5000-6000. Governments attempted to centrally regulate markets in the past, but government regulation has reduced since economic liberalization in the 1990s and the system overall appears to run efficiently supplying different groups of consumers of different status and tastes (Anonymous, 1992a). In general, larger more valuable fish are traded into cities and also out of the basin, especially to Bangkok and Ho Chi Minh City, capture fish command higher prices than culture fish, and smaller fish continue to be competed for by poorer rural people and those wishing to supply the trash fish market for aquaculture. Markets are characterized by the high proportion of live fish which are transported by river or canal in floating cages, in barges, in trucks, or on motorcycles.

Marine products are also sold throughout the basin; mackerel and dried squid are staples throughout northeast Thailand and are also traded into Laos, and seafood of all types is common throughout the delta, whereas marine products are relatively unimportant in Cambodia.

Export markets are poorly documented, largely clandestine, and may operate along a chain that includes many unofficial taxes or fees. For example, fish traders transporting their products from the Tonle Sap Lake to the Thai border by pickup truck typically made 27 different fee payments to 15 institutions in 16 different places, with the sum of fees plus costs associated with weight loss and spoilage exceeding the profit margin on the shipment by more than three times (Yim and McKenney, 2003). Export of high-value fish from Laos (including fish caught in northern Cambodia) to Thailand is also a large and underregulated industry (Baird, 2006b). The export industry is, in general, relatively "inefficient" compared

to the domestic marketing industry. The largest export industries for Pangasiid catfish and shrimp in Viet Nam are well regulated; however, markets are maintained by consistent export quality.

3.6. Fisheries in the Real Economy

In the MB countries, much of the economy is based on nonmarket (or nonmonetized) goods and services, which in less "developed" parts of the basin may provide most or all of the needs of households. For example, along the floodplains of the Se Bang Hieng in southern Laos, households collected on average about 2.4 tonnes (t) year^{-1} of nontimber forest products (NTFPs), which included fish (704 kg household (hh)$^{-1}$ year^{-1}) and OAAs (97 kg hh^{-1} year^{-1}), as well as medicines, fruits, herbs, resins, bamboo shoots, and mushrooms, none of which is visible in the cash economy Mollot et al., (2005). Even where development and agricultural intensification have greatly reduced the yield from "nature," most people still benefit from open-access to wetland resources. For example, in northeast Thailand, virtually all households around Nong Han wetland utilized its resources, collecting on average 711 kg hh^{-1} year^{-1} of NTFPs, which included 61 kg hh^{-1} year^{-1} of fish and 8 kg hh^{-1} year^{-1} of OAAs, as well as a variety of plants including lotus (Nelumbo spp.) and water morning glory (Ipomoea aquatica), with the wetland provided the equivalent of 32% of an average household's income (Pagdee et al., 2007). In rice fields (the most extensive aquatic habitats) fish and OAAs are usually significant but unaccounted open-access resources (Hortle et al., 2008).

Within the cash economy, the informal sector is significant in each MB country. Although little accurate information is available, in

Cambodia about 85% of the workforce and 62% of the estimated Gross Domestic Product (GDP) is within the informal sector (Nuth, 2005). Fisheries are also well-represented within the informal economy, because of the importance of small-scale unregulated middleman and traders, as well as payments that are made to governmental officials for licenses or to allow continuation of illegal fishing activities (Touch and Todd, 2003; Yim and McKenney, 2003). The lack of accurate data on the real economy and the poor representation of fisheries within the official statistics suggest that most official figures on the economic contribution of fisheries should be regarded skeptically (e.g., see compilations by Baran *et al.*, 2007a).

Because the nonmonetary sector is officially nonexistent, and the informal sector is usually underrepresented in official statistics, an apparent improvement in economic conditions (e.g., rising GDP) may actually reflect a loss of natural resources (including fisheries) and a simplification and monetization of production systems, rather than reflecting any actual improvement in living conditions for rural people. Fisheries are particularly likely to suffer from misguided interventions, because their size and importance are grossly underestimated (Coates, 2002) and are particularly sensitive to the impacts of development in other sectors. "Official" (but erroneous) figures may be extremely misleading, for example, "apparent consumption" figures published by the FAO, have been used to show that "highly urbanized populations consume more meat (including fish) per capita than those with less urbanized populations" (York and Gossard, 2004). A more plausible interpretation is that urbanization is correlated with improved data collection systems since the analysis is based on "official" consumption figures of: Cambodia 9, Laos 9, Thailand 34, and Viet Nam 17 kg person^{-1} year^{-1}, values which are much less than the survey-based figures presented in the following section.

4. LMB FISHERY YIELD AND VALUE

4.1. LMB Fisheries Yield

Yield can be estimated directly (from catches plus aquaculture yield) either from consumption and market sales, or indirectly based on estimates of yield per area. Official data on catches are likely underestimates (Coates, 2002). Accurate data have been collected for catches from some well-defined areas (such as rice fields or reservoirs), but large-scale surveys are often unreliable because of the difficulty of obtaining coverage of the huge diversity of gears (especially illegal gears) and fishers, and the likelihood that fishers routinely lie about catches to avoid taxes. A major catch survey in the 1990s in Cambodia produced national estimates that have been widely quoted (Baran and Myschowoda, 2008; Hortle *et al.*, 2004), despite the original authors explaining that the data were simply "expert guesses" (van Zalinge and Touch, 1996) with subjective correction factors applied (Sensereivorth *et al.*, 1999). Data from markets are not widely available and would not include the large proportion of fish and OAAs which are either directly consumed or traded off market. Therefore, approaches using yield per unit area (discussed below) or household consumption are most promising for accurately estimating the system yield.

Total consumption of fish and OAAs within the LMB was estimated based on 20 studies (Hortle, 2007). Various adjustments were made to standardize units to fresh weight, to fill in missing data in some studies (e.g., for preserved fish or OAAs), and to extrapolate to provinces where data were collected for only part of a province. The estimated total consumption of about 2.6 Mt year^{-1} (Table 9.1) is consistent with the range based on areal yield estimates. The MB's fisheries provide a significant proportion of the world fisheries yield, estimated

TABLE 9.1 Estimated consumption of fish and OAAs in the LMB, Year 2000 after Hortle (2007)

Country	Population (millions)	Inland fish	Inland OAAs	Inland fish plus OAAs	Marine products	Total aquatic
Cambodia	11.4	482	105	587	11	598
Laos	4.9	168	41	209	2	211
Thailand	22.5	721	191	911	130	1042
Viet Nam	17.5	692	161	853	129	982
Delta	15.1	640	149	789	119	908
Highland	2.4	52	12	64	10	74
Total	56.3	2062	498	2560	273	2833

Units are ×000 t year^{-1} as FWAEs, that is, converted to the weight of animals "as landed."

at 95 Mt year^{-1} from all capture fisheries and 59.4 Mt year^{-1} from aquaculture in 2004 (FAO, 2007). About 10% of the MB yield was assumed to be derived from aquaculture based on figures compiled by Phillips (2002), but these do not include much of the small-scale household yield. Household surveys suggest that aquaculture contributed about 20% and is probably increasing in importance, especially in the Viet Nam delta.

Consumption-based estimates may be biased to some extent by differences between responses in interviews and actual consumption figures. On the other hand, the consumption figures are conservative as basis for a yield estimate, because they do not take into account at least 120,000 t year^{-1} of small fish from the capture fishery that are used as "trash fish" to feed aquaculture fish and other animals, including crocodiles; they exclude wastage prior to consumption of at least 10% (Hortle, 2007); and they do not cover local clandestine exports which could be quite significant.

A large proportion of catches is consumed locally, but improvements in roads and marketing systems increasingly allow redistribution of catches and products from aquaculture, particularly from rural areas to cities (Baird and Flaherty, 2005; Phonvisay, 2003). Hence,

consumption within a province or country may include significant local (intrabasin) imports. It can be assumed that the highly productive provinces around the Tonle Sap-Great Lake system and in the upper delta are nett exporters, with Laos and northeast Thailand probably nett importers of both fresh and dried and fermented fish. Cambodian fisheries products are sold throughout the Viet Nam delta, but the export of aquaculture fish, including snakeheads and catfish to Cambodia probably balances this trade.

The consumption-based estimates are higher than official estimates, which grossly underestimate the entire LMB yield (Table 9.2). The discrepancy suggests that either the official figures are too low, as is likely (Coates, 2002) or that the consumption figures are too high, which seems less likely. Catches may vary significantly between years, and the proportion contributed by aquaculture is increasing, so the basinwide figure should be considered indicative.

Official aquaculture exports from the basin (mainly from Viet Nam) are now about 1 Mt year^{-1} and increasing (Section 6.1). Hence, an updated summary of the total basinwide yield is about 3.6 Mt year^{-1}, with aquaculture contributing about 40% of the total.

TABLE 9.2 LMB Year 2000 consumption-based estimates (k year^{-1}) compared with some "official" figures after Hortle (2007)

Country	Consumption-based estimate	Official estimate	Ratio (%)	Reference for "official" estimate
Cambodia	587	338	174	Sam *et al.* (2003): consumption and catch estimates
Laos	209	62	334	Souvanaphanh *et al.* (2003): areal yield times areas of habitat
Thailand	911	76	1204	Pawaputanon *et al.* (2003): commercial figures, mainly reservoirs
Viet Nam (delta only)	789	682	116	GSO (2003): yield minus sea catches
Total	2560	1157	221	

Units are ×000 t year^{-1} as FWAEs. Note that Cambodia recently adopted the consumption-based estimate as its official estimate (Nao Thuok, personal communication).

4.2. The Value of LMB Fisheries

Fisheries have both indirect and direct values. Indirect values include biodiversity values (see Chapter 8) and cultural values (Baran *et al.*, 2007a). Direct values are usually based on first-sale or market prices. Sverdrup-Jensen (2002) provided figures on total yield and value which equated to an average first-sale price of about US$0.73 kg^{-1} based on limited data, and which are now out-of-date. A preliminary appraisal of unpublished MRC data from detailed monitoring during 2003-2004 showed that weighted average prices across inland species in representative markets were lowest in the Viet Nam delta (US $1.40 kg^{-1}), slightly higher in Phnom Penh, Cambodia (US$1.60 kg^{-1}), higher again in some small city markets in northeast Thailand (US $1.90 kg^{-1}) and were highest in Vientiane, Laos ($2.60 kg^{-1}). These prices are biased towards larger, high-value species which are selectively traded in markets; on the other hand the market price of all foods has risen significantly since the survey, for example, by about 40% based on late 2008 figures in Vientiane. Figures quoted by Truong *et al.* (2008) suggest a first-sale price across all capture species in the Viet Nam delta

in 2004 of about US$0.62, and prices of at least $1.20 kg^{-1} for freshwater fish exports and US $3.10 for brackish water (mostly shrimp) exports. Phonvisay and Bush (2001) reported that retail selling prices of fish traded from the south of Laos to Vientiane were approximately double the first-sale price, a markup ratio reasonably consistent with the difference between the market and first-sale prices noted above for Viet Nam ($0.62/$1.40).

An up-to-date estimate of typical retail market prices (allowing for recent increases) is about US $2-$3.60 kg^{-1}, and for first-sale prices is about US$1-1.80 kg^{-1}. With a total yield of about 3.6 Mt, the LMB fishery is "worth" about US $3.6-6.5 billion dollars as a first-sale value. The economic value of fisheries also includes the downstream processing of products (Truong *et al.*, 2008), or associated industries. Of course the true worth of the fishery could be judged in many other ways, for example, based on its replacement cost, its value to livelihoods, its relative profitability or by including its benefits to and impacts on the environment, or perhaps in terms of opportunity costs. Nevertheless the price-based estimates provide a crude yardstick which is useful in comparisons with the often-quoted benefits from other sectors, such as hydroelectricity and irrigated agriculture. The uncertainties in the estimates illustrate the need for better data collection.

4.3. The Value of Associated Industries

Many industries support fisheries, including gear-making, boatbuilding, and salt and ice production, all of which are carried out at scales varying from individual households to large commercial operations. Most rural households near waterways own a small boat, typically a two-person wooden sampan 9-10 m in length, which costs about US$1000-2000 or about US$3000-5000 with an engine. It is likely that there are several million sampans in the MB with a combined value of several billion dollars. In 2004, there were 105,055 registered fishing boats (generally larger powered vessels) in the Viet Nam delta alone (Truong *et al.*, 2008). Salt is produced both by evaporation of sea salt (e.g., at Kampot in Cambodia) and also from salt mines on the Korat Plateau. About one-third of the 2.6 Mt of fish and OAAs that are consumed each year is preserved (Hortle, 2007), most by fermentation or salt-drying, which typically requires salt in the proportion

of one-fifth of the weight. Hence, the industry probably consumes about 170,000 tonnes of salt each year. The lack of ice, until recently, has led to a strong preference for sale of live animals, but ice is now produced in many small factories and transported widely to landing sites and to markets; however, there are no accurate statistics on the industry.

4.4. Fisheries and Nutrition

Fishery products (fish and OAAs) are the main source of animal protein throughout the basin (Hortle, 2007), as well as a key source of micronutrients. National average values for per capita consumption (Table 9.3) are within the ranges generally reported for developed countries (about 13-67 kg person^{-1} year^{-1}), and actual consumption on average is about 75% of consumption expressed as Fresh Weight Animal Equivalent (FWAE) (Hortle, 2007). Per capita consumption varies widely from each country, being highest near productive water bodies and lowest in drier or mountainous regions. A large proportion of fish are small and are eaten whole, so their skeletons are an important source of calcium in a region where dairy foods are rare; the iron provided by fish is particularly important for women, and fish

TABLE 9.3   Per capita fish and OAA consumption estimates

Country	Inland			Marine products	Total aquatic
	Fish	OAAs	Fish plus OAAs		
Cambodia	42.2	9.2	51.4	1.0	52.4
Laos	34.6	8.4	43.0	0.5	43.5
Thailand	32.0	8.5	40.5	5.8	46.2
Viet Nam	39.5	9.2	48.7	7.4	56.1
Delta	42.5	9.9	52.4	7.9	60.3
Highland	21.2	4.9	26.1	4.1	30.2
TOTAL	36.6	8.8	45.5	4.9	50.3

Units are kg per capita year^{-1} as FWAEs, not actual consumption, after Hortle (2007).

eyes and some of the other organs contain high concentrations of vitamin A (Mogensen, 2001; Roos *et al.*, 2007a,b). The health benefits of fish have been widely promoted, but in the LMB the habit of eating raw or partly preserved fish and OAAs has led to high rates of infection with certain parasites, including liver and intestinal flukes (Hortle, 2008); educating people to cook fish properly should be a key element of public health education. Fisheries development is often promoted as a way to increase the intake of nutrient-dense foods to combat malnutrition, which remains common in the LMB (MRC, 2003). However, the benefits of any dietary improvement may be negated by the effects of common water and food-borne parasites that infect most people in the basin (Nguyen *et al.*, 2006b, Phathammavong *et al.*, 2007; Tep *et al.*, 2006; Wongstitwilairoong *et al.*, 2007). Improvements in basic hygiene and sanitation must be made at the same time as improvements in the food supply, which should include fisheries within a general development framework.

4.5. Habitats and Yield

Aquatic habitats are most extensive in the lowlands; of most importance to fisheries are (1) large river-floodplain systems, (2) rice fields and associated habitats, (3) large man-made reservoirs, and (4) the estuary and brackish-water zone. Marine fisheries (coastal and offshore) also depend upon nutrients in the Mekong's plume and the division of the fishery into inland and marine by a line across the mouth is quite arbitrary in an ecological sense. Numerous smaller streams and rivers run from uplands of northern Laos, northern Thailand and the Annamite mountain chain (which roughly follows the western border of Viet Nam), as well as from the smaller mountains that delimit the catchment in southeast and northern Cambodia; uplands form a fifth habitat that is of relatively minor direct importance in fisheries. Land-use data (Table 9.4

and Figure 9.1) indicate that wetlands—land that is covered permanently or seasonally by water—cover about 30% of the LMB. Most (about 80%) of the wetland area is rice field habitat. Virtually all wetland habitats are in lowland areas, whereas forest and swidden characterize uplands.

At least 90% of the wetland area is seasonal as is evident from the low proportion of permanent water bodies. Seasonality is more extreme than suggested by the GIS data, because digitizing the area of permanent water bodies (with the exception of the Great Lake) gives a value about 80-90% of their maximum area.

Major floods cover up to 58,000 km^2 or about 30% of the flood area (Table 9.5); this figure includes 6300 km^2 (or about 10%) of permanent water bodies. Other wetland areas are not subject to flooding by the major rivers but are inundated by local rainfall or by local diversions from smaller watercourses. About 60% of the wetland areas in Cambodia and Viet Nam lie within the major flood zone, which includes the highest quality habitats for fisheries production. About half of the land covered by floodwaters is rice fields, about one-quarter is more natural land such as forests, shrubland, or swamps, and the remainder includes plantations (such as flood-tolerant paperbarks (*Melaleuca*)) in the delta.

The estimates of wetland areas are simplified in Table 9.6 and combined with estimates for areal yields from each habitat (as discussed below) to show that the total yield of the inland capture fishery of the LMB is about 1.3-2.7 Mt year^{-1}, a range that is consistent with the estimate from consumption studies. The two main habitats contributing most to the total catch are likely to be river-floodplain habitats (high areal yield over a moderate total area) and rice field habitats (low-moderate yield over a very large area). However, the extent of wetlands and the areal yield figures are subject to considerable uncertainty which should be reduced by further research.

TABLE 9.4 Land use and wetlands in the LMB

Land use type	Laos	Thailand	Cambodia	Viet Nam delta	Viet Nam highlands	Total LMB	% of total
Rice field habitats	10.6	98.3	28.5	20.2	1.6	159.2	25.1
Forest within flood zones and grassland/shrub/swamps	2.6	1.4	11.7	1.0	0	16.8	2.6
Large-scale aquaculture	0	0.1	0	2.4	0	2.4	0.4
Permanent water bodies	2.4	4.5	5.0	1.8	0.2	13.8	2.2
Other land in flood zone—includes plantations and crops	0.1	0.3	1.5	3.7	0	5.6	0.9
Total wetlands	**15.7**	**104.5**	**46.7**	**29.1**	**1.8**	**197.8**	**31.3**
Wetland as % of total	7.6	51.5	29.9	84.6	5.6	31.3	
Forest and degraded forest	182.3	56.2	102.0	1.7	25.7	367.8	58.1
Field crop	1.2	35.4	3.5	0.4	1.8	42.3	6.7
Plantation	0	5.6	0.7	3.0	2.1	11.5	1.8
Swidden	5.2	1.4	3.1	0	0	9.6	1.5
Urban and other	2.2	0	0.4	0.2	1.1	3.9	0.6
Total nonwetlands	**190.9**	**98.6**	**109.7**	**5.3**	**30.6**	**435.1**	**68.7**
Nonwetlands as % of total	92.4	48.5	70.1	15.4	94.4	68.7	
Total	**206.6**	**203.1**	**156.4**	**34.4**	**32.4**	**632.9**	**100**

Areas are ×000 km^2. Land use is based on MRC GIS data post-2000, which is more current and comprehensive than MRC wetland data cited by Hortle (2007). GIS data does not resolve small areas of a particular class, so for example, aquaculture areas are underestimated.

5. HABITATS AND THEIR FISHERIES

5.1. Lowland River-Floodplain Systems and Their Ecology

The Mekong River and large tributaries such as the Tonle Sap-Great lake system include a great variety of in-stream habitats as well as a range of seasonal and permanent water bodies (lakes and marshes) on their floodplains, of which the Great Lake is the largest. The flow of rivers is highly seasonal with much of the total annual discharge during the wet season (see Chapter 4). Although storage in reservoirs and abstraction for irrigation are significant in

Thailand, overall the river system is relatively unregulated, with a fairly predictable monotonic flood pulse each year. In Cambodia and the upper Viet Nam delta floodplains are less than 20 m ASL and most of the wetland area is flooded by river water in most years (Table 9.5); the Great Lake depth increases by up to 9 m during each flood season (Chapter 11), and much of the floodplain is covered by several metres of floodwater for 3-4 months each year. Because of the extensive, deep, and prolonged flooding, this part of the LMB is generally considered to support the most productive fisheries (Lamberts, 2006; van Zalinge et al., 2004). In Thailand and Laos, a small proportion of the total wetland area comprises

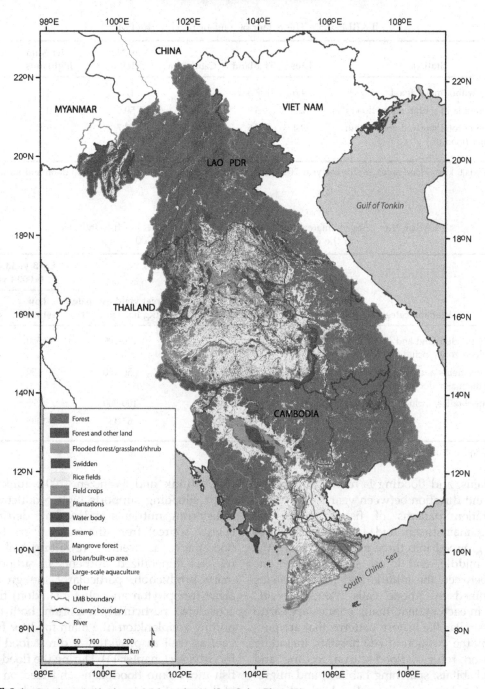

FIGURE 9-1 Land-use in the lower Mekong basin (See Color Plate 20).

TABLE 9.5   The extent of a major flood in the LMB

Statistic	Laos	Thailand	Cambodia	Viet Nam delta	Viet Nam highlands	Total LMB
Land area within major flood	4.6	7.8	28.3	17.3	0	58.0
Proportion of land within major flood (%)	2.2	3.8	18.1	50.5	0	9.2
Proportion of total wetland area within the major flood (%)	29.4	7.5	60.5	59.7	0	29.3

Units are $\times000$ km^2. Flood areas are for the year 2000 estimated from Radarsat imaging, elevation data, and modeling.

TABLE 9.6   Approximate extent of wetlands and likely yield of fish and OAAs in the capture fishery of the LMB, Year 2000

Wetland category	Area ($\times000$ km^2)	Areal yield estimate (kg ha^{-1} year^{-1})	LMB yield estimate ($\times000$ t year^{-1})	
			Low estimate	High estimate
Seasonally flooded land and water within the major flood zone, including some rice fields	58.0	100-200	580	1160
Rain-fed rice fields and associated habitats not within the major flood zone	129.9	50-100	650	1299
Large water bodies, including reservoirs	7.5	100-300	75	225
Total	**195.4**	**67-137**	**1305**	**2684**

Excludes aquaculture.

floodplains, and flooding is highly variable in extent and duration between years.

Migration patterns of fishes within the Mekong mainstream and lowland tributaries can be grouped into three main "systems": the upper, middle, and lower systems, with the break between the middle and lower systems well-defined by Khone Falls (Poulsen et al., 2002a). In each system, many species appear to exhibit similar life history patterns that are governed by the position of key habitats, including dry-season refuges, flood-season feeding and rearing habitats, spawning habitats, and migration routes. However, the entire lowland river-floodplain system can be regarded ecologically as a single unit under the flood pulse concept (FPC) (Junk and Wantzen, 2004; Junk et al., 1989). Flooding supports high productivity by transferring nutrients and organic detritus (an energy source) from the main rivers to their floodplains, as well as rewetting soil which releases mineralized nutrients to support primary production, particularly the growth of algae (periphyton and phytoplankton) that are considered particularly important both in supporting zooplankton (the main food for fish larvae) as well as providing a direct food source for fish (see Chapter 11). During the flood, larger fish move onto floodplains and feed on terrestrial vegetation, fruits, detritus, and terrestrial animals (such as insects and other arthropods), and many fish species also spawn in flooded areas. The fry of black and gray fish originate from spawning on the floodplain, whereas white fish fry—from spawning upstream in rivers—arrive as drift in the rising floodwaters (Chea et al., 2003; Nguyen et al., 2006c, 2008; Thach et al., 2006). Spawning in upper Cambodia results in enormous quantities of fry drifting downstream with rising floodwaters; one study made a crude estimate of 120 million fish per day, most of which are cyprinids or Pangasiid catfishes. Invertebrates such as shrimps are also a large component of the early wet-season drift (Hortle et al., 2005a).

Most of the larger "white fish" leave floodplains soon after water levels peak and then migrate via rivers to dry-season refuges, such as deep pools, which may be distant from the floodplains, for example, in northern Cambodia and near Khone Falls (Baran et al., 2005; Poulsen et al., 2002b). Most of the smaller white fish follow later towards the end of the flood recession, moving with the falling waters in migration "waves," which seem to be timed around lunar cycles (Baird et al., 2003a). Black and gray fishes move short distances to dry-season refuges on or near floodplains. Catches of small-scale fishers may comprise mostly black or gray fish, especially where there are significant floodplain water bodies (Dubeau et al., 2001), whereas large-scale commercial fisheries tend to target migratory white fish, as shown by summary data for Cambodia by van Zalinge and Nao (1999). However, black fish productivity may depend partly on feeding upon white fish which have migrated onto floodplains, so there is a need both for better data on catch composition as well as on the basis for production. Fishing pressure is very heavy during the flood recession when fish are migrating, as they are caught in large quantities by gears that filter the water (such as dais) or that use large fences or barriers to divert them into traps. Many small-scale fishers are active only during the flood recession when fish are highly concentrated and are most catchable; areas where fishing is particularly intense include the Tonle Sap, and Khone Falls, just upstream of the border of Laos and Cambodia (Baran et al., 2005).

World data (Welcomme, 1985) and recent detailed studies in other Asian systems (de Graaf et al., 2001) show that total catches (kg ha^{-1} year^{-1}) are well-correlated with the extent of annual flooding, as is also evident in the only long-term dataset of a significant fishery in the MB, where catches are well-correlated with flood height or the area and duration of flooding (Halls et al., 2008). Where fishing effort and other conditions are constant, larger catches result from larger numbers of fish or higher mean size or both (e.g., Halls et al., 2008). In general, larger floods of longer duration allow more fish to survive and grow to a larger size, so flood amplitude and duration have a direct effect on fish catches, other factors being equal. Other features of the flood pulse such as its timing, continuity, smoothness, and rapidity of change (Welcomme and Halls, 2004) may affect productivity (Lamberts, 2008), but there is no information which would allow a precise prediction of the importance for fisheries production of changes in the shape or timing of the flood pulse. Retention of water on floodplains during the dry season also enhances fish production, but the "optimal" balance between wet- and dry-season flows is unclear, because production does not depend upon hydrology alone, but other factors such as nutrient release from exposed sediments during the flood pulse (Welcomme and Halls, 2004).

The Mekong and its tributaries transport processed allochthonous (terrestrial) organic material and nutrients to the lowlands and the estuary in a more or less predictable way, with a succession of organisms associated with habitat changes down the system and dependent upon the flow of materials and energy broadly conforming to the river continuum concept (RCC) (Vannote et al., 1980). A downstream transport of fine particulate organic matter and dissolved forms of organic carbon takes place continuously, and the quantity and variety of visible coarse particulate matter moved downstream by the Mekong during major floods is quite extraordinary—many thousands of tonnes of trees, logs, sticks, leaves, bark, and detritus are flushed through each day. Such organic material supports detritus-based food chains on floodplains as well as the many species of "mud-eating" animals in the estuary and coastal waters. The RCC has not been studied in the Mekong system, but as a generally accepted concept it is useful in predicting, for example, that reservoirs are likely to reduce productivity downstream in rivers and the estuary by trapping sediment, organic material and nutrients, and that the ongoing logging of the catchments is likely to have a wide range of effects downstream.

Although the FPC and the RCC have some predictive value, the Mekong system differs greatly from the river systems on which these concepts were developed, mainly in North and South America, where rice fields are not a common land use. The FPC and RCC also primarily relate to the ecology of "natural" systems, hardly appropriate for the Mekong where, for example, nutrient flows in some areas may be dominated by the inputs from artificial fertilizers (MRC, 2003), as well as untreated sewage disposal for 70 million people; 24 million livestock (cattle, buffalo, and pigs) (Nesbitt et al., 2004); and many more ducks, chickens, and other domestic animals. It seems obvious that the river system (including its fisheries) cannot be managed sensibly without a much better understanding of its ecology, particularly the relative importance of the various flows of nutrients and energy.

5.1.1. Yield from Floodplain-River Systems

Other factors being equal, a larger river system—as indicated by catchment area or river length—tends to produce a larger total catch

of fish, as discussed in detail by Welcomme (1985). Hoggarth et al. (1999a) suggest that typical floodplain yields are 70-140 kg ha^{-1} year^{-1}, and a recent review of world data by Halls et al. (2006) suggest that catches are sustained around 100-150 kg ha^{-1} year^{-1} over a range of fishing effort of 5-30 fishers km^{-2}. Comprehensive data from Bangladesh, where hydrology and the fauna are similar to the Mekong, include a range of yields of up to 574 kg ha^{-1} year^{-1} for wild fish (Ali, 1997), with some higher yields recorded where floodplain water bodies were stocked. The main environmental factors which affect yields have been well-documented in a number of studies that systematically compared catches from a range of river-floodplain systems (Ali, 1997; de Graaf et al., 2001; Halls et al., 1999; Hoggarth et al., 1999a). The highest yields are consistently associated with systems where floodplains are (1) open to fish access from adjacent rivers, (2) have deep and extended flooding, and (3) have depressions that allow water to persist through the dry season. Conversely, low yields are recorded where floodplains are (1) isolated from rivers by levees, (2) relatively shallow, and (3) dry out quickly. These findings are likely to apply to the Mekong system and are consistent with much anecdotal information, although they have not been documented in systematic studies. The theoretical bases for fish production in floodplain fisheries are discussed in Halls et al. (2001), Halls and Welcomme (2004) and Welcomme (1985), and are generally supported by empirical studies, which show that most of the variation in catches between years is directly related to the size of the annual flood, the retention of water in the dry season, and fishing pressure.

5.1.2. Yield from LMB Floodplains

Only four studies provide estimates of yield from floodplains in the Mekong system based on actual catch measurements.

Dubeau *et al.* (2001) estimated yield from a well-defined "floodplain" area adjacent to the Tonle Sap. About 82% of the area would be flooded during the highest flood and perhaps half of the surveyed area is active floodplain in an average year; most is rain-fed or recession rice fields and water persists through the dry season in permanent lakes which cover about 5% of the area. Fishers were surveyed throughout the year using logbooks and commercial catches were monitored accurately. Several approaches were used to estimate the total catch of 243-532 kg ha^{-1} year^{-1} (mid-value 388 kg ha^{-1} year^{-1}) from the entire area. Small-scale artisanal fishers caught 94% of the catch. The total estimate includes only fish and OAAs caught within the study area, that is, not including fish which swam into the Tonle Sap and were caught there (or elsewhere), therefore an estimate of 300-400 kg ha^{-1} year^{-1} may be conservative, and would be consistent with the study site having the attributes that favor high yield, as mentioned earlier for Bangladesh.

In the Mekong delta over a 90-day flood fishing season, a site in a deepwater but acid-affected floodplain produced 63 kg ha^{-1} and a deepwater, nonacid site produced 119 kg ha^{-1} (de Graaf and Chinh, 2000). In this region, rice is grown throughout the dry season by irrigating from canals, so dry-season catches in rice fields and associated habitats would increase the total yield.

Troeung *et al.* (1999) compared the yields from three large commercial fishing lots on forested, partly cleared (31% forest), and completely cleared floodplains. This study appeared to show little effect of partial clearing (yield falling from 95 to 92 kg ha^{-1} year^{-1}) and a relatively large reduction (to 55 kg ha^{-1} year^{-1}) in completely cleared areas, with the total value of the catch declining disproportionately as large high-value species such as snakeheads became less abundant. However,

because artisanal catches were not included it is likely that total catches were much higher in all three study areas, as in the study by Dubeau *et al.* (2001) discussed previously, and would perhaps have been highest in the completely cleared area, where the largest numbers of artisanal fishers would be operating, and where catches would be made from recession rice fields during the dry season.

In Cambodia, but outside the Mekong system, a yield of 630 kg ha^{-1} year^{-1} was reported from a largely artificial coastal floodplain created by polders (Lim *et al.*, 2005). Although this yield seems high, the study area is deeply flooded (to 2 m) each year and a large proportion is permanent water bodies which support year-round fishing. The flooded area may also have been underestimated by a factor of about 2, so a yield of 300-400 kg ha^{-1} year^{-1} may be reasonable.

Crude estimates for the Tonle Sap-Great Lake system include 139-190 kg ha^{-1} year^{-1} (Lieng and van Zalinge, 2001) and 230 kg ha^{-1} year^{-1} (Baran *et al.*, 2001), but these were apparently based on estimates of catch which are subject to great uncertainty.

Overall, the limited data from the LMB, and studies elsewhere, indicate that a likely yield from river-floodplain systems is 100-200 kg ha^{-1} year^{-1}, with possibly higher yields in more productive parts of the system in Cambodia and the upper parts of the Viet Nam delta, and lower yields where flooding is of relatively short duration and depth, such as in Laos and Thailand.

5.2. Rice Fields and Associated Habitats

Rice originated in Asia, and the dominant lowland cultures in the LMB are often regarded as "rice-fish" societies. Modification of landscapes for rice farming has been proceeding for over 2000 years in the LMB, and traditionally managed rice fields teem with life. The abundant black fish are carnivores (or mainly carnivorous), feeding on tadpoles, frogs, crabs, shrimps, and insects, which themselves are also significant elements of fisheries (Halwart, 2006; Hortle et al., 2008). Rice field fertility and the food chain depend upon the rapid turnover of a nonrice aquatic biomass of a few hundred kg (dry weight) ha^{-1}, with blue green algae particularly important in biological nitrogen fixation (Roger and Kurihara, 1988); artificial fertilization and conversion of rice stubble to manure in the dry season by cattle grazing increases available nutrients. The managed inundation of rice fields stimulates nutrient release from sediments, initiating primary production, with a consequent succession of plants and animals adapted to the temporary environment (Bambaradeniya and Amarasinghe, 2004; Heckman, 1974, 1979). In many respects rice fields are miniature floodplains, with a short-managed "flood pulse," similar to that described for larger natural systems (Junk et al., 1989), despite the different origin of the floodwater. Rice fields are, however, generally more stressful environments, with a lower diversity than natural floodplains, but with relatively high productivity and yield when the shorter duration of inundation is taken into account.

Fish and OAAs typically breed several times during the wet season, with fish fry or immature stages colonizing newly flooded fields. Most large rice field animals, including the common black fishes, can move over land, bypassing the many barriers that are built through the agroecosystem. Growth is rapid and many species reproduce continuously, so that fishers harvest aquatic animals throughout the wet season along the edges of rice fields and in nearby water bodies (Hortle et al., 2008; Meusch et al., 2003). In the dry season, some animals survive in refuges (canals, ponds, and nearby streams) while some aestivate deep below the dried surface of the fields, where they may be dug up by "mud-fishing." Insects are a major component of the fishery, with large quantities of dragonfly larvae (Manning

and Lertprasert, 1973) caught with small seines, and large water beetles and bugs (Hortle et al., 2005b) caught by light trapping.

As well as black fishes, river/stream fishes may migrate into rice fields for short periods to spawn and feed, with the proportion of white fish in catches dependent on the connectivity to nearby floodplain water bodies, rivers, or streams (Meusch et al., 2003). However, rice fields are at times hostile environments for river fishes, because the water underlying actively growing rice is typically anoxic and contains limited palatable biomass (Vromant et al., 2004). Fish and OAAs which spend significant periods of time in rice fields must be tolerant of anoxia, elevated temperatures and fluctuating water levels. Snakeheads, walking catfish, and climbing perch dominate rice field catches throughout the LMB (Hortle et al., 2008) and are probably the most important fishery species in the LMB, given the vast extent of rice fields and associated habitats, and the fact that these species may also be the most abundant on floodplains. As the environment of floodplains is modified for rice cultivation the proportion of blackfishes increases. In five provinces of northeast Thailand, over half of the weight of fish that were eaten by people comprised black-fish, consistent with the recent expansion of rice-field habitats (Prapertchob et al., 1989). In Long An province, Viet Nam, an intensively farmed rice field landscape, snakeheads, climbing perch, and walking catfish also made up more than half of the catch (Pham and Guttman, 1999). These amphibious carrnivores, together with frogs, may also be important in controlling some of the pests of rice, so reducing the need for use of pesticides. The dominance of carnivores is of some interest and contrasts with river-floodplain fisheries, where omnivorous and herbivorous fish are most important in catches.

At higher, colder locations in the north of the basin, rice fields are often stocked with temperate or subtropical fishes, such as carp and goldfish, which have formed wild populations. The need for stocking arises because diversity, biomass, and richness in terraced rice fields decline with altitude (Margraf *et al.*, 1996).

Apart from rain-fed rice which accounts for about 80% of all rice fields by area (Nesbitt *et al.*, 2004), rice is also grown on floodplains as floodwaters are rising (long-stemmed flood rice) and in paddies as floodwaters are receding (recession rice). Within the floodplains, recession rice is the dominant land use; rice fields retard the flood recession and with associated water bodies support a dry-season fishery. Hence, floodplains support a double-pulse of fisheries productivity—predominantly white and gray fish are caught as they retreat from the floodplain during the recession, whereas a fishery during the dry season in rice fields and floodplain water bodies catches predominantly black and gray fish. In the Viet Nam delta, most rice fields are double cropped for rice (each crop requiring about 100 days), with small quantities of fish and OAAs also caught, and then the fields are inundated by floodwaters which support fisheries production. Rice-fish systems are widely promoted and rice is also grown in rotation with shrimps in coastal areas.

5.2.1. *Yield from Rice Fields*

The yield of wild fish and OAAs from unstocked rain-fed rice field habitats can be conservatively estimated as 50-100 kg ha^{-1} year^{-1}, based on several studies in the LMB and elsewhere, as discussed by Hortle *et al.* (2008). Yields are likely to be favored by inundation of rice fields to greater depths and for longer duration (Khoa *et al.*, 2005), and where farmers maintain ponds as dry-season refuges (Angporn *et al.*, 1998). Small water bodies including ponds and reservoirs up to about 100 ha in area are usually intimately connected with surrounding rice fields, and fish and fishers tend to move through the landscape;

their yield cannot generally be separately accounted, but is part of the "rice field landscape" yield. As discussed below, small water bodies may be very productive, which may compensate for losses of fishery production in intensively farmed landscapes.

Expansion of the area of rice fields may impact river-floodplain fisheries by depriving them of water through many small-scale diversions into fields, but losses to river fisheries may be compensated for by the additional catches of the more restricted suite of fish and OAAs from rice fields, as well as by capture in small reservoirs (Khoa *et al.*, 2005). The extent of compensation for any losses depends upon management; highly intensive cultivation—high yielding varieties with shallow, short-duration flooding, and high pesticide use may support very limited fisheries, whereas rice-fish culture is likely to produce the highest yields.

5.3. Reservoir Fisheries

In recent decades, dams have been built on all MB river systems in northeast Thailand, several large dams have been built on major tributaries in Laos and Viet Nam, and many more dams are being built or planned. In the MB in Thailand, Virapat and Mattson (2001) estimated there were 1872 reservoirs (mostly built for irrigation) with individual surface areas greater than 100 ha, their total area was 2120 km^2, which may be an underestimate, because the combined area of the 13 largest Thai dams alone is 1665 km^2 (Table 9.7). The 22 largest existing LMB dams have a combined surface area of 2737 km^2, so the total area of LMB reservoirs (larger than 100 ha) is probably 4000-5000 km^2. Most dams were built for irrigation, with relatively few until recently built for hydroelectricity production; some dams—classed as multipurpose—may have small hydroelectric plants while also providing irrigation water. Most dams are designed to store wet-season flows for release during

TABLE 9.7 Some key features of the largest dams in the LMB

Country	Dam name	Code	Status	River system	Completion	Purpose	Elevation (m ASL)	Wall height (m)	Wall length (m)	Inundated area (km²)	FSL volume (million m³)	Mean depth (m)	Catchment (km²)	Inflow (m³ s⁻¹)
China	Xiaowan		Construction	Mekong	2012	Hydro	1236	292	nd	190	14,550	76.6	113,300	1220
	Manwan			Mekong	1996	Hydro	994	132	418	nd	920	nd	114,500	1230
	Dachaoshan			Mekong	2003	Hydro	895	111	460	nd	880	nd	121,000	1340
	Mengsong			Mekong	2008?	Hydro	519	28	nd	nd	nd	nd	160,000	2020
Laos	Houay Ho	HH		Se Kong	1999	Hydro	883	79.5	400	42	620	14.8	192	9.5
	Nam Leuk	NL		Leuk	2000	Hydro	405	45.5	800	13	185	14.5	274	16.4
	Nam Theun 2	NT2	Filling	Theun	2010	Multi	538	45	48	450	3680	8.2	4013	245.3
	Nam Ngum (1)	NN		Ngum	1971/84	Multi	212	75	468	370	7000	18.9	8460	427
Thailand	Nam Pung	NP		Pung	1966	Multi	284	41	1720	22	165	7.7	296	4
	Lam Nam Rong	NR		Rong	1991	Irrig	143	23.5	1500	25	218	8.7	453	5
	Chulaphorn	CP		Phrom	1972	Multi	759	70	700	12	188	15.7	545	5
	Huai Luang	HL		Huai Luang	1973	Irrig	198	12.5	1400	31	113	3.6	666	Small
	Nam Pra Phloeng	PP		Pra Phloeng	1968	Irrig	272	50	575	19	220	11.6	807	6
	Nam Oon	NO		Oon	1973	Irrig	178	29.5	3300	85	520	6.1	1100	12
	Kwan Phayao	KP		Ing	1941	Fisheries	405	5	10	24	11	0.5	1161	Small
	Lam Ta Khong	TK		Ta Khong	1969/2001	Irrig	227	40.3	527	44	445	10.1	1430	8

Name	Code	River	Year	Status	Purpose								
Nong Han	NH	Kam	1953		Fisheries	157	5	135	200	64	0.5	1653	Small
Sirindhorn	SI	Dom Noi	1971		Multi	142	42	288	940	1966	6.8	2097	53
Lam Pao	LP	Pao	1968		Irrig	160	33	400	7800	2640	6.6	5964	45
Ubolratana	UR	Pong	1966		Multi	182	35.1	410	800	2264	5.5	12,104	71
Pak Mun	PM	Mun	1994		Multi	108	17	60	324	350	5.8	117,040	759
Rasi Salai	RS	Mun	1994		Irrig		9	110	nd	~440	~4	~48,000	~310
Viet Nam Buon Tua Srah	BT	Sre Pok	2008?	Construction	Multi	488	83	37	1035	787	21.2	2930	100
Plei Krong	PK	Se San	2006		Hydro	570	71	53	495	1049	19.7	3216	128
Yali Falls	YF	Se San	2000		Multi	515	69	53	1190	1037	19.5	7455	270
Se San 4	S4	Se San	2010	Construction	Hydro	215	74	54	850	893	16.5	9326	329

Showing only those with surface area >20 km² or volume >150 million m³.

the dry season, and as such reduce the flood pulse downstream, inevitably impacting river-floodplain fisheries. The combined storage of the existing large dams (excluding those under construction) (Table 9.7) is about 25 km^3, so as yet they have only a minor effect on the seasonality in flow of the entire Mekong, which discharges about 475 km^3 year^{-1} (MRC, 2005).

Many of the reservoirs behind dams support productive fisheries for both indigenous and exotic fishes as well as some other animals, including shrimps (Bernacsek, 1997; Phan and Sollows, 2001; Pholprasith and Sirimongkontha-worn, 1999; Sricharoendham et al., 1999). Reservoirs are not suitable for many river species, which may decline or disappear, but up to about 100 species persist in the largest reservoirs such as Ubolratana (Pawaputanon, 1986) and about 55 species are common in catches in Nam Ngum (Mattson et al., 2001). Fewer species persist in smaller reservoirs, a "species-area effect" caused inter alia by fewer niches and less spawning habitat. Not surprisingly, the reservoir fauna includes black fishes (such as snakeheads, gouramies, and spiny eels) which live in standing waters, as well gray fishes (such as some cyprinids, some catfishes, glass-fishes, leaf-fishes, and some gudgeons) which are floodplain spawners. However, larger reservoirs with large tributary rivers continue to support many species that require running water for spawning, including many migratory cyprinids (especially carps) and catfishes (e.g. Mattson et al., (2001) for Nam Ngum reservoir). The larger reservoirs have an extensive pelagic zone that seems to be well-suited to at least one indigenous species of planktivorous freshwater herring, *Clupeichthys aesarnensis* (Jutagate et al., 2003), which may form a large component of catches; for example, about one-third of total catches in Nam Ngum (Mattson et al., 2001) and 50-60% of catches in Sirindhorn Reservoir (Sricharoendham et al., 1999).

There are no fish in the Mekong system that have evolved within large lakes, so fish from other continents are often stocked to fill particular niches. These include African tilapias (*Oreochromis* spp.), hardy fish which are adapted to the littoral zone, herbivorous grass carp (*Ctenopharyngodon idella*), and planktivores, including silver carp (*Hypophthalmichthys molitrix*) and bighead carp (*Hypophthalmichthys nobilis*), which are large fast-growing species that can support commercial fisheries. Tilapias and silver carp are particularly useful in eutrophic systems as they can feed on and digest and assimilate blue-green algae (Piyasiri and Perera, 2001; Xie and Liu, 2001), which are thought either not to be eaten or to be inefficiently processed by indigenous fishes.

5.3.1. Productivity and Yield of Reservoirs

Secondary production in reservoirs is often high initially, when nutrients are released from inundated terrestrial vegetation, and then may decline over a period of some years, later stabilizing at a lower level. However, in the MB, the natural decline in productivity may be offset where other sources of nutrients or organic material compensate, for example, where large tributary rivers supply organic material from a forested catchment, as postulated for Nam Ngum (Mattson et al., 2001). An expansion of population into the catchment area of reservoirs may lead to increased nutrient inputs which can increase productivity, but at the risk of sedimentation. In the few cases where catches have been accurately monitored after a reservoir filled, small-scale artisanal catches appear to increase, whereas catches of larger predatory species decline as they are fished down (Mattson et al., 2001; Pholprasith and Sirimongkonthaworn, 1999). Official data, such as those compiled by Bernacsek (1997), appear to be the estimates of commercial catches and understate yields in comparison with studies where catches have been accurately monitored (Table 9.8). Artisanal catches in Ubolratana

TABLE 9.8 Estimated fisheries yields from reservoirs and small water bodies in the LMB

Waterbody	Location	Year constructed	Year(s) of survey	Area (km²)	Mean depth (m)	Catch (t year⁻¹)	Yield (kg ha⁻¹ year⁻¹)	Species makeup	Dominant fishes	Source
Ho 31 Reservoir	Viet Nam Highlands	nd	1997-1999	0.0537	~1	6	1139	99% stocked exotic	Silver carp, bighead carp, common carp, and Indian carps	Tran et al. (2001)
Yang Re Reservoir	Viet Nam Highlands	1984	1997-1999	0.56	6.1	32	575	87% stocked exotic	Silver carp, bighead carp, common carp, and Indian carps	Tran et al. (2001)
Ea Kar Reservoir	Viet Nam Highlands	1978	1997-1999	1.41	5.2	55	388	98% stocked exotic	Silver carp, bighead carp, common carp, and Indian carps	Tran et al. (2001)
Ea Kao Reservoir	Viet Nam Highlands	1979	1997-1999	2.1	5.1	123	588	77% stocked exotic	Silver carp, bighead carp, common carp, and Indian carps	Tran et al. (2001)
Ea Soup Reservoir	Viet Nam Highlands	1980/2002	1997-1999	2.4	6.1	51	214	98% self-recruiting indigenous	Indigenous fish	Tran et al. (2001)
Lak Lake	Viet Nam Highlands	Natural	1997-1999	6.58	1.0	83	126	97% self-recruiting indigenous	Indigenous fish	Tran et al. (2001)
Huai Muk Reservoir	NE Thailand	nd	2002?	2.0	~1	13	66[a]	79% exotic	Common carp	Nachaipherm et al. (2003)
Kaeng La Wa Reservoir	NE Thailand	1983	2002?	19	1.4	512	269	~62% exotic	Common carp and Nile tilapia	Nachaipherm et al. (2003)
Huai Luang Reservoir	NE Thailand	1973	2000	31	3.6	781	252	63% exotic	Nile tilapia and common carp	Nakkaew et al. (2002)

Continued

TABLE 9.8 Estimated fisheries yields from reservoirs and small water bodies in the LMB—Cont'd

Waterbody	Location	Year constructed	Year(s) of survey	Area (km²)	Mean depth (m)	Catch (t year⁻¹)	Yield (kg ha⁻¹ year⁻¹)	Species makeup	Dominant fishes	Source
Nam Oon Reservoir	NE Thailand	1973	2002?	85	6.1	1032	121	50% exotic	Common carp	Nachaipherm et al. (2003)
Nam Ngum Reservoir	Laos	1971/84	1998	370	18.9	6833	185	All self-recruiting indigenous	*Clupeichthys aesarnensis* (28%), cyprinids	Mattson et al. (2001)
Ubolratana Reservoir	NE Thailand	1965	1992	410	16.0	3714	61	97% self-recruiting indigenous	Cyprinids	Polprasith and Sirimong-konthaworn (1999)
16 Village Ponds	Thailand		1994-1996	1.8-20 ha	~2		26-2881, med. 652	Most stocked	Silver carp, bighead carp, common carp, Indian carps, silver barb, and Nile tilapia	Lorenzen et al. (1998a)
17 natural and reservoirs	Laos		1995-1997	1-60 ha	39,569.0		60-690	Various	Stocked and indigenous	Lorenzen et al. (1998b)

[a] Reservoir was silted and weed-choked.

Reservoir make up about one-half of total catches according to Pholprasith and Sirimong-konthaworn (1999) (Table 9.4).

The features that favor high productivity per unit area in reservoirs include shallow depth, small size, and optimal nutrient status. Stocking and higher fishing pressure lead to higher catches, as is evident from systematic catch assessment in highland reservoirs in Viet Nam (Phan and Silva, 2000; Tran *et al.*, 2001) and from smaller water bodies in Thailand (Lorenzen *et al.*, 1998b). Accurate data on a few reservoirs (Phan and Sollows, 2001) and official data from all Vietnamese reservoirs (Ngo and Le, 2001) show that yield declines exponentially with reservoir size.

The highest yields are recorded from small, shallow, stocked reservoirs or other small water bodies (Table 9.8). Smaller reservoirs are more productive for fisheries for several reasons: (1) smaller reservoirs have more shoreline relative to area and shorelines are more productive and accessible than deeper water, (2) they have a lower proportion of deep water, so are less likely to stratify and lockup nutrients, and (3) the fish in small reservoirs are more easily caught, avoiding wastage of productivity in a large standing stock of fish that are not growing. Fisheries production also appears to be correlated with the proportion of a reservoir that is "drawn-down" each year during the dry season (Nissanka, 2001); probably through a similar process of nutrient release from reflooding of exposed shoreline sediments as described for the flood pulse. Most large reservoirs in the highly seasonal Mekong system are highly drawn-down each year, a factor that contributes to the apparently high measured yields.

An estimate of total annual catches of 25,428 tonnes from LMB Thai reservoirs larger than 100 ha by Virapat and Mattson (2001) equates to a yield of 120 kg ha^{-1} year^{-1}, which is probably an underestimate based on the figures in Table 9.8. As the mean size of reservoirs in

Thailand is much smaller than Nam Ngum or Nam Oon, and as areal yields are higher in smaller reservoirs, the yield of reservoirs is likely to average at least 200 kg ha^{-1} year^{-1}. Reservoir yields in the LMB, therefore, appear to be quite significant and sustainable, and should be considered in any balanced assessment of dam impacts, which should also take into account that most of the yield is taken by the many unmonitored small-scale fishers. However, the figure of 240,000 t year^{-1} from reservoirs quoted by Sverdrup-Jensen (2002) and others from Virapat and Mattson (2001) appears to be a misquote. The origin of the estimate of reservoir catches quoted by van Zalinge *et al.* (2004) is not clear.

5.3.2. Assessing the Impacts of Dams

The type and scale of impacts of dams vary greatly depending on baseline conditions at a location, the nature and size of the scheme, and mitigation and management measures adopted. Negative effects on downstream fisheries include the direct effects on productivity caused by trapping of nutrients and detritus, release of hypolimnetic water which may be anoxic and toxic due to the presence of hydrogen sulphide, rapid downstream water-level fluctuations caused by hydroelectric releases, and blockage of spawning migrations (Jackson and Marmulla, 2001; Kruskopf, 2006; Schouten, 1998). Fishing activities may also be impacted by water-level fluctuations. Because dams and weirs are so numerous and variable in the LMB, it would be unwise to generalize about the extent of the impacts of all existing dams, particularly as useful pre- and postproject data are almost nonexistent. However, the severity of the negative impacts is likely to depend on the size of the dam, the size of the dammed watercourse, and the location of the dam relative to fish migration routes, valuable habitats, and settlement patterns. The Pak Mun dam was particularly controversial as it cut off Thailand's largest river system from the Mekong as well as submerging important spawning and feeding habitat for fish migrating from the Mekong (Amornsakchai et al., 2000; Jutagate et al., 2001). The Rasi Salai Dam, further upstream on the Mun, was also particularly poorly conceived, flooding extensive floodplain wetlands and the villages that depended upon them, as well as being sited on a salt dome (Sretthachau et al., 2000). After long periods of protests by affected villagers the gates of both of these dams were opened, which allowed riverine fish to migrate past the dams (Jutagate et al., 2005), and in the case of Rasi Sali reduced the negative impacts on the diverse floodplain wetlands that had been submerged by the reservoir. The severe negative impacts of both dams could have been reasonably foreseen from their location on the mainstream of the Mun River in heavily settled locations.

The proposed mainstream Mekong dams are also likely to have very significant negative impacts. The dams proposed in upper Cambodia would be particularly damaging, causing direct interference to migration and spawning and recruitment to floodplains. Mainstream dams in the LMB would be >30 m high and would form reservoirs 75-200 km long (Roberts, 1995) transforming large reaches of the Mekong into standing water. Apart from the issues of location and size, the operational regime of dams varies greatly; fluctuating flows from hydroelectric releases are particularly damaging, as are accidental or emergency releases (Wyatt and Baird, 2007). Damming for interbasin diversions is also likely to be particularly damaging for river fisheries, because the flow of the dammed river is cut off while the channel and banks of the receiving river must erode to adjust to the increased flow. Recent interbasin schemes include Nam Theun to Hinboun, and Nam Theun 2 to Se Bang Fai, and Nam Song and Nam Leuk to Nam Ngum (Koizumi, 2006; Roberts, 2004; Warren, 2000; Watson and Schouten, 2001). Dams on the tributary rivers of major reservoirs (such as several that are

being built upstream of Nam Ngum) are likely to impact spawning habitat and the inflow of nutrients and detritus that support reservoir fisheries.

The impacts of large dams on fish migrations are often discussed in environmental impact assessments (EIAs), and fish passes have been built in many countries (and on the Pak Mun Dam in the Mekong system) in an attempt to mitigate impacts. Apart from the technical difficulty of passing large numbers of fish, fishways are likely to be ineffective on large dams that modify the environment, so that conditions are no longer suitable for migrating fishes (Roberts, 2001a), as for example, when fish migrate upstream to spawn in rapids but find themselves in a large reservoir. Should fish manage to swim into tributaries and spawn, fish fry or larvae are unlikely to negotiate the downstream passage through the reservoir and turbines (Agostinho et al., 2007).

Large dams are designed to cause permanent changes in the environment, so efforts to maintain the existing fish migrations are likely to be futile. If a large dam is to be constructed, mitigation should focus on those aspects where demonstrable benefits are likely to provide a reasonable return on investment, for example, maintaining water quality in the reservoir and in the discharge, and reregulation of the fluctuating flows which characterize hydropower dams. Impacts of dams and technical mitigation measures are discussed in detail by Jackson and Marmulla (2001) and Kruskopf (2006). Dams cause an unavoidable loss of production on downstream floodplains, which is likely to be roughly proportional to the reduction in the extent of annual flooding. The compensating effect of fisheries development in reservoirs varies depending upon the characteristics of the reservoir previously noted and management measures, such as fish stocking. The productivity of the large deep dams such as those built in China (Table 9.6) and those proposed on the mainstream and on some tributaries is unlikely to compensate for lost productivity on floodplains, because the water held in such deep reservoirs is spread over a relatively small area, and such dams stratify, with nutrients depleted as seston settles below a thermocline which develops at about 10-20 m depth (Sitthichaikasem, 1990).

Although the socioeconomic impacts of large dams are well documented (Lawrence, 2008), small dams and weirs that attract little publicity may have a large overall effect on system yield, and deserve more attention because they may also provide better opportunities for mitigation of negative impacts and for enhancing positive impacts, as well as being less socially disruptive than large dams. The impacts of small dams on capture fisheries may be a nett benefit in relatively arid lowland parts of the basin. For example, Schouten (1999) found, based on fisher interviews, that more species of fish (70 cf. 46) were present upstream of a dammed Tonle Sap tributary than upstream of an adjacent undammed tributary (the Stung Chinit), and he concluded that the shallow reservoir provided an important dry-season refuge. Subsequent rehabilitation of a reservoir on the Stung Chinit has led to perceptions of various negative impacts on fisheries by villagers as discussed by Arthur et al. (2006), but no actual catch data were collected, and increased access, population, and fishing pressure could also be causing changes; moreover, the entire system around the reservoir would take time to adjust.

In Thailand, two of the large LMB reservoirs were built to raise the water level (0.5-1 m) of natural swamps to support fisheries production (Table 9.6). Although no accurate data are available for total catches of these "fisheries" reservoirs, Nong Han is accessed by almost all nearby households for subsistence, and provides very large catches of fish and OAAs as well as a range of other wetland products (Pagdee et al., 2007). Villagers in the lower Songkhram basin included improved water flows to swamps or deepening or raising small weirs as their first choice for improved fisheries management (Hortle and Suntornratana, 2008). In Laos, Khoa et al. (2005) found that catches in small irrigation reservoirs compensated for lost catches on rice fields. Local fisheries management of a small (28 ha) floodplain waterbody in southern Laos included installation of a sluice gate to maintain water levels (Tubtim and Hirsch, 2005). Overall, it is reasonable to conclude that dams that are small and shallow and weirs offer significant opportunities for conservation and enhancement of fisheries, because they provide dry-season refuges, potentially high productivity, an extended fishing season, and allow efficient fishing activities by small-scale fishers. As mentioned earlier, smaller schemes can be considered part of the rice field landscape, indicating the possible benefits of an integrated approach to development of fisheries and agriculture.

5.4. Mekong Delta and Estuary

The Vietnamese refer to the delta as *Cuu Long*, "Nine Dragons," because the river system is divided into nine main distributaries. The rivers have been interconnected to form a hydraulic network by large canals, mostly 50-60 m wide, the first of which were built in the early nineteenth century (Bourdeaux, 2005). The levees formed by canal spoil provide space for roads and settlements, and the canal construction allowed drainage of marshes, human penetration and settlement, and rice farming and fishing. Secondary canals (typically 10 m wide) and small tertiary canals further subdivide the landscape. Many of the canals have watergates that when opened allow drainage and ingress of water for irrigation but are closed to prevent floodwater intrusion. Pumps are used to drain canals when watergates are closed; hence, canals may be isolated from the river systems for months, reducing access for fish and fish fry. The canal system functions as a vast reservoir and distribution system for irrigation, and is also the basis for much transportation, including of fishery products and ice. The total length of canals is many thousand kilometres, for example, in Can Tho Province alone there are 4032 km of canals (Akira, 2005); and they occupy perhaps 3% of the delta's area, making them extremely important as an artificial habitat. A large fishery operates year-round in the canals and in the main river systems. Common commercial gears include trawlers (typically operated by two people), dai nets, lift nets, and brush parks. On many permanent watercourses, fish are held in cages underneath floating houses, and most commercial fishers transfer some part of their catch to cages for holding or grow-out. Fence net culture is also becoming common, and the edges of rivers are in many cases effectively enclosed and their use privatized for aquaculture. Most fish are transported alive to markets in locally made wooden sampans, either in containers or in the hull of vessels, through which water is exchanged.

Fisheries in the delta are increasingly integrated with agriculture and forestry. The VAC (Vuong-Ao-Chuong/Garden-Pond-Barn) system is one of the most common systems that integrates fruit trees, aquaculture, and pig farming. Rice-fish farming is also widespread in the delta, and rice alternated with shrimp farming is common near the coast, and *Melaleuca* trees are grown together with rice crops to improve soils in acid sulfate soil area. About 60% of the Vietnamese part of the delta is irrigated (MRC, 2003) from the canal system. During the flood season and for most of the dry season, water is fresh throughout the estuary, but late in the dry season river inflows into the canal network are less than irrigation withdrawals, so saline water is drawn inland up to 60 km, the limits of the "estuary."

The Mekong's freshwater plume extends up to 500 km into the south China Sea, so marine productivity is thought to be heavily influenced by river-derived nutrients, detritus, and sediment (Lagler, 1976a). Plankton and fish larvae biomass are highest at the edge of the plume during the flood season (Lagler, 1976a), consistent with more detailed studies that show the general importance of river plumes to nearshore productivity (e.g., Grimes and Finucane, 1991). Coastal fisheries depend upon this productivity, but nutrients are also recycled back to the river system, through food to the basin's population, as well as in the form of trash fish for aquaculture feed. Marine and estuarine fishes and other animals are likely to follow favored salinity levels, as well as the food chain that depends upon plankton that typically blooms where turbid nutrient-rich river waters meet clear, nutrient-poor salt waters. There is little quantitative information available on the estuarine fisheries of the Mekong delta; monitoring data are lacking and in socioeconomic surveys, it is not possible to separate the estuarine component of the inland catch (Sjorslev, 2002) or, in coastal areas, the marine component from catches (Phanh *et al.*, 2002).

About 60% of the area of the Viet Nam delta is underlain by pyrite-rich soils (Hanhart, 1997). Clearing and burning of *Melaleuca* forests by early farmers caused exposure and oxidation of pyrite generating sulphuric acid; soils became so acidic over a vast area in the northeast corner of the delta that only a single species of reed could survive, leading to its description as the Plain of Reeds (Hanhart, 1997). Birds and fish were scarce throughout the acid-affected areas, and the construction of new canal systems since the 1970s initially worsened acidity by exposing pyritic soil to oxidation, but eventually allowed flushing by alkaline river water (Anonymous, 1992b). Gradual improvement in pH and improved access via roads and canals have allowed development of agriculture, capture fisheries, and aquaculture. Acid drainage is still a major issue in many parts of the delta, and no doubt influences fish production, but specific details are lacking.

Most of the smaller coastal estuarine chan-nels have been blocked by watergates, and large dykes have been built inland and near the coast (parallel to the sea) to control saline water intrusion which facilitates farming of rice as well as other crops, such as fruit trees and freshwater aquaculture. The most easterly of the nine main Mekong distributaries has been completely dammed by the Ba Lai dam at Ben Tre, closed in May 2002 to limit saltwa-ter intrusion (Anonymous, 2002); the dam would also prevent migration of fish and OAAs, but no data are available on the impact on fisheries of this structure, which would include impacts on fish and shrimp migrating upstream as far as Laos. Since the 1990s, there has been an ongoing evolution of farming sys-tems in the delta. In some areas, there were conflicts between farmers who wished to maintain freshwater conditions and those who preferred to allow the seasonal ingress of saline water so they could grow shrimp (Penaeid species) in brackish-water ponds or flooded rice fields.

5.4.1. Yield from the Delta

The upper delta is part of the lowland river-floodplain system, whereas the remainder of the delta comprises irrigated rice fields, planta-tions of fruit trees, brackish-water aquaculture ponds, and the estuarine and canal systems. Capture fisheries in the major flood zone are likely to yield 100-200 kg ha^{-1} year^{-1} as dis-cussed in Section 5.1.2. Yields from rice fields have probably been reduced greatly in some areas by intensification, but small-scale aquacul-ture and intensive fishing in canals would tend to compensate for losses. Marine coastal capture fisheries in the delta provinces were estimated to yield about 726,000 t year^{-1} in 2004 (Truong et al., 2008) of which a large proportion is trash fish that supports aquaculture production. As discussed in the following section, the highest yielding aquaculture systems are in the delta.

5.5. Upland Fisheries

Uplands that are undulating, hilly, or moun-tainous land at altitudes greater than about 300-400 m, make up about 40% of the LMB corresponding approximately to land classed as forest in Table 9.1. Much of the forest is sec-ondary regrowth, gardens and, particularly in Thailand, includes some land converted to plantations. People are mainly concentrated along streams or highland valleys, where rice and plantation crops are grown, with swidden or shifting cultivation common on hillsides. Major tributaries flow from the northern high-lands in Laos and the Annamite chain in Laos, Cambodia, and Viet Nam while smaller high-land tributaries run from the southern uplands in Cambodia toward the Tonle Sap. Rainfalls in mountainous areas are generally higher than the average of the Mekong catchment, so for example, the Se San-Se Kong-Sre Pok, the larg-est tributary system, contributes about 21% of the total annual flow of the Mekong from only 10% of its catchment area (Pantulu, 1986a). In general, tributaries with large upland catch-ments have less seasonally variable flows than many purely lowland systems as a result of more even rainfall, and forested and often karstic catchments; they are also relatively clear and cold with a range of in-stream habitats that favor specialized species.

The distribution of fish and other aquatic animals within tributaries is not well-described, but the results of several surveys suggest that lowland Mekong species are found in their lower reaches (which can be considered part of the entire lowland river-floodplain system), some generalist species are found throughout tributaries, and some species (including ende-mics) are restricted to particular reaches (espe-cially headwaters) or particular habitats, for example, rocky fast-flowing reaches, or caves as for the Nam Theun and Xe Bang Fai Rivers (NT2PC, 2006; Roberts, 2004). In some Lao tribu-taries, lowland species migrating upstream may be blocked by natural barriers such as water-falls or cascades (Noraseng and Warren, 2001; Warren, 1999) and some karst rivers flow through cave systems that may also block migrations (Shoemaker *et al.*, 2001). Fish species diversity usually declines with altitude because of the smaller size of streams, lower tempera-tures, and a more restricted range of habitats (Welcomme, 1985), but the relationship between altitude or stream size and fish diversity in Mekong tributaries has not been well docu-mented. In the Nam-Theun-Xe Bang Fai system, Kottelat (1996) recorded 38-41 species per low-land site near the Mekong and 3-16 species per site in highland upstream of the Nam Theun 2 dam site.

Swidden agriculture is expanding and fal-low periods are decreasing, increasing erosion and probably affecting the water quality of riv-ers (Sjorslev, 2000; Sprenger, 2006). People who live along streams typically erect barriers to divert fish and other aquatic animals into traps, clear streamside vegetation, and build small weirs to divert water into ponds and rice fields (Choulamany, 2005). As temperature declines with altitude, stocking of exotic temperate or subtropical fishes (such as carp or goldfish) in ponds in highland areas is also common (Choulamany, 2005; Funge-Smith, 1999b). There are no systematic studies on the overall effects of such small-scale but very extensive environmental modifications, but it can be assumed that indigenous species and stream and river fisheries would generally be nega-tively impacted by the environmental changes. On the other hand, fisheries production is enhanced by weirs and ponds functioning as dry-season refuges (Choulamany, 2005), and ponds and rice fields provide new habitats for fisheries production. Stocked highland rice fields, for example, may produce about 199 kg of fish per hectare per year (Funge-Smith, 1999b).

Some isolated and relatively unpopulated upland tributaries are lightly fished because access is poor or restricted by unexploded ordi-nance, and fishing is difficult in some large, rapid, and inaccessible rivers, particularly in the wet season, which may lead fishers to use explosives or longlines (NT2PC, 2006). Upland rivers are also inherently less productive for fisheries than the large lowland rivers that are connected to active floodplains, which provide a greatly increased area for aquatic production during the annual flood pulse.

There are few specific details on the fisheries of upland sections of most tributaries; even fish distribution records are found mainly within EIA-related studies for hydroelectric dams, and most information is not readily available (Warren, 2000). One ground-breaking survey covered upland fisheries of Luang Phabang, a rugged mountainous province (247-1600 m) in northern Laos (Sjorslev, 2000). Although rice and livestock farming were the most important activities, 83% of households (42% of people) engaged in capture fisheries, with the focus of most fishing effort from rivers and small streams and producing 90% of reported catches, the remainder being caught in rice fields (7%) and ponds (3%). Although villages in this province are concentrated in valleys, even highlanders go fishing (including collect-ing aquatic animals). Recent catches included at least 73 taxa covering a wide range of mainly white fishes, and only two exotic species—Nile tilapia and common carp—were reported, both escapees from pond aquaculture. As is com-monly found in more well-studied lowland fisheries, villagers used a wide range of gears—cast nets, gill nets, scoop nets, hooks, traps as well as collecting by hand, with the methods varying with habitat. Fishing effort varies about twofold through the year, peaking at the driest time when fish and other aquatic animals are most concentrated. This highland fishery also showed the usual gender division in which about two-thirds of fishing trips are made by men. Fish and other aquatic animals provided around 20% of total animal protein intake, ranking equally with beef and pork. Other aquatic animals represented about 15% of the weight reported for recent catches. Choulamany (2005) and Meusch (2005) described various aspects of the fisheries of highland rice fields in Laos and northern Viet Nam, respectively, and list fishes, amphibians, reptiles, crustaceans, molluscs, and insects caught or cultured in rice field farming systems. Fisheries are important in these more settled areas, based predominantly on rice fields and associated habitats. The perceptions of villagers are similar to those often noted in lowland surveys; for example, declining catches per fisher are attributed to habitat changes and increasing numbers of fishers (Choulamany, 2005). In response to the perceived decline in fisheries, communities have implemented local regulations controlling fishing, and they stock exotic species in ponds.

In the highlands of Viet Nam, there is little information on the fisheries within the river systems, except for that of Lak Lake, the largest waterbody in the province (658 ha). Lak Lake lies at a relatively low altitude (~300 m) within a natural floodplain of a Sekong River tributary and supports an intensive fishery, with catches comprising approximately 50% indigenous floodplain and river fishes, one-quarter shrimps and the remainder mainly self-recruiting introduced species (Thai et al., 2001). In this region, there are many small irrigation reservoirs—about 500 with a total surface area of 105 km² in the main province, Dak Lak (Ly et al., 2006), and more than half the reservoirs are less than 10 ha in area (Phan and Sollows, 2001); smaller reservoirs have larger catch per unit area.

Hydroelectric dams have been built on several tributaries in Laos and Viet Nam, and because of their elevation and large size all major tributaries are currently the subject of development or feasibility studies for further

hydroelectric dams (Chantawong, 2006; Lawrence, 2008). Reports of negative impacts of tributary dams on downstream fisheries have been widely documented, with transboundary impacts of particular concern (e.g., Wyatt and Baird, 2007). Upland people are disproportionately swidden farmers from ethnic minorities and are generally poorer, less literate and less educated than people from the dominant lowland groups, who immigrate to work on dam projects and who may stay and capture ongoing benefits from the project, including the new reservoir fisheries that require capital and technical know-how.

5.6. Fisheries in the Upper Mekong Basin in China

The upper Mekong, or Lancang, runs from tributaries on the Tibetan Plateau in Qinghai and the Tibetan Autonomous Region (TAR) (formerly Xizang) through Yunnan (Chapter 14). The catchment is relatively narrow and steep, highly dissected and erosional; contributing about 16% of river flow but a high sediment load. Southern Qinghai, where the Mekong rises, is a high grassy plateau, mostly at about 3500 m ASL, with peaks to 6500 m, with glacier fields which are an important source of melt-water during the dry season (MRC, 2005). Six large high-altitude natural fold-fault lakes in Yunnan along the main eastern Mekong tributary, the Ehr (or Yangpi) River, have a combined surface area of 273 km²; the largest, Lake Erhai covers about 259 km².

Although mean population density of the Upper Mekong Basin (UMB) is relatively low, settlements are concentrated on the limited areas of flat land near lakes and along rivers, and in forested areas in southern Yunnan, where swidden cultivation extends onto steep hillsides (Heinonen and Vainio-Mattila, 1997). Terraced rice paddies are a dominant land use on flat or gently sloping land throughout the province; rice fields and associated habitats generally support productive fisheries, as is documented in southern Yunnan by Luo (2005).

High altitude and latitude lead to low temperatures in the Mekong and tributaries, which, together with swift currents and elevated sediment concentrations restrict colonization by the diverse lowland tropical fauna of the Lower Mekong. Conversely, most animals adapted to these elevated conditions are not found in the lowlands downstream. About 140 fish species are now recorded from the Mekong system in Yunnan (Chapter 8); an earlier review by Yang (1996) recorded 130 species. The fish fauna is allied to both the Chinese and Mekong (lowland Indochinese) faunas, but the fauna is less diverse and more specialized than that of the Lower Mekong. Fishes of the rivers and streams of Yunnan's mountainous environment are adapted to high and variable flows and include specialized fast-swimming cyprinids (*Cyprinus*, *Garra*, and *Onychostoma* spp.) and montane species of loaches (Balitoridae) and catfishes (Siluridae and Sisoridae), which are adapted to live on or within stony substrata (Yap, 2002). Yunnan's lakes have been colonized by river fishes, some of which have evolved into specialized endemics, including pelagic "snow trouts" (Cyprinidae, Schizothoracinae) and spring fish, *Cyprinus* species. These endemic fishes were the target of important fisheries, but they have been affected by eutrophication and competition from exotic fishes that were introduced to boost fisheries productivity since the 1950s. Temperate herbivorous species introduced to most Yunnan lakes include silver carp (*H. molitrix*), bighead (*Aristichthys nobilis*), and grass carp (*C. idella*) (Xie *et al.*, 2001). Lake Erhai was formerly oligotrophic (Hsiao, 1946), but is now eutrophic as a result of pollution from a large paper plant and other industries; its indigenous fishes are extinct or close to extinction, and have been replaced by pollution-tolerant introduced aquaculture species, including cyprinids such as crucian carp, *Carassius auratus*; common carp, *Cyprinus carpio*; the Chinese false gudgeon, *Abbottina rivularis*; as well as the small short-lived ice fish, *Neosalanx pseudotaihuensis* (Salangidae) (Yan and Chen, 2007). Pollution, eutrophication, and replacement of indigenous fish by exotics appear to be common phenomena in lakes in Yunnan (Kong *et al.*, 2006; Whitmore *et al.*, 1997; Xie *et al.*, 2001).

Four small hydroelectric dams have been built on the Yangpi River, and four major dams have been built on the Mekong (Table 9.7), with several more dams under construction or planned. Large dams on the Mekong mainstream have limited value for fisheries because they are in narrow, steep-sided gorges, which form deep but unproductive water bodies. Moreover, the sides are prone to landslides and slumping into the reservoirs which together with natural sediment loads and erosion caused by settlement lead to high turbidity as has occurred at Manwan (Fu and He, 2007). As Roberts (2001b) predicted, Manwan is filling much faster with sediment than the dam builders expected (Fu and He, 2007), which is reducing its functionality and depriving the downstream Mekong of sediment. Naturally unfavorable conditions for reservoirs on this part of the Mekong are exacerbated by an influx of settlers that leads to increased erosion and pollution from industries (He, 2004).

Upstream of Yunnan in the Lancang River system in the TAR and Qinghai, Walker and Yang (1999) list only eight species of fish, of which six are also found in Yunnan (Table 9.9). Few fishes can tolerate the extreme conditions at high altitudes, and some of the recorded species may be found upstream of Yunnan only at certain times. All those recorded are small species (<31 cm maximum length) and only the snow trout, *Scizothorax lantsangensis* is noted as important in fisheries of the Lancang in Qinghai; steep gorges make access difficult, and

TABLE 9.9 Fishes recorded from the Upper Lancang River in Qinghai, China (from Walker and Yang, 1999)

Family	Subfamily	Species		Note	Maximum length (cm)
Balitoridae	Nemacheilinae	*Triplophysa microps*	(Steindachner, 1866)	Also in Yunnan	6.8
(River loaches)	(Stone loaches)	*Triplophysa orientalis*	(Herzenstein, 1888)		14.2
Cyprinidae	Schizothoracinae	*Gymnocypris potanini*	(Herzenstein, 1891)	Also in Yunnan	16
(Barbs, carps)	(Snow trouts)	*Diptychus kaznakovi*	(Nikolskii, 1903)	Also in Yunnan	30?
		Schizothorax lantsangensis	(Tsao, 1964)		31.1
		Schizothorax lissolabiatus	(Tsao, 1964)	Also in Yunnan	30
Sisoridae	Glyptosterninae	*Pareuchiloglanis gracilicaudata*	(Wu and Chen, 1979)	Also in Yunnan	15?
(Sucker catfish)	(Hillstream catfishes)	*Pareuchiloglanis kamengensis*	(Jayaram, 1966)	Also in Yunnan	15?

ethnic Tibetans do not have a fishing tradition. In all of the water bodies of the Qinghai-Xizang plateau, there are 112 native and 17 introduced species, so more species may be found to be present in the Lancang, which is poorly surveyed (Walker and Yang, 1999). Lake fisheries are important in the north of Qinghai, outside the Mekong catchment.

There is little published information on fisheries activities in the UMB in China. Heinonen and Vainio-Mattila (1997) visited villages near nature reserves in Yunnan and observed that hunting was more important than fishing, but that most people went to fish in nearby rivers when they had time, and that nets, electricity, and poisoning were all used. Baran *et al.* (2007a) concluded from personal observations that fishing is "not a dominant activity in this region." This contrasts with the findings of Luo (2005), based on a 3-month survey in Xishuangbanna, the most densely populated part of Yunnan, that wild fish and OAAs caught in rice fields and rivers are important throughout the region in people's diet and as a source of income. In rice fields alone, 60 fish species were recorded, as well as frogs, snails, crabs, and shrimps. Overfishing and illegal

fishing using electrofishing, poisoning, and explosives, as well as impacts of agricultural intensification appear to be impacting fish production. The most important species caught or cultured now in the villages surveyed by Luo were exotics, including the African tilapias *Oreochromis mossambicus* and *Oreochromis niloticus* as well as the golden apple snail *Pomacea canaliculata*. Aquaculture is expanding, even at altitudes up to 1200 m ASL, reportedly producing over 12,000 t year^{-1}, mostly from ponds, but also with significant production from rice-fish systems.

According to official statistics, the Chinese provinces of the MB contribute a low proportion of the total Chinese inland fisheries production of about 17 Mt year^{-1} (2.5 Mt year^{-1} from capture and 15 Mt year^{-1} from culture). According to Xie and Li (2003), in all of Yunnan, capture fisheries produce about 20,000 t year^{-1}, but this figure is likely to be an underestimate, as is usual for capture fishery statistics which may understate commercial catches and which do not include catches by small-scale family and artisanal fishers (Coates, 2002). About 80,000 ha in Yunnan is devoted to aquaculture; assuming a mid-range productivity of 2.5 t ha^{-1} (based on fig. 9 of Xie and Li), aquaculture might add 200,000 t year^{-1} to the province's production. Most of Yunnan lies outside the MB, so the total fisheries production in the Chinese part of the MB is likely to be less than 100,000 t year^{-1} or less than 4% of the 2.6 Mt year^{-1} estimated as the total yield from the LMB by Hortle (2007). In Qinghai and in the TAR of MB, fishery production figures were not reported by Xie and Li, but presumably would add little to the total estimate.

Although river/floodplain fisheries may be of limited direct importance, the Chinese portion of the UMB is of great significance to fisheries in the downstream countries. The effects on hydrology and water quality of the existing and planned cascade of dams are discussed in Chapter 16. Such changes will have direct effects on all components of aquatic ecosystems, and a wide range of negative effects on fisheries has been predicted (Roberts, 2001a), but no field data are available which would allow an objective evaluation of the effects of these dams. Of particular concern are direct and indirect effects on fish migrations and spawning. In northern Laos, many fish species migrate upstream along the Mekong and presumably some would have moved into China each year to spawn (Poulsen and Viravong, 2002); annual upstream migrations would be consistent with the patterns observed elsewhere in the basin and in other well-studied river fisheries. The longest postulated migration is that of the giant catfish (*Pangasianodon gigas*) which Smith (1945) believed, based on observations by Pavie in 1904, migrated annually from the Cambodian floodplains upstream via the Mekong into China to spawn in Lake Erhai. The existing dams have already cut off the route for all such migrations before they could be documented in any systematic way, so it will never be known which species migrated into China or to what extent their recruitment downstream has been affected. Operation of the four completed dams causes dramatic diurnal variability in flows, which based on many studies elsewhere would be expected to affect migration triggers for fishes downstream (Baran, 2006). Apart from the likely impacts of dams, much concern has been expressed about the effects of clearing of shoals to improve navigation; based on other studies, a wide range of negative effects has been predicted on fisheries in the vicinity and downstream (Roberts, 2001a). The EIA work associated with the project is of little value in predicting impacts (Finlayson, 2002; McDowall, 2002; Roberts, 2001a), and no field data to quantify the effects have been collected.

Apart from fish, Yunnan supports many species of other aquatic animals, such as crabs, shrimps, molluscs, frogs, and turtles; diversity is likely to be high, as has been noted for many groups of land vertebrates by Yang *et al.* (2004). For example, in the MB in southern Yunnan, 13 species of shrimp are recorded (Cai and Dai, 1999), but there is little information on most groups of aquatic animals (Campbell, 2001) or on their importance in fisheries in the region. Significant resources should be devoted to monitoring the effects of ongoing hydrological and land-use changes on fishery species and fisheries to document their value and to evaluate the effectiveness of any management measures.

6. AQUACULTURE

Aquaculture in the LMB includes production and sale of fry or fingerlings, raising of fry produced by hatcheries, raising of wild-caught fry, and grow-out of wild-caught fish. Freshwater aquaculture is practiced in ponds, rice fields, and in cages or fenced-off areas in rivers. Brackish-water aquaculture is practiced along the coast in the Viet Nam delta in ponds and tidal flats.

Aquaculture was not common when populations were small and wild fish were abundant. Aquaculture typically develops along a path of "intensification and enclosure" which parallels population growth and the modification of landscapes, including the expansion of agriculture, especially rice farming and water management. Aquaculture is rarely a totally separate activity from capture fisheries. In its simplest form, wild fish or fry are caught and "grown-out" or reared in cages or ponds, a common practice of commercial fishers, particularly in the delta in Cambodia and Viet Nam. Typically catches are divided; large high-value fish are sold, small low-value fish are eaten in the fishers' households, and "trash fish" are fed to other fish that are being grown-out. Feed may also include leftover food or offal from animal or fish processing as well as farm wastes. Where aquaculture is promoted as a separate or even "new" activity for farmers, the linkages with wild capture fisheries remain: "self-recruiting" fish are usually present in ponds, broodstock or fry may be caught from the wild, "trash" fish are used to feed captive fish, and wild fisheries are impacted by competition for space and water, the enrichment or pollution of water by feed and waste, and by escapes of aquaculture fish. Aquaculture is unlikely to ever substitute for the capture fishery, because much of the aquaculture production is based on conversion of wild-caught fish, with a proportion lost in the process. Moreover, aquaculture requires much greater investments in capital and labor than capture fisheries, so increased production by a household may not translate to a proportional increase in profitability.

Exotic species were widely adopted during the 1960s as techniques for their culture were well-known. They include primarily herbivorous species such as the Indian and Chinese major carps, large fast-growing fishes which include species that efficiently convert phytoplankton to edible biomass. Some of these carps are likely to be substituted by native species as culture techniques are refined. Other stocked species include Nile tilapia *O. niloticus* and its hybrids, African walking catfish (*Clarias gariepinus*) and its hybrids, and South American pirapitinga *Piaractus brachypomus* and pacu *Colossoma macropomum*, which originate from dilute acidic blackwater areas making them suitable for aquaculture in some parts of the delta.

Domestication of some important Mekong species has been achieved quite rapidly and research now centers on improving strains for maximum yield (Froese and Pauly, 2008; MRC, 2003). Pangasiids are particularly promising as several species are air-breathers that tolerate extremely high densities under intensive culture (Trong *et al.*, 2002; van Zalinge *et al.*, 2002). Three giant species (giant carp *Catlocarpio siamensis*, giant catfish *P. gigas*, and Jullien's barb *Probarbus jullieni*) are regularly stocked in ponds and small reservoirs (Mattson *et al.*, 2002). Progress is also being made in techniques for culturing a range of other indigenous species (Cacot and Phengarouni, 2006; Dang *et al.*, 2005; Nuanthavong and Vilayphone, 2006; Prasertwattana *et al.*, 2006; Singsee *et al.*, 2003; Trinh *et al.*, 2006).

Aquaculture development has been most intense in the Mekong delta, where flat terrain, year-round availability of water from the canal system, and settlement patterns favor intensification of production. Recently, integrated agriculture-aquaculture (IAA) systems have been promoted to support smaller scale aquaculture within agricultural systems. IAA aims to minimize nutrient losses from a farm, through concentrating wastes and manures in fish ponds or trenches, where they support a crop of fish, and from which sludge is periodically removed and recycled onto fruit trees or other crops (Nhan *et al.*, 2007). By contrast, larger commercial systems are usually flow-through, which leads both to inefficient use of nutrients and pollution of downstream water bodies.

6.1. Aquaculture Yield

Aquaculture systems range from extensive stocking with no or little feeding at low intensity to intensive systems involving very high densities of aquatic animals that are fed frequently to maximize growth rates. Small-scale aquaculture typically produces less than 1 tonne ha^{-1} from ponds, while intensive systems for shrimp produce up to 9 t ha^{-1} of ponds, and intensive *Pangasius* catfish culture produces up to 1000 t ha^{-1} (Anonymous, 2007). Reliable and up-to-date figures for the actual extent of aquaculture and production figures are only available for official aquaculture for the Viet Nam delta (Truong *et al.*, 2008) and indicate an average yield across all systems (including cage culture) of 1.2 t ha^{-1} $year^{-1}$.

Phillips (2002) estimated that aquaculture production in freshwaters of the LMB had grown from about 60,000 t $year^{-1}$ in 1990 to 260,000 t $year^{-1}$ in 1999/2000, with a further 135,000 tonnes estimated to be produced in brackish-water aquaculture, of which about 72,000 tonnes was high-value shrimp. A significant proportion of brackish-water production was exported, so aquaculture appeared to account for about 10% of the total estimated LMB consumption of 2.6 Mt $year^{-1}$ in 2000 (Hortle, 2007). However, the official figures used by Phillips include commercial production and understate or ignore household production. Small-scale or artisanal aquaculture is common throughout the basin, most often in ponds, and commercial aquaculture is most developed in Thailand and Viet Nam. About 6-8% of rural households are engaged in aquaculture in Thailand and Laos (Phillips, 2002); in Cambodia up to 5% of households and in the delta in Viet Nam 14-61% of households are engaged in aquaculture (Hortle, 2007).

Household surveys suggest that aquaculture contributes less than 10% of the total yield in Cambodia and Laos (e.g., Ahmed *et al.*, 1998;

Sjorslev, 2000). In the lower Songkhram basin (northeast Thailand), where a large river-floodplain fishery persists, about 10% of total production was from aquaculture (Hortle and Suntornratana, 2008); elsewhere in Thailand, the proportion may be higher, but is not known because reservoirs and rice field habitats may be the main sources of production. However, in the Viet Nam delta, aquaculture's contribution to total yield varies between about 30% in Long An (Pham and Guttman, 1999) up to 80% in Tien Giang (Setboonsarng *et al.*, 1999), with a mean contribution of about 50%, including data from Tra Vinh and An Giang (Phanh *et al.*, 2002; Sjorslev, 2002). People in the delta consume about 30% of the basin's total inland fish and OAAs (Table 9.3), implying a minimum basinwide contribution to consumption of about 15% from aquaculture, so the overall contribution could be up to 20%, allowing for about 10-20% of Thai consumption (basinwide proportion 35%) being aquaculture-derived.

The area under aquaculture systems in the delta expanded from about 5% in the early 1990s to about 15% by 2004 and about 75% of this area is devoted to shrimp farming (Truong *et al.*, 2008). The total official aquaculture production in the delta increased to about 800,000 tonnes by 2004, of which about 60% was freshwater (mostly catfish) and of the remainder about 70% was shrimp. Most of the production was exported and shrimp provided the most revenue (Truong *et al.*, 2008). Aquaculture continues to expand, and production of Pangasiid catfishes has grown to about 1 Mt $year^{-1}$ (Anonymous, 2007), implying that total aquaculture production in the delta could now be about 1.5 Mt $year^{-1}$. Most of the recent increase in catfish production is based on intensive systems which provide high yields per unit area. Pond culture yields 50-300 t ha^{-1}, net pens along river banks yield 1000 t ha^{-1}, and fish may be grown in cages at densities of 100-300 kg m^{-3} (Anonymous, 2007).

7. MANAGEMENT OF FISHERIES

Fisheries laws within each LMB country (in draft in Laos) provide national governments with ownership of "living aquatic resources" on public land and waters, and the ability to regulate and license their exploitation. National objectives of fisheries management include conservation, promotion of aquaculture, improved efficiency, disease control, and "sustainable fisheries development."

Preventing conflicts within the sector requires, however, more focused objectives that clarify the balance between potentially competing (or interfering) activities, for example, capture versus culture fisheries, small- versus large-scale fishers, and internal consumption versus export. The importance of transboundary migrating species and the large-scale ecological linkages in the system imply a need for much more specific objectives than simply promoting "sustainable development." Technical measures include managing fish (e.g., by stocking), managing the environment—water and habitat, or managing fishers and their activities. The specific measures depend upon local conditions and particular objectives; for example, to maximize catches and spread the benefits among poor people, few restrictions may be applied in river-floodplain fisheries, as such systems can withstand heavy fishing pressure. If maximum value is desired, various restrictions (i.e., managing the fishers) may reduce "fishing down," thereby reducing the total catch but increasing the catch of large and valuable fish. Stocking of some species may augment catches in both situations, particularly if certain niches can be filled. In most situations, however, managing the environment is of primary importance as is well-known to MB peoples, who apply various aphorisms to describe the dependence of fish on water and habitat, for example, *where there is water there is fish* in Khmer. Many recent surveys record

that people throughout the basin perceive an ongoing decline in fish catches, and generally wish to correct the situation by applying measures that restrict certain fishing activities or modify aquatic systems to increase fish abundance and growth. Until recently, however, government policies have largely not taken into account the opinions of the primary resource users. Governments assumed that capture fisheries will inevitably decline, and have promoted aquaculture as a solution.

Large or middle-scale gears and small trawlers are licensed in well-defined and productive fisheries—the Cambodian floodplains, some reservoirs, and rivers and canals in the Viet Nam delta, but the licensing system collects very little revenue in comparison to the value of the resource (Touch and Todd, 2003). In practice, most fisheries have become "open-access" to small-scale fishers, with very limited enforcement of regulations or active management of habitat for fish. Open-access tends to lead to the well-known "tragedy of the commons," in which individuals maximize their catches because any benefits from conservation are unlikely to accrue to them personally, leading to ever-increasing fishing pressure (Hardin, 1968). Ownership of fisheries resources on seasonally flooded land (the largest area of aquatic habitat) is ambiguous under national laws. Landholders have little incentive to improve habitats (such as refuge ponds or small watercourses) on their land if others derive the benefits, so they generally prefer to invest in measures that increase agricultural output (such as higher yielding plant varieties supported by pesticides), even if fisheries yield is reduced.

Given the huge variation in ecological and socioeconomic conditions, national governments cannot define objectives or management measures which will be appropriate at each locality and which can be adjusted in a timely manner to suit changing conditions. Comanagement—governments and communities working together—has been increasingly promoted throughout the basin (Hartmann, 2000). Comanagement delegates authority to users (often by formalizing traditional ownership) to set local objectives and to devise and test the best mixture of technical measures through a process of adaptive management. Government's role is to legislate some commonly agreed rules and create mechanisms for local management, as well as to provide technical and financial support to communities and to assist with enforcement and conflict resolution. For comanagement to succeed, many conditions must be met; these include clearly defined boundaries, clearly defined membership, the benefits to individuals must exceed costs, management rules must be simple and enforceable by the community, and cooperation at leadership and community levels is necessary (Pomeroy et al., 1998).

Comanagement has been quite successful in some reservoirs, partly because boundaries are clear and fishing activities are observable (Hoang et al., 2006; Nguyen et al., 2003; Niphonkit et al., 2008). Similarly, in the Viet Nam delta, comanagement has recently had success in a rice-shrimp-farming area (Nguyen et al., 2006d). Successful comanagement in the complex and less well-defined river-floodplain fisheries is more problematic (Khumsri et al., 2006), but has been achieved where traditional family and social structures are largely intact, traditional ownership has been recognized by government, and locally formulated regulations "make ecological sense" to fishers (Baird et al., 2003b). Commonly adopted measures include fish conservation zones (FCZs) to protect broodstock in deep pools (and in some areas of Laos in caves), banning of destructive methods and methods that catch spawning fish or frogs and juveniles of some species, closed seasons, and protection of inundated forest habitat along the Mekong (Baird, 2006a; Baran et al., 2005; Hogan, 1998). Community-based management may cause conflicts as exclusion

is not only of "outsiders" but also of nearby communities who may have traditionally shared the resource (Tubtim and Hirsch, 2005). Community-based management may lead to a better return on effort or more economically efficient harvesting when outsiders are excluded (Lorenzen et al., 1998a), but the total yield may be reduced, an outcome that could be inconsistent with government objectives for food security. In Cambodia, community fisheries management is being heavily promoted, but with varying success (Kaing et al., 2005). Conflicts between subsistence and commercial fishers in fishing lots continue, the legitimacy of community-appointed inspectors is questioned (Ly, 2003; Ratner, 2006; Resurreccion, 2006), more people are fishing, there is increased use of illegal gears, especially fine-mesh nets and electrofishers, and catches appear to have declined despite increasing individual effort. The apparent problems in Cambodia are largely a result of general social breakdown caused by recent wars and civil unrest, and should not discourage a continuing focus on community-level management.

Some issues within the fisheries sector require involvement of government at regional, national, or international levels. These include the conservation of long-distance migratory species, including endangered giant fishes, regulation of exports and imports of live fish, fish products and fish feed, and exchange of research and technical information. Some approaches to establishing management structures at different levels are discussed by Hoggarth et al. (1999b) and in the MB, the MRC has a key role to play in coordinating international efforts.

Although there have been some examples of successful fisheries management in the LMB, the major threats to fisheries arise from activities in other sectors, particularly as national policies tend to favor large irrigation and hydroelectric projects that may cause major impacts. For such projects, all LMB countries now implement EIAs, which should fairly evaluate all impacts, but the EIA process has been heavily criticized and has been generally ineffective in evaluating alternatives or leading to mitigation of impacts (Blake, 2005; Schouten, 1998; Warren, 2000).

Even if the EIA process can be improved, assessments are not required for smaller developments (e.g., land conversion, small weirs, roads, housing), resulting in an accumulation of many small changes that eventually completely alter landscapes and their aquatic environments. For example, there are at least 11,000 MRC-registered irrigation schemes in the LMB, most in northeast Thailand, with most including several barriers across watercourses many of which would obstruct fish passage, and if built now the vast majority would not be required to undergo any EIA. Many other barriers are formed by levees and roads. Baran *et al.* (2007b) identified 14,459 "built structures" in the Tonle Sap basin, most having some degree of barrier effect and many designed to store or divert water. Each barrier potentially reduces fishery production by preventing migration, especially of fish moving upstream to spawn, and of fry drifting downstream. Measures other than conventional EIA are needed to improve outcomes for fisheries and to encourage the use of "fish-friendly" designs. Fisheries agencies could be given more control over the environment, as for example, under Thai fisheries law where the Department of Fisheries may require construction of fishways over barriers, or under Cambodian fisheries law which includes clauses that aim to protect flooded forest around the Tonle Sap-Great Lake system. In general, however, conservation of fisheries should be supported through engineering codes of practice for fish-friendly structures, such as "overshot" weirs (Baumgartner *et al.*, 2006) and by environmental guidelines and regulations which apply to all sectors and which if implemented could produce a range of benefits. A number of other approaches including catchment management could be effective if appropriate management structures and systems are established.

Although it is likely that the Mekong system can continue to sustain current yields, as populations grow pressure on fisheries will increase. Providing alternative livelihood opportunities in other labor-intensive industries would reduce the number of people in the fishery, which would increase per capita returns and provide greater flexibility for different management approaches.

8. CONCLUSIONS

The fisheries of the Mekong River basin are vital for the nutrition and livelihoods of millions of people and appear to be particularly diverse and productive as a result of the persistence of the natural flood pulse through a wide range of habitats.

Most households in the basin are engaged in fisheries to some degree and rely upon fish and other aquatic animals to provide food security and support their livelihoods. Based on available data, the yield from the LMB is now estimated to be about 3.6 Mt year^{-1}, of which aquaculture accounts for about 1.5 Mt year^{-1}, most of which is exported. The main habitats which support the LMB's yield are rivers and their floodplains (which cover a moderate area but have a high yield per unit area) and ricefields and associated habitats (which cover most of the total wetland area and provide a low-to-moderate yield per unit area). In the upper Mekong basin fisheries are poorly described but appear to be significant, and the inputs of nutrients and organic detritus in the Mekong's plume support a large coastal fishery.

Although the importance of fisheries is increasingly recognized, management information should be improved by systematic basin-wide research on aspects such as: biodiversity, productivity, attributes of the flood pulse, the size and value of fisheries, the contribution from aquaculture, the possibilities for conserving or increasing production, and ways to mitigate the impacts of water resources developments. The actual management systems and processes to be applied also require a great deal more trial and refinement for local conditions, and should be the subject of basinwide appraisal to document the key factors in success and failure.

However, a shortage of detailed technical information should not be a barrier to increased efforts to apply adaptive management at a variety of scales within fisheries, as research can be included within a management framework. Similarly, the general impacts of developments in other sectors on fisheries are well known, so the lack of specific information should not prevent efforts to take fisheries into account in planning and EIA processes.

While "overfishing" may continue to be a problem for individual fishers and could even eliminate some of the largest species, fishing pressure alone is unlikely to cause irreversible effects on system productivity and fisheries yield. The greatest threats to the Mekong's fisheries arise from large-scale environmental modifications made to support other sectors, particularly hydroelectricity and irrigated agriculture. Experiences in many developed countries (and also from the Mekong) show that single-sector approaches are often shortsighted and may lead to great deal of environmental damage which can only be undone at great cost, if at all.

Despite the threats to river-floodplain fisheries, there are also opportunities, which include improving habitat management, controls on fishing in deep pools to protect broodstock, reinstating fish passage across the many existing barriers, improvement of design in water-management structures and creation of refuges on floodplains. Fisheries opportunities also exist within rice field habitats (e.g.,

development of refuge ponds and IPM to reduce pesticide use), within reservoirs (e.g., management of stocking and fishing pressure), and in aquaculture through integration with agricultural systems and domestication of Mekong species. Improving management would, however, require general improvements in governance and ownership of living aquatic resources, to ensure that some of the value derived from fisheries is reinvested in their conservation and development.

ACKNOWLEDGMENTS

Preparation of this review was supported by funding from the Fisheries Programme of the Mekong River Commission. I thank the Water Studies Centre of Monash University for providing office space and access to library resources. Ms Penroong Bamrungrach kindly provided GIS data and assisted in map production. Ms Pham Mai Phuong, Dr Suchart Ingthamjitr, Mr Ngor Pengby, and Mr Phan Than Lam are also thanked for providing helpful information. Drs Chris Barlow and Ian Campbell kindly provided comments on a draft text.

References

Agostinho, A. A., Marques, E. E., Agostinho, C. S., de Almeida, D. A., de Oliveira, R. J., and de Melo, J. R. B. (2007). Fish ladder of Lajeado Dam: Migrations on one-way routes? *Neotropical Ichthyology* 5(2), 121-130.

Ahmed, M., Hap, N., Ly, V., and Tiongco, M. (1998). "Socio-Economic Assessment of Freshwater Capture Fisheries in Cambodia. Report on a Household Survey," Danish International Development Assistance and Mekong River Commission, Phnom Penh, Cambodia, 138pp.

Akira, Y. (2005). Zoning for risk assessment of water-related natural disasters in the Mekong delta, Unpublished Masters Thesis. Can Tho University, CanTho, Vietnam, 130pp.

Ali, Y. (1997). "Fish, Water and People. Reflections on Inland Openwater Fisheries Resources of Bangladesh," The University Press Ltd., Dhaka, Bangladesh, 126pp.

Allan, J. D., Abell, R., Hogan, Z., Revenga, C., Taylor, B. W., Welcomme, R. L., and Winemiller, K. (2005). Overfishing of inland waters. *BioScience* **55**(12), 1041-1051.

Amornsakchai, S., Annez, P., Vongvisessomjai, S., Choowaew, S., Thailand Development Research Institute, Bangkok, Kunurat, P., Nippanon, J., Schouten, R., Sripatprasite, P., Vaddhanaphuti, C., Vidthayanon, C., Wirojanagud, W., and Watana, E. (2000). "Pak Mun Dam Mekong River Basin Thailand," Submission to the World Commission on Dams. www.dams.org, 170pp.

Angporn, P., Guttman, H., Turongruang, D., Mingkano, P., and Demaine, H. (1998). "Survey of Trap Pond Owners in Sisaket and Roi Et Provinces, Thailand," AIT Aqua Outreach Working Paper T-6. AIT Aqua Outreach, Bangkok, Thailand, 45pp.

Anonymous. (1992a). "Fisheries in the Lower Mekong Basin. Review of the Fishery Sector in the Lower Mekong Basin. Main Report," Interim Committee for Coordination of Investigations of the Lower Mekong Basin, Bangkok, Thailand, 92pp.

Anonymous. (1992b). "Fisheries in the Lower Mekong Basin. Review of the Fishery Sector in the Lower Mekong Basin. Annexes," Interim Committee for Coordination of Investigations of the Lower Mekong Basin, Bangkok, Thailand, 11pp Annexes + Plates.

Anonymous. (2002). Cuu Long Delta's largest irrigation project erects dam on Ba Lai River. *Viet Nam News* 27 March 2002, 4.

Anonymous. (2007). Catfish processing takes off in delta as global demand soars. *Catch and Culture* **13**(2), 16-17.

Arthur, R., Baran, E., So, N., Leng, S. V., Prum, S., and Kura, Y. (2006). "Influence of Built Structures on Tonle Sap Fish Resources," Cambodia National Mekong Committee and WorldFish Center, Phnom Penh, Cambodia, 55pp.

Baird, I. G. (2006a). Strength in diversity: Fish sanctuaries and deep-water pools in Lao PDR. *Fisheries Management and Ecology* **13**, 1-8.

Baird, I. G. (2006b). *Probarbus jullieni* and *Probarbus labeamajor*: The management and conservation of two of the largest fish species in the Mekong River in southern Laos. *Aquatic Conservation: Marine and Freshwater Ecosystems* **16**, 517-532.

Baird, I. G. and Flaherty, M. S. (2005). Beyond national borders: Important Mekong River medium sized migratory carps (Cyprinidae) and fisheries in Laos and Cambodia. *Asian Fisheries Science* **17**, 279-298.

Baird, I. G. and Phylavanh, B. (1998). "Preliminary Survey of Aquatic Resources, with a Special Emphasis on Freshwater Fish and Fisheries, in the Xe Lamphao River Sub-basin (Dong Kanthoung proposed NBCA) in Mounlapmok District, Champasak Province, Southern Lao PDR," Environmental Protection and Community Development in Siphandone Wetland Project, Pakse, Laos, 31pp.

Baird, I. G., Flaherty, M. S., and Phylavanh, B. (2003a). Rhythms of the river: Lunar phases and migrations of small carps (Cyprinidae) in the Mekong River. *Natural History Bulletin of the Siam Society* **51**(1), 5-36.

Baird, I. G., Haggan, N., Brignall, C., and Wood, L. (2003b). Local ecological knowledge and small-scale freshwater fisheries management in the Mekong River in southern Laos. *In* "Putting Fishers Knowledge to Work," Haggan, N., Brignall, C., and Wood, L. (eds.), pp 87-89. Fisheries Center, University of British Columbia, Vancouver, BC, Canada.

Bambaradeniya, C. N. B. and Amarasinghe, F. P. (2004). "Biodiversity Associated with the Rice Field Agroecosystem in Asian Countries: A Brief Review," International Water Management Institute, Colombo, Sri Lanka, 24pp.

Baran, E. (2006). Fish migration triggers in the Lower Mekong Basin and other tropical freshwater systems. *MRC Technical Paper* **14**, 1-56.

Baran, E. and Myschowoda, C. (2008). Have fish catches been declining in the Mekong River Basin? *In* "Modern Myths of the Mekong," Kummu, M., Keskinen, M., and Varis, O. (eds.), pp 55-64. Helsinki University of Technology, Helsinki, Finland, 187pp.

Baran, E., van Zalinge, N. P., and Ngor, P. B. (2001). Floods, floodplains and fish production in the Mekong Basin: Present and past trends. *In* "The Asian Wetlands Symposium 2001," Universiti Sains Malaysia, George Town, Pinang. Malaysia, 27–30 August 2001, 11pp.

Baran, E., Baird, I. G., and Cans, G. (2005). "Fisheries Bioecology at the Khone Falls (Mekong River, Southern Laos)," WorldFish Center, Penang, Malaysia, 84pp.

Baran, E., Jantunen, T., and Kieok, C. C. (2007a). "Values of Inland Fisheries in the Mekong River Basin," WorldFish Center, Phnom Penh, Cambodia, 76pp.

Baran, E., Starr, P., and Kura, Y. (2007b). "Influence of Built Structures on Tonle Sap Fisheries," Cambodia National Mekong Committee and the WorldFish Center Phnom Penh, Cambodia, 44pp.

Baumgartner, L. J., Reynoldson, N., and Gilligan, D. M. (2006). Mortality of larval Murray cod (*Maccullochella peelii peelii*) and golden perch (*Macquaria ambigua*) associated with passage through two types of low-head weirs. *Marine and Freshwater Research* **57**, 187-191.

Bernacsek, G. M. (1997). "Large Dam Fisheries of the Lower Mekong Countries. Review and Assessment. Vol. 1 Main Report,". Mekong River Commission, Bangkok, Thailand, 118pp + Annexes.

Blake, D. J. H. (2005). "A Review of the Nam Theun 2 Environmental Assessment and Management Plan (EAMP) As It Pertains to Impacts on Xe Bang Fai Fisheries,"

International Rivers Network, Berkeley, California, 23pp.

Bouakhamvongsa, K., Viravong, S., and Hortle, K. G. (2006). Monitoring river fishers along the Mekong River in Lao PDR. *MRC Conference Series* 6, 295-300.

Bourdeaux, P. (2005). Reflections on the notion of the "riverine civilization" and on the History of the Mekong delta seen through some aspects of the settlement of the village of Sóc Són (1920-1945). *In* "Proceedings of a Conference: Water in Mainland Southeast Asia," Centre for Khmer Studies, Siem Reap, Cambodia, and the International Institute for Asian Studies, Amsterdam, the Netherlands. 30 November-2 December 2005.

Bush, S. R. (2008). Contextualising fisheries policy in the Lower Mekong Basin. *Journal of Southeast Asian Studies* 39(3), 329-353.

Cacot, P. and Phengarouni, L. (2006). Artificial reproduction of the carp *Cirrhinus microlepis* performed in the south of Laos by using LHRHa implant. *MRC Conference Series* 6, 171-196.

Cai, Y. and Dai, A. Y. (1999). Freshwater shrimps (Crustacea: Decapoda: Caridea) from the Xishuangbanna region of Yunnan Province, southern China. *Hydrobiologia* 400, 211-241.

Campbell, I. (2001). Invertebrates, biodiversity and the Upper Lancang-Mekong Navigation Project. *In* "Biodiversity Management and Sustainable Development. Lancang-Mekong River in the New Millenium." (Min Cao, Kevin Woods, Huabin Hu and Liming Li, eds) pp 97-105. China Forestry Publishing House, Kunming, China, 245pp.

Chandler, D. (2003). "A History of Cambodia," Silkworm Books, Thailand, 296pp.

Chantawong, M. (2006). The Mekong's changing currency. *Watershed* 11, 12-25.

Chea, T., Lek, S., and Thach, P. (2003). Fish larvae and juvenile drift at the confluence of four rivers near Phnom Penh: The Mekong upstream and downstream, the Tonle Sap and the Bassac River, June-September 2002. *MRC Conference Series* 4, 21-28.

Choulamany, X. (2005). Traditional use and availability of aquatic biodiversity in rice-based ecosystems. III. Xieng Khouang and Houa Phanh Provinces, Lao PDR. *In* "Aquatic Biodiversity in Rice-based Ecosystems," Halwart, M., Bartley, D., and Margraf, J. (eds.). FAO Inland Water Resources and Aquaculture Service, Rome, Italy, 18pp.

Claridge, G., Sorangkhoun, T., and Baird, I. (1997). "Community Fisheries in Lao PDR: A Survey of Techniques and Issues," Technical Report 1. IUCN, The World Conservation Union, Vientiane, Laos, 69pp.

Coates, D. (2002). "Inland Capture Fishery Statistics of Southeast Asia: Current Status and Information Needs." RAP Publication 2002/11: 1-114. FAO Regional Office for Asia and the Pacific, Bangkok, Thailand.

Dang, V. T., Nguyen, M. T., Hoang, Q. B., Thi, T. V., Pham, D. K., Nguyen, T. H. V., and Trinh, Q. T. (2005). Artificial propagation of Hoeven's slender carp (*Leptobarbus hoevenii*). *MRC Conference Series* 5, 89-96.

Deap, L., Degen, P., and van Zalinge, N. P. (2003). "Fishing Gears of the Cambodian Mekong," *Inland Fisheries Research and Development Institute of Cambodia Technical Paper Series* 4, 1-269.

de Graaf, G. J. and Chinh, N. D. (2000). "Floodplain Fisheries in the Southern Provinces of Vietnam," www .nefisco.org, 9pp.

de Graaf, G. J. and Xuan, T. T. (1998). Extensive shrimp farming, mangrove clearance and marine fisheries in the southern provinces of Vietnam. *Mangroves and Salt Marshes* 2, 159-166.

de Graaf, G., Born, B., Uddin, A. M. K., and Marttin, F. (2001). "Floods Fish and Fishermen," The University Press, Dhaka, Bangladesh, 110pp.

Dubeau, P., Ouch, P., and Sjorslev, J. G. (2001). "Estimating Fish and Aquatic Animal Productivity/Yield per Area in Kampong Tralach: An Integrated Approach." *Inland Fisheries Research and Development Institute of Cambodia Technical Paper Series* 3, 143-164.

FAO. (2007). "The State of World Fisheries and Aquaculture 2006," Food and Agriculture Organization of the United Nations, Rome, Italy, 162pp.

Finlayson, B. (2002). "Report on Environmental Impact Assessment: The Navigation Improvement Project of the Lancang-Mekong River from China-Myanmar Boundary Marker 243 to Ban Houei Sai of Laos," Mekong River Commission, Phnom Penh, Cambodia, 12pp.

Fox, J. and Ledgerwood, J. (1999). Dry-season flood-recession rice in the Mekong Delta: Two thousand years of sustainable agriculture? *Asian Perspectives* 38(1), 37.

Froese, R. and Pauly, D. (eds.). (2008). "FishBase." World Wide Web Electronic Publication Version (04/2008), www.fishbase.org.

Fu, K. D. and He, D. M. (2007). Analysis and prediction of sediment trapping efficiencies of the reservoirs in the mainstream of the Lancang River. *Chinese Science Bulletin* 52(Supp. II), 134-140.

Funge-Smith, S. J. (1999a). Small-scale rural aquaculture in Lao PDR (Part 1). *FAO Aquaculture Newsletter* 22, 1-7.

Funge-Smith, S. J. (1999b). Small-scale rural aquaculture in Lao PDR (Part 2). *FAO Aquaculture Newsletter* 23, 1-4.

Garnier, F. (1996). "Travels in Cambodia and Part of Laos," Volume 1, The Mekong Exploration Commission Report. 1866-1868. Translated by W. E. J. Tips. White Lotus Co., Bangkok, Thailand, 346pp.

Gorman, C. (1971). The Hoabinhian and after: Subsistence patterns in Southeast Asia during the late Pleistocene and early Recent Periods. *World Archaeology* 2(3), 300-320.

Green, J. R., Corbet, S. A., Watts, E., and Lan, O. B. (1976). Ecological studies on Indonesian lakes. Overturn and restratification of Ranu Lamongan. *Journal of Zoology* **180**, 315-354.

Grimes, C. B. and Finucane, J. H. (1991). Spatial distribution and abundance of larval and juvenile fish, chlorophyll and macrozooplankton around the Mississippi River discharge plume, and the role of the plume in fish recruitment. *Marine Ecology Progress Series* **75**, 109-119.

Halls, A. S. and Welcomme, R. L. (2004). Dynamics of river fish populations in response to hydrological conditions: A simulation study. *River Research and Applications* **20**, 985-1000.

Halls, A. S., Hoggarth, D. D., and Debnath, K. (1999). Impacts of hydraulic engineering on the dynamics and production potential of floodplain fish populations in Bangladesh. *Fisheries Management and Ecology* **6**, 261-285.

Halls, A. S., Kirkwood, G. P., and Payne, A. I. (2001). A dynamic pool model for floodplain-river fisheries. *Ecohydrology and Hydrobiology* **1**(2), 323-339.

Halls, A. S., Welcomme, R. L., and Burn, R. W. (2006). The relationship between multi-species catch and effort: Among fishery comparisons. *Fisheries Research* **77**, 78-83.

Halls, A., Lieng, S., Ngor, P., and Tun, P. (2008). New research reveals ecological insights into dai fishery. *Catch and Culture* **14**(1), 8-12.

Halwart, M. (2006). Biodiversity and nutrition in rice-based aquatic ecosystems. *Journal of Food Composition and Analysis* **19**, 747-751.

Hanhart, K. (1997). "Management of Acid Sulphate Soils Project," Review and Assessment Report For Phase 1 (1989–1995). Environment Unit, MRC Secretariat, Bangkok, Thailand, 45pp.

Hardin, G. (1968). The tragedy of the commons. *Science* **162**, 1243-1248.

Hartmann, W. (2000). It's co-management or no management. *Catch and Culture* **5**(4), 1-4, 11–12.

Hatha, P., Narath, S., and Gregory, R. (1995). "Stocks and Shares" A Study of the Roles and Responsibilities of Cambodian Women and Children in Small-Scale Aquaculture," Working Paper C-2. http://www.aquainformation.ait.ac.th/aarmpage/documents/outreach/workingpapers/WP-C-2.htm. Asian Institute of Technology, Bangkok, Thailand, 5pp.

He D., Zhao W., Chen L. (2004). The ecological changes in Manwan Reservoir area and its causes. *In* 'International Conference on "Advances in Integrated Mekong River Management" 25-27 October 2004. Lao Plaza Hotel, Vientiane, Lao PDR. pp. 123-131. 360 pp.

Heckman, W. C. (1974). The seasonal succession of species in a rice paddy in Vientiane, Laos. *Internationale Revue der Gesamten Hydrobiologie* **59**, 489-507.

Heckman, W. C. (1979). Rice field ecology in Northeastern Thailand. *Monographiae Biologicae* **34**, 1-227.

Heinonen, J. and Vainio-Mattila, K. (1997). "Biodiversity/ Ecotourism Assessments in Yunnan, China." Special

Report. ADB RETA 5771 Poverty Reduction & Environmental Management in Remote Greater Mekong Subregion Watersheds Project (Phase I). Asian Development Bank, Manila, Philippines.

Hoang, T. T., Phan, P. D., Truong, P. H., Ly, T. N., Nguyen, T. T., Duong, P. T., and Sollows, J. D. (2006). Established process and activities of fisheries co-management model in Lak Lake, Daklak Province, Viet Nam. *MRC Conference Series* **6**, 311-320.

Hogan, Z. (1998). Aquatic conservation zones: Community management of rivers and fisheries. *Watershed* **3**(2), 29-33.

Hoggarth, D. D., Cowan, V. J., Aeron-Thomas, M., McGregor, J. A., Garaway, C. A., Payne, A. I., and Welcomme, R. L. (1999a). "Management Guidelines for Asian Floodplain River Fisheries. Part 2: Summary of DFID Research,". Food and Agricultural Organisation of the United Nations, Rome, Italy, 117pp.

Hoggarth, D. D., Cowan, V. J., Halls, A. S., Aeron-Thomas, M., McGregor, J. A., Garaway, C. A., Payne, A. I., and Welcomme, R. L. (1999b). "Management Guidelines for Asian Floodplain River Fisheries. Part 1: A Spatial, Hierarchical and Integrated Strategy for Adaptive Comanagement," Food and Agricultural Organisation of the United Nations, Rome, Italy, 63pp.

Hortle, K. G. (2007). Consumption and the yield of fish and other aquatic animals from the lower Mekong basin. *MRC Technical Paper* **16**, 1-88.

Hortle, K. G. (2008). Liver and intestinal flukes: An underrated health risk in the Mekong basin. *Catch and Culture* **18**(2), 14-18.

Hortle, K. G. and Suntornratana, U. (2008). Socio-economics of the fisheries of the lower Songkhram River Basin, northeast Thailand. *MRC Technical Paper* **17**, 1-85.

Hortle, K. G., Lieng, S., and Valbo-Jorgensen, J. (2004). An introduction to Cambodia's inland fisheries. *Mekong Development Series* **4**, 1-41.

Hortle, K. G., Chea, T., Bun, R., Em, S., and Thach, P. (2005a). Drift of fish juveniles and larvae and invertebrates over 24-hour periods in the Mekong River at Phnom Penh, Cambodia. *MRC Conference Series* **5**, 19-34.

Hortle, K. G., Roth, T., Garrison, J., and Cans, G. (2005b). Harvesting insects—A growing industry in Cambodia. *Catch and Culture* **11**(1), 9-10.

Hortle, K. G., Troeung, R., and Lieng, S. (2008). Yield of the wild fishery of rice fields in Battambang Province, near the Tonle Sap Lake, Cambodia. *MRC Technical Paper* **18**, 1-62.

Hsiao, S. C. (1946). A limnological study of Erh Hai, Yunnan, China: Physico-chemical characteristics. *Journal of Animal Ecology* **15**, 1-8.

Islam, M. S., Rahman, M. M., Halder, G. C., and Tanaka, M. (2006). Fish assemblage of a traditional fishery and the seasonal variations in diet of its most abundant species

Wallago attu (Siluriformes: Siluridae) from a tropical floodplain. *Aquatic Ecology* 40, 263-272.

Jackson, D. and Marmulla, G. (2001). "The Influence of Dams on River Fisheries," Report to the World Commission on Dams. www.dams.org, 17pp.

Junk, W. J. and Wantzen, K. M. (2004). The flood pulse concept: New aspects, approaches and applications—An update. RAP Publication 2004/16, pp 117-140.

Junk, W., Bayley, P., and Sparks, R. (1989). The flood pulse concept in river-floodplain systems. *Canadian Special Publications in Fisheries and Aquatic Sciences* 106, 110-127.

Jutagate, T., Lamkom, T., Satapornwanit, K., Naiwinit, W., and Petchuay, C. (2001). Fish species diversity and ichthyomass in Pak Mun Reservoir, five years after impoundment. *Asian Fisheries Science* 14, 417-424.

Jutagate, T., Silva, S. S. D., and Mattson, N. S. (2003). Yield, growth and mortality rate of the Thai river sprat, *Clupeichthys aesarnensis*, in Sirinthorn Reservoir, Thailand. *Fisheries Management and Ecology* 10, 221-231.

Jutagate, T., Krudpan, C., Ngamsnae, P., Lamkom, T., and Payooha, K. (2005). Changes in the fish catches during a trial opening of sluice gates on a run-of-the river reservoir in Thailand. *Fisheries Management and Ecology* 12, 57-62.

Kaing, K., Sung, S., and Un, K. (2005). Participation in fisheries co-management in Kandal Province, Cambodia. *MRC Conference Series* 5, 143-153.

Khay, D. and Hortle, K. G. (2005). Monitoring sales of fish and other aquatic animals at retail markets in Phom Penh, Cambodia. *MRC Conference Series* 5, 3-18.

Khoa, S. N., Lorenzen, K., Garaway, C., Chamsinhg, B., Siebert, D., and Randone, M. (2005). Impacts of irrigation on fisheries in rain-fed rice-farming landscapes. *Journal of Applied Ecology* 42, 892-900.

Khumsri, M., Sriputinibondh, N., and Thongpun, W. (2006). Fisheries co-management in Lower Songkhram River Basin: Problems and challenges. *MRC Conference Series* 6, 121-126.

Ko-Anandakul, K. (2004). "Fishing Gears of the Songkhram River Basin (in Thai)," Department of Fisheries (Thailand) and Mekong River Commission, Bangkok, 87pp.

Koizumi, H. (2006). "Nam Ngum Dam after 30 Years of Operation," Koei Research Institute International Corporation, Japan, 68pp + Annexes.

Kong, D.-P., Chen, X.-Y., and Yang, J.-X. (2006). Fish fauna status in the Lugu Lake with preliminary analysis on cause and effect of human impacts. *Zoological Research* 1, 94-97.

Kottelat, M. (1996). "Potential Impacts of Nam Theun 2 Hydropower Project on the Fish and Aquatic Fauna of the Nam Theun and Xe Bangfai Basins, Lao PDR,"

Report to NTEC Development Group, Vientiane, Lao PDR, 57pp.

Kruskopf, M. (2006). "Influence of Built Structures on Tropical Floodplains Worldwide," Cambodia National Mekong Committee and WorldFish Center, Phnom Penh, Cambodia, 88pp.

Lagler, K. F. (1976a). "Fisheries and Integrated Mekong River Basin Development," Field Investigations Appendix Volume 1. Annex E: Coastal Fishery Field Investigations, Terminal Report of the Mekong Basinwide Fishery Studies. University of Michigan School of Natural Resources, Michigan, USA, 39pp.

Lagler, K. F. (1976b). "Fisheries and Integrated Mekong River Basin Development," Executive Volume, Terminal Report of the Mekong Basinwide Fishery Studies. University of Michigan School of Natural Resources, Michigan, USA, 367pp.

Lamberts, D. (2006). The Tonle Sap Lake as a productive ecosystem. *International Journal of Water Resources Development* 22(3), 481-495.

Lamberts, D. (2008). Little impact, much damage: The consequences of Mekong River flow alterations for the Tonle Sap ecosystem. *In* "Modern Myths of the Mekong," Kummu, M., Keskinen, M., and Varis, O. (eds.), pp 3-18. Helsinki University of Technology, Helsinki, Finland, 187pp.

Lawrence, S. (2008). "Power Surge. The Impacts of Rapid Dam Development in Laos," International Rivers Network, Berkely, CA, USA, 88pp.

Lieng, S. and van Zalinge, N. P. (2001). Fish yield estimation in the floodplains of the Tonle Sap Great Lake and River, Cambodia. *Cambodia Fisheries Technical Paper Series* 3, 23-26.

Lim, P., Villavueva, M. C., Chhouk, B., Chay, K. K., Brun, J. M., and Moreau, J. (2005). Fish assessment in the rehabilitated polders of Prey Nup (Cambodia). *Asian Fisheries Science* 18, 241-253.

Lorenzen, K., Garaway, C. J., Chamsingh, B., and Warren, T. J. (1998a). Effects of access restrictions and stocking on small water body fisheries in Laos. *Journal of Fish Biology* 53, 345-357.

Lorenzen, K., Juntana, J., Bundit, J., and Tourongruang, D. (1998b). Assessing culture practices in small waterbodies: A study of village fisheries in north-east Thailand. *Aquaculture Research* 29, 211-224.

Luo, A. (2005). Traditional use and availability of aquatic biodiversity in rice-based ecosystems II. Xishuangbanna, Yunnan, People's Republic of China. *In* "Aquatic Biodiversity in Rice-based Ecosystems," Halwart, M., Bartley, D., and Margraf, J. (eds.), pp 1-11. FAO Inland Water Resources and Aquaculture Service, Rome, Italy. A CD-ROM.

Ly, S. (2003). Cambodia fisheries conflicts in the post war era and the creation of fishing communities: A case of fishing lot conflicts around Tonle Sap River and the Great Lake. *Journal of Regional Fisheries*, **43**(2), 89-106.

Ly, N. T., Sollows, J., Hartmann, W., Phan, D. P., Hoang, T. T., Nguyen, T. T., Truong, H. P., and Duong, T. P. (2006). Is there fisheries co-management in Daklak province, Central Highlands of Viet Nam? *MRC Conference Series* **6**, 147-158.

Manning, G. S. and Lertprasert, P. (1973). Studies on the life cycle of *Phaneropsolus bonnei* and *Prosthodendrium molenkampi* in Thailand. *Annals of Tropical Medicine and Parasitology* **67**(3), 361-365.

Margraf, J., Voggesberger, M., and Milan, P. P. (1996). Limnology of Ifugao rice terraces, Philippines. *In* "Perspectives in Tropical Limnology," Schiemer, F. and Boland, T. (eds.), pp 305-319. SPB Academic Publishing BV, Amsterdam, Netherlands, 347pp.

Mattson, N., Nilsson, H., and Phounsavath, S. (2000). "The Fishery of Nam Ngum reservoir Lao PDR," Mekong River Commission, Vientiane, Lao PDR, 72pp.

Mattson, N. S., Balavong, V., Nilsson, H., Phounsavath, S., and Hartmann, W. (2001). Changes in fisheries yield and catch composition at the Nam Ngum Reservoir, Lao PDR. *ACIAR Proceedings* **98**, 48-55.

Mattson, N. S., Buakhamvongsa, K., Sukumasavin, N., Nguyen, T., and Ouk, V. (2002). Mekong giant fish species: On their management and biology. *MRC Technical Paper* **3**, 29.

McDowall, R. M. (2002). "Evaluation of: Report on Environmental Impact Assessment: The Navigation Channel Improvement Project of the Lancang-Mekong River from China-Myanmar Boundary Marker 243 to Ban Houei Sai of Laos," Mekong River Commission, Phnom Penh, Cambodia, 9pp.

Meusch, E. (2005). Traditional use and availability of aquatic biodiversity in rice-based ecosystems. IV. Lai Chau and Hoa Binh provinces, Viet Nam. *In* "Aquatic Biodiversity in Rice-based Ecosystems," Halwart, M., Bartley, D., and Margraf, J. (eds.). FAO Inland Water Resources and Aquaculture Service, Rome, Italy, 14pp.

Meusch, E., Yhoung-Aree, J., Friend, R., and Funge-Smith, S. (2003). The role and nutritional value of aquatic resources in the livelihoods of rural people: A participatory assessment in Attapeu Province, Lao PDR. *In* "Dialogue on Water, Food and the Environment." Food and Agriculture Organization of the United Nations, Regional Office for Asia and the Pacific and IUCN—The World Conservation Union, Bangkok, Thailand, 34pp.

Mogensen, M. T. (2001). The importance of fish and other aquatic animals for food and nutrition security in the lower mekong basin, MSc Thesis. Department of Human Nutrition, The Royal Agricultural and Veterinary University, Denmark, 129pp + Appendices.

Mollot, R., Phothitay, C., and Kosy, S. (2005). Hydrology, habitat and livelihoods on the floodplains of southern Lao PDR. *MRC Conference Series* **5**, 155-174.

MRC. (2003). "State of the Basin Report 2003," Mekong River Commission, Phnom Penh, Cambodia, 300pp.

MRC. (2005). "Overview of the Hydrology of the Mekong Basin," Mekong River Commission, Vientiane, Lao PDR, 73pp.

Nesbitt, H. J., Johnston, R., and Solieng, M. (2004). Mekong River water: Will river flows meet future agriculture needs in the Lower Mekong Basin. *ACIAR Proceedings* **116e**, 86-104.

Ngo, S. V. and Le, T. L. (2001). Status of reservoir fisheries in Vietnam. *ACIAR Proceedings* **98**, 235-245.

Ngor, P. B., Aun, S., Deap, L., and Hortle, K. G. (2005). The *dai trey linh* fishery on the Tonle Touch (Touch River), southeast Cambodia. *MRC Conference Series* **5**, 35-56.

Ngor, P., Aun, S., and Hortle, K. G. (2006). The *dai bongkong* fishery for giant river prawns, *Macrobrachium rosenbergii*, in southeastern Cambodia. *MRC Conference Series* **6**, 3-32.

Nguyen, X. V. and Wyatt, A. B. (2006). "Situation Analysis: Plain of Reeds, Viet Nam," United Nations Development Programme, Vientiane, Lao PDR, 60pp.

Nguyen, T. N., Vo, T. D., Truong, H. P., Phan, T. H., Phan, D. P., Nguyen, N. V., Tran, M. H., Nguyen, T. B., Hoang, D. T. V., and Sollows, J. D. (2003). The future of fisheries co-management in six reservoirs in the Central Highlands of Viet Nam. *MRC Conference Series* **4**, 195-213.

Nguyen, N. D., Smallwood, C., Nguyen, V. H., Nguyen, X. T., and Nguyen, T. T. (2006a). "Fishing Gears of the Mekong Delta," Research Institute for Aquaculture 2 and Mekong River Commission, Ho Chi Minh City, Viet Nam (in Vietnamese), 351pp.

Nguyen, P. H., Nguyen, K. C., Nguyen, T. D., Le, M. B., Bern, C., Flores, R., and Martorell, R. (2006b). Intestinal helminth infections among reproductive age women in Viet Nam. *Southeast Asian Journal of Tropical Medicine and Public Health* **37**(5), 865-874.

Nguyen, T. T., Nguyen, N. D., Vu, V. A., and Truong, T. T. (2006c). Monitoring of fish larvae during the annual flood of the Mekong and Bassac rivers, Mekong Delta, Viet Nam. *MRC Conference Series* **6**, 71-88.

Nguyen, V. H., Mai, T. T. C., Pham, B. V. T., Pham, T. B. H., Sollows, J., and Davidson, J. (2006d). Strengthening co-management of water resources for rice-shrimp farming in Soc Trang Province—Viet Nam. *MRC Conference Series* **6**, 127-146.

Nguyen, N. D., Vu, V. A., Doan, V. T., Lam, N. C., and Nguyen, V. P. (2008). Monitoring of fish larvae in the Mekong and Bassac Rivers, Viet Nam, 2006. *MRC Conference Series* **7**, 31-40.

Nhan, D. K., Phong, L. T., Verdegem, M. J. C., Duong, L. T., Bosma, R. H., and Little, D. C. (2007). Integrated freshwater aquaculture, crop and livestock production in the Mekong delta, Vietnam: Determinants and the role of the pond. *Agricultural Systems* **94**, 445-458.

Niphonkit, N., Sriputinibondh, N., and Buttaprom, W. (2008). Co-management of reservoir fisheries in Huay Luang Reservoir, Udonthani, Thailand. *MRC Conference Series* **7**, 93-98.

Nissanka, C. (2001). Effect of hydrological regimes on fish yields in reservoirs of Sri Lanka. *ACIAR Proceedings* **98**, 93-100.

Noraseng, P. and Warren, T. J. (2001). "Fish Migration and Species Composition in a Major Mekong Tributary: A Rapid Assessment of Lower Sedone River Fisheries using CPUE," The Small-scale Wetlands Indigenous Fisheries Management Project, Pakse, Lao PDR, 17pp.

NT2PC. (2006). "Fish and Aquatic Habitats Survey in NTPC Area, 2006," Nam Theun 2 Hydroelectric Project. Final Report. Nam Theun 2 Power Company, http://www.namtheun2.com/Fish%20and%20Aquatic%20Habitats%20Survey%20-%20Aug06.pdf, 96pp.

Nuanthavong, T. and Vilayphone, L. (2006). Reproduction and nursing of *Cirrhinus molitorella* in a small fish farm in Luang Prabang Province, Lao PDR. *MRC Conference Series* **6**, 197-204.

Nuth, M. (2005). "The Informal Economy in Cambodia. An Overview," Economic Institute of Cambodia, Phnom Penh, Cambodia, 5pp.

Pagdee, A., Homchuen, S., Sangpradab, N., Hanjavanit, C., and Uttharak, P. (2007). Biodiversity and economic value of wetland resources at Nong Han, Udonthani Province, Northeast Thailand. *Natural History Bulletin of the Siam Society* **55**(2), 323-339.

Pantulu, V. R. (1986a). The Mekong River system. In "The Ecology of River Systems," Davies, B. R. and Walker, K. F. (eds.), pp 695-719. Dr W. Junk Publishers, Dordrecht, The Netherlands, 793pp.

Pantulu, V. R. (1986b). Fish of the Lower Mekong Basin. In "The Ecology of River Systems," Davies, B. R. and Walker, K. F. (eds.), pp 721-741. Dr W. Junk Publishers, Dordrecht, The Netherlands.

Pauly, D., Christensen, V., Dalsgaard, J., Froese, R., and Torres, F. (1998). Fishing down marine food webs. *Science* **279**, 860-863.

Pawaputanon, O. (1986). Fisheries and fishery management of large reservoirs in Thailand. In "The First Asian Fisheries Forum," McLean, J. L., Dixon, L. B., and Hosillos, L. V. (eds.), pp 389-392. Asian Fisheries Society, Manila, The Philippines.

Pham, V. N. and Guttman, H. (1999). "Aquatic Resources Use Assessment in Long An Province, Vietnam (Results from 1997 Survey)," College of Agriculture and Forestry/Asian Institute of Technology, Ho Chi Minh City, Viet Nam/Bangkok, Thailand, 32pp + Appendices.

Phan, P. D. and Silva, S. S. D. (2000). The fishery of the Ea Kao reservoir, southern Vietnam: A fishery based on a combination of stock and recapture, and self-recruiting populations. *Fisheries Management and Ecology* **7**, 251-264.

Phan, D. P. and Sollows, J. D. (2001). Status and potential of reservoir fisheries in Dak Lak Province, Vietnam. *ACIAR Proceedings* **98**, 36-42.

Phanh, T. L., Pham, M. P., Nguyen, T. T., Visser, T., and Hortle, K. G. (2002). "Tra Vinh Fisheries Survey. Tra Vinh Province-Viet Nam," RIA2, Department of Fisheries, Tra Vinh; Department of Statistics, Tra Vinh; AMFC of the MRC Fisheries Programme, Ho Chi Minh City, Viet Nam, 69pp + Annexes.

Phathammavong, O., Moazzam, A., Xaysomphoo, D., Phengsavanh, A., and Kuroiwa, C. (2007). Parasitic infestation and nutritional status among schoolchildren in Vientiane, Lao PDR. *Journal of Paediatrics and Child Health* **43**, 689-694.

Phillips, M. J. (2002). Freshwater aquaculture in Lower Mekong Basin. *MRC Technical Paper* **7**, 62.

Phithakpol, B., Varanyanond, W., Reungmaneepaitoon, S., and Wood, H. (1995). "The Traditional Fermented Foods of Thailand," ASEAN Food Handling Bureau, Kuala Lumpur, Malaysia, 35pp.

Pholprasith, S. and Sirimongkonthaworn, R. (1999). The fish community of Ubolratana Reservoir, Thailand. In "Fish and Fisheries of Lakes and Reservoirs in Southeast Asia and Africa," Densen, W. L. T. (ed.), pp 103-115. Westbury Publishing, Otley, UK, 432pp.

Phonvisay, A. (2003). "Monitoring of Fish Trade Study of the Siphandone Fishery, Champassak Province, 2003," Living Aquatic Resource Research Centre, Ministry of Agriculture and Forestry, Lao PDR, 60pp.

Phonvisay, A. and Bush, S. (2001). Baseline study of fish trade in Champasack. *LARReC Research Report* **4**, 1-70.

Piyasiri, S. and Perera, N. (2001). Role of *Oreochromis* hybrids in controlling *Microcystis aeruginosa* blooms in the Kotmale Reservoir. *ACIAR Proceedings* **98**, 137-148.

Pomeroy, R. S., Katon, B. M., and Harkes, I. (1998). Fisheries co-management: Key conditions and principles drawn from Asian experiences. In "Crossing Boundaries," The Seventh Annual Conference of the International Association for the Study of Common Property," Vancouver, British Columbia, Canada, 10–14 June 1998, p 15. 23pp.

Poulsen, A. F. and Viravong, S. (2002). Fish migrations and maintenance of biodiversity in the Mekong River Basin. In "International Symposium on Biodiversity Management and Sustainable Development in the Lancang-Mekong River Basin," Xishuangbanna, Yunnan Province of China, 4–7 December 2001.

Poulsen, A. F., Ouch, P., Viravong, S., Suntornratana, U., and Nguyen, T. T. (2002a). Fish migrations of the Lower

Mekong River Basin: Implications for development planning and environmental management. *MRC Technical Paper* **8**, 1-62.

Poulsen, A. F., Ouch, P., Viravong, S., Suntornratana, U., and Nguyen, T. T. (2002b). Deep pools as dry season fish habitats in the Mekong River basin. *MRC Technical Paper* **4**, 1-24.

Prapertchob, P., Kachamart, P., Pakuthai, W., Viratchakul, J., Hornak, A., Thiranggon, P., and Kamsrakaeo, P. (1989). "Summary Report on Analysis of Freshwater Fish Consumption and Marine Product Marketing in Northeast Thailand," Khon Kaen University, Khon Kaen, Thailand, 35pp.

Prasertwattana, P., Manee, N., and Namtum, S. (2006). Culture of Red-tail Mystus, *Hemibagrus wyckioides*, in earthen ponds with different stocking densities. *MRC Conference Series* **6**, 213-216.

Ratner, B. D. (2006). Community management by decree? Lessons from Cambodia's fisheries reform. *Society and Natural Resources* **19**, 79-86.

Resurreccion, B. P. (2006). Rules, roles and rights: Gender, participation and community fisheries management in Cambodia's Tonle Sap region. *Water Resources Development* **22**, 433-447.

Roberts, T. R. (1993). Just another dammed river? Negative impacts of Pak Mun Dam on fishes of the Mekong Basin. *Natural History Bulletin of the Siam Society* **41**, 105-133.

Roberts, T. R. (1995). Mekong mainstream hydropower dams: Run-of-the-river or ruin-of-the-river? *Natural History Bulletin of the Siam Society* **43**, 3-13.

Roberts, T. R. (2001a). On the river of no returns: Thailand's Pak Mun Dam and its fish ladder. *Natural History Bulletin of the Siam Society* **49**, 189-230.

Roberts, T. R. (2001b). Killing the Mekong: China's fluvicidal hydropower cum-navigation development scheme. *Natural History Bulletin of the Siam Society* **49**, 143-159.

Roberts, T. R. (2002). Fish scenes, symbolism, and kingship in the bas-reliefs of Angkor Wat and the Bayon. *Natural History Bulletin of the Siam Society* **50**, 135-193.

Roberts, T. R. (2004). "Fluvicide: An Independent Environmental Assessment of Nam Theun 2 Hydropower Project in Laos, with Particular Reference to Aquatic Biology and Fishes," http://internationalrivers.org/files/tysonfluvicide0904.pdf, 46pp.

Roger, P. A. and Kurihara, Y. (1988). Floodwater biology of tropical wetland ricefields. *In* "The First International Symposium on Paddy Soil Fertility," International Soil Science Society, Chiang Mai, Thailand, pp 275-300.

Roos, N., Chamnan, C., Loeung, D., Jakobsen, J., and Thilsted, S. H. (2007a). Freshwater fish as a dietary source of vitamin A in Cambodia. *Food Chemistry* **103**, 1104-1111.

Roos, N., Thorseng, H., Chamnan, C., Larsen, T., Gondolf, U. H., Bukhave, K., and Thilsted, S. H. (2007b). Iron content in common Cambodian fish species: Perspectives for dietary iron intake in poor, rural households. *Food Chemistry* **104**, 1226-1235.

Schouten, R. (1998). Effects of dams on downstream reservoir fisheries, case of Nam Ngum. *Catch and Culture* **4**(2), 1-5.

Schouten, R. (1999). "Rehabilitation of Stung Chinit Diversion Weir in Cambodia and Impacts on Fish Migration and Fisheries," Asian Development Bank, Phnom Penh, Cambodia, 31pp.

Sensereivorth, T., Diep, L., and Nao, T. (1999). Freshwater capture fisheries data collection in 1998. *In* "Present Status of Cambodia's Freshwater Capture Fisheries and Management Implications, 19–21 January 1999," Van Zalinge, N. P., Nao, T., and Deap, L. (eds.), pp 40-53. Mekong River Commission and Department of Fisheries, Phnom Penh, Cambodia, 170pp.

Setboonsarng, S., Le, H. H., and Pham, C. T. (1999). "Report of Baseline Survey of Tien Giang Province," Mekong River Commission Fisheries Programme, Rural Extension for Aquaculture Development in the Mekong Delta (Phase 1) Cambodia and Vietnam, Phnom Penh, Cambodia, 113pp + Annexes.

Shoemaker, B., Baird, I. G., and Baird, M. (2001). "The People and Their River. A Survey of River-Based Livelihoods in the Xe Bang Fai River Basin in Central Lao PDR," Lao PDR/Canada Fund for Local Initiatives, Vientiane, Laos, 79pp.

Singsee, S., Udomkarn, C., Prasertwattana, P., and Sukumasavin, N. (2003). Development of techniques for induced oocyte maturation and ovulation in *Pangasius bocourti. MRC Conference Series* **4**, 151-154.

Sitthichaikasem, S. (1990). Thermal and chemical stratification phenomena in Srinagarind Reservoir. *In* "2nd National Seminar on Water and Wastewater Technology," Panwad, T. and Nayomtoon, I. (eds.), pp 93-107. College of Engineering, Chulalongkorn University, Chulalongkorn, Thailand.

Sjorslev, J. G. (2000). "Fisheries Survey, Luangprabang Province, Lao PDR," NAFRI and MRC Fisheries Program AMFC Component, Vientiane, Lao PDR, 45pp.

Sjorslev, J. G. (ed.). (2002). "An Giang Fisheries Survey. An Giang Province—Viet Nam," Draft Report. RIA2, Department of Fisheries An Giang, Fisheries Department, Can Tho University, AMFC of MRC Fisheries Programme, Vientiane, Laos, 70pp.

Smith, H. M. (1945). The fresh-water fishes of Siam, or Thailand. *Bulletin of the Smithsonian Institution* **188**, 1-577.

So, N., Ouch, V., Hao, V., and Nandeesha, M. C. (1998). Women in small-scale aquaculture development in Cambodia. *Aquaculture Asia* **3**, 20-22.

Soukhaseum, V., Viravong, S., Sinhanouvong, D., Phounsavath, S., and Warren, T. (2007). Fish migration studies and CPUE (catch per unit effort) data collection in southern Lao PDR from 1994 to 2006. Paper presented in "8th Asian Fisheries Forum," Kochi, India, 20-23 November 2007, Asian Fisheries Society, Manila, Philippines, 19pp.

Sprenger, G. (2006). Out of the ashes. Swidden cultivation in Laos. *Anthropology Today* **22**, 9-13.

Sretthachau, C., Nungern, K., and Olsson, A. (2000). "Social Impacts of the Rasi Salai Dam, Thailand: Loss of Livelihood Security and Social Conflict," Submission to the World Commission on Dams, East/South-East Asia Regional Consultation, Hanoi, Vietnam. 26–27 February 2000. Southeast Asia Rivers Network (SEARIN), Thailand, 14pp.

Sricharoendham, B., Leelapatra, W., Ratanachamnong, D., Kaewjaroon, P., and Aimsab, M. (1999). "Variation on Fish Community and Catch of Sirindhorn Reservoir, Ubon Ratchathani Province," National Inland Fisheries Institute, Department of Fisheries, Ministry of Agriculture and Cooperatives, Thailand, Bangkok, Thailand, 30pp.

Suwannapeng, N. (2002). Induced spawning of Jullien's mud carp, *Henicorhynchus siamensis*. *MRC Conference Series* **2**, 253-256.

Sverdrup-Jensen, S. (2002). Fisheries in the lower Mekong Basin: Status and perspectives. *MRC Technical Paper* **6**, 1-95.

Taki, Y. (1978). An analytical study of the fish fauna of the Mekong Basin as a biological production system in nature. *Research Institute of Evolutionary Biology Special Publications* **1**, 1-77.

Tep, C., Sinuon, M., Doung, S., and Odermatt, P. (2006). Intestinal parasites in school-aged children in villages bordering Tonle Sap Lake, Cambodia. *Southeast Asian Journal of Tropical Medicine and Public Health* **37**(5), 859-864.

Thach, P., Chea, T., and Hortle, K. G. (2006). Drift of fish fry and larvae in five large tributaries of the Tonle Sap-Great Lake system in Cambodia. *MRC Conference Series* **6**, 289-294.

Thai, N. C., Sollows, J. D., Nguyen, Q. A., Phan, D. P., Nguyen, Q. N., and Truong, H. P. (2001). Is Lak Lake overfished? *ACIAR Proceedings* **98**, 71-80.

Touch, S. T. and Todd, B. H. (2003). "The Inland and Marine Fisheries Trade of Cambodia," Economic, Social and Cultural Observation Unit, Royal Government of Cambodia and Oxfam America, Phnom Penh, Cambodia, 147pp.

Tran, T. V., Do, T. L., Nguyen, N. V., Phan, D. P., Phan, T. H., Thai, N. C., Nguyen, Q. A., and Sollows, J. D. (2001). An assessment of the fisheries of four stocked

reservoirs in the Central Highlands of Vietnam. *ACIAR Proceedings* **98**, 81-92.

Trinh, Q. T., Huynh, H. N., Thi, T. V., Nguyen, M. T., and Hoang, Q. B. (2006). Preliminary results of domestication of *Pangasius krempfi*. *MRC Conference Series* **6**, 217-222.

Troeung, R., Aun, S., Lieng, S., Deap, L., and van Zalinge, N. (1999). A comparison of fish yields and species composition between one fishing lot in Battambang province and two fishing lots in Prey Veng Province. *MRC Conference Series* **4**, 9-16.

Trong, T. Q., Nguyen, V. H., and Griffiths, D. (2002). Status of Pangasiid aquaculture in Viet Nam. *MRC Technical Paper* **2**, 1-16.

Truong, T. T., Do, Q. T. V., and Nguyen, V. H. (2008). Fisheries and aquaculture statistics for the Mekong River Delta in Viet Nam. *MRC Conference Series* **7**, 65-81.

Tubtim, N. and Hirsch, P. (2005). Common property as enclosure: A case study of a backswamp in southern Laos. *Society and Natural Resources* **18**, 41-60.

Vannote, R. L., Minshall, G. W., Cummins, K. W., Sedell, J. R., and Cushing, C. E. (1980). The River Continuum Concept. *Canadian Journal of Fisheries and Aquatic Sciences* **37**, 130-137.

van Zalinge, N. and Nao, T. (1999). Summary of project findings: Present status of Cambodia's freshwater capture fisheries and management implications. *In* "Present Status of Cambodia's Freshwater Capture Fisheries and Management Implications," Van Zalinge, N. P., Nao, T., and Deap, L. (eds.), pp 11-20. Mekong River Commission and Department of Fisheries, Phnom Penh, Cambodia, 150pp.

van Zalinge, N. and Touch, S. T. (1996). Catch assessment and fisheries management in the Tonle Sap Great Lake and River. *In* Paper presented in "Workshop on Fisheries Statistics," 18 September 1996, Department of Fisheries, Phnom Penh, Cambodia, 21pp.

van Zalinge, N. P., Lieng, S., Ngor, P. B., Heng, K., and Valbo-Jorgensen, J. (2002). Status of the Mekong *Pangasianodon hypophthalmus* resources, with special reference to the stock shared between Cambodia and Viet Nam. *MRC Technical Paper* **1**, 1-29.

van Zalinge, N. P., Degen, P., Pongsri, C., Nuov, S., Jensen, J. G., Nguyen, V. H., and Choulamany, X. (2004). The Mekong River system. *In* "Proceedings of the Second International Symposium on the Management of Large Rivers for Fisheries," Phnom Penh, Cambodia, 11–14 February 2003, Volume 1, pp 335-357. RAP Publication 2004/16. FAO Regional Office for Asia and the Pacific, Bangkok, Thailand, 357pp.

Virapat, C. and Mattson, N. (2001). Inventory of reservoir fisheries in Thailand. *ACIAR Proceedings* **98**, 43-47.

Voeun, V. (2001). "Fishbones from Archaeological Remains from Angkor Borei," Heinrich-Boli-Foundation, Phnom Penh, Cambodia, 1p.

Vromant, N., Nam, C. Q., HChau, N. T., and Ollevier, F. (2004). Survival rate and growth performance of *Cyprinus carpio* L. in intensively cultivated rice fields. *Aquaculture Research* **35**, 171-177.

Walker, K. F. and Yang, H. Z. (1999). Fish and fisheries in western China. *FAO Fisheries Technical Paper* **385**, 237-278.

Warren, T. J. (1999). "A Monitoring Study to Assess the Localized Impacts Created by the Nam Theun-Hinboun Hydro-scheme on Fisheries and Fish Populations," Theun-Hinboun Power Company, Vientiane, Laos, 68pp.

Warren, T. (2000). Impacts to fish populations and fisheries created by the Nam Theun-Hinboun Hydropower Project, Lao PDR. *In* "Summary of Presentation at Conference on "Accounting for Development," University of Sydney, 23–24 June 2000, 7pp.

Watson, S. P. and Schouten, R. (2001). "Nam Song Diversion Project." Draft Impact Analysis Report and Action Plan. Asian Development Bank, Vientiane, Laos.

Welcomme, R. L. (1985). River fisheries. *FAO Fisheries Technical Paper* **262**, 1-358.

Welcomme, R. L. (2001). "Inland Fisheries Ecology and Management," Fishing News Books, Oxford, UK, 358pp.

Welcomme, R. L. and Halls, A. (2004). Dependence of tropical river fisheries on flow. RAP Publication 2004/16, pp 267-283.

Westing, A. H. (1976). "Ecological Consequences of the Second Indochina War," Stockholm International Peace Research Institute and Amqvist and Wiksell, Stockholm, Sweden, 119pp.

Whitmore, T. J., Brenner, M., Jiang, Z., Curtis, J. H., Moore, A. M., Engstrom, D. R., and Wu, Y. (1997). Water quality and sediment geochemistry in lakes of Yunnan Province, southern China. *Environmental Geology* **32**, 45-55.

Wongstitwilairoong, B., Srijan, A., Serichantalergs, O., Fukuda, C. D., McDaniel, P., Bodhidatta, L., and Mason, C. J. (2007). Intestinal parasitic infections among pre-school children in Sangkhlaburi, Thailand. *American Journal of Tropical Medicine and Hygiene* **76**, 345-350.

Wyatt, A. B. and Baird, I. G. (2007). Transboundary impact assessment in the Sesan River Basin: The case of the Yali Falls Dam. *International Journal of Water Resources Development* **23**(3), 427-442.

Xie, P. and Chen, Y. (2001). Invasive carp in China's Plateau lakes. *Science* **224**, 999-1000.

Xie, S. and Li, Z. (2003). "Inland Fisheries Statistics in China." RAP Publication 2003/01: 20–26. FAO Regional Office for Asia and the Pacific, Bangkok, Thailand.

Xie, P. and Liu, J. (2001). Practical success of biomanipulation using filter-feeding fish to control cyanobacteria blooms. *The Scientific World* **1**, 337-356.

Xie, Y., Li, Z., Gregg, W. P., and Li, D. (2001). Invasive species in China—An overview. *Biodiversity and Conservation* **10**, 1317-1341.

Yan, Y. Z. and Chen, Y. F. (2007). Changes in the life history of *Abbottina rivularis* in Lake Fuxian. *Journal of Fish Biology* **70**, 959-964.

Yang, J. X. (1996). The alien and indigenous fishes of Yunnan: A study on impact ways, degrees and relevant issues. *In* "Conserving China's Biodiversity," Wang, S., Peter, J. S., and Xie, Y. (eds.), pp 129-138. China Environmental Science Press, Beijing (in Chinese).

Yang, Y., Tian, K., Hao, J., Pei, S., and Yang, Y. (2004). Biodiversity and biodiversity conservation in Yunnan, China. *Biodiversity and Conservation* **13**, 813-826.

Yap, S. Y. (2002). On the distributional patterns of Southeast-East Asian freshwater fish and their history. *Journal of Biogeography* **29**, 1187-1199.

Yim, C. and McKenney, B. (2003). "Domestic Fish Trade: A Case Study of Fish Marketing from the Great Lake to Phnom Penh—Working Paper 29," Cambodia Development Resource Institute, Phnom Penh, Cambodia, 24pp.

York, R. and Gossard, M. H. (2004). Cross-national meat and fish consumption: Exploring the effects of modernisation and ecological context. *Ecological Economics* **48**, 293-302.

Zhou, D. (2001). "The Customs of Cambodia," Edited and translated by Michael Smithies. The Siam Society, Bangkok, 147pp.

Tonle Sap Lake, the Heart
of the Lower Mekong

I.C. Campbell,[1] S. Say,[2] and J. Beardall[2]

[1]Principal Scientist, River Health, GHD, 180 Lonsdale Street, Melbourne 3000,
Australia & Adjunct Research Associate, School of Biological Sciences,
Monash University, Wellington Rd, Victoria 3800, Australia
[2]School of Biological Sciences, Monash University, Wellington Rd, Victoria 3800, Australia

1. INTRODUCTION

Tonle Sap Great Lake, or Boeng Tonle Sap in Cambodian terminology, is the largest natural freshwater lake in southeast Asia and arguably the most important. It plays a critical role in Cambodian Khmer culture, in the economy of Cambodia, and in the ecology and hydrology of the Mekong River basin. The annual pulse of water and fish that arises from the lake and its floodplain constitutes the heartbeat of the lower Mekong, supporting the life of the river, and contributing to the productivity of the floodplains and the agricultural systems from Kratie in Cambodia to the South China Sea.

2. PHYSICAL DESCRIPTION OF TONLE SAP LAKE

Boeng Tonle Sap comprises a permanent lake surrounded by an extensive floodplain (Fig. 10.1). The permanent lake consists of two basins—a large northwest basin and a smaller southeast

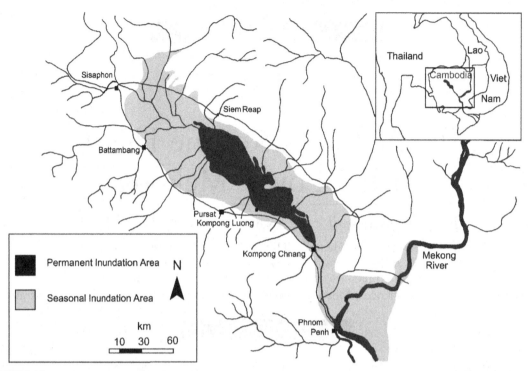

FIGURE 10.1   Location map indicating the extent of the permanent lake and the seasonally inundated floodplain.

basin—linked by a relatively narrow strait. During the dry season, the lake is about 120 km long and 35 km wide with an area of about 2500 km². A bathymetrical survey of the lake proper, conducted between 1997 and 1999, revealed a relatively flat bottom, with a maximum depth of about 3.3 m (Campbell *et al.*, 2006). During the flood phases the system enlarges to about 250 km long and 100 km wide with an area of about 17,500 km², and the depth reaches 8-10 m.

The floodplain surrounding the lake extends 20-40 km and is dominated by seasonally inundated forests and rice fields. A natural levee and distinct vegetation change marks the margin of the permanent lake. While the vegetation of the floodplain is still poorly mapped, the dominant taxa are *Barringtonia, Diospyros, Terminalia, Homalium*, and *Hydnocarpus*, and diversity is not particularly high (Campbell *et al.*,

2006; Junk *et al.*, 2006). The phenology of the plant communities is primarily governed by the flood cycle (McDonald *et al.*, 1997a,b). In some areas of the floodplain, forest is rapidly being cleared for agriculture (i.e., Siem Reap Province), while in other areas a largely intact *Barringtonia* community remains.

Tonle Sap Great Lake, like much of the rest of the Mekong River system of which it forms a component, is a monotonal flood-pulsed system (Fig. 10.2). The dry season permanent lake, begins to enlarge as flood waters from the Mekong back up the Tonle Sap river. The sediment-laden water flows through channels in the low levees surrounding the dry season lake margins, and extends over the extensive surrounding floodplain, depositing its sediment as if flows (Koponen *et al.*, 2003).

Much of the sediment is deposited close to the dry season lake margin as the floodplain,

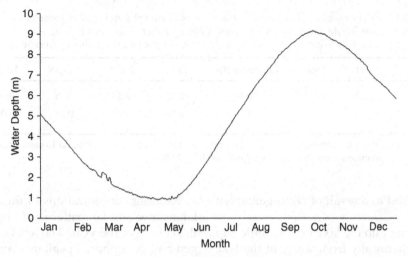

FIGURE 10.2 Annual flood cycle of Tonle Sap Great Lake. Average daily depth at the gauge at Kompong Luong for the 4 years from 1999 to 2002.

presumably because the inundated floodplain vegetation reduces turbulence within the water column and emergent vegetation reduces surface wind velocity. (Koponen *et al.*, 2003).

Water from the catchment of the Great Lake also contributes to the flood pulse. Monthly rainfall in the catchment exceeds 100 mm from at Siem Reap May to October with a maximum of 276 mm in September (World Meteorological Organization, 2007), while Mekong flood flows peak in September at Kratie. So the two water sources reach their peak at about the same time upstream of the junction of the Tonle Sap River.

Garsdal (2004) calculated that about 50% of the water flowing into the Lake was derived from the Mekong while about 40% was derived from the catchment, although this figure underestimates the catchment contribution (Campbell *et al.*, 2006).

The lake is very shallow in the dry season with only a few very small localities exceeding 3 m in depth. The mean depth, even at the height of the flood, is less than 7 m. This has a number of consequences for lake chemistry and ecology. As the water level reduces during the dry season, turbidity increases (Fig. 10.3), presumably because of resuspension of sediment

FIGURE 10.3 Mean monthly values for total suspended solids (solid line) and conductivity (dashed line) measured at Kompong Luong from July 1995 to December 2005. Error bars represent standard errors. Data from Mekong River Commission water quality database.

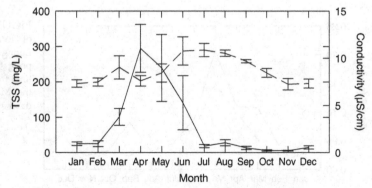

TABLE 10.1 Comparisons of mean values of a range of chemical water quality parameters measured in July-September in the Tonle Sap River at Kompong Chnang (when water was flowing into the Great Lake) and in June at Kompong Luong in the Great Lake prior to the influx of river water

Site	pH	TSS	Conductivity	Ca^{2+}*	Na^+*	NO_3-N	Total P	COD*
Kompong Luong	7.0	141	10.8	0.42	0.33	0.25	0.087	4.8
Kompong Chnang	7.3	102	11.1	0.59	0.23	0.24	0.053	2.5

Data from the Mekong River Commission water quality database for the period July 1995 to December 2005.
*Differences between the means were statistically significant ($p < 0.05$).

from the lake bed as a result of wind-generated turbulence.

Water flowing into the lake up the Tonle Sap River differs chemically from water in the lake (Table 10.1). There are large (and statistically significant) differences in conductivity, calcium ion concentration, and Chemical Oxygen Demand (COD). Although the incoming river water has a slightly lower level of suspended solids, it is clearly not so low as to account for the decline in Total Suspended Solids (TSS) in the lake which drops from about 300 mg l^{-1} in April and May to about 20 mg l^{-1} in July (Fig. 10.3). Presumably, the large change results mainly from the increased lake depth resulting in less resuspension of bed material. No chemical data are available for water entering the lake from tributary streams within its own catchment, so these may also be a contributing factor, but it would be expected that they would also be relatively turbid during the wet season.

Nutrient concentrations in the river and the lake show similar patterns of change through the course of the year. The levels of both nitrogen and phosphorus peak in May at both locations (Figs. 10.4 and 10.5), as water levels start to rise from their minima, and a little after the TSS peak in the lake. Concentrations are similar in both the lake and the river, so the river water is neither diluting nor boosting lake nutrient levels.

3. SIGNIFICANCE OF TONLE SAP LAKE

Tonle Sap Great Lake is of international significance culturally, hydrologically, and ecologically. Culturally it was the center of the great Khmer civilization from at least 800 AD to the 1440s (Chandler, 1996). Hydrologically

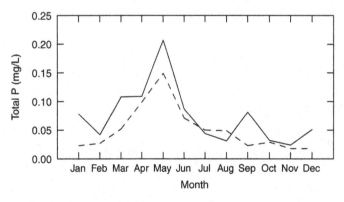

FIGURE 10.4 Mean monthly concentrations of total phosphorus measured at Kompong Luong in Tonle Sap Lake (solid line) and Kompong Chnang on the Tonle Sap River (dashed line) from July 1995 to December 2005. Data from the Mekong River Commission water quality database.

FIGURE 10.5 Mean monthly concentrations of nitrate + nitrite measured at Kompong Luong in Tonle Sap Lake (solid line) and Kompong Chnang on the Tonle Sap River (dashed line) from July 1995 to December 2005. Data from the Mekong River Commission water quality database.

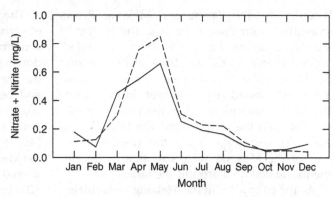

the lake provides a regulator for the flow of the Mekong River into the delta of Viet Nam. Ecologically it is a key component of one of the world's largest fisheries, and certainly the largest subsistence fishery.

3.1. Cultural Significance

The Khmer civilization is the subject of intense archaeological interest and as understanding grows, so interpretations change. A part of the attraction for non-Cambodians is the Angkor cultural site on the shores of Tonle Sap Great Lake, one of the wonders of the world. There is debate about the nature of Angkorean society and to what extent it was a hydraulic society (e.g., see Higham, 2001 and Coe, 2003 for differing viewpoints). More recently, Fletcher and coworkers have provided more detailed support for hydraulic dependence of the city of Angkor (Evans et al., 2007; Fletcher, 2001).

It is clear that to support the population required to build so many labor-intensive structures the Khmers required a substantial food surplus. From the images on the temple walls, and from our knowledge of the present fishery in Tonle Sap Lake, it is evident that fish from the lake formed a significant component of the food surplus. Daguan (1312) referred to Khmers from Angkor growing flood recession rice and floating rice around the Great Lake, and also to the great variety of fish consumed.

The integration between the fisheries and the society were so deep that the present-day Cambodian currency—the *Riel*—was named after the key commercial fish species termed trey riel in Cambodian (*Henicorhynchus lobatus* and *H. siamensis*).

3.2. Hydrological Significance

Hydrologically the lake plays a key role in the lower Mekong as the regulator of the Mekong flood. Fujii et al. (2003) used a combination of hydrological measurement, modeling and remote sensing to investigate the hydrological role of the Cambodian floodplain, of which the Tonle Sap Great Lake floodplain forms the largest component. Based on the year 2002 wet season inundation they estimated the total area of the floodplain as 26,150 km^2, of which the Great Lake floodplain comprises about 17,500 km^2 or 67%.

Not only is the Tonle Sap Lake floodplain more than half the Cambodian floodplain area, but it represents an even higher percentage of the total inundated area in any flood. In March 2002, at the end of the dry season, Fujii et al. (2003) estimated that 16% of the Great Lake floodplain was inundated (including the dry season permanent lake area), while at the peak of the flood in late September that increased to 68%. Two other zones of the floodplain had similar proportions inundated, the area bounded by the Bassac and Mekong Rivers extending from Phnom Penh to the border of Viet Nam increased from 1% to 63% inundated, while the area on the east bank of the Mekong River, opposite Phnom Penh, increased from 3% to 77% inundated. However, in both cases the total areas inundated at the peak were far smaller than the area around the lake. At 720 and 2560 km^2 respectively the flooded areas east of the Mekong were less than a quarter the inundated area around the lake.

As the river level in the Mekong rises during the annual wet season, it eventually exceeds the water level in the Tonle Sap Great Lake causing the famous flow reversal in the Tonle Sap River (Fig. 10.6). As water levels rise further in the Mekong upstream of Phnom Penh water also begins to flow overland across the floodplain toward the Great Lake. Based on simulations, Garsdal (2004) estimated that this overland flow constituted about 20% of the water flowing into the lake between 1985 and 2003.

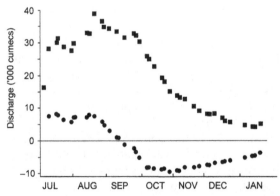

FIGURE 10.6 Measured discharge in the Mekong upstream of the junction with the Tonle Sap River (square symbols) and in the Tonle Sap River at Phnom Penh port (round symbols) from July 2002 to January 2003, based on data from Garsdal (2003). The negative discharge in the Tonle Sap River indicates a flow from the Mekong toward the Tonle Sap Lake.

The storage of water around the lake and elsewhere on the floodplain during the wet season ameliorates flood levels from Phnom Penh downstream. Fujii *et al.* (2003) estimated that, during the 2002 flood peak, between August and September, out of a total flow of about 50,000 m^3 s^{-1} entering their study area at Kompong Cham upstream of Phnom Penh, up to 5400 m^3 s^{-1} flowed across the floodplain from the Mekong to the Tonle Sap system, 8700 m^3 s^{-1} flowed up the Tonle Sap River channel and up to 8200 m^3 s^{-1} flowed to the east across the floodplain rejoining the river downstream of Phnom Penh. Thus, about 45% of the flow was "diverted" from around Phnom Penh. Similar amelioration occurs with the flows passing into Viet Nam during the peak of the flood. Fujii *et al.* (2003) estimated that flows into Viet Nam are only about 40% of the flows at Kompong Cham on the same date.

As water levels fall following the end of the wet season, the water stored on the floodplain during the flood season augments the dry season flows. Following the end of the wet season rains, and as the Mekong River recedes from its flood levels, water again begins to flow from the Great Lake and the floodplains down the Tonle Sap River toward the sea, supplementing water flowing down from the upper Mekong. For example, on December 5, 2002 the flow in the Mekong at Kompong Cham was measured at 5900 m^3 s^{-1}, while the flow at the Viet Nam border, including flows in the Mekong main stream, the Bassac and several smaller streams draining the Cambodian floodplains, was almost double that at 11,000 m^3 s^{-1}.

Both flooding and dry season low flows are perceived as significant transboundary environmental problems in the lower Mekong (Campbell, 2007). The Cambodian floodplains, of which those around the Great Lake are the largest component, play an important role in ameliorating both problems.

4. LIMNOLOGY OF TONLE SAP LAKE

Although its ecology has been poorly studied, it is clear that Tonle Sap Lake has enormous local, regional, and global ecological significance. Locally it supports a productive fishery and a substantial human population. Regionally it acts as a nursery for fish which migrate to contribute to the overall Mekong fishery, as well as providing habitat for many bird, mammal, and plant species. Globally, parts of the lake are now recognized as having World Biosphere significance.

The energy which supports most ecosystems is derived from sunlight harvested by plants, primary producers, through photosynthesis. The rate at which sunlight is harvested is a measure of the primary production of the system. In most ecosystems the amount of solar energy used is substantially less that the amount available. So, to understand how an ecosystem functions it is necessary to know which plants are the most important source of energy for higher consumers, how much of the sun's energy they trap and pass on to consumer organisms, and what limits the rate at which they trap energy from sunlight.

Although many authors have claimed that Tonle Sap Lake is highly productive (Kummu *et al.*, 2005; Lamberts, 2001, 2006; Rainboth, 1996), these opinions were based solely on the fish catch, not on any measures of energy flow. The only previous data which could indicate productivity of the lake was some limited chlorophyll data presented by Campbell *et al.* (2006) and Song *et al.* (2004).

Campbell *et al.* (2006) believed that chlorophyll levels were highest during the wet season flood, and lowest during the low water period. They used an *in situ* fluorometric probe technique to measure chlorophyll in the water column and found that levels peaked at about 20 µg l^{-1} in February and then crashed and were unmeasurable in May. However, they

noted that the high level of turbidity present in May probably interfered with the readings at that time.

In contrast, Song *et al.* (2004) presented data that showed maximal chlorophyll levels in May, during the dry season, with far lower levels during the flood period. They collected surface water samples from 10 sites in the lake in May 2003 and from 10 sites in the lake and 10 in the flooded forest in October 2003. Chlorophyll concentrations were higher in the lake in May (mean: 54 µg l^{-1}, SD: 33) than in October (mean: 33 µg l^{-1}, SD: 17), and the difference was significant ($p = 0.013$). Chlorophyll was higher in the flooded forest (mean: 44 µg l^{-1}, SD: 26) than in the open lake, but the difference was not significant. Both the lake and flooded forest chlorophyll values were quite variable, as can be seen from the standard deviation values. The highest levels recorded were 117 µg l^{-1} in the lake in May, and 89 µg l^{-1} in the flooded forest in September.

More recent detailed data collected by Say (unpublished data) indicate a spatially and temporally complex pattern of chlorophyll variation in the lake. Remote sensing data show surface chlorophyll peaking during the dry phase when turbidity is high, reducing as the river water begins to fill the lake, remaining generally low but with hot spots fringing the open water during the high water period, and then rising again as the water drains out. Peak surface concentrations reach levels in excess of 100 µg l^{-1}, dropping to as low as 10–15 µg l^{-1} in the central part of the lake during the high water phase. The depth-averaged chlorophyll values calculated by Say (unpublished data) showed the same trend as surface measurements. Both measurements are consistent with those of Song *et al.* (2004).

However, depth-integrated chlorophyll data collected by Say (unpublished data) show a quite different pattern to the surface chlorophyll. These data show minimal levels of chlorophyll *a*, as little over 20 mg m^{-2}, during the dry season while the lake level is falling and is low, but increasing to over 35 mg m^{-2} while the lake is rising and peaking at over 45 mg m^{-2} when the water level is highest. This is the opposite of the trend shown by the remote sensing results but consistent with the results reported by Campbell *et al.* (2006).

The key to reconciling the two apparent opposite interpretations lies in the data on algal assemblages included in Song *et al.* (2004). The high level of chlorophyll reported in surface samples during the dry season by Song *et al.* (2004) and others consists largely of species of cyanobacteria from the genus *Anabaena*. *Anabaena* spp. have gas vacuoles, so they float near or on the surface of the water at a time when the turbidity of the lake is high. Their buoyancy, combined with an ability to fix nitrogen in nitrogen-limited waters, give them an advantage over other algal species present, and based on the data in Song *et al.* (2004) they comprised as much as 60% of the algal biomass between March and April 2002. Because these species accumulate at the surface they show up in the remote sensing data and also in chlorophyll samples collected at or near the water surface. With the influx of river water, and the reduction in turbidity, *Anabaena* loses its advantage and other algal species, mainly diatoms, grow rapidly in the nutrient-rich transparent water. The surface water sampling conducted by Song *et al.* (2004) collected the *Anabaena* bloom, while the probe sampling by Campbell *et al.* (2006) missed it.

Although there are noticeable differences in turbidity between the lake basin and the floodplain waters, this does not appear to impact chlorophyll levels greatly. Song *et al.* (2004) cite a mean TSS value in the open lake during the high water period of 1992 at 77 mg l^{-1} while the mean level in the flooded forest was only 23 mg l^{-1}. However, neither Song *et al.* (2004) nor Say (unpublished data) found a statistically significant difference in chlorophyll concentrations between the two, although in both cases the floodplain means were higher

than those of the open lake, and in the case of the data of Song *et al.* (2004) the difference was almost significant ($p = 0.051$).

The more detailed studies of Say (unpublished data) indicate that phytoplankton production in the lake is high, and peaks during rising and high water levels, when fish larva are actively feeding and growing in the lake. Chlorophyll *a* levels in Tonle Sap Lake are well below those in many temperate and tropical shallow saline lakes (Hammer, 1981; Payne, 1986) as well as those of some well known freshwater shallow tropical lakes, such as Lake George in Uganda with levels above 200 µg l^{-1} (Ganf, 1974) and Lake Apopka in Florida which averages 105 µg l^{-1}. But chlorophyll levels in Tonle Sap Lake are comparable with many other examples including Lake Okeechobee in Florida at 15–50 µg l^{-1} (Phlips *et al.*, 1995).

Certainly, chlorophyll levels in Tonle Sap Lake are well above the levels generally associated with eutrophication in temperate lakes. Thomas *et al.* (1997) cite annual mean concentrations of total chlorophyll of 25 µg l^{-1} and maxima of 75 µg l^{-1} as the levels above which lakes are classified as hypereutrophic. In Tonle Sap Lake, Say (unpublished data) measured the annual mean for total chlorophyll as 51 µg l^{-1} with a maximum of 89 µg l^{-1} in the lake basin and a mean of 70 µg l^{-1} and a maximum of 103 µg l^{-1} on the floodplain.

It is not yet known what limits phytoplankton productivity in Tonle Sap Lake. Phytoplankton production in Northern Hemisphere temperate regions is most often limited by phosphorus (Wetzel, 1975), while lakes in many tropical regions, such as those of east Africa, are often limited by nitrogen (Payne, 1986; Talling and Lemoalle, 1998). Highly turbid waters are also often light limited. In the absence of any experimental data testing for nutrient limitation it is not possible to come to a definite conclusion. Sarkkula and Koponen (2003) stated that the lake was phosphorus limited, while Sarkkula *et al.* (2003) claimed that the lake was mostly phosphorus limited but occasionally nitrogen limited. However, neither publication gave any evidence for phosphorus limitation, and the evidence for nitrogen limitation was limited to the occurrence of blooms of nitrogen-fixing cyanobacteria in the March-April period.

One method of assessing the relative significance of phosphorus or nitrogen as a limiting nutrient is to examine the ratio between them. Generally N/P ratios in excess of 10 are considered indicative of phosphorus limitation, while ratios below 7 are considered to indicate nitrogen limitation (Meybeck *et al.*, 1989; Thomas *et al.*, 1997). Using the MRC water chemistry data for total phosphorus and nitrate + nitrite from 1995 to 2005, we calculated mean ratios for each month (Fig. 10.7). Based on the ratios the lake would appear to be nitrogen limited throughout the year except during periods of light limitation. Nitrogen is most often the limiting nutrient for agriculture in Cambodian soils (MRC, 2003b).

Although the annual inflow from the Mekong flood and catchment streams are undoubtedly important nutrient sources for the Tonle Sap Lake, nutrient dynamics in the lake must certainly be dominated by internal loading. As previously noted, phosphorus and nitrogen concentrations in the inflowing water at Kompong Chnang are similar (Figs. 10.4 and 10.5). It is

well known that mixing processes in shallow lakes play an important role in nutrient release from sediments, especially for phosphorus (e.g., Jensen and Andersen, 1992; Sondegaard *et al.*, 2003). Even if all external sources of nutrients ceased, internal loading would be sufficient to maintain the lake in its present trophic state for many years, if not decades.

5. FISHERIES IN TONLE SAP LAKE

As previously noted, the lake has been claimed to be "among the most productive freshwater ecosystems in the world," or words to that effect, by a large number of authors. However, as Lamberts (2006) has recently pointed out, these statements are based on the large fish catch, and not on any data on production for any component of the lake ecosystem, although his comment that there has been no research to determine primary production in the lake is not correct.

Production, in an ecological sense is the rate at which biomass is produced per unit area. It is usually expressed either in terms of energy or, more often, as grams of carbon per unit area per year. Organic C produced by plants using sunlight to convert water and carbon dioxide into plant biomass is termed primary production, while production by animals, including

FIGURE 10.7 Mean monthly ratio of N/P at Kompong Luong based on data from July 1995 to December 2005. Data from the Mekong River Commission water quality database.

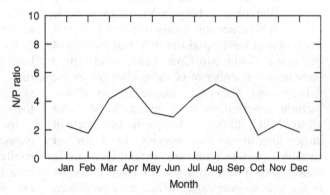

fish, is referred to as secondary production. Clearly, the rate of secondary production in an ecosystem depends on the rate of primary production, although there may be energy subsidies from allochthonous sources—primary production from outside the system. Examples would include algae produced in a reservoir that are washed into the river downstream (e.g., Ward and Stanford, 1979) and terrestrial plant material that falls, or is washed, into headwater streams (e.g., Cummins, 1986).

In the case of Tonle Sap Lake, the chlorophyll data cited previously indicate that autochthonous primary production is high, but certainly not among the highest in the world. Lake productivity would only be among the highest in the world if there was a very substantial input of allochthonous production from the floodplain vegetation at least equivalent to the autochthonous production level.

The Tonle Sap Lake is noted for its large fish catch, which is critical to those who live near it, but catch is not the same as production. This has been most spectacularly illustrated in recent years by the collapse of the Grand Banks cod fishery in the Atlantic Ocean (see Casey and Myers, 1998; Pilkey and Pilkey-Jarvis, 2007), a result of fish catch exceeding fish production. Equally, fish catch may be substantially lower than fish production in water bodies that are rarely fished. There is simply no obligate relationship between the two.

Where there has been long-term heavy exploitation of a fishery with no observable impact, it is reasonable to assume that fish production is at least equal to catch. But that is not the case in Tonle Sap Great Lake, where there have been a number of obvious changes to the fishery over the past decade or so. These include reported declines in catch per unit effort (MRC, 2003a), a decline in the catch of large long-lived fish species in favor of smaller species with an annual life cycle (MRC, 2003a) and an apparent large variation in the year-to-year catch, which may be driven

by climatic factors or human activities (Hortle et al., 2004a,b).

Many tropical flood-pulse rivers or river floodplain systems are notable for having high biodiversity and high productivity (Junk, 1999; Junk and Wantzen, 2004; Junk et al., 1989). One of the factors often cited as playing an important role in contributing to the high productivity levels was the contribution of terrestrial organic material, which was available for consumption during the inundation phase (Junk and Howard-Williams, 1984). Studies using stable carbon isotopes have since shown that terrestrial carbon sources, including flooded forest trees, inundated grasses, and macrophytes contribute surprisingly little to fish populations. Results have been generally consistent between tropical flood pulse systems in the Amazon and Orinoco river systems in South America (Forsberg et al., 1993; Hamilton et al., 1992) and the equatorial Fly River system in Papua New Guinea (Bunn et al., 1999).

In the Amazon, phytoplankton was estimated to contribute less than 10% of floodplain primary production, but an average minimum of 37% of fish carbon, and for a number of commercially important species it was the predominant carbon source (Araujo-Lima et al., 1986; Forsberg et al., 1993). In the Orinoco, the ultimate carbon sources for both invertebrates and fish were reported to be phytoplankton and periphyton, even though macrophytes and terrestrial litter fall composed 98% of the potentially available carbon for the river-floodplain system (Hamilton et al., 1992; Lewis et al., 2001). In the Fly River, algal carbon (including both periphyton and phytoplankton) accounted for 32–72% of the carbon in floodplain fish species and 31–46% of the carbon in fish from the main channel (Bunn et al., 1999).

In the case of Tonle Sap Lake it might be expected that the relative contribution of terrestrially derived carbon would be less than the case for at least the Amazon and Orinoco rivers. The Tonle Sap inundation area flooded

FIGURE 10.8   Aerial photograph of flooded forest vegetation in the Battambang river area. Photograph by Joe Garrison. (See Color Plate 21)

forest is a relatively open canopy (Fig. 10.8), while those along the Amazon and Orinoco are not. Thus, plant litter fall in the flooded forest areas of the South American rivers should be higher than in the Tonle Sap flooded forest, and with less light penetration algal growth should be lower. However, both South American rivers, and the Fly River in PNG, like the Tonle Sap Lake have large unforested areas on the floodplain. In these areas, in the absence of floating vegetation which could shade the water column, algal production should be high.

Algal material is known to be a richer food source for aquatic animals than detritus or even growing macrophyte material (Forsberg *et al.*, 1993). Therefore, when algal food is available in aquatic ecosystems it is often utilized to an apparently disproportionate level.

Stable carbon isotope data from Tonle Sap Lake indicate that, as for the other systems, algal carbon from phytoplankton and periphyton (algae growing attached to surfaces) is the predominant energy source. Say (unpublished data) investigated the stable carbon and nitrogen isotope signatures for six commercially important fish species from the lake that included *Barbonymus gonionotus*, *Channa striata*, *Cirrhinus microlepis*, *Cyclocheilichthys enoplos*, *Henicorhychus lobatus*, and *Pangasius larnaudii*. Four of the six, which comprise 30% of Cambodia's annual fish catch (Hortle *et al.*, 2004a,b), clearly derived most of their energy from phytoplankton or periphyton. For two others, both of which are carnivores and comprising 11% of the Cambodian annual catch, the results were less clear.

Although the flood pulse has commonly been cited as a key contributing factor to Tonle Sap productivity, through the transfer of terrestrial primary production into the aquatic phase (Kummu *et al.*, 2006; Lamberts, 2001; Mizino and Mori, 1970; Pantulu, 1986b; Sarkkula *et al.*, 2005), in reality this is not the case. In fact the flood contributes to the lake productivity by providing a large shallow, warm, nutrient-rich, and brightly illuminated water body that allows rapid algal growth. The dry period is also critical because it is during the low flow periods that sediments are resuspended, making their nutrients available for the next wet season production pulse. If the water was maintained at the wet season depth permanently the sediment-associated nutrients would settle to the lake floor, becoming far less available to support algal production and the invertebrate and fish production dependent on it.

6. THREATS TO THE LAKE

The high production levels and diverse habitat resulting from the mosaic of riparian vegetation bordering the lake have resulted in a diverse fauna. Although many of the faunal elements are as yet poorly known, Campbell *et al.* (2006) documented more than 200 species of higher plants, 149 fish species, at least 220 species of birds, 16 species of snakes, and at least 10 large mammal species including 3 species of primates, 2 otters, fishing cats, Asian jackals, and the Asian mongoose. There is also a small population of Siamese crocodile remaining in the lake. Many of the animal species are rare, including 9 globally threatened bird species.

However, Tonle Sap Lake is coming under increasing environmental pressure from changing conditions within Cambodia as well as elsewhere in the Mekong Basin. Challenges can be grouped broadly as those arising within the Lake and its floodplain, within the lake catchment, and elsewhere within the Mekong Basin or outside it.

6.1. Changes in the Lake and Floodplain

Within the lake and its inundation area there are two pressing problems, clearing of the vegetation on the inundation area and fishing pressure. Both are results of the same cause—rapid growth of the population in Cambodia following the political stability and freedom that followed the demise of the Khmer Rouge in the early 1990s. Following decades of war, political unrest, and repression, the peaceful conditions that are now established have led to a baby boom. In 1998 it was estimated that 54% of Cambodians were aged 19 or younger (MRC/ UNDP, 1998) and the annual rate of population increase was estimated in 2003 as around 2–3% (Hook *et al.*, 2003). Most Cambodians are poor subsistence farmers, dependent on rice and wild-caught fish, so as the population has grown, forest has been cleared to provide land for farming and firewood for cooking.

The area around Tonle Sap Lake is relatively flat, with good access to water and close to good fishing areas, so it has tended to attract populations from elsewhere in Cambodia. MRC/UNDP (1998) estimated that there were about 1.2 million people living within the area enclosed by the roads circling Tonle Sap Lake, with a further 1.4 million living in the zone extending 15 km outward from the roads. Population growth within the basin was estimated to be about 4.8% (Leang, 2003).

As a consequence, there has been concern expressed about the impacts on forest- and lake-based resources. MRC (1997) claimed that, based on remote sensing data, there had once been about 1 million ha of forest around the lake and that this had been reduced to about 360,000 ha of flooded forest and 157,000 ha of degraded forest and associated vegetation types. They suggested that clearing had recently intensified but was unquantified. These claims should be treated cautiously, since the vegetation surrounding the lake is, and always has been, a complex mosaic, and no information was provided on the actual data source or processing and ground truthing. Nevertheless it is widely believed that forest cover surrounding the lake has been, and continues to be, reduced by human clearing.

The fishery of Tonle Sap Lake is linked to the broader lower Mekong fishery, which is clearly under pressure (Hortle, Chapter 9). Fishing pressure within the lake is particularly intense,

FIGURE 10.9   Fish traps at the mouth of the Battambang River in 2004. (See Color Plate 22)

with large fish traps fringing the lake margin, and some tributaries, such as the Battambang River (Fig. 10.9), almost completely occluded at low flows.

Fish are not the only aquatic resource intensively exploited within the lake. Also important are "snake fisheries" (Brooks *et al.*, 2007). There is an intensive harvest of water snakes, with upward of 8500 snakes per day being captured during the wet season (Stuart *et al.*, 2000). The major market appears to be the crocodile farms in Siem Reap and Battambang provinces, where they are used for food but some are consumed by humans either locally or in export markets in Thailand, Viet Nam, and China. Waterbirds and their eggs are also intensively harvested, especially in areas near villages (McDonald *et al.*, 1997a,b). Most of these harvests are uncontrolled and certainly not sustainable.

There have been a number of proposals to build a dam or barrage across the entrance to Tonle Sap Lake, possibly near Kompong Chnang (Hori, 2000). While these proposals have not received much attention recently

they need to be discussed because undoubtedly they will be raised again in the future. The proposal, in its most recent form, was to build a large (700 m long) relatively low-level barrage that would fill during the Mekong flood and then retain the lake at its highest water level for some months after the normal flood recession. The water would then be used partly for irrigation and partly for release later in the dry season to increase dry season flows.

Numerous potential benefits were identified as arising from such a project (Hori, 2000). These included an increase in the fish catch in the lake, a reduction in wet season flooding in Cambodia and Viet Nam, improvement in navigation in the lake and in Tonle Sap River, facilitation of pump irrigation in Cambodia and Viet Nam, and a reduction in salt water intrusion into the delta. These were in addition to the promotion of stable irrigation around the lake. In its most recent appearance, mathematical modeling indicated that the project would be less effective at reducing flooding than expected, so it was allocated a low priority.

In fact it is difficult to envisage any project within the Mekong that would have been as environmentally, economically, and socially destructive as a Tonle Sap Lake barrage. The effect of retaining water after the end of the wet season would be to block fish migration down the Tonle Sap River. Fish migrate down the river to breed in the Mekong during the low flow period, with their larva being washed back up the river in the following wet season to grow in the shallow waters around the lake. Thus the dam, in a single year, would have virtually wiped out the fishery for migratory white fish in Cambodia and Viet Nam in a single year. The 10 million or so people dependent on the fishery would have lost their protein source, and the Cambodian government would have lost a USD $2 billion a year fishery.

Within the broader Tonle Sap catchment, many of the challenges are similar to those on the lake floodplain. There is rapid population growth, with pressure on natural resources. The major pressure that has been often identified is the potential for deterioration in water quality in streams flowing into the lake, although diversion of inflowing streams for irrigation development is also a potential issue.

There is extraordinary growth in the city of Siem Reap, the city from which tourists visit Angkor. Campbell et al. (2006) cited annual increases in hotel rooms of 112% between 2003 and 2004. The city at present has no waste treatment facilities, so hotels build their own package treatment plants or septic systems, and there is increasing concern about contamination of the Siem Reap River which drains to the lake.

A number of other towns around the lake are also likely to experience rapid growth. Battambang and Pursat are located on the main road between Phnom Penh and Thailand. The road has recently been upgraded and both are likely to grow with the increasing in commercial traffic. Both are also located on rivers which drain into the lake. On the northern side of the lake the road from Phnom Penh to Siem Reap has also been upgraded, so that a journey which formerly took about 18 h can now be completed in 4 h. As a result it is likely that much of the traffic which formerly went by boat up the Tonle Sap River, a 5-h journey, will now go by the quicker road, and the towns along that road will also grow more rapidly, along with the wastes that they will discharge into Tonle Sap tributaries.

One of the concerns has been that forest clearing in the catchment is causing sedimentation in the tributary streams and thence in the lake. Sedimentation is alleged to have become so severe that the lake will shortly cease to exist as an aquatic system. Campbell et al. (2006) traced these beliefs back to an unpublished report by Csavas (1990), which suggested that the lake would cease to exist within a decade, but he may have been repeating, and adding doubtful quantitative estimates to, accepted beliefs (Pantulu, 1986a; Welcomme, 1979). The story has unfortunately become entrenched in both the popular (e.g., Earth Report, 2000; Richardson, 1999) and scientific literature (Douglas, 2005; Dudgeon, 1999; Gupta, 2005). It has recently been shown to be completely without foundation (e.g., Penny et al., 2005). The low slopes of streams on the Tonle Sap Lake floodplain will ensure that any sediment movement along them will be extremely slow, and the flatness of the land in-between, and rapid growth of vegetation at the start of the wet season, will reduce soil erosion rates.

6.2. Changes in the Mekong Basin

Changes in the broader Mekong River basin have more recently been seen as constituting environmental threats to Tonle Sap Lake. Two types of activities are seen as potentially significant threats. Projects that divert water from the river could alter the flow regime to the lake, and construction of reservoirs, especially reservoirs on the main stream, could impact the hydrology, water chemistry, and fisheries in the lake.

Water diversion projects upstream of the Chaktomuk junction at Phnom Penh could result in altered flow regimes in the Mekong, and thus in Tonle Sap Lake. Modeled projections of possible irrigation diversions, such as the Chi-Mun proposal, indicate that their effect on Tonle Sap Lake is likely to be negligible (Campbell, Chapter 16; Podger *et al.*, 2004). In view of the limited land suitable for irrigation available in upstream areas, and the fact that water would almost certainly be extracted during the wet season flows, there is not likely to be sufficient demand to make a noticeable impact on the hydrology of Tonle Sap Lake.

Of more concern are the possible impacts of main stream dams. Of these there are two scenarios that are of greatest concern. One is a proposal to build a hydropower dam at Sambor rapids in Cambodia, and the other is the Lancang cascade of dams in China.

The Sambor rapids dam proposal has recently been revived by Cambodian government officials. The Sambor dam was included as one of the projects in the Mekong Committee's 1970 indicative basin plan (Hori, 2000; Mekong Committee, 1970). As originally conceived it was to be a 54-m-high, 26-km-wide dam, with a storage pool stretching to Stung Treng, 147 km upstream (Hori, 2000), and its purpose was to produce hydroelectricity. Cambodia has chronic electricity shortages, with few high-gradient streams suitable for hydropower development, and no significant fossil fuel deposits discovered so far. As a result most of the electricity generated for Phnom Penh is produced by diesel generators, so any large-scale hydropower project within the country's borders superficially seems attractive. The project could also provide water for irrigation of 34,000 ha of land immediately downstream on both banks of the river.

The problem is that a dam at Sambor rapids, like the Tonle Sap Lake barrage, would devastate the lower Mekong fishery. The impact would be twofold. Firstly, the fish which migrate upstream each year to breed in the

river between Sambor and Stung Treng, or even further upstream, would be unable to pass the dam, and thus would perish without reproducing. Fish ladders are unlikely to be successful because they need to be tailored to the needs of particular fish species. That is relatively easy in rivers where there are few fish of interest, as in the salmonid rivers in North America, but far more difficult in a river such as the Mekong where there is a high diversity of fish species, all with unique requirements. Riverine fish also often have difficulty with passage through the impoundment, further reducing the effectiveness of the ladder.

Apart from blocking fish passage, a dam at Sambor would submerge key fish breeding areas. Within the pondage of the dam, the lake bed would develop quite different physical characteristics to the river bed. Fine silt would accumulate on areas that were formerly gravel and boulders and deoxygenation would probably occur near the river bed during the dry season. Thus, fish which lay their eggs on the river bed would be unable to find suitable habitat. For fish which produce planktonic eggs which float downstream the situation would be worse. In the pondage in the dry season there would be insufficient current to carry them, and most larva would perish in the relatively food-free pondage water rather than being carried by the current to the shallow fertile waters of the floodplain.

The impact of the Lancang cascade on Tonle Sap Lake is also likely to be appreciable (Campbell, Chapter 16). Two sets of potential impacts have been identified in the literature, hydrological impacts and chemical impacts. The hydrological potential impacts have been discussed elsewhere (Campbell, Chapter 16; Podger *et al.*, 2004), and so will not be discussed much further here. The key hydrological issues are to what extent the dams will reduce the level of the flood, to what extent they will raise the level of the dry season flows, and to what extent will the flow changes interfere with fish reproduction.

Predicting the impact of the hydrological changes is difficult—at best there would be multiple small changes which may or may not add up to more than the sum of their parts. The reduction in wet season flows is likely to be relatively small. Combined with the increase in dry season flows Campbell *et al.* (2006) suggested a decrease in the area of seasonally inundated area around the lake of about 10%, while Podger *et al.* (2004) suggested generally smaller changes for the lower basin as a whole.

The increase in dry season water depth is of concern for two reasons. One is the reason previously discussed, the seasonally inundated area is determined by the difference between dry season low water levels and wet season flood levels. It is therefore subject to change either through a decrease in flood levels but also through an increase in dry season low water levels. But the dry season water level may also be important because it is during the dry season that lake bed sediment is resuspended, providing the nutrient pulse that is one of the drivers of primary production. The deeper the dry season water level, the less sediment suspension there will be, and the weaker the nutrient pulse. So changes to dry season water levels may have a range of as yet unforeseen impacts.

Finally, there is a potential hydrological-impact of flow changes on fish reproduction. Many of the fish species that make up a large part of the Cambodian fishery migrate to the lake and breed in the Mekong River, mainly between Phnom Penh and Khone falls. The fish leave the lake during the beginning of the dry season, contributing to the dai fishery in the Tonle Sap river upstream of Phnom Penh, and lay their eggs in the Mekong during the low flow period (Hortle, Chapter 9). Increases in the dry season flows have the potential to interfere with breeding, or the passive dispersal of larva downstream and back on to the floodplain and back to the lake in the following wet season.

One other concern that has been raised is that the Lancang cascade, or other main stream dams, could reduce the sediment and nutrient transport into the Tonle Sap Lake (Kummu and Varis, 2006; Kummu *et al.*, 2004; Sarkkula *et al.*, 2003). Concerns about sediment trapping by Chinese dams were first raised by MRC (2003a), who noted a drop in concentration in total suspended solids at Chiang Saen in Thailand following the closure of Manwan dam. Kummu and Varis (2006) conducted more detailed analyses calculating sediment loads based on suspended sediment. However, more comprehensive data on Mekong sediment loads using depth integrated sediment samples show no indication of such an impact (Walling, Chapter 6) a contradiction for which there is no obvious explanation.

Kummu and coworkers (Kummu and Varis, 2006; Kummu *et al.*, 2004) have suggested that dams under construction and proposed for the upper Mekong or Lancang River in China will reduce the sediment load entering the Tonle Sap Lake during the flood season, thereby reducing the nutrients which the flood pulse sediments supply to the lake. They have modeled these impacts assuming an admitted extreme reduction of 50% in the flood pulse sediment load, and suggest that impacts would include a drop in productivity due to reduced nutrient availability.

There are several difficulties with this scenario. The first is that, although several of the dams in China are very large, and will cause major reductions in sediment load in downstream sections of the river, they are located more than 1000 km upstream of the junction of the Tonle Sap River and the Mekong. By that distance there is unlikely to be any noticeable sediment trapping impact of the dams. Secondly, the models ignore internal sediment and nutrient loading within the lake. The high nutrient levels in the lake would be maintained for decades even if nutrient and sediment inputs from the Tonle Sap River ceased. This has been exactly the problem for a number of shallow eutrophic lakes in developed countries, where reducing nutrient inputs has not necessarily been sufficient to improve nutrient status of the water because of internal loading (Sondegaard *et al.*, 2001, 2003). In those cases chemical treatment to "lock" nutrients in the sediments in insoluble form, or sediment removal by dredging, was necessary to reduce water column nutrient concentrations.

6.3. Global Changes

One key global factor is global climate change, but it is difficult to make predictions about the future impact of climate change on Tonle Sap Lake. Different climate change models make contradictory predictions about the possible direction of future climate change, but it is possible that climate change has already played a significant role in the history of the lake.

Two prominent modeling exercises are attempting to make predictions about the consequences of climate change for the Mekong basin. One of these, conducted by the SEA START project based in Bangkok, predicts that rainfall in southeastern Lao will increase by between 500 and 1000 mm per year (Chinvanno, 2004). The other, the Adapt project run through the CGIAR centers predicts more droughts for the same area. (Adapt, 2003). Other researchers have predicted that rising temperatures will lead to loss of glaciers and snowmelt from the Himalaya, which would reduce the wet season flood pulse (WWF, 2005). Whether any or none of these scenarios is correct, it is clear that there may be substantial repercussions from climate change within the Mekong Basin, including Tonle Sap Lake, and these changes will add to the existing environmental pressures.

Climate change may well have played a role in the decline of the Angkorean empire. The decline of Angkor is usually dated at as commencing around the mid-1300s (Chandler,

1996). It is not clear to what extent the decline was rapid or slow, although slow seems more likely. Evans *et al.* (2007) have attributed the decline to silting up of the hydraulic system as a result of erosion brought about by forest clearing, however this seems to us to be simply an extension of the misconception that Tonle Sap Lake is presently silting up. The land around Angkor is relatively flat, so sediment movement would be slow.

In China the medieval warming period ended from some time after 1300 AD, to be followed by the Little Ice Age (De'er, 1994, but also see Hughes and Diaz, 1994). Undoubtedly this was a period of rapid global climate change, which raises the possibility that climatic change, or prolonged drought, was one of the pressures that led to the decline of the Angkorean period. A decline in fish production due to reduced wet season flows, or reduced rice production if the amplitude of the annual flood was reduced, may have produced a positive feedback spiraling a society short of irrigable land into decline.

7. MANAGEMENT OF TONLE SAP LAKE

The Royal Government of Cambodia has given increasing attention to Tonle Sap Lake over the past 15–20 years. There has been a growing recognition within the government of the importance of the lake as a fishery resource, as a cultural icon, and as a valuable biodiversity resource. The temple complex at Angkor has had World Heritage listing since 1992 (Campbell *et al.*, 2006), but protection for the lake and floodplain have been slower in coming.

Developing a management structure for the Lake has been complicated, as is often the case with biological resources, by the number of government agencies with real or potential interests within the lake and its catchment (Bonheur and Lane, 2002). The major players are the Ministry of Agriculture Forests and Fisheries (MAFF), the Ministry of Environment (MoE), and the Ministry of Water Resources and Meteorology (MOWRAM), but other government agencies including the Cambodian National Mekong Committee (CNMC) and tourism and minerals agencies also have interests in the lake and its management.

In November 1993 the Royal Cambodian Government, by Royal Decree, designated the lake as a multiple use protected area and in 1997 the Lake was successfully nominated as a biosphere reserve under the UNESCO Man and the Biosphere Reserve Programme. The reserve is intended to serve three functions: the conservation of landscapes, ecosystems, and species; culturally, socially, and ecologically sustainable development; and research and education.

The reserve, as established, is divided into three zones: core areas, buffer zones, and transition zones (Fig. 10.10). The three core areas are Prek Toal (31,280 ha) Boeng Tonle Chhmar/Moat Khla (32,970 ha), and Stung Sen (6590 ha). They are considered to be unique ecosystems of high conservation value with a level of protection similar to national parks. However, there are an estimated 2000 people living within the core areas, who fish, hunt, and collect firewood within them. The buffer zones amount to some 540,000 ha and include mainly flooded forest and open lake. With an estimated human population of about 100,000, the buffer zones incorporate some cropping including dry season rice farming, and growing lotus and vegetables in the terrestrial parts of the buffer zones as well as community

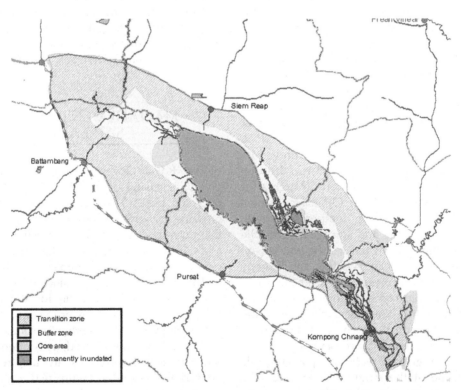

FIGURE 10.10 Map showing the three types of conservation areas around Tonle Sap Lake. Based on material in www.cambodiaatlas.com.

subsistence and commercial fishing. It is intended to develop community management of these areas. The transition zone is primarily farmland where rain-fed and floating rice are cultivated.

Following the listing of Tonle Sap Lake as a Man and Biosphere site in 1997, the Royal Cambodian Government established a Tonle Sap Biosphere Reserve secretariat within the CNMC in 2001 (Sokhem and Sunada, 2006). The CNMC has the advantage of including representatives of many of the ministries with potential interests in the Lake, including MAFF, MoE, MOWRAM, and seven others, but it has very limited financial and human resources (Sokhem and Sunada, 2006).

More recently, the Asian Development Bank (ADB) has funded and supported the development of a lake basin management organization for Tonle Sap Lake under the Tonle Sap Initiative. There has been argument about whether the final structure should be as a coordinating organization working to encourage cooperation between other government agencies, or as an authority with overriding veto power (Sokhem and Sunada, 2006). Not surprisingly, the latter approach is strongly resisted by existing players but seems to have support of the Prime Minister (Sen, 2007). It remains to be seen what sort of organization finally emerges and whether it will, in the long term, have the financial and human resources, and sufficient political power, to function as an effective management agency, locally, nationally, and internationally.

In August 2007, a proposed Boeng Tonle Sap Authority was established by the Royal Cambodian Government sub-decree. This move by the government was intended to minimize inter-ministerial conflicts, which are often plagued by unclear authority and responsibility. At the same time, each government agency often initiates activity or management policy that replicates those of other agencies. Thus, this new authority will be a sole agency and provide a tool in coordinate management activities, conservation and development of Boeng Tonle Sap that are being initiated and undertaken by various government agencies. However, the effectiveness of the new authority remains to be seen.

ACKNOWLEDGMENTS

Thanks to the Mekong River Commission for the water quality data. The possible link between the end of Angkor and the Little Ice Age was suggested to us by Robyn Johnston.

References

Adapt. (2003). Mekong basin in southeast Asia: Climate change and challenges. http://www.waterandclimate.org/dialaogue/Adapt/documents/Adapt%Mekong%2010203.pdf. Accessed 4 February 2008.

Araujo-Lima, C. A. R. M., Forsberg, B. R., Victoria, R., and Martinelli, C. (1986). Energy sources for detritivorous fishes in the Amazon. *Science* **234**, 1256-1258.

Bonheur, N. and Lane, B. D. (2002). Natural resources management for human security in Cambodia's Tonle Sap Lake biosphere reserve. *Environment Science and Policy* **5**, 33-41.

Brooks, S. E., Allison, E. H., and Reynolds, J. D. (2007). Vulnerability of Cambodian water snakes: Initial assessment of the impact of hunting at Tonle Sap Lake. *Biological Conservation* **139**, 401-414.

Bunn, S. E., Tenakanai, C., and Storey, A. (1999). "Energy Sources Supporting Fly River Fish Communities," Report to Ok Tedi Mining Limited, Environment Department, 38pp.

Campbell, I. C. (2007). Perceptions, data and river management: Lessons from the Mekong River. *Water Resources Research* **43**, doi:10.1029/2006WR005130.

Campbell, I. C., Poole, C., Giesen, W., and Valbo-Jorgensen, J. (2006). Species diversity and ecology of the Tonle Sap Great Lake, Cambodia. *Aquatic Sciences* **68**, 355-373.

Casey, J and Myers, R. A. (1998). Near extinction of a large widely distributed fish. *Science* **281**, 690-692.

Chandler, D. (1996). "A history of Cambodia," 2nd edition, Silkworm Books, Chiang Mai, Thailand, 288pp.

Chinvanno, S. (2004). Information for sustainable development in light of climate change in Mekong River basin. http://203.159.5.16/digital_gms/Proceedings/A77_SUPPAKORN_CHINAVANNO.pdf. Accessed 4 February 2008.

Coe, M. D. (2003). "Angkor and the Khmer Civilisation," Thames and Hudson, London, 240pp.

Csavas, I. (1990). "Report of the Mekong Secretariat Mission to Cambodia," Appendix 3d, 3pp.

Cummins, K. W. (1986). Riparian influence on stream ecosystems. In "Stream Protection, the Management of Rivers for Instream Uses," Campbell, I. C. (ed.), pp 45-55. Water Studies Centre, Chisholm Institute of Technology, East Caulfield, Australia.

Daguan, Z. (1312)."The Customs of Cambodia," Smithies, M. (ed.). (2001). Published in translation by the Siam Society, Bangkok, 147pp.

De'er, Z. (1994). Evidence for the existence of the medieval warm period in China. Climatic Change 26, 289-297.

Douglas, I. (2005). The Mekong River basin. In "The Physical Geography of Southeast Asia," Gupta, A. (ed.), pp 193-218. Oxford University Press, Oxford, 440pp.

Dudgeon, D. (1999). "Tropical Asian Streams. Zoobenthos, Ecology and Conservation," Hong Kong University Press, Hong Kong, 830pp.

Earth Report. (2000). Take the money and run. http://www.tve.org/earthreport/archive/28Jul2000.html. Accessed 3 February 2008.

Evans, D., Pottier, C., Fletcher, R., Hensley, S., Tapley, I., Milne, A., and Barbetti, M. (2007). A comprehensive archaeological map of the world's largest preindustrial settlement complex at Angkor, Cambodia. Proceedings of the National Academy of Sciences 104(36), 14277-14282.

Fletcher, R. (2001). Seeing Angkor: New views of an old city. Journal of the Oriental Society of Australia 32, 1-25.

Forsberg, B. R., Araujo-Lima, C. A. R. M., Martinelli, L. A., Victoria, R. L., and Bonassi, J. A. (1993). Autotrophic carbon sources for fish of the Central Amazon. Ecology 74, 643-652.

Fujii, H., Garsdal, H., Ward, P., Ishii, M., Morishita, K., and Boivin, T. (2003). Hydrological roles of the Cambodian floodplain of the Mekong River. International Journal of River Basin Management 1(3), 1-14.

Ganf, G. G. (1974). Phytoplankton biomass and distribution in a shallow eutrophic lake (Lake George, Uganda). Oecologia 16, 9-29.

Garsdal, H (2004). "Study on the Natural Reverse Flow in the Tonle Sap River," Report to the Mekong River Commission, Vientiane, 25pp.

Gupta, A. (2005). Accelerated erosion and sedimentation in Southeast Asia. In "The Physical Geography of Southeast Asia," Gupta, A. (ed.), pp 239-249. Oxford University Press, Oxford, 440pp.

Hamilton, S. K., Lewis, W. M., and Sippel, S. J. (1992). Energy sources for aquatic animals in the Orinoco River floodplain: Evidence from stable isotopes. Oecologia 89, 324-330.

Hammer, U. T. (1981). Primary production in saline lakes. A review. Hydrobiologia 81, 47-57.

Higham, C. (2001). "The Civilization of Angkor," Wiedenfeld and Nicholson, London, 192pp.

Hook, J., Novak, S., and Johnston, S. (2003). "Social Atlas of the Lower Mekong Basin," Mekong River Commission, Phnom Penh, 154pp.

Hori, H. (2000). "The Mekong, Environment and Development," United Nations University Press, Tokyo, 398pp.

Hortle, K. G., Lieng S., and Valbo-Jorgensen, J. (2004a). "An Introduction to Cambodia's Inland Fisheries," Mekong Development Series No. 4. Mekong River Commission, Phnom Penh, 41pp.

Hortle, K. G., Pengbu, N, Rady, H., and Sopha, L. (2004b). Trends in the Cambodian dai fishery: Floods and fishing pressure. Catch and Culture 10(1), 7-9.

Hughes, M. K. and Diaz, H. F. (1994). Was there a "medieval warm period," and if so, where and when? Climatic Change 26, 109-142.

Hun, S. (2007). Keynote address. In "National Forum on the Tonle Sap Initiative," Hotel Intercontinental, Phnom Penh, 5 March 2007. http://www.env.org.kh/2007_releases/05mar07_tonlesap_initiative_forum.htm.

Jensen, H. S. and Andersen, F. (1992). Importance of temperature, nitrate, and pH for phosphorus release from aerobic sediments of four shallow, eutrophic lakes. Limnology and Oceanography 37, 577-589.

Junk, W. J. (1999). The flood pulse concept of large rivers: Learning from the tropics. Archiv für Hydrobiologie Supplement 1153, 261-280.

Junk, W. J. and Howard-Williams, C. (1984). The ecology of aquatic macrophytes in Amazonia. In "The Amazon," Sioli, H. (ed.), Dr W. Junk, Publishers, The Hague, pp 269-293. 763pp.

Junk, W. J. and Wantzen, K. M. (2004). The flood pulse concept: New aspects, approaches and applications— An update. In "Proceedings of the Second International Symposium on the Management of Large Rivers for Fisheries," Volume II, Welcomme, R. L. and Petr, T. (eds.), pp 117-140. FAO Regional Offices for Asia and the Pacific, Bangkok.

Junk, W. J., Bayley, P. B., and Sparks, R. E. (1989). The flood pulse concept in river-floodplain ecosystems. Canadian Special Publications in Fisheries and Aquatic Sciences 106, 110-127.

Junk, W., Brown, M., Campbell, I. C., Finlayson, M., Gopal, B., Ramberg, L., and Warner, B. G. (2006). Comparative biodiversity values of large wetlands: A synthesis. Aquatic Sciences 68, 400-414.

Koponen, J., Josza, J., Lauri, H., Sarkkula, J., and Virtanen. (2003). "Modelling Tonle Sap Watershed and Lake Processes for Environmental Change Assessment," MRCS/WUP-FIN Model Report. Mekong River

Commission and Finnish Environment Institute, Helsinki, 172pp.

Kummu, M. and Varis, O. (2006). Sediment related impacts due to upstream reservoir trapping in the lower Mekong River. *Geomorphology* **85**, 275-293.

Kummu, M., Koponen, J., and Sarkkula, J. (2004). Upstream impacts on Lower Mekong Floodplains: Tonle Sap case study. *In* "Proceedings of the International Conference on Advances in Integrated Mekong River Management," Vientiane, Lao PDR, pp 347-352.

Kummu, M., Koponen, J., and Sarkkula, J. (2005). Assessing impacts of the Mekong Development in the Tonle Sap Lake. *In* "Proceedings of the International Symposium of the Role of Water sciences in Transboundary River Basin Management," Ubon Ratchathani, Thailand, pp 1-10.

Kummu, M., Sarkkula, J., Koponen, J., and Nikula, J. (2006). Ecosystem management of the Tonle Sap Lake: An integrated modelling approach. *Water Resources Development* **22**, 497-519.

Lamberts, D. (2001). "Tonle Sap Fisheries: A Case Study on Floodplain Gillnet Fisheries in Siem Reap, Cambodia," FAO Regional Office for Asia and the Pacific, Bangkok, Thailand, 133pp. RAP Publication 2001/11.

Lamberts, D (2006). The Tonle Sap Lake as a productive ecosystem. *Water Resources Development* **22**, 481-495.

Leang, P. (2003). "Sub-Area Analysis, The Tonle Sap Sub-Area. Report for the Basin Development Plan," Mekong River Commission, Phnom Penh, 79pp.

Lewis, J., Hamilton, W. M., Rodriguez, S. K., Saunders, M. A., and Lasi, M. A. (2001). Foodweb analysis of the Orinoco floodplain based on production estimates and stable isotope data. *Journal of the North American Benthological Society* **20**, 241-254.

McDonald, A. J., Bunnat, P., Virak, P., and Bunton, L. (1997a). "Plant communities of the Tonle Sap Floodplain. Final Report as a Contribution to the Nomination of Tonle Sap as a Biosphere Reserve of UNESCO's Man in the Biosphere Program," Ministry of Environment, Phnom Penh, 90pp.

McDonald, A., B. Pech, V. Phauk, and B. Leeu, (1997b). "Plant communities of the Tonle Sap Floodplain. Final Report in Contribution to the Nomination of Tonle Sap as a UNESCO Biosphere Reserve," UNESCO, IUCN, Wetlands International and SPEC (European Commission), Phnom Penh, 30pp.

Mekong Committee. (1970). "Report on Indicative basin plan," Mekong Committee, Bangkok, 73pp.

Meybeck, M., Chapman, D., and Helmer, R. (eds.). (1989). "Global Freshwater Quality. A First Assessment," Blackwell Reference, Oxford, 306pp.

Mizino, T. and Mori, S. (1970). Preliminary limnological survey of some S.E. Asian lakes. *In* "Proceedings of the Regional Meeting of Inland Water Biologists in South

East Asia," pp 105-107. UNESCO Field Science Office, Jakarta, Indonesia.

MRC. (1997). "Mekong River Basin Diagnostic Study," Final Report. Mekong River Commission and UNEP, Bangkok, Thailand, 200pp.

MRC. (2003a). "State of the Basin Report 2003," Mekong River Commission, Phnom Penh, 300pp.

MRC. (2003b). "Water Used for Agriculture in Lower Mekong Basin," Basin Development Plan Report BDP 017. Mekong River Commission, Phnom Penh, 56pp.

MRC/UNDP. (1998). "Natural Resources-Based Development Strategy for the Tonle Sap Area, Cambodia," Volume 2, Part B. Sectoral Studies, (CMB/95/003). Final Report. Mekong River Commission Secretariat/UNDP, Phnom Penh, 278pp.

Pantulu, V. R. (1986a). The Mekong River system. *In* "The Ecology of River Systems," Davies, B. R. and Walker, K. F. (eds.), pp 695-719. Dr. W Junk, Publishers. The Hague.

Pantulu, V. R. (1986b). Fish of the lower Mekong Basin. *In* "The Ecology of River Systems," Davies, B. R. and Walker, K. F. (eds.), pp 721-741. Dr W Junk, Publishers. The Hague, 793pp.

Payne, A. J. (1986). "The Ecology of Tropical Lakes and Rivers," John Wiley and Sons, Chichester, 301pp.

Penny, D., Cook, G., and Sok, S. I. (2005). Long term rates of sediment accumulation in the Tonle Sap, Cambodia: A threat to ecosystem health. *Journal of Palaeolimnology* **33**, 95-103.

Phlips, E. J., Aldridge, F. J., Schelske, C. L., and Crisman, T. L. (1995). Relationships between light availability, chlorophyll *a*, and tripton in a large, shallow subtropical lake. *Limnology and Oceanography* **40**, 416-421.

Pilkey, O. H. and Pilkey-Jarvis L. (2007). "Useless Arithmetic. Why environmental scientists Can't Predict the Future," Columbia University Press, New York, 230pp.

Podger, G, Beecham, R., Blackmore, D., Perry, C., and Stein, R. (2004). "Modelled Observations of Development Scenarios in the Lower Mekong Basin," World Bank, Vientiane, 122pp.

Rainboth, W. J. (1996). "Fishes of the Cambodian Mekong," FAO, Rome, 265pp.

Richardson, M. (1999). Ecology: Cambodia's next man-made disaster. International Herald Tribune 15 January 1999. http://www.iht.com/articles/1999/01/15/Mekong.t.php?page=1. Accessed 3 February 2008.

Sarkkula J. and Koponen J., (2003). "Modelling Tonle Sap for Environmental Impact Assessment and Management Support," Draft Final Report. Finnish Environment Institute, Helsinki, 109pp.

Sarkkula, J., Kiirikki, M., Koponen, J., and Kummu, M. (2003). "Ecosystem processes in Tonle Sap Lake," Full Paper for the 1st Workshop of Ecotone Phase II, http://users.tkk.fi/u/mkummu/publications/sarkkula&al_Ecotone_Siem-Reap_2003.pdf. Accessed 5 August 2009. 14pp.

Sarkkula, J., Baran, E., Chheng, P., Keskinen, M., Koponen, J., and Kummu, M. (2005). Tonle Sap pulsing system and fisheries productivity. *Verhandlungen die Internationale Vereinigung für Theoretische und Angewandte Limnologie* **29**, 1099-1102.

Sokhem, P. and Sunada, K. (2006). The governance of the Tonle Sap Lake, Cambodia: Integration of local, national and international levels. *Water Resources Development* **22**, 399-416.

Sondegaard, M., Jensen, J. P., and Jeppesen, E. (2001). Retention and internal loading of phosphorus in shallow, eutrophic lakes. *The Scientific World* **1**, 427-442.

Sondegaard, M., Jensen, J. P., and Jeppesen, E. (2003). Role of sediment and internal loading of phosphorus in shallow lakes. *Hydrobiologia* **506**, 135-145.

Song, S. L., Sopha, L., Sarkkula, J., and Kummu, M. (2004). "Technical Report on the Relationship between Primary Productivity and Fish Production in Tonle Sap Lake (RE-CBD-002)," Report to the ASEAN Regional Centre for Biodiversity Conservation (ARCBC) and European Commission (EU), May 2004, Cambodia, 45pp.

Stuart, B. L., Smith, J., Davey, K., Din, P., and Platt, S. G. (2000). Homalopsine watersnakes. The harvest and trade from Tonle Sap, Cambodia. *Traffic Bulletin* **18**, 115-124.

Talling, J. F. and Lemoalle, J. (1998). "Ecological Dynamics of Tropical Inland Waters," Cambridge University Press, Cambridge, 441pp.

Thomas, R, Meybeck, M., and Beim, A. (1997). Lakes. *In* "Water Quality Assessments. A Guide to the Use of Biota, Sediments, and Water in Environmental Monitoring," 2nd edition, Chapman, D. (ed.), pp 319-368. E & F.N. Spon, London, 626pp.

Ward, J. V. and Stanford, J. A. (eds.). (1979). "The Ecology of Regulated Streams," Plenum Press, New York, 398pp.

Welcomme, R. L. (1979). "Fisheries Ecology of Floodplain Rivers," Longman, London, 317pp.

Wetzel, R. G. (1975). "Limnology," W.B. Saunders, Philadelphia, USA, 743pp.

World Meteorological Organization. (2007). www.world-weather.org/145/c00347.html. Acessed on 19 August 2007.

WWF. (2005). "An Overview of Glaciers, Glacier Retreat, and Subsequent Impacts in Nepal, India and China," WWF Nepal Program, Kathmandu, 79pp.

11

Mekong Schistosomiasis: Where Did It Come from and Where Is It Going?

S.W. Attwood

State Key Laboratory of Biotherapy, West China Hospital, West China Medical School, Sichuan University, Chengdu 610041, PR China

OUTLINE

1. INTRODUCTION

1.1. The Mekong Schistosome, the Mekong River, and a Remarkable Malacofauna

The remarkably high fish diversity of the Mekong River system is well known as the third largest after the Amazon and the Zaire (see Chapter 8). The endemics include celebrated species such as the giant catfish (*Pangasionodon gigas*), the giant Mekong carp (*Catlocarpio siamensis*), and the saee (*Mekongina erythrospila*). Other taxa, such as the giant stingray (*Himantuira chaophraya*) and the endangered Irrawaddy dolphin (*Orcaella brevirostris*), are similarly well documented.

Less well-known and less surveyed are the large numbers of molluscs also endemic to the Mekong drainage system. The most reliable and complete account was published by Davis (1979) and revealed an endemic fauna that included over 90 species of Triculinae Annandale, 1924 (Caenograstropoda: Rissooidea: Pomatiopsidae Stimpson, 1865), 32 species of Stenothyra (Rissooidea: Stenothyridae), 42 Viviparidae (Caenogastropoda: Rissooidea), and approximately 121 species of Buccinidae (Neogastropoda: Muricoidea).

Among this fauna, the Triculinae are most noteworthy in that, with over 90 endemic species occurring within a single subfamily and along only a 300-km stretch of the Mekong River, they represent a unique freshwater molluscan biodiversity. Other radiations of Triculinae also occur, each based on a different tribe. The Yunnan/Sichuan triculine radiation (in Southwest China) comprises nine species of the tribe Triculini and two of Pachydrobiini. The Hunan radiation (Southeast China) comprises 13 taxa of Pachydrobiini (mainly Neotricula Davis, 1968). The Mekong River Triculinae belongs mostly to the tribe Jullieniini.

The triculine radiation would have been even less well-known were it not for their importance as intermediate hosts for certain species of digenean blood flukes (Trematoda: Digenea), which infect humans and/or their livestock in China and Southeast Asia. The most publicized of these parasites in the Mekong region is the endemic *Schistosoma mekongi*. Like many of the endemic taxa discussed in this volume, *S. mekongi* shows an unusual distribution that has been the focus of much study. Questions as to the origins, current human population at risk, and the likelihood of the future spread of disease have prompted research into both the parasites and the snails involved in their transmission. It is such questions that are the focus of this chapter.

1.2. Mekong Schistosomiasis: A Persistent Public Health Problem

In Cambodia and Laos, schistosomiasis in humans is caused by infection with *S. mekongi*. Historically, Mekong schistosomiasis has caused severe morbidity and prior to the first 10-year control program, deaths due to ascites, jaundice, or gastrointestinal bleeding were not uncommon in southern Laos (Gang, 1991). The disease develops after free-swimming parasite larvae (cercariae), released from a snail intermediate host, penetrate the skin during bathing. Only snails of the correct species (or a few related species) can act as intermediate hosts for a particular *Schistosoma* species. The snails become infected by similarly mobile and penetrative larvae (miracidia) that hatch from eggs usually passed in the feces of definitive hosts, which may be humans, rodents, livestock, or other mammals—depending on the species of *Schistosoma*.

The only species of snail known to be compatible with *S. mekongi* is the triculine *Neotricula aperta* (Temcharoen, 1971) (Triculinae: Pachydrobiini). Three strains of *N. aperta* have been described (α, β, and γ) (Davis *et al.*, 1976), and all are to varying degrees compatible with *S. mekongi* in laboratory experiments, but only the γ-strain is known to transmit the parasite in nature (Attwood *et al.*, 1997). Like the Mekong Triculinae, *S. mekongi* is endemic to the lower Mekong basin and is a member of the *Schistosoma japonicum* group (see Rollinson and Southgate, 1987).

The first cases of Mekong schistosomiasis were reported among Laotian immigrants presenting with chronic symptoms at the Paris hospital (Dupont-Vic *et al.*, 1957). Epidemiological surveillance for schistosomiasis in Laos began in 1960 coordinated by the WHO, but the first direct evidence of disease transmission was not found until 1967 (Iijima and Garcia, 1967). The endemic focus was in the Mekong River at Khong Town, on Khong Island, Champassac Province, southern Laos, close to the border with Cambodia (Fig. 11.1).

The first (10-year) national control program began in Laos in 1989, when the prevalence of schistosomaisis in children under 15 years of age at Khong Island was 42.1% and the prevalence by village ranged from 15.3% to 94.0% (Gang, 1991). A similar survey by the Lao-PDR Project for Parasite Eradication in 2003, near the end of the second (1999-2006) national control program, recorded a mean prevalence of 11% for 63 sampled villages of Khong District in Champassac Province (Samlane *et al.*, 2007). The target population of the program was children and the oral anthelminthic praziquantel was administered with 88% coverage (Ohmae *et al.*, 2004).

In Cambodia, the prevalence prior to the period of integrated control (1966-1969) was 13.6% in the Stung-Treng district and 35.9% in the Kratié district (Ohmae *et al.*, 2004). Political unrest in Cambodia prevented any coordinated control efforts until 1994, by which time the prevalence was 72.9% among primary school children (Stich *et al.*, 1999). In 1996, ultrasound examination was introduced to the surveillance program; these investigations revealed pathological changes consistent with *Schistosoma* infection in 84% of the 299 villagers tested (Ohmae *et al.*, 2004). The Cambodian National Center for Parasitology, Entomology and Malaria Control began coordinated antischistosomiasis efforts in 1994, initially covering 20 villages around Kratié, and in 1997 this was extended to Stung-Treng. The target population was thereby expanded from 45,000 in 1994 to 80,000 by 2005. Diagnosis was primarily by Kato-Katz test, but the more sensitive ELISA test was introduced in 1999. The surveys were conducted in an average of 20 randomly selected villages (2000-3000 people) each year, with praziquantel (40 mg kg^{-1}) given to all people in all villages except for children under 2 years of age and pregnant women (Sinuon *et al.*, 2007). In 2004, the prevalence by Kato-Katz was 0% in both Kratié and Stung-Treng districts. A few new cases were reported in 2005, 2006, and 2007, most notably at Sa Dao village on the Xe Kong River near Stung-Treng (Fig. 11.1A); however, the prevalence remained <1%. The prevalence in Kratié Province by ELISA was found have decreased from around 97% in 1997 to 30% in 2007 (Sinuon *et al.*, 2007).

The 8-year national control program in Cambodia and 17 years of coordinated control efforts in Laos have achieved great reductions in the prevalence of infection and have been even more effective in reducing morbidity. No new cases of severe morbidity have been reported in Cambodia since 2003 (Sinuon *et al.*, 2007). Such successes have led some to regard Mekong schistosomiasis as almost eradicated (Freiermuth *et al.*, 2005), or at least readily controllable (Ohmae *et al.*, 2004), but the control campaigns appear to have a tenuous hold on transmission.

In spite of many years of concerted control efforts, high prevalence is still observed in some villages and infections reoccur in areas previously apparently clear of the disease. Kato-Katz surveillance in eight villages (548 people) of Khong Island revealed a prevalence of 28.1% (Samlane *et al.*, 2007). At Sa Dao, no

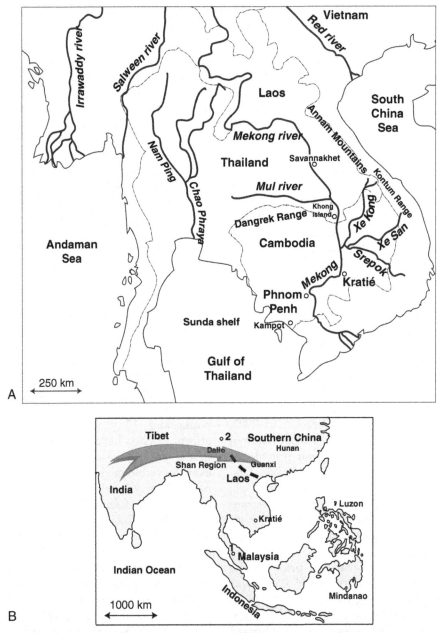

FIGURE 11.1   (A) The lower Mekong Basin (Central Sundaland), showing the major rivers draining the region. The bold lines indicate river courses, whereas the broken lines show approximate international boundaries. The thick broken line near Guanxi Province (China) represents the Zhaotong-Ling Mountains. Reprinted from Attwood S. W., Fatih F. A. & Upatham E. S., 2008 DNA-sequence variation among *Schistosoma mekongi* populations and related taxa; phylogeography and the current distribution of Asian schistosomiasis. *PLoS Negl Trop Dis* 2(3): e200. doi:10.1371/journal.pntd.0000200. (B) Eastern Asia showing the current locations of key *Schistosoma japonicum* group taxa mentioned in the text. (1) *S. malayensis* Baling, Pahang, West Malaysia, (2) *S. sinensium* Mianzhu, Sichuan, PR China; scale approximate.

human infections were detected in 2004, but the disease re-emerged in 2005 (Sinuon *et al.*, 2007). In addition, in 2004 the prevalence of infection among *N. aperta* collected at Sa Dao was 0.14% (Attwood *et al.*, 2004a). Similarly, despite an almost eight-fold reduction in the prevalence in the human population at Khong Island in Laos (1969-2003), the estimated prevalence in the local *N. aperta* populations had changed little (Attwood *et al.*, 2001).

These observations suggest that there may be a significant zoonotic component to the transmission of *S. mekongi*, a reservoir of infection in animal populations that escapes the control efforts. Further evidence for the importance of reservoir hosts is the observation that dogs (in Cambodia) and pigs (in Laos) are known to be infected at prevalence of 3.6% and 12.2%, respectively (Matsumoto *et al.*, 2002; Strandgaard *et al.*, 2001). Alternatively, Mekong River *N. aperta* populations may be maintained by colonization rather than by recruitment, with foci of infection, probably animal mediated, in Mekong River tributaries of the Annam highland region (Fig. 11.1) acting as sources or seeder populations for the Mekong River (Attwood *et al.*, 2008a,b).

2. WHERE DID *SCHISTOSOMA MEKONGI* COME FROM?

2.1. Mekong Schistosomiasis Is Not Limited to the Mekong River

The almost complete absence of *S. mekongi* from Laos north of the Khong Island, and its restriction in the Mekong River has traditionally been attributed to the absence of *N. aperta* and/or unsuitable ecological conditions for transmission in this area. Most of the control and study of Mekong schistosomiasis have focused on the 300-km section of the Mekong River, between Khong Island in Laos and Pong

Ro in Cambodia (25 km downstream of Kratié), along which all reliably documented, persistent, foci of transmission were located (Attwood 2001; Biays *et al.*, 1999; Ohmae *et al.*, 2004; Urbani *et al.*, 2002). The first report of *N. aperta* outside the Mekong River was from the Khammouanne Province of Central Laos, in the middle Xe Bang-Fai River, a tributary of the Mekong (Attwood and Upatham, 1999). The first transmission focus outside the Mekong River was found at Sa Dao (lower Xe Kong, Stung-Treng Province), in the human population in 1997 (Sinuon *et al.*, 2007) and in the snail population in 2004 (Attwood *et al.*, 2004a). Subsequently, 11 new *N. aperta* populations have been reported, involving 6 new river systems in Cambodia and Laos (Attwood *et al.*, 2004a). More recently, 15 new populations, including a population of *S. mekongi* infected snails in the middle Sre Pok River of the Rattanakiri Province in eastern Cambodia (Fig. 11.1A), were reported (Attwood *et al.*, 2008b), substantiating earlier reports of *S. mekongi* in Rattanakiri (Iijima, 1970). Consequently, there are now 31 populations of *N. aperta* known, and 7 foci of transmission involving 4 river systems (Fig. 11.2; Table 11.1).

This extension of the known range of *N. aperta* and the conditions under which the disease can be transmitted, raises the estimated population at risk from the 140,000 people considered by Urbani *et al.* (2002) to approximately 1.5 million. Further, it seems less likely that the distribution of Mekong schistosomiasis is being limited primarily by ecological conditions. So, why is *S. mekongi* not a more serious public health problem in Laos?

Studies of the evolutionary history of both *S. mekongi* and other relevant schistosome species (Blair *et al.*, 1997; Le *et al.*, 2001; Littlewood and Johnston, 1995; Littlewood *et al.*, 2006; Woodruff *et al.*, 1987; Yong *et al.*, 1985) and the intermediate host (Attwood *et al.*, 2004a; Davis, 1980, 1992) shed some light on

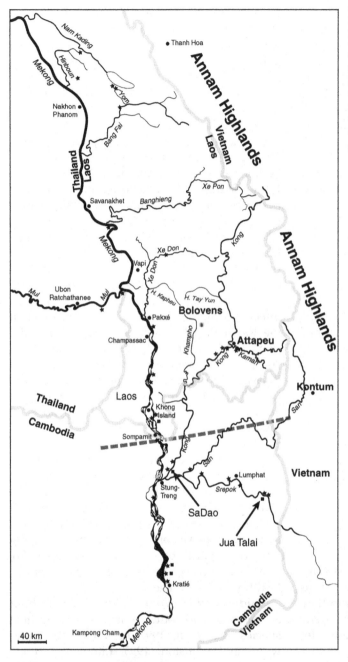

FIGURE 11.2 Sketch map of the Annam and Kontum region of Southeast Asia showing the locations of the *Neotricula aperta* populations sampled (denoted by star symbols) and transmission foci for *Schistosoma mekongi* (denoted by black box symbols, see also Table 11.1). Important geographical features or locations are given in large bold type. The thick broken line along the Cambodian border approximately marks the uplifted edge of the Khorat Basin. Adapted from figure 5 of Attwood *et al.* (2008a).

TABLE 11.1 Details of published foci of *Schistosoma mekongi* transmission (i.e., where infected snails have been found)

River	Territory	Province/district	Village	Location (GPS)	Reference
Mekong	Cambodia	Sambour (ST)	San Dan	12°44′20″; 106°00′30″	Attwood *et al.* (2004a)
Mekong	Cambodia	Sambok (KR)	**Xai-Khoun**	12°37′15″; 106°01′15″	Sinuon *et al.* (2007)
Mekong	Laos	Champassac (KI)	**Don Khong**	14°06′30″; 105°51′15″	Kitikoon *et al.* (1973)
Mekong	Laos	Champassac (KI)	**HXK**	14°06′00″; 105°52′00″	Kitikoon *et al.* (1973)
Sre Pok	Cambodia	Rattanakiri	Jua Talai	13°28′30″; 106°59′45″	Attwood *et al.* (2008b)
Xe Kong	Cambodia	Stung-Treng	**Sa Dao**	13°36′45″; 106°06′00″	Attwood *et al.* (2004b)
Xe Kong	Cambodia	Stung-Treng	Srae Khoean	13°36′30″; 106°22′15″	Ohmae *et al.* (2004)

HXK, Ban Hat-Xai-Khoun; KI, Khong Island; KR, Kratié; ST, Stung-Treng. Sites with confirmed human infections are given in bold.

the origins of the *S. japonicum* group, including *S. mekongi*, which may offer some answers to the public health question.

2.2. The Origin of the *Schistosoma japonicum* Group

Early biogeographical studies of *Schistosoma* and pomatiopsid snails suggested a primarily southern continental distribution (Fig. 11.3). This led Davis (1979) to ascribe an African-Gondwanan origin to *Schistosoma* and the Pomatiopsidae, with rafting on the Indian craton, via "continental drift," as the route into southern China and East Asia. The idea of Indian rafting remained the consensus until Snyder and Loker (2000) proposed an Asian origin for *Schistosoma*, evidenced by an rRNA gene molecular-based phylogeny for the Schistosomatidae; these authors found that Asian *Schistosoma*, together with *Orientobilharzia*, also restricted to Asia, appeared basal to the clade containing African *Schistosoma*. These findings were subsequently corroborated by several authors using different combinations of taxa/ genes (Agatsuma *et al.*, 2002; Attwood *et al.*, 2002; Lockyer *et al.*, 2003).

The Pomatiopsidae comprises two subfamilies, the Pomatiopsinae, with an apparent Gondwanan distribution (Fig. 11.3), and the Triculinae, which are found from northern India into southern China and Southeast Asia (see Fig. 11.1B). The Pomatiopsinae include *Oncomelania hupensis* subspecies that transmit *S. japonicum* in China, in the highlands of Yunnan and Sichuan and in the lowlands across the marshland flats of the middle and lower Yangtze plain (Fig. 11.3). The Triculinae are freshwater snails and several species transmit schistosomiasis to humans and/or animals in mainland Southeast Asia (Attwood *et al.*, 2004a). The Pomatiopsinae are amphibious to terrestrial and freshwater to brackish or marine, which is probably the reason for their much wider distribution.

Davis (1979) proposed that the Pomatiopsidae arose in a region of Gondwana centered on South Africa and spanning parts of modern South America, southeast India, southwestern Australia (i.e., the current range of the southern Pomatiopsinae), and Antarctica. The various pomatiopsine genera were then isolated by the break-up of Gondwana approximately 100 million years ago (Ma). Rafting of Pomatiopsinae

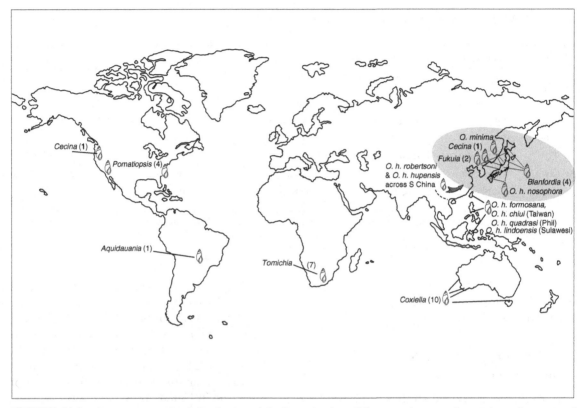

FIGURE 11.3   Current geographical distribution of the Pomatiopsinae. Where species name is not given, the names of genera are followed by total number of species in parentheses (in bold if not endemic to region). Taxa shown in the shaded area refer to Japan only (not also to China). Where appropriate, *Oncomelania hupensis* subspecies names are given. After Davis (1979), with updated distributional data supplied by the Biodiversity Center of Japan. The broken line near Hainan indicates the Zhaotong-Ling Mountains (Indosinian fold-belt).

to Cathaysian Asia on the Indian craton occurred until the Eocene, when India collided with Asia and initiated the major Himalayan orogeny. Proto-Pomatiopsidae were assumed to have colonized southern China from northeast India during the mid-Miocene. The Indian pomatiopsid stock was proposed to have given rise to the Triculinae in the highlands of Tibet, southwest China (Yunnan Province), and Southeast Asia, and to *Oncomelania* Gredler, 1881 in northern Burma. Fossils interpreted as *Oncomelania* by Davis (1979) have been reported from northern India and Burma, but

records of the Triculinae are restricted to the Shan region.

Davis's (1979) hypothesis provides a parsimonious explanation for the "southern continental" distribution of the Pomatiopsinae; however, there are several problems. First, the absence of Pomatiopsinae and *S. japonicum* from India or the valleys of Burma and most of Yunnan must be explained because this is the proposed route of entry for these taxa into China. Pliocene fossil hydrobioid taxa, interpreted as *Oncomelania* by Davis, cannot be taken as reliable evidence. Hydrobioids are an artificial grouping of small basal, plesio-morphic, prosobranch snails; these include Hydrobiidae, Assimineidae, or Stenothyridae (Rissooidea). According to Davis (1980), one cannot distinguish such taxa on the basis of shell alone. The Indian proto-Pomatiopsidae may have became extinct through competition with pulmonates (Gastropoda: Pulmonata) or climatic changes. However, this requires that isolated pomatiopsine populations have remained viable since the Cretaceous, in places such as South Africa's Cape Province (in spite of Plio-Pleistocene xerification of the region; Dupont *et al.*, 2005; Leroy and Dupont, 1994).

The absence of *Oncomelania* from mainland Southeast Asia is another problem. *Oncomelania* is common across southern China East of the Zhaotong-Ling Mountains (Fig. 11.1B), but is not found to the southwest. Small foci of *Onco-melania* are found in isolated valleys around Dali Lake in Yunnan, but these are most likely due to a recent introduction by humans.

The population phylogeny for the *S. indicum* group shown in Fig. 11.4 agrees with the phylo-genies published by Lockyer *et al.* (2003) and Webster *et al.* (2006). The early divergent posi-tion of *S. nasale* and *S. incognitum* in these phy-logenies and their close affinity with African species is consistent with the hypothesis of Barker and Blair (1996). These authors also pro-posed that *Schistosoma* arose in Asia with a subsequent radiation into Africa (producing the three main lineages of *Schisotosoma*, includ-ing *S. mansoni* (see Morgan *et al.*, 2005), with a final reintroduction to Asia as the *S. indicum* group radiation. Attwood *et al.* (2007) dated

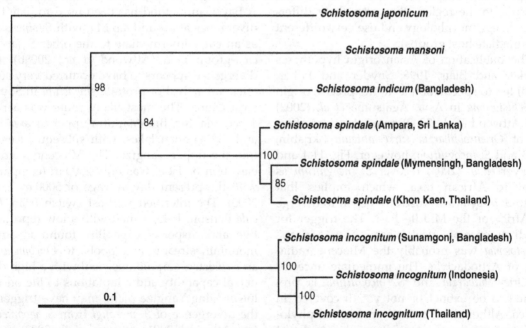

FIGURE 11.4 Phenogram summarising the phylogeny estimated from combined 12S and 16S DNA-sequence data (mitochondrial rRNA genes) using a Bayesian approach (outgroup *Schistosoma japonicum*). Numbers assigned to each node represent the posterior probability that the hypothesis represented by this bi-partition, and under all parameters of the model, is true given the observed data. Reprinted with permission from Cambridge University Press. Originally pub-lished in Parasitology 134(14):2009-2020, Copyright 2007.

the radiation of the *S. indicum* group sensu stricto at around 250,000 years before present (YBP) and linked this to lowered (interpluvial) sea levels; these authors proposed an unrelated radiation of *S. incognitum* dated at 165,000 YBP and linked it to the dispersal of rodents across the Pleistocene wetlands of the then exposed Sunda shelf. In reflection of their different phylogeographies, many authors now regard the *S. indicum* group as paraphyletic, with *S. incognitum* as the odd one out (Agatsuma *et al.*, 2002; Attwood *et al.*, 2002; Morgan *et al.*, 2003; Webster *et al.*, 2006). The recolonization of Asia (from Africa) probably began ca. 2 Ma with the Plio-Pleistocene large mammal migration along the Sinai-Levant dispersal tract (Matthee *et al.*, 1997, 2001) (see Fig. 11.5). Consequently, although *S. incognitum* and *S. spindale* are found in the lower Mekong Basin, they have no phylogeographical affinity with *S. mekongi* and are not endemic to the region (they also show differences in egg morphology and use very different intermediate host groups).

The publication of Asian origin hypotheses (Barker and Blair, 1996; Snyder and Loker, 2000) led to speculation as to the precise origin of *Schistosoma* in Asia. Agatsuma *et al.* (2002) and Attwood *et al.* (2002) noted the close affinity of *Orientobilharzia turkestanicum* Skrjabin, 1913 and *S. incognitum*; however, Fig. 11.4 and Lockyer *et al.* (2003) recover *S. incognitum* as basal to African taxa, which implies that proto-*S. incognitum* arose and began to diverge in Africa or the Middle East. The trigger for the divergence of *Orientobilharzia* and "African" *Schistosoma* was probably the Miocene radiation of Artiodactyla. The immediate ancestor of *Orientobilharzia* and *S. incognitum* is presumed to be extinct or not yet discovered in Africa. Although all recently published phylogenies show the *S. japonicum* group to lie at the root (Agatsuma *et al.*, 2002; Attwood *et al.*, 2002; Morgan *et al.*, 2003; Webster *et al.*, 2006) and support an Asian origin, there are few published estimates of the date and pattern of the

initial radiation. All *Schistosoma* species arising early in the phylogeny are restricted to, or significantly utilize, rodents, whereas those species arising later tend to be much more dependent on Artiodactyla, especially Bovidae (cattle). Murid rodents first appeared in China in the late Miocene, around 15 Ma (Qiu and Li, 2003); this implies a mid-Miocene origin for *Schistosoma* in the region of Tibet or Central Asia (Attwood *et al.*, 2007), prior to the major Himalayan uplift in 5 Ma (Li, 1991; Mitchell, 1981; Rost, 2000).

Attwood *et al.* (2002) proposed that the *S. sinensium* lineage arose from proto-*S. incognitum*/*Orientobilharzia* as an adaptation to transmission in the small river valleys of uplifted highland areas (Fig. 11.5). Host switches from cattle and pulmonate snails to rodents and Triculinae would facilitate transmission in the rising highland areas of the Shan region. A Bayesian method has been used to date this divergence at around 4.6 Ma, with *S. sinensium* as an early intermediate to the other *S. japonicum*-group taxa (Attwood *et al.*, 2008b); the divergence appears to have occurred very soon after the arrival of proto-*S. sinensium* in Southwest China. The most likely route was across Tibet, via the Brahmaputra-upper Irrawaddy and the Yunnan lakes, with subsequent entry into the upper Yangtze. The Miocene average elevation of Tibet was only 2500 m (compared with the present day average of 5000 m; Rost, 2000). The intermediate host switch from *Tricula* Benson, 1843, a snail with a low reproductive and dispersal capability found in small mountain streams and pools, to *Oncomelania*, an *r*-selected amphibious snail with a high dispersal capability and adaptations to life on the lower lying Yangtze plain, may have triggered the divergence of *S. mekongi* from *S. japonicum* (see Attwood, 2001; Attwood *et al.*, 2002).

Several authors have described the evolution of *Schistosoma* as tightly linked, or even coevolved, with that of the snails (Davis, 1980; Davis *et al.*, 1998; Lockyer *et al.*, 2004; Webster

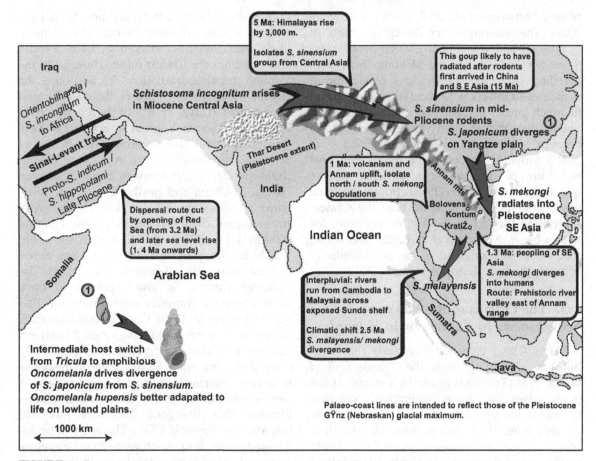

FIGURE 11.5 Semischematic summarizing a phylogeography for *Schistosoma*. The genus is assumed to have arisen in rodents of Miocene Central Asia/Tibet (prior to major uplift there), with an antecedent typified by modern *Schistosoma incognitum*. The *Schistosoma sinensium* lineage next radiates in rodents in China and Southeast Asia. *Orientobilharzia* acquires artiodactyls as hosts in Central Asia; this lineage (or one based on proto-*S. incognitum*) enters Africa in late Miocene, as the Tethyan seaway closes, to form the *Schistosoma haematobium* and *Schistosoma mansoni* lineages. The Pliocene large mammal radiation in Africa triggers the divergence of several lineages of *Schistosoma* utilizing Artiodactyla. The Plio-Pleistocene large mammal divergence into Asia (probably via the Sinai-Levant tract), and the emergence of Bovidae, then drives the divergence of a proto-*Schistosoma indicum/Schistosoma spindale* lineage from one of the African artiodactyle *Schistosoma* clades. *Schistosoma* of bovids establish on Indian subcontinent and in Southeast Asia. *S. malayensis* and *S. mekongi* colonize mainland Southeast Asia (via rodents), becoming isolated in Vietnam following the later Indosinian orogeny. *S. mekongi* then establishes in the larger rivers of Cambodia after diverging into humans.

and Davies, 2001). Davis (1979) proposed the same India-Tibet-Myanmar-Yunnan radiation for the Asian Pomatiopsidae, but with a much earlier (18 Ma) date, and the phylogeography of the *Schistosoma* species was considered to be linked to this. This sort of linkage may be

true for some *Schistosoma* species (Webster *et al.*, 2004), but it is unlikely to be true for *S. japonicum*-group species because of their low prevalence in intermediate host populations (Attwood *et al.*, 2004a) and evidence for frequent host switches in their evolutionary history (Attwood *et al.*, 2002, 2008b; Blair *et al.*, 2001). The Mekong River dwelling strains of *N. aperta* (β and γ) are relatively compatible with *S. mekongi*, while only the Mekong River pool dwelling strain (α) shows signs of significant resistance to infection. Miracidial success would be much greater in these pools, which cattle frequent, than in the main channel of the river because of the lack of water flow. Cribb *et al.* (2001) provided evidence for long-range host switching by digeneans in general. Consequently, although Davis (1979) applied the same phylogeography and dates to both the schistosomes and the snails, there is no necessity for this to be so. Agatsuma *et al.* (2002) also stated that there was unlikely to be any similarity between the two phylogeographies.

2.3. The Origins of the Mekong Schistosome: In Yunnan or in Hunan?

Davis (1992) regarded *Neotricula* (Pacydrobiini) as diverging from the *Tricula bollingi* Davis, 1968 (Triculini) group in Yunnan, at the point where the more "primitive" Triculini colonized the evolving Mekong River system. At this time, these rivers were cutting their way southwards through Laos and Cambodia to the sea in Vietnam, as Central Sundaland continued to emerge from the sea. The evolution of these rivers, the Pliocene orogeny, the Pleistocene changes in sea level and basin tilting with the associated flow reversals, lake formation/extinction, and stream captures, are considered to be the driving force behind speciation in the Triculinae and is used to explain the high biodiversity and endemism (Davis, 1979). Davis (1980) saw Dali in Yunnan (Fig. 11.1A) as a center of radiation, with *S. japonicum* and *Oncomelania* diverging into the Yangtze River, and *S. mekongi* and *N. aperta* into the Mekong River.

Although attractive, this model fails to explain several biogeographical features of the group. At least five species of *Neotricula* are known from Hunan, but only one from Laos and none from Yunnan; therefore, it is more likely that *Neotricula* arrived in Laos directly and not from the *Tricula bollingi* lineage of the Yunnan triculine radiation. In addition, the closest relatives of *Robertsiella*, *Guoia*, and *Wuconchona* (Davis *et al.*, 1992) are found in Hunan and not in Yunnan or Southeast Asia. *Robertsiella* are the intermediate hosts of *S. malayensis*. Much of the Annam mountain chain (which today forms a barrier between Hunan in China and northern Laos and Vietnam) is Mesozoic, and at 1.3 Ma the only trans-Annam dispersal corridor would be the 900-km long valley of the Red River fault, which in the past ran up to 400 km closer to Laos than today (Lacassin *et al.*, 1993). The Pliocene Yangtze is also reported to have flowed along a common course with the Red River (Brookfield, 1998). Consequently, proto-*S. malayensis/mekongi* could have entered northern Laos from Hunan via the Red River (Fig. 11.5), there diverging from *S. japonicum* in response to a new intermediate host (*N. aperta*) and a new environment (a return to mountainous terrain). This divergence has been dated to late Miocene (around 4 Ma). The entry of proto-*S. malayensis* into Southeast Asia therefore appears to have occurred just prior to the second major uplift on the Indosinian block; this would have been a significant event in the region, isolating taxa and driving divergence (Attwood *et al.*, 2008b).

The evolution of the Mekong River and associated Plio-Pleistocene tectonic events in central Sundaland is unlikely to have been involved in the divergence of *S. malayensis/mekongi* from *S. japonicum* and the establishment of these parasites in Southeast Asia. Although earlier palaeogeographical studies (Attwood *et al.*, 2002) dated *S. malayensis* diverging from *S. mekongi* to <1.5 Ma (Attwood *et al.*, 2005), recent molecular dating estimates (Attwood *et al.*, 2008b) suggest a divergence around 2.5 Ma. Consequently, an *S. malayensis*-like schistosome, almost certainly infecting mainly or exclusively rodents, probably diverged from *S. japonicum* in Guanxi, China around 4 Ma following isolating events linked to the second major uplift of the Himalaya.

Cambodia was the most likely point of entry for proto-*S. malayensis* into central Sundaland from Vietnam. A preliminary DNA-sequence-based phylogeny for *S. mekongi* populations of Khong Island, Sambour (both Mekong River), Sa Dao (Xe Kong River), and Lumphat (Sre Pok River, Rattanakiri) showed the Rattanakiri population to be early divergent (Attwood et al., 2008b), which is consistent with a Vietnam to Cambodia route of entry. The Pleistocene Mekong River flowed further west along the Dangrek escarpment, then southwards along the modern Tonlé Sap, and across the Sunda shelf from Kampot (Cambodia Fig. 11.1A) to the present day west Malaysia (Hall, 1998). The lower Mekong River in the area of Stung-Treng and Kratié probably did not occupy its present course until 5-6 KYBP (Hutchinson, 1989). Consequently, populations of schistosomes in these regions must have been established in the Holocene well after the divergence of *S. malayensis/mekongi*.

The Pliocene Dong-Ngai-Mekong River, which once flowed north to south on the exposed shelf east of the Annam mountains (Workman, 1972) could have introduced a proto-*S. malayensis/mekongi* to the whole Sundaland region, with later range contraction, fragmentation, and divergence. Attwood *et al.* (2008a) dated the divergence of *S. mekongi* from *S. malayensis* at around 2.5 Ma, which coincides with a major intensification of monsoon winds affecting rainfall and flow patterns in the rivers of the region (Xiao and An, 1999); this would have impacted the distribution of the intermediate hosts and could have isolated Cambodian proto-*S. mekongi* from Malaysian *S. malayensis*. In this case, *S. malayensis* appears to have retained the ancestral (*S. sinensium*-like) habit of transmission by spring or primary stream dwelling triculinine snails and parasitism of

rodents, whereas *S. mekongi* has switched to transmission in large rivers by *N. aperta* and the use of humans as the main definitive host.

The place of origin of *S. mekongi* is probably on a part of the Sunda shelf, east of Vietnam, which is now under the sea and the Khong Island population represents a more recent colonization of Laos. Besides the transmission focus at Khong Island, which lies on the Dangrek range, *S. mekongi* has not been found north of the Dangrek Mountains. At the time of the dispersal of proto-*S. malayensis/mekongi* from China, a vast plain of wetlands and rivers existed to the east of present day Vietnam (and between west Malaysia and Cambodia); these alkaline wetlands would be ideal habitats for triculine snails. Rising sea levels would have reduced the habitat area down to only the rivers southwest of the Annam range in Cambodia and to those of West Malaysia.

2.4. An Australian Origin for the Asian Pomatiopsidae?

Today, *Oncomelania* is found down the Japanese, Philippine, and Indonesian island chain. *Oncomelania minima* is found on Honshu, Japan, and the only other recognized species (*O. hupensis*) has diverged into at least three subspecies on the Chinese mainland, two on Taiwan, and one each on Japan, the Philippines, and Sulawesi (Fig. 11.3). *Oncomelania* and Pomatiopsinae are absent from Southeast Asia. It is possible that a pomatiopsid snail had evolved on Gondwana by the Jurassic. The proto-pomatiopsid was pomatiopsine and probably occupied the marine/brackish to semiamphibious niche of most modern Pomatiopsinae.

The breakup of Gondwana initiated the divergence of the Pomatiopsinae into the genera seen on the southern continents today (Fig. 11.3); but rather than rafting via India, a *Coxiella*-like pomatiopsine could have rafted from northwest Australia (Jurassic) on the small continental fragments of Gondwanan origin, which today form part of Borneo and eastern Indonesia (Metcalfe, 1998). This proto-*Oncomelania* could then have dispersed from Borneo and Sulawesi by island hopping across the extensive island complex that formed North of Borneo during the Oligocene (Fig. 11.6). The dispersal would have been facilitated by the more southerly position of the southern Philippines, and by Mio-Pliocene northwards rafting and rotation, which brought southerly terrains to abut the northern Philippines (Hall, 2002). Dispersal would have

continued with *Oncomelania* colonizing China and the then emergent parts of Miocene Japan. Tertiary tectonic instability in Japan and an island-complex environment could have driven cladogenesis and dispersal in the Japanese Pomatiopsinae to produce the cluster of genera found there today. Prior to the opening of the Sea of Japan (23-16 Ma) (Hall, 1998), colonization of Japan from China would have involved much shorter distances than today. *Oncomelania* would have readily colonized west into Sichuan towards the Shan Massif area.

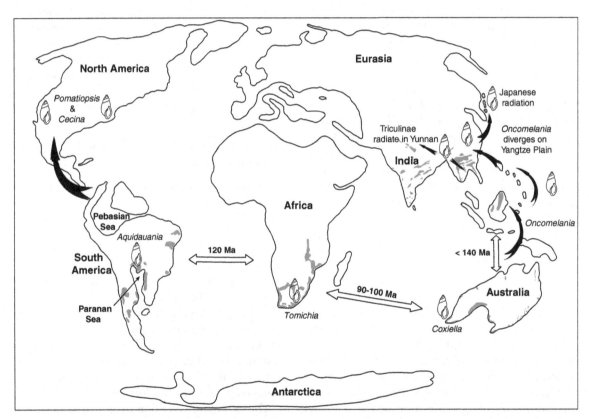

FIGURE 11.6 Tracts of dispersal for a possible Miocene radiation of the Pomatiopsinae. Unfilled (block) arrows show rafting events with approximate dates. Dark gray arrows show the overland (or island hopping) radiation of the Oncomelania subspecies complex from Indonesia into Japan and the Philippines. The approximate distributions of major deposits of calcareous rocks (most are Permian to mid-Mesozoic) are shown as shaded areas for India, Indochina, Australia (karst only), Japan, Borneo and coastal South Africa, and South America. Coastlines are rough approximations for 10-15 Ma. Compiled from multiple sources (Dey, 1968; Hall, 1998, 2002; Hartt, 1870; Hutchinson, 1989; Ministry of Geology, 1947; Van Royen, 1952; Yamada *et al.*, 1987).

Although now mountainous, the Himalaya of eastern Tibet and the lands eastwards were an area of low hills and shallow seas until the Pliocene (Hutchinson, 1989).

The Pomatiopsidae probably diverged into the two modern subfamilies mid-Miocene in what is central southern China today; the emerging complex of islands and hills in the shallow seas of the region may have triggered this divergence. Chinese *Oncomelania* dispersing westwards in the late Miocene would have been caught among rapidly uplifting terrain in the Shan and north Indosinian mountains (Yunnan/northern Laos). The early Triculinae (like extant triculines) were probably adapted to highland streams and a cooler climate. However, *Oncomelania* with its adaptations to life on the lower, flatter terrain of the Yangtze plain, probably became extinct (pre-Pliocene) west of the orogenic belts on the Sichuan/Yunnan border, before they had time to colonize suitable habitats in Yunnan and mainland Southeast Asia.

This hypothesis agrees with dispersal tracts noted for a wide range of terrestrial fauna, all appearing to have reached the Philippines from the south, via a South China Sea crossing (W-E), and not from the north (Heaney, 1986). The absence of Pomatiopsinae and *S. japonicum* from India and mainland Southeast Asia is accommodated. It is consistent with the isolated foci of *S. japonicum* and *Oncomelania* on Sulawesi being representative of the distribution of islands (now mountain peaks) during periods of higher sea level. The hypothesis is also consistent with an *S. japonicum*-like schistosome infecting the macaques of northeast Borneo (Kuntz, 1978). It also allows more time for the divergence of the east Asian Pomatiopsinae (11 species/subspecies) relative to that of the lesser Chinese radiation, with only 4 subspecies, and agrees with molecular systematics showing the Triculinae as derived relative to *Oncomelania*.

The main drawback of the hypothesis is that the evidence for the drift of a Gondwanan fragment from Australia to Borneo is weak, though there are stratigraphic and some palaeomagnetic data (Metcalfe, 1998). It is interesting to note that the distribution of the southern continental Pomatiopsinae falls within that of southern continental coastal limestone (see Fig. 11.6); this alone could explain the apparent Gondwanan distribution if the ancestral form was a marine calciphilic snail. Nevertheless, a refugial Gondwanan distribution seems likely and probably reflects range contraction following increasing aridity in South Africa and Australia. Even for a calciphilic snail, the Mesozoic Bau limestone would have provided habitats in Borneo. As *S. sinensium* is phylogenetically basal to *S. japonicum* and has a Tibet/Yunnan origin, the alternative hypothesis implies that *S. japonicum* colonized *Oncomelania* more recently than the latter colonized its current range.

Davis (1979) suggested that *Pomatiopsis* and *Cecina* in North America were part of the late Tertiary faunal influx to North America, from Hokkaido and Manchuria via the Bering land bridge. This is unlikely under a Gondwanan hypothesis because *Cecina* (part of the Tertiary radiation) appears closely related to *Tomichia* and *Coxiella* (see Davis, 1980) and these taxa supposedly diverged in 120 Ma. The North American taxa may have colonized from South America, with convergence explaining similarities between these taxa and Asian *Oncomelania* and *Cecina*. Misassignment of taxa cannot be ruled out because published descriptions date from the 1950s and were not sufficiently detailed or were based on taxonomically less reliable characters (e.g., conchology, see Temcharoen, 1971). Davis (1980) explained the occurrence of a single species inland in South America as vicariance; however, *Aquidauania* can alternatively be explained as refugial on the limestone coast of the mid-Miocene Paranan Sea (Lundberg *et al.*, 1998) (Fig. 11.6). The adoption of an East-to-West phylogeography for the Chinese Pomatiopsidae would support the Red River hypothesis (Attwood *et al.*, 2003), a Hunan/Guanxi origin for the antecedent Mekong schistosome. The possibility of an east-to-west dispersal is yet to be tested for Chinese *Oncomelania*.

2.5. Historical Biogeography and Demographic History of *Neotricula aperta*

Although early studies suggested that the radiation of the Pomatiopsidae and the *S. japonicum* group was isochronous (e.g., Attwood *et al.*, 2004a; Davis, 1992), more recent work indicates that this was not the case. Attwood *et al.* (2008b) dated the divergence of *S. mekongi* among the different river systems of Cambodia, and up to Khong Island at 1.3 Ma and linked it to Pleistocene tectonic upheavals, such as the tilting of the Khorat Basin and volcanism in the Bolovens and eastern Cambodia. More recently, modeled gene trees identified two ecotypes of *N. aperta*, one lentic spring dwelling not found in second-order streams or above, and one lotic found in small or large rivers. Genetically, each was characterized by unique haplotypes. The ecotypes, which all belonged to the γ-strain, were estimated to have diverged about 10 Ma (Attwood *et al.*, 2008a), suggesting that the radiation of *N. aperta* began long before that of *S. malayensis* and *S. mekongi*. The *N. aperta* populations with the greatest internal genetic diversity were found in the Attapeu region, suggesting that *N. aperta* entered central Sundaland from Vietnam via the Xe Kong or upper Sre Pok (similar to the tract inferred for *S. mekongi*). Coalescent simulations of population history also favored an ancestral clade in Attapeu, diverging to produce first a Bolovens clade and then a Kontum clade (i.e., the rivers of eastern Cambodia, see Fig. 11.1A); this agrees with the Red River hypothesis of a Vietnamese dispersal followed by a radiation in the Annam hills before invasion of the Mekong River.

Major divergence events inferred across the range of *N. aperta* between 4 and 6.5 Ma may be attributed to the final Indosinian orogeny at around 5 Ma. The coalescent-based analyses of gene flow and the nested clade analysis (NCA) indicated a Pliocene (or pre-Pliocene) colonization from North Central Laos so that *N. aperta* was present across most of its range by 4 Ma; however, this was followed by a more recent (early Pleistocene) switch in the direction of prominent gene flow from south to north (mostly from the Cambodian lower Mekong into Laos). A major dispersal tract was detected between Cambodia (Xe Kong and Xe San rivers) and Attapeu (Fig. 11.7). The findings are consistent with an earlier (6-7 Ma) expansion of the lotic ecotype in the Cambodian lower Mekong River, followed by re-invasion of the larger rivers of Central Laos by this ecotype. The lentic ecotype is assumed to have inhabited the spring-fed pools of North Central Laos throughout its history and not dispersed far, though it may have been displaced from Attapeu and southern Khammouanne through competition with the lotic form and/or by habitat destruction caused by human activities.

A northward migration tract is against the flow of the modern rivers in the region (except for Fig. 11.7 tract 2, which could be more recent); however, the Pliocene river courses were very different from those of the modern rivers. Marked tectonic upheavals during the Pleistocene effected major course changes, stream captures and flow reversals (Hutchinson, 1989). Consequently, by the time that *S. mekongi* appears to have been introduced to Cambodia, snail and human migration rates between Cambodia and Laos would have been much reduced from their pre-Pleistocene levels. Indeed, the nested clade analysis of Attwood *et al.* (2008b) showed a biogeographical divide between western (Lao/Thai) and eastern (eastern Lao/Cambodian) clades; this divide runs along the Bolovens uplift and the uplifted edge of the Khorat Basin (Fig. 11.7). The barriers affecting this genetic discontinuity include

FIGURE 11.7 The approximate geographical extent of key clades for *Neotricula aperta* as identified by nested clade analysis (NCA). The depth of shading of each clade is proportional to its effective population size (as estimated by a coalescent method). The major tracts of gene flow among clades are shown as arrows whose width is proportional to the level of gene flow measured using a coalescent model. The inferences of the NCA are also shown for each shaded area: CRE, contiguous range expansion; PAN, panmixis; RGE, restricted gene flow. The thick broken line along the Cambodian border approximately marks the uplifted edge of the Khorat Basin. Historical processes inferred are an ancestral spring dwelling *N. aperta* clade in the Annam region diverges to produce the river ecotype (10-9 Ma); river ecotype enters Lao and Cambodian Mekong rivers in independent migrations (6-5 Ma); after major population growth in the Mekong river, Cambodian *N. aperta* recolonizes larger rivers of Annam region (<5 Ma); spring ecotype is relict; Pleistocene uplifts break connections between Cambodia and Laos (<1 Ma). Adapted from figure 6 of Attwood *et al.* (2008a).

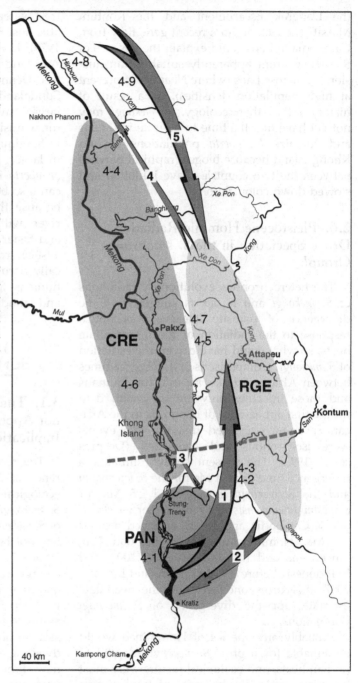

the Dangrek Escarpment and the Kontum Massif. The lack of more recent gene flow from Cambodia to Laos could explain the absence of *S. mekongi* from apparently suitable transmission foci across Laos (where *N. aperta* is present at high population densities), as a result of history rather than ecology. *S. mekongi* may not yet have had the time to migrate into Laos and colonize *N. aperta* populations beyond Khong Island because biogeographical barriers between the two countries have inhibited and slowed down colonization.

2.6. Pleistocene Hominid Radiation Drove Speciation in the *S. sinensium* Group?

The heterochronous evolutionary radiations of *S. mekongi* and *N. aperta* suggest that the divergence of parasite lineages was not a response to the radiation of the intermediate hosts; so what could have driven the radiation of *S. mekongi*? Comparisons of divergence times between African *Schistosoma* infecting humans and those infecting only animals, with data for the first appearance of hominids in the African Savanna, suggested that the two events were isochronous and connected (Despres *et al.*, 1992). The present study estimated a divergence time of ca. 4 Ma for the *S. japonicum* and the *S. mekongi* lineage, and 2.5 Ma for *S. mekongi/malayensis*. The radiation of *S. mekongi* across Cambodia apparently occurred around 1.3 Ma. Hominins appear in Sichuan Province at around 1.3 Ma (Zhu *et al.*, 2003) and in Indonesia before 0.9 Ma (Moore and Brumm, 2007); this corresponds well with the mean date estimates for the divergence of *S. mekongi* populations.

Probably any species of Homininae would be suitable for a proto-*S. mekongi* adapted to human hosts, and ecological factors were most likely the main determinants of transmission success. *Homo erectus*, at least, appears to have lived in fairly stable camps located close to

freshwater and as such would have been a suitable host for *Schistosoma* (Clark and Kurashina, 1979). *Homo ergaster* is known from Southeast Asia and is considered similar to *H. erectus* (see Dennell and Roebroeks, 2005). As central Sundaland continued to uplift, emerge, and tilt, larger rivers evolved and host switching (or acquisition), from stream associated rodents to hominins, would have allowed transmission in larger, faster flowing, rivers. The switch to *N. aperta* from *Robertsiella* by proto-*S. mekongi* can also be seen as an adaptation to hominins because this snail is epilithic in parts of the river with shallow, well-oxygenated, clean, and faster flowing water. The snail also selects a sheet rock substratum; these habitats typically occur in places also favored by modern humans for fishing, boat moorings, bathing, and laundry.

3. WHERE IS MEKONG SCHISTOSOMIASIS GOING?

3.1. The Distribution of *S. mekongi* Does not Appear to be Limited by Ecology: Implications for Control Strategies

The recently discovered widespread occurrence of *N. aperta* across Laos suggests that ecological conditions in Laos are suitable for *S. mekongi* transmission and that the absence of *S. mekongi* from most of Laos is because there has not been sufficient time for *S. mekongi* to invade Laos. The implication is that the future emergence of *S. mekongi* across Laos is possible, and that disease surveillance efforts in Laos need to be enhanced in areas previously assumed to be free from the risk of schistosomiasis. In addition, the recognition of a spring-dwelling ecotype of *N. aperta* in Laos implies a new dimension to the *S. mekongi* control program. If *N. aperta* can inhabit small streamlets and pools, the potential range of *S. mekongi* will be much greater than previously thought.

The challenge for disease surveillance is also greater as habitats may be hidden around villages or in forests, and the exact site of transmission where human cases are discovered will be much more difficult to determine. Human exposure patterns will also differ; exposure along the Mekong River is concentrated among those involved in fishing, laundry, etc. However, transmission via the spring-dwelling ecotype would affect the whole community, with an exposure pattern similar to that of *S. malayensis*, which is transmitted by *Robertsiella* spp. living in springs and seepages (Attwood *et al.*, 2005; Greer *et al.*, 1984).

3.2. The Future of Schistosomiasis Control in the Face of Expanding Snail Populations and Marked Gene-Flow into Laos

Coalescent-based (population genetic) estimates indicated that the lower Mekong River clades (South of Khong Island) had the highest *N. aperta* population sizes and historical population growth rates (approximately four times greater than other populations). The next largest values were for the eastern Cambodian rivers, with much smaller values for the Thai Mekong and Xe Bang-Fai rivers. The northernmost clades had the smallest estimated population sizes, only 5-1.5% of those of the Cambodian clades (the smallest values were for the lentic spring populations). Growth rates for the spring populations were also small, being <10% of those for the lotic populations (Attwood *et al.*, 2008a).

The observation of a marked expansion in the snail populations at the *S. mekongi* transmission foci along the Mekong River in Cambodia highlights the need for snail control and suggests that control of *S. mekongi* will require continual efforts in the face of a large and historically fast-growing, panmictic, snail population.

3.3. Summary of the Public Health Implications and Directions for Future Work

The reconstructions of population history reviewed here were based on preliminary data for *N. aperta* (Attwood *et al.*, 2008b), and further data are required using additional DNA-sequence loci and samples for geographically intermediate populations, if they exist. Nevertheless, the general features summarized here appear to be well-established: lower Mekong snail populations have been growing rapidly, most of the recent gene flow is out of the Cambodian lower Mekong, there was a post-Pliocene biogeographical divide between Cambodia and Laos, there are two ecotypes of *N. aperta*, and the radiations of *S. mekongi* and *N. aperta* are heterochronous. Furthermore, although the lentic *N. aperta* populations were generally declining, one population near to the construction site of the Nam Theun II dam project showed a high growth rate. In addition, high levels of gene flow (in a region of otherwise restricted gene flow) were reported between this population and those along the Xe Bang-Fai and Nam Yom rivers (the discharge channel for the dam). Such reports require further investigation; the possible detection of unusual population expansion and migration rates among *N. aperta* at the dam site has public health implications (Attwood *et al.*, 2008b).

Based on recent literature, an expansion of *S. mekongi* into Laos is possible from Khong Island to Savannakhet. So increased and sustained surveillance is required in Central Laos. There is also an increased possibility of transmission in nonriparian communities because the lentic ecotype of *N. aperta* is able to live in tiny seepages and streamlets or pools. The future control of Mekong schistosomiasis may not be as easy to achieve and maintain as previously assumed; the parasite may even emerge in unexpected areas. Certainly, there is no case for a winding down of schistosomiasis control efforts in the lower Mekong; rather, this currently relatively neglected disease warrants increased study and monitoring.

References

Agatsuma, T., Iwagami, M., Liu, C. X., Rajapakse, R. P. V. J., Mondal, M. M. H., Kitikoon, V., Ambu, S., Agatsuma, Y., Blair, D., and Higuchi, T. (2002). Affinities between Asian non-human *Schistosoma* species, the *S. indicum* group, and the African human schistosomes. *Journal of Helminthology* **76**, 7-19.

Attwood, S. W. (2001). Schistosomiasis in the Mekong region: Epidemiology and phylogeography. *Advances in Parasitology* **50**, 87-152.

Attwood, S. W. and Upatham, E. S. (1999). A new strain of *Neotricula aperta* found in Khammouanne Province, central Laos, and its compatibility with *Schistosoma mekongi*. *Journal of Molluscan Studies* **65**, 371-374.

Attwood, S. W., Kitikoon, V., and Southgate, V. R. (1997). Infectivity of a Cambodian isolate of *Schistosoma mekongi* to *Neotricula aperta* from Northeast Thailand. *Journal of Helminthology* **71**, 183-187.

Attwood, S. W., Upatham, E. S., and Southgate, V. R. (2001). The detection of *Schistosoma mekongi* infections in a natural population of *Neotricula aperta* at Khong Island, Laos, and the control of Mekong schistosomiasis. *Journal of Molluscan Studies* **67**, 400-405.

Attwood, S. W., Upatham, E. S., Meng, X. H., Qiu, D.-C., and Southgate, V. R. (2002). The phylogeography of Asian *Schistosoma* (Trematoda: Schistosomatidae). *Parasitology* **125**, 1-13.

Attwood, S. W., Ambu, S., Meng, X. H., Upatham, E. S., Xu, F.-S., and Southgate, V. R. (2003). The phylogenetics of triculine snails (Rissooidea: Pomatiopsidae) from south-east Asia and southern China: Historical biogeography and the transmission of human schistosomiasis. *Journal of Molluscan Studies* **69**, 263-271.

Attwood, S. W., Campbell, I., Upatham, E. S., and Rollinson, D. (2004a). Schistosomes in the Xe Kong River of Cambodia: The detection of *Schistosoma mekongi* in a natural population of snails and observations on intermediate host distribution. *Annals of Tropical Medicine & Parasitology* **98**, 221-230.

Attwood, S. W., Upatham, E. S., Zhang, Y.-P., Yang, Z.-Q., and Southgate, V. R. (2004b). A DNA-sequence based phylogeny for triculine snails (Gastropoda: Pomatiopsidae: Triculinae), intermediate hosts for *Schistosoma* (Trematoda: Digenea): Phylogeography and the origin of *Neotricula*. *Journal of Zoology* **262**, 47-56.

Attwood, S. W., Lokman, H. S., and Ong, K. Y. (2005). *Robertsiella silvicola*, a new species of triculine snail (Caenogastropoda: Pomatiopsidae) from peninsular Malaysia, intermediate host of *Schistosoma malayensis* (Trematoda: Digenea). *Journal of Molluscan Studies* **71**, 379-391.

Attwood, S. W., Fatih, F. A., Mondal, M. M. H., Alim, M. A., Fadjar, S., Rajapakse, R. P. V. J., and Rollinson, D. (2007). A DNA-sequence based study of the *Schistosoma indicum* (Trematoda: Digenea) group: Population phylogeny, taxonomy and historical biogeography. *Parasitology* **134**, 2009-2020.

Attwood, S. W., Fatih, F. A., Campbell, I. C., and Upatham, E. S. (2008a). The distribution of Mekong schistosomiasis, past and future: Preliminary indications from an analysis of genetic variation in the intermediate host. *Parasitology Int.*, **57**, 256-270.

Attwood, S. W., Fatih, F. A., and Upatham, E. S. (2008b). A study of DNA-sequence variation among *Schistosoma Mmekongi* (Trematoda: Digenea) populations and related taxa; phylogeography and the current distribution of Asian Schistosomiasis. *PLoS Neglected Tropical Diseases* **2**(3): e200. doi:10.1371/journal.pntd.0000200.

Barker, S. C. and Blair, D. (1996). Molecular phylogeny of *Schistosoma* species supports traditional groupings within the genus. *Journal of Parasitology* **82**, 292-298.

Biays, S., Stich, A. H. R., Odermatt, P., Long, C., Yersin, C., Men, C., Saem, C., and Lormand, J.-D. (1999). Foyer de bilharziose à *Schistosoma mekongi* redécouvert au Nord du Cambodge: I. Perception culturelle de la maladie; description et suivi de 20 cas cliniques graves. *Tropical Medicine & International Health* **4**, 662-673.

Blair, D., van Herwerden, L., Hirai, H., Taguchi, T., Habe, S., Hirata, M., Lai, K., Upatham, E. S., and Agatsuma, T. (1997). Relationships between *Schistosoma malayensis* and other Asian schistosomes deduced from DNA sequences. *Molecular & Biochemical Parasitology* **85**, 259-263.

Blair, D., Davis, G. M., and Wu, B. (2001). Evolutionary relationships between trematodes and snails emphasizing schistosomes and paragonimids. *Parasitology* **123**, S229-S243.

Brookfield, M. E. (1998). The evolution of the great river systems of southern Asia during the Cenozoic India-Asia collision: Rivers draining southwards. *Geomorphology* **22**, 285-312.

Clark, J. D. and Kurashina, H. (1979). Hominid occupation of the east-central highlands of Ethiopia in the Plio-Pleistocene. *Nature* **282**, 33-39.

Cribb, T. H., Bray, R. A., and Littlewood, D. T. J. (2001). The nature and evolution of the association among digeneans, molluscs and fishes. *International Journal for Parasitology* **31**, 997-1011.

Davis, G. M., Kitikoon, V., and Temcharoen, P. (1976). Monograph on "*Lithoglyphopsis*" *aperta*, the snail host of Mekong River schistosomiasis. *Malacologia* **15**, 241-287.

Davis, G. M. (1979). The origin and evolution of the gastropod family Pomatiopsidae, with emphasis on the Mekong River Triculinae. *Academy of Natural Sciences of Philadelphia, Monograph* **20**, 1-120.

Davis, G. M. (1980). Snail hosts of Asian *Schistosoma* infecting man: Evolution and coevolution. *In* "The Mekong Schistosome," Bruce, J. I., Sornmani, S., Asch, H. L., and Crawford, K. A. (eds.), pp 195-238. *Malacological Review*, Suppl. 2. Society for Experimental and Descriptive Malacology, The University of Michigan, Ann Arbor, Michigan.

Davis, G. M. (1992). Evolution of prosobranch snails transmitting Asian *Schistosoma*; coevolution with *Schistosoma*: A review. *Progress in Clinical Parasitology* **3**, 145-204.

Davis, G. M., Chen, C.-E., Wu, C., Kuang, T.-F., Xing, X.-G., Li, L., Liu, W.-J., and Yan, Y.-L. (1992). The Pomatiopsidae of Hunan, China (Gastropoda: Rissoacea). *Malacologia* **34**, 143-342.

Davis, G. M., Wilke, T., Spolsky, C., Zhang, Y., Xia, M.-Y., and Rosenberg, G. (1998). Cytochrome Oxidase I-based phylogenetic relationships among the Hydrobiidae, Pomatiopsidae, Rissoidae, and Truncatellidae (Gastropoda: Prosobranchia: Rissoacea). *Malacologia* **40**, 251-266.

Dennell, R. and Roebroeks, W. (2005). An Asian perspective on early human dispersal from Africa. *Nature* **438**, 1099-1104.

Despres, L., Imbert-Establet, D., Combes, C., and Bonhomme, F. (1992). Molecular evidence linking hominid evolution to recent radiation of schistosomes (Platyhelminthes: Trematoda). *Molecular Phylogenetics & Evolution* **1**, 295-304.

Dey, A. K. (1968). "Geology of India." National Book Trust, New Delhi.

Dupont -Vic, B. E., Soubrane, J., Halle, B., and Richir, C. (1957). Bilharziose à forme hépato-splénique révélée par une grande hématémèse. *Bulletin Mémoires de la Société MédicaleHôpitaux Paris* **73**, 933-994.

Dupont, L. M., Donner, B., Vidal, L., Pérez, E. M., and Wefer, G. (2005). Linking desert evolution and coastal upwelling: Pliocene climate change in Namibia. *Geology* **33**, 461-464.

Freiermuth, J. P., Sauvet, F., and Morillon, M. (2005). Bilharziasis caused by *Schistosoma mekongi*: An almost eradicated endemic disease. *Medécin Tropicale* **65**, 421-422.

Gang, C. M. (1991). "Mission Report: Schistosomiasis Control Programme in Khong District." WHO/ICP/PDP/004, WHO, Geneva.

Greer, G. J., Ambu, S., and Davis, G. M. (1984). Studies on the habitat, distribution and schistosome infection of *Robertsiella* spp., snail hosts for a *Schistosoma japonicum*-like schistosome in Peninsular Malaysia. *Tropical Biomedicine* **1**, 85-93.

Hall, R. (1998). The plate tectonics of Cenozoic Asia and the distribution of land and sea. *In* "Biogeography and Geological Evolution of SE Asia," Hall, R. and Holloway, J. D. (eds.), pp 99-131. Backbuys, Leiden.

Hall, R. (2002). Cenozoic geological and plate tectonic evolution of SE Asia and the SW Pacific: Computer-based reconstructions, model and animations. *Journal of Asian Earth Sciences* **20**, 353-431.

Hartt, F. (1870). "Geology and Physical Geography of Brazil." Boston, Fields, Osgood & Co., Cambridge.

Heaney, L. R. (1986). Biogeography of mammals in Southeast Asia: Estimates of rates of colonization, extinction and speciation. *Biological Journal of the Linnean Society* **28**, 127-165.

Hutchinson, C. S. (1989). "Geological Evolution of South-East Asia." Clarendon Press, Oxford.

Iijima, T. (1970). "Enquête sur la Schistosomiase dans la Bassin du Mékong: Cambodge." Rapport de Mission 13/11/68 - 08/05/69. WPR/059/70. Organisation Mondiale de la Santé, Genève.

Iijima, T. and Garcia, E. G. (1967). "Preliminary Survey for Schistosomiasis in South Laos." WHO Assignment Report, WHO/BILH/67.64. World Health Organisation, Geneva.

Kitikoon, V., Schneider, C. R., Sornmani, S., Harinasuta, C., and Lanza, G. R. (1973). Mekong schistosomiasis: II. Evidence of the natural transmission of *Schistosoma japonicum*, Mekong strain at Khong Island, Laos. *South East Asian Journal of Tropical Medicine & Public Health* **4**, 350-358.

Kuntz, R. E. (1978). North Borneo (Malaysia): A new locality for *Schistosoma japonicum*. *American Journal of Tropical Medicine & Hygiene* **27**, 208-209.

Lacassin, R., Leloup, P. H., and Tapponier, P. (1993). Bounds on strain in large Tertiary shear zones of SE Asia from boudinage restoration. *Journal of Structural Geology* **15**, 677-692.

Le, T. H., Humair, P.-F., Blair, D., Agatsuma, T., Littlewood, D. T. J., and McManus, D. P. (2001). Mitochondrial gene content, arrangement and composition compared in African and Asian schistosomes. *Molecular & Biochemical Parasitology* **117**, 61-71.

Leroy, S. and Dupont, L. M. (1994). Development of vegetation and continental aridity in northwestern Africa during the Late Pliocene: The pollen record of ODP Site 658. *Palaeogeography, Palaeoclimatology, Palaeoecology* **109**, 295-316.

Li, J. (1991). The environmental effects of the uplift of the Qinghai Xizang Plateau. *Quarterly Science Review* **10**, 479-483.

Littlewood, D. T. J. and Johnston, D. A. (1995). Molecular phylogenetics of the four *Schistosoma* species groups determined with partial 28S ribosomal RNA gene sequences. *Parasitology* **111**, 167-175.

Littlewood, D. T. J., Lockyer, A. E., Webster, B. L., Johnston, D. A., and Le, T. H. (2006). The complete mitochondrial genomes of *Schistosoma haematobium and Schistosoma spindale* and the evolutionary history of mitochondrial genome changes among parasitic flatworms. *Molecular Phylogenetics & Evolution* **39**, 452-467.

Lockyer, A. E., Olson, P. D., Østergaard, P., Rollinson, D., Johnston, D. A., Attwood, S. W., Southgate, V. R., Horak, P., Snyder, S. D., Le, T. H., Agatsuma, T., McManus, D. P. *et al.* (2003). The phylogeny of the Schistosomatidae based on three genes with emphasis on the interrelationships of *Schistosoma* Weinland, 1858. *Parasitology* **126**, 203-224.

Lockyer, A. E., Jones, C. S., Noble, L. R., and Rollinson, D. (2004). Trematodes and snails: An intimate association. *Canadian Journal of Zoology* **82**, 251-269.

Lundberg, J. G., Marshall, L. G., Guerrero, J., Horton, B., Malabarba, M. C. S. L., and Wesselingh, F. (1998). The stage for neotropical fish diversification: A history of tropical South American rivers. *In* "Phylogeny and Classification of Neotropical Fishes," Malabarba, L. R., Reis, R. E., Vari, R. P., Lucena, Z. M. S., and Lucena, C. A. S. (eds.), pp 13-48. EDIPUCRS, Porto Alegre, Brazil.

Matsumoto, J., Muth, S., Socheat, D., and Matsuda, H. (2002). The first reported cases of canine Schistosomiasis mekongi in Cambodia. *South East Asian Journal of Tropical Medicine & Public Health* **33**, 458-461.

Matthee, C., Catzeflis, F. M., and Douzery, E. J. P. (1997). Phylogenetic relationships of artiodactyls and cetaceans as deduced from the comparison of cytochrome *b* and 12S rRNA mitochondrial sequences. *Molecular Biology & Evolution* **14**, 550-559.

Matthee, C., Burzlaft, J. D., Taylor, J. F., and Davis, S. K. (2001). Mining the mammalian genome for Artiodactyl systematics. *Systematic Biology* **50**, 367-390.

Metcalfe, I. (1998). Palaeozioc and Mesozoic geological evolution of the South East Asian region: Multidisciplinary constraints and implications for biogeography. *In* "Biogeography and Geological Evolution of SE Asia," Hall, R. and Holloway, J. D. (eds.), pp 25-41. Backbuys, Leiden.

Ministry of Geology. (1947). "Carta Geologico-Economica de la Republica Argentina." Direccion General de Minas y Geologia, Buenos Aires.

Mitchell, A. H. G. (1981). Phanerozoic plate boundaries in mainland S.E. Asia, the Himalayas and Tibet. *Journal of the Geological Society of London* **138**, 109-122.

Moore, M. W. and Brumm, A. (2007). Stone artifacts and hominins in island Southeast Asia: New insights from Flores, eastern Indonesia. *Journal of Human Evolution* **52**, 85-102.

Morgan, J. A. T., Dejong, R. J., Kazibwe, F., Mkoji, G. M., and Loker, E. S. (2003). A newly identified lineage of *Schistosoma*. *International Journal for Parasitology* **33**, 977-985.

Morgan, J. A. T., Dejong, R. J., Adeoye, G. O., Ansa, E. D. O., Barbosa, C. S., Bremond, P., Cesari, I. M., Charbonnel, N., Correa, L. R., Coulibaly, G., D'andrea, P. S., De Souza, C. P. *et al.* (2005). Origin and diversification of the human parasite *Schistosoma mansoni*. *Molecular Ecology* **14**, 3889-3902.

Ohmae, H., Sinuon, M., Kirinoki, M., Matsumoto, J., Chigusa, Y., Socheat, D., and Matsuda, H. (2004). Schistosomiasis mekongi: From discovery to control. *Parasitology International* **53**, 135-142.

Qiu, Z.-D. and Li, C. (2003). Chapter 22: Rodents from the Chinese Neogene: Biogeographic relationships with Europe and North Africa. *Bulletin of the American Museum of Natural History* **279**, 586-602.

Rollinson, D. and Southgate, V. R. (1987). The genus *Schistosoma*: A taxonomic appraisal. *In* "The Biology of Schistosomes: From Genes to Latrines," Rollinson, D. and Simpson, A. J. G. (eds.), pp 1-49. Academic Press, London.

Rost, K. T. (2000). Pleistocene paleoenvironmental changes in the high mountain ranges of central China and adjacent regions. *Quaternary International* **65/66**, 147-160.

Samlane, P., Phetsouvanh, R., Hongvanthong, B., and Sengsavath, V. (2007). The situation of *Schistosoma mekongi* in Khong District. *In* "Proceedings of the 7th Workshop of the Regional Network on Asian Schistosomiasis and Other Helminth Zoonoses (RNAS+)," September 5-7, 2007, pp 20-22. RNAS, Lijiang, Yunnan.

Sinuon, M., Tsuyuoka, R., Socheat, D., Odermatt, P., Ohmae, H., Matsuda, H., Montresor, A., and Palmer, K. (2007). Control of *Schistosoma mekongi* in Cambodia: Results of eight years of control activities in the two endemic provinces. *Transactions of the Royal Society of Tropical Medicine & Hygiene* **101**, 34-39.

Snyder, S. D. and Loker, E. S. (2000). Evolutionary relationships among the Schistosomatidae (Platyhelminthes: Digenea) and an Asian origin for *Schistosoma*. *Journal of Parasitology* **86**, 283-288.

Stich, A. H. R., Biays, S., Odermatt, P., Men, C., Saem, C., Sokha, K., Ly, C. S., Legros, P., Philips, M., Lormand, J.-D., and Tanner, M. (1999). Foci of schistosomiasis mekongi, northern Cambodia: II. The distribution of infection and morbidity. *Tropical Medicine & International Health* **4**, 674-685.

Strandgaard, H., Johansen, M. V., Pholsena, K., Teixayavong, K., and Christensen, N. O. (2001). The pig as a host for *Schistosoma mekongi* in Laos. *Journal of Parasitology* **87**, 708-709.

Temcharoen, P. (1971). New aquatic molluscs from Laos. *Archiv für Molluskenkunde* **101**, 91-109.

Urbani, C., Sinuon, M., Socheat, D., Phoisena, K., Strandgaard, H., Odermatt, P., and Hatz, C. (2002). Epidemiology and control of mekongi schistosomiasis. *Acta Tropica* **82**, 157-168.

Van Royen, W. (1952). "Mineral Resources of the World." Prentice-Hall, New York.

Webster, J. P. and Davies, C. M. (2001). Coevolution and compatibility in the snail-schistosome system. *Parasitology* **123**, S41-S56.

Webster, J. P., Gower, C. M., and Blair, L. (2004). Do hosts and parasites coevolve? Empirical support from the *Schistosoma* system. *American Naturalist* **164**, S33-S51.

Webster, B. L., Southgate, V. R., and Littlewood, D. T. J. (2006). A revision of the interrelationships of *Schistosoma* including the recently described *Schistosoma guineensis*. *International Journal for Parasitology* **36**, 947-955.

Wilke, T., Davis, G. M., Gong, X., and Liu, H.-X. (2000). *Erhaia* (Gastropoda: Rissooidea): Phylogenetic relationships and the question of *Paragonimus* coevolution in Asia. *American Journal of Tropical Medicine & Hygiene* **62**, 453-459.

Workman, D. R. (1972). "Geology of Laos, Cambodia, South Vietnam and the Eastern Part of Thailand: A Review." Institute of Geological Science Overseas Division Report, London.

Xiao, J. and An, Z. (1999). Three large shifts in East Asian monsoon circulation indicated by loess-paleosol sequences in China and late Cenozoic deposits in Japan. *Palaeogeography, Palaeoclimatology, Palaeoecology* **154**, 179-189.

Yamada, N., Teraoka, Y., and Hata, M. (1987). "Geological Atlas of Japan." Geological Survey of Japan, Tokyo.

Yong, H. S., Greer, G. J., and Ow-Yang, C. K. (1985). Genetic diversity and differentiation of four taxa of Asiatic blood flukes (Trematoda: Schistosomatidae). *Tropical Biomedicine* **2**, 17-23.

Zhu, R., An, Z., Potts, R., and Hoffman, K. A. (2003). Magnetoshperic dating of early humans in China. *Earth-Science Reviews* **61**, 341-359.

Water Quality of the Lower Mekong River

Edwin D. Ongley

4951 Connaught Ave., Montreal, QC, Canada H4V 1X4

1. INTRODUCTION

The Mekong River rises in the Tibetan Plateau of western China and flows southward through China, Myanmar, Laos, Thailand, Cambodia, and Viet Nam, where it flows into the South China Sea through the Mekong main stem and its principal distributary, the Bassac River that branches from the main stem at Phnom Penh (see Chapter 5 Figure 1). The Mekong is longest river in southeast Asia and the 10th largest world river by discharge according to Dai and Trenberth (2002). The basin is comprehensively described in the Mekong River Basin Diagnostic Study (MRC, 1997), the State of the Basin Report (MRC, 2003), and in this volume.

The hydrology of the Mekong basin is dominated by monsoon rains that occur in the period June-November and by the dry period from December to May. China contributes only 16% of the mean annual discharge, whereas Laos contributes some 35% and up to 60% during the wet season (MRC, 2005). In contrast, China contributes 50% of the total sediment load of the Mekong (MRC, 2005). For water quality monitoring, the period of high flows creates problems at monitoring sites in tributaries that are close to the main stem due to extensive back flooding into the tributaries.

The hydrology of the lower Mekong basin is particularly complex in the area from Kratie (Cambodia) downstream due to the extremely low gradients. During the dry season, the Mekong drains through the Mekong and Bassac distributary channels through the Mekong Delta to the South China Sea. Salt water influences are felt some 70 km upstream from the distal part of the delta and tidal influences can extend upstream as far as Phnom Penh. During the rainy season, Mekong water flows "upstream" in the Tonle Sap River into the Great Lake of Cambodia, causing expansion of the surface area of the lake by up to sixfold (Campbell et al., Chapter 10), creating extensive wetlands around the entire lake. The water drains out of the Great Lake into the Mekong system during the dry season, thereby adding to low flow discharges in the region downstream of Phnom Penh. This hydrological pattern is the source of the abundance of biodiversity of this region, but makes hydrological and water quality monitoring and interpretation difficult, especially in main stem stations below Kratie. Reverse flows occur during parts of the tidal cycle in delta stations and during wet season reverse flow in the Tonle Sap River.

The tropical Tonle Sap Lake (the Great Lake of Cambodia) and the Tonle Sap River are a unique lacustrine and wetland complex. The water quality of this lake has been monitoring by the Mekong River Commission (MRC) since 1993 and as part of a special study on nutrient and sediment fluxes as part of the Water Utilization Project. However, there has been no systematic or substantial scientific study of the nutrient dynamics of the Great Lake and it is not known if the lake is N and/or P limited. Indeed, there is relatively little research on nutrient limitation in tropical systems. Various authors have suggested that tropical lake Victoria (Africa) may be N limited (Lehman and Branstrador, 1993; Moss, 1969). Guildford et al. (2003) found that N is limiting in tropical Lake Malawi when there is sufficient light. In contrast, they found that in eutrophic Lake Victoria there has been a shift from diatom dominance to N-fixing cyanobacteria (blue-green algae), which is not N limited. Hecky (2007, personal communication), on the basis of his work in African tropical lakes, is of the opinion that in the absence of extensive light shading (due to dense algal blooms or turbidity), algal populations may exhaust soluble reactive phosphorus (SRP) in which case P is limiting. Algal blooms are not widely reported in Tonle Sap Lake (but see Chapter 10), but the water is turbid and supports abundant water hyacinth. Therefore, without research it cannot be determined if N or P should be the focus of water quality assessment or of control in the Tonle Sap Lake. It is known that there is extensive anoxia in the wetlands surrounding the lake probably due to oxygen consumption by intensive bacterial decay of organic matter in this zone; however, there is no monitoring in this zone. It is not known if, or how much, of the nutrient loadings from the surrounding land are transported through the wetlands into the open lake, or whether these loads are consumed within the wetlands. Monitoring has been restricted to open lake locations and it is not known if these data are representative of the lake basin as a whole and are responding to nutrient loadings from the basin, or are only representative of a limited area around the monitoring site.

2. POLLUTION SOURCES

2.1. Upper Mekong Basin

The provincial government of Yunnan Province, through which the upper Mekong (Lancang River) flows, is reported to have inspected 1042 industrial enterprises in the basin in 2000, and shut down four of these (CIIS, 2002). Since 1986, the Simao Paper Plant and the Lanping Lead-Zinc Mine have been built by the river. In addition to industrial enterprises, hydropower stations such as those at Manwan, Dachaoshan have been established on the Lancang River since 1986 with five more hydropower stations planned in the next 20 years. Xiaowan dam, which is now under construction, is located 550 km upstream of the China/Lao border, and will be the largest hydro-electric dam in China after the Three Gorges Dam on the Yangtze River (CERN, undated). Two other dams, the Jinhong station and the Nuozhadu station were scheduled to commence construction in the period 2005-2006. According to China.Org (n.d.), there is an anticipated 2.7 °C reduction in temperature in the water flowing from the Xiaowan dam. There is the danger of release of anoxic bottom waters from reservoirs, although Chinese sources claim that these dams will have no impact on water quality in the lower Mekong basin. Nevertheless, there are concerns about potential impacts in the lower Mekong basin due to rapid development in Yunnan Province and in view of the serious and increasing pollution in Chinese rivers and of the Chinese acknowledgment of failure in water quality enforcement (Ongley and Wang, 2004; Xinhua New Agency, 2005). Chinese news sources (e.g., CIIS, 2002) claim that the water of the Lancang meets international drinking water standards (for those parameters for which Chinese agencies routinely monitor), but data for water quality of the Lancang are not accessible. Additional information on the Upper Mekong is contained in Chapter 14.

2.2. Lower Mekong Basin

In the lower Mekong basin, pollution sources contributing directly to the Mekong main stem are few. Thailand's contribution to pollution in the Mekong is mainly limited to salt leaching from the Korat Plateau—part of the Nam Mun watershed. However, as noted below, data indicate that this is diluted to the point where there is no visible change in main stem water quality where the Nam Mun joins the Mekong River. There are no data that suggest that irrigated agriculture or the limited industrial areas in Thailand that are within the Mekong basin are significant contributors of pollution to the main stem of the Mekong.

The two largest urban areas, Vientiane in Laos and Phnom Penh in Cambodia, are of potential concern as they lie on the banks of the Mekong. Currently Vientiane, a city of approximately 500,000 inhabitants, discharges its municipal sewage into the That Luang Swamp—a wetland that discharges into the Mekong River downstream of Vientiane. This discharge is small at this time and there is no evidence that suggests that it poses any immediate risk to the Mekong main stem. However, development of Vientiane with substantial land reclamation in the That Luang Swamp for urban purposes is a concern, and may pose greater threats to the Mekong main stem in the future. Phnom Penh, a city of approximately 1.7 million inhabitants, is built around the junction of the Mekong, Bassac, and Tonle Sap rivers. Urban sewage is untreated and much is discharged into wetlands that drain into the Bassac River or into the Tonle Sap River. Some industrial and municipal discharges and storm water runoff discharge directly into the Tonle Sap River. There is local pollution of an industrial nature in the Tonle Sap River at Phnom Penh (MRC, 2007). There is a substantial riparian population in Phnom Penh that occupies housing located on piles along the margin of the river. There are also a number of floating villages in the Tonle Sap Lake. These populations discharge domestic waste directly into the water column, however the loading and its environmental significance is not known.

There are some well-known water quality issues in the Mekong Delta, including saline intrusion and acidity caused by acid-sulfate soils. While these are widespread, their impact on water quality is restricted to canals and distributaries that dissect the delta. They do not affect the main stems of the Mekong or Bassac, and are not discussed in this chapter. In the delta, the Vietnamese cities of Tan Chau and Chau Doc on the Mekong and the Bassac rivers, respectively, are major urban centers and are affected by tidal influences. Identified river pollution at these locations is probably attributable to local sources; however, there has been no definitive work on transboundary transport of pollutants from upstream. Analysis of transboundary risk concluded that the current data could neither support nor deny the presence of transboundary pollution between Cambodia and Viet Nam (Hart et al., 2001) Extensive caged fish culture lining the banks of the Mekong and Bassac rivers downstream of the Cambodian-Viet Nam boundary are likely sources of pollutants. In-stream caged fish culture occurs throughout much of the lower basin, however elsewhere it is not on such a large scale.

Until recently, there has been very limited research or other data on organic contaminants or of nonpoint sources of pollutants in the Mekong River basin. Evidence presented at the 2nd Asia Pacific International Conference on Pollutants Analysis and Control held in HoChiMinh City, Viet Nam in December of 2003 indicates that there is little evidence of persistent organic pollutants in the basin even in locations where there are known to have been high levels of use (e.g., TCDD dioxin from Agent Orange used during Viet Nam American war; agricultural pesticides used intensively in parts of Thailand). In 2004, MRC commissioned a large diagnostic study of contaminants in the lower basin (MRC, 2007) and, as noted below, that study found little evidence of the presence of contaminants in the basin. Recent work by Agusa et al. (2005) in Cambodia, a country where fish is the main source of dietary protein, indicates that mercury in some species of freshwater and marine fish is above dietary guidelines. However, their work does not imply large-scale mercury contamination in the freshwater system, although there is anecdotal evidence of mercury usage in artisanal placer gold extraction upstream in the main stem in Laos and possibly in some tributaries of the Mekong. Other nonpoint sources include rapid expansion of caged fish culture throughout the Mekong and its tributaries, discharge of human wastes from all river vessels plying the Mekong, especially tour boats in the middle reaches extending upstream from Luang Prabang (Laos) to close to the China border, and accidental spills from river boat traffic. Increasing commercial river traffic arising from channel "improvements" in the sector between China and northern Thailand and Laos will add additional stress to the upper reaches.

Other water quality issues exist in the basin. One is the potential for elevated levels of natural arsenic in drinking water wells in some countries. Another is the presence of a variety of water-borne diseases arising from endemic parasites (e.g., schistosomiasis). Sediment is a water quality issue, however this is dealt with in detail elsewhere in this volume.

MRC (2003) assessed a limited number of water quality attributes within the basin, but has never comprehensively assessed the entire database, although this is now in progress. Other observers have concluded, albeit without detailed analysis, that the water quality of the main stem of the Mekong is good, but with high levels of suspended sediment, especially during wet season runoff. Using MRC data, Campbell (2007) found that riparian concerns about water quality were largely perceptions and not based on empirical evidence.

3. WATER QUALITY DATA

Prior to 1985, there was little systematic ambient water quality monitoring in any of the Mekong countries with the exception of Thailand. In 1985, in response to riparian concerns over potential water pollution and its transboundary implications, the Mekong River Committee, the forerunner of the MRC, commenced a water quality monitoring program in Laos, Viet Nam, and Thailand; Cambodia joined this program in 1993. With financial and technical assistance from the Government of Sweden, the program was implemented with approximately 100 permanent stations on the main stem and important tributaries of the Mekong River (Fig. 12.1). In the period 2001-2005, MRC extensively revised the monitoring program to ensure the program met specific data needs of the countries and of MRC. About one-third of the original stations were eliminated and 23 new stations added. Some stations' locations were moved to reflect current land use activities. The network was divided into those stations having mainly international or transboundary significance (primary network of 55 stations) and those having mainly local or national significance (secondary network). This revised network was put into operation in 2004 and is essentially the same for the mainstream stations as shown in Fig. 12.1, but with the addition of a transboundary station on the Mekong River at the Lao-China border. The stations used in this chapter are shown in Fig. 12.1.

Data collection is implemented by designated national agencies in each country with technical and financial assistance from MRC. As a facilitating agency, MRC does not have its own laboratories. The basic parameters collected by all countries are noted in Table 12.1. NH_4 is the sum of $NH_3 + NH_4$; however at the neutral pH values of the Mekong system, most of the ammonia is in the less toxic NH_4 form. Other parameters such as total coliform, other metals, and some pesticides, were collected by one or more countries but without external QA/QC. These were abandoned by MRC in 2001 due to concerns over data quality and inconsistencies in methodologies. Coliforms and chlorophyll-a were added to the program in 2007.

In this chapter, we look specifically at the water quality of the main stem of the Mekong and the Bassac using the period of record from 1986 to 2005 (1994-2005 for Cambodia). Parameters chosen for analysis represent those most significant for (1) aquatic life (pH, dissolved oxygen (DO), and ammonia), (2) water supply and agricultural use (conductivity), (3) eutrophication (total phosphorus and NO_{2-3} [Total-N is not measured in two of the countries]), and (4) human impact (chemical oxygen demand (COD) and ammonia).

Sampling is generally carried out at the midpoint of each month, usually at 0.5 m depth (up to 1 m in deep rivers), at low tide in tidally affected reaches such as the delta, and from a point either in the middle of the river or in the thalweg where there is maximum flow. Samples are preserved and/or kept in coolers (depending on the parameter) and transported to the laboratory, usually within 24 h. In some countries, some sites are not easily accessible and there is delay before the samples reach the laboratories, in which case some sample deterioration may be unavoidable. Obtaining ice for field coolers is often a major problem. Field staff are well aware of these problems and do their best; however, the Mekong basin is tropical and much of it has poor access. Logistics are often extremely difficult, especially during the monsoon period. There was no systematic QA/QC program until 2005; however, analytical capacity building since 1985 included standard QA/QC protocols such as sample blanks, replication and use of reference materials, and occasional interlaboratory comparisons (round-robin). Historical data quality has been assessed by MRC using a suite

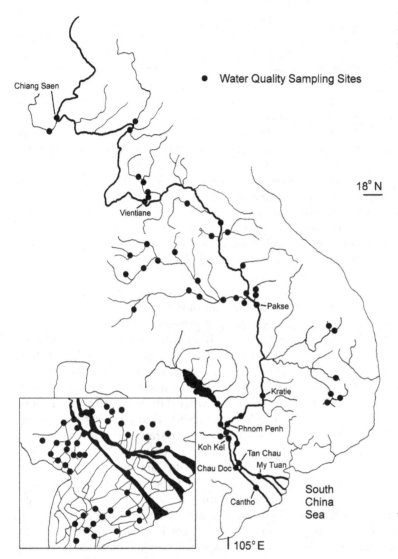

of internal checks (e.g., ion balance) and they have been integrated into a "Reliability Score" shown in Table 12.2. The higher the score, the more "reliable" are the data. As this score is unique to MRC, it is not possible to evaluate these scores relative to peer performance elsewhere. The score can be biased (usually upward) in cases of nonperformance of some parameters for which the reliability criteria are automatically omitted from the calculation of the score.

The data used in this analysis cover the period 1986-2005, except for Cambodia which is 1994-2005. Monitoring for stations in Thailand, Laos, and Viet Nam began in 1985, but as the year is incomplete these records have been omitted. Similarly, Cambodia began sampling in 1993; however, records for that year are omitted for

TABLE 12.1 Parameters of the MRC water quality program

Temperature	Sodium	Nitrite + nitrate	Phosphate ($PO_4{}^{3-}$)	COD_{mn}
Conductivity (mS m^{-1})	Potassium	Total ammonia	Total-P	Aluminum
TSS (mg l^{-1})	Calcium	Total nitrogen[a]	Silica	Iron
pH	Magnesium			
DO (mg l^{-1})	Chloride			Chlorophyll-a[b]
	Sulfate			Fecal coliforms[b]
	Alkalinity			

Shaded cells denote the parameters used in this chapter.
[a]Two of the countries do not measure this.
[b]Added in 2007.

the same reason. As a result of optimization of the network, the station at Pakse was moved a short distance upstream in 2004 and the records combined for this analysis (#13900 with former 13901). The Mekong is represented by seven benchmark stations and the Bassac by three benchmark stations as noted in Table 12.2. The sequence begins upstream at Chiang Saen (Thailand), which was the most upstream station in the network until 2004. The most

TABLE 12.2 Benchmark stations used in this chapter (upstream to downstream)

	Station code	Station name	River	Comment	Reliability score[a] (%)
Mainstream stations					
1	010501	Chiang Saen	Mekong	Most upstream, long-term stations, Thailand	81
2	011901	Vientiane	Mekong	Laos	79
3	013901	Pakse	Mekong	Laos	78
4	014901	Kratie	Mekong	Cambodia	85
5	019801	Phnom Penh (now named Chrouy Changvar)	Mekong	Cambodia, 3 km upstream of Phnom Penh City	87
6	019803	Tan Chau	Mekong	Viet Nam	76
7	019804	My Thuan	Mekong	Most downstream station above distributaries, Viet Nam	81
8	033402	Koh Khel	Bassac	Cambodia, also spelled as Koh Kel	87
9	039801	Chau Doc	Bassac	Viet Nam	74
10	039803	Can Tho	Bassac	Most downstream, long-term station, Viet Nam	79

Shaded cells denote transboundary stations. Station code is MRC station number.
[a]Data reliability index developed by MRC, reflecting internal checks (e.g., ion balance) of historical database; the higher the score, the more reliable the data.

downstream stations are My Tuan on the Mekong and Can Tho on the Bassac. Transboundary flux is examined using the sequence Phnom Penh, Chau Doc, and My Tuan on the Mekong and Koh Khel, Tan Chau, and Can Tho on the Bassac. The Koh Khel station is located in Cambodia, downstream of the city of Phnom Penh, which mainly influences the Bassac River. The Phnom Penh station is some 3 km upstream of the city of Phnom Penh.

4. WATER QUALITY GUIDELINES AND SUGGESTED THRESHOLDS

Technical criteria for water quality for various beneficial uses have not yet been adopted by the MRC. Therefore, the following discussion is based on technical material prepared by the author for the MRC (Ongley, 2006). While two of the countries have guidelines, they are not always the same and it is not known if these were developed mainly from the literature or through study of aquatic life in their territories. In part, the problem of technical criteria reflects the limited information available on, in particular, protection of aquatic life in tropical systems in Asia.

Water quality guidelines are discussed here for three types of beneficial uses/impacts—aquatic life, agricultural use of river water, and human impact. Additionally, eutrophication is discussed.

Table 12.3 compares guidelines used by a number of countries for protection of aquatic life and compares these with threshold values suggested by this author for use in the Mekong. The values used by most developing countries reflect research in North America, and to a lesser extent research from Europe and Australia, and attempt to adapt these results to local conditions. The two sets of guidelines in the Mekong basin are those of Thailand and Viet Nam; however, the technical rationale for the Thai and Viet Nam systems is unavailable. Research in the Northern Hemisphere that focuses on salmonids as the most sensitive species is not very relevant in the Mekong, where fish species are adapted to living in conditions of high suspended sediment concentration and, in some locations, in low oxygen conditions. The values suggested here for the Mekong are noted in Table 12.3. These are based upon review of guidelines used by other countries and their related technical basis. Some published values are clearly not appropriate to the Mekong and other values are suggested. Because there is little technical information on the relationship between Mekong aquatic life and water quality requirements, the suggested values of Table 12.3 are "best estimates" of this author following limited peer review.

4.1. Dissolved Oxygen (Aquatic Life)

Although DO is held as mg l^{-1} in the Water Quality Monitoring Network (WQMN) database, the critical issue for aquatic life is the percentage oxygen saturation as the physiological process of oxygen uptake by diffusion across a gill membrane is a direct function of the difference in oxygen saturation between the blood and the surrounding water. Much of the temperature data required to calculate percentage oxygen saturation is missing from Viet Nam stations. Therefore, for this assessment, DO as mg l^{-1} is used. Oxygen guidelines from North America or Europe are not suitable due to the very different temperature regimes and their use of salmonids to derive oxygen guidelines. The Australian values in Table 12.3 are taken from the ANZECC and ARMCANZ (2000) for tropical rivers in Australia. The values, expressed by ANZECC, are in percentage saturation and have been converted in Table 12.3 to O_2 concentration using published conversion factors using the median value of temperature of 27.2 °C for the entire database. Generally, water quality for aquatic life is considered good if saturation is $\geq 90\%$. For average temperature

TABLE 12.3 Comparative water quality criteria

MRC parameters (mg l⁻¹ except as noted)	Mekong suggested[a]	Australia[b]	US-EPA (CCC)[e]	Canada	Viet Nam	Thai. (Class 2)	France (aquatic life)	Malaysia (Class 2)	Taiwan	Hong Kong	China (Class 2)
Conductivity (mS m⁻¹) (agric)	<70	200-2500									
pH (aquatic life)	6.5-8.5	6.5-8.0			6.5-8.5	5.0-9.0		6-7	6.0-9.0	6.5-8.0	
DO (aquatic life)	≥5	6.6 - 8.5[h]		>5.5	>5.0	>6.0	>8	5.0-7.0	>5.5	4	6
NH4-N[c] (aquatic life)	<0.1				0.05						
NO3-2-N (eutrophication)	<0.7	<0.2			13[d]		5.0[d]	2[d]			
Total-P (eutrophication)	<0.13	0.01	0.126[f]	0.01[g]			0.05		<0.05		0.1[i]
NH4-N (human impact)	<0.05					0.05					
COD_mn (human impact)	<4				5[k]	2.5[l]	3	j			4

[a]Not for delta stations off the main stream, for which different values may apply due to salinity and pH conditions.

[b]Values are for tropical rivers.

[c]NH4 is the sum of NH3 (toxic) and NH4 (less toxic).

[d]This is for NO3.

[e]CCC is chronic concentration criteria and has more meaning for the Mekong than CMC (criteria maximum concentration).

[f]Nutrient guidelines are dependent on temperature, turbidity, color, and many other factors.

[g]The threshold between oligotrophic and mesotrophic waters is 0.01 mg l⁻¹ in the Canadian system (CCME, 2004).

[h]Country Guidelines are expressed as O2 saturation. The threshold values for O2 concentration are derived be converting saturation to concentration at 27.2 °C. Australian data are drawn mainly from North American studies and may not be entirely suitable.

[i]Chinese values are for lakes and reservoirs.

[j]Malaysia uses COD in their classification system; however, it is not known if this is COD_cr or COD_mn.

[k]Viet Nam has proposed a COD standard of 5 mg l⁻¹ for first grade water (excellent) for the MRC Basin Report Card.

[l]The Thai value is for BOD5 (1.5 mg l⁻¹). Assuming this as approximately 60% of COD, the derived COD value would be 2.5.

(27.2 °C) of the Mekong system, this corresponds to a DO level of approximately 7 mg l^{-1}. However, some fish species of the Mekong system are adapted to breathing at the surface at times of low oxygen, therefore the values used require adjustment downwards. If *Tilapia*, exotic fish widely cultivated in the Mekong system, is used as an indicator of oxygen requirements, research (Tsadik and Kutty, 1987) indicates that below 90% oxygen saturation, there is measurable decline in fish growth. It should be noted that WQMN values are obtained during the daytime and do not account for diurnal variations in oxygen, which can impact aquatic life, especially at night when O_2 is lower. However, it is reasonable to assume that aquatic life is adjusted to the natural diurnal variations in oxygen of this river system. As a result of discussion within MRC technical groups, a value of ≥ 5.0 mg l^{-1} was proposed as the limit for aquatic life. In Table 12.3, the suggested value for the Mekong is ≥ 5.0 mg l^{-1}, which conforms with classifications used in the region. This is slightly less than the threshold reported for nonsalmonid fishes (Lake Access, 2006).

4.2. Dissolved Oxygen (Human Impact)

The recommended DO guideline value for human impact is ≥ 6 ppm, which conforms to regional guidelines and is a measure of departure from "average" conditions that is attributable to human activity.

4.3. Conductivity (Aquatic Life and Water Use in Agriculture)

There is no specific threshold for salinity insofar as the effects of salinity are incremental. Chapman (1996) reports that conductivity in freshwater ranges from 1 to 100 mS m^{-1}. According to Food and Agriculture Organization, FAO (1985), values above 70 mS m^{-1} are somewhat impaired, even for irrigation. Table 12.3 notes the Australian guidelines of tropical rivers (ANZECC

and ARMCANZ, 2000) for conductivity for protection of aquatic life. However, salinity is a significant factor in many Australian rivers; therefore, these values are probably too high for use in the freshwater part of the Mekong. According to Hart *et al.* (1991), salinity above 1500 mSm^{-1} causes direct degradation of aquatic life in Australian rivers. Period of record values in the WQMN database for conductivity are highly positively skewed as they include historical values from tributaries and delta locations that have high salinity. The median value is 20.3. It seems reasonable to use a threshold value that is above the mean and median values for the Mekong, but which is lower than the Australian value to ensure protection of Mekong aquatic life. Therefore, the value of 70 mS m^{-1} suggested in Table 12.3 seems suitable to ensure that salinity is below the threshold for freshwater (70 mS m^{-1}, FAO). Mainstream stations will always be less than this guideline value. For example, My Thuan's highest recorded value is only 29.6 with a mean of 14.6 mS m^{-1}.

4.4. Phosphorus (Eutrophication)

Most guidelines use only Total-P insofar as PO_4 is unstable and difficult to analyze accurately. P is the limiting nutrient in northern hemispheric waters; however, the role of P as a limiting nutrient in the Mekong system has not been established. Some tropical aquatic systems may be N limited (Guildford *et al.*, 2003; Lehman and Branstrator, 1993; Moss, 1969). However, as Mugidde *et al.* (2003) show in African Lake Victoria, although the low TN: TP ratio indicates N limitation, eutrophication of this lake has shifted phytoplankton composition to dominance by N-fixing (and potentially toxic) cyanobacteria. It not thought that Tonle Sap lake has reached this level of eutrophication; however, there is little research on phytoplankton composition and nutrient cycling and, in any case, determination of nutrient limitation in tropical systems is greatly complicated by a wide range of light and other geochemical factors (Guildford *et al.* 2003). In the Mekong system, the high sediment concentration and low light penetration, especially during flood periods, depress the eutrophication effects of N and P. Also, the high suspended sediment concentration sequesters phosphorus into particulate-associated phosphorus that is not readily available in a fluvial environment. Algal blooms have been observed in the Mekong River in the south of Laos, however these are not documented and the chemistry at that time is not known. The US-EPA (2001) found 0.126 mg l^{-1} at the 25th percentile for Region X, which includes the Gulf Coast and lower Mississippi. Seventy-five percent of the Region X data are, therefore, more than this value. The EPA does suggest that this value may be too high (not sufficiently restrictive); however, the turbid conditions of the Mekong would probably accommodate a higher level of Total-P without greatly increasing the risk of algal blooms. Under these circumstances, a guideline of 0.13 mg l^{-1} for Total-P is recommended.

4.5. Nitrogen (Eutrophication)

Total-N is not evaluated here because of absence of data from Laos, Cambodia, and early Thai data. However, evaluation of nutrient status of the Mekong must consider Total-N so that threshold values established for NO_{3-2} are reasonable relative to Total-N and to ensure that the natural N:P ratio in the Mekong is maintained. Nitrogen is linked to phosphorus in the development of phytoplankton (algae) and in eutrophication; however, most of the research has been conducted in lakes and reservoirs and not in rivers. There appears to be no published research into trophic status of the Mekong relative to N and P and algal species/abundance. N:P ratios that are less than 10 may indicate nitrogen deficiency and ratios greater than 20 can indicate phosphorus deficiency (UNEP, 1999); however,

as noted for phosphorus (above), this relationship is extremely complicated, reflecting other variables and site-specific conditions.

The MRC data for both N and P are highly positively skewed. After eliminating stations for which there are no Total-N values and using recalculated median values, the weight ratio of N to P is approximately 10.6:1. This ratio is biased downward due to the larger amounts of phosphorus at many delta stations. If the objective is to minimize the possibility of occurrence of cyanophyte abundance (blue-green algae), a 13:1 ratio may be used as the basis for determining a basin-wide guideline for Total-N. This corresponds with the oligotrophic-mesotrophic boundary (UNEP, 1999; but is for lakes and reservoirs). Therefore, based on the suggested P value of 0.13, it seems reasonable to use a guideline value of 1.7 mg l^{-1} Total-N ($0.13 \times 13 = 1.69$) to maintain the existing balance between N and P in the Mekong. Turbidity levels of the Mekong would probably allow for a higher value of Total-N insofar as the Mekong main stem is probably light limited and not nutrient limited for algal growth.

MRC reports NO_{3-2} as a combined measure; however, at ambient pH of the Mekong it is believed that almost all is NO_3 in most locations. CCME (1995) notes that NO_3 values of 90 mg l^{-1} protect most warm-water fishes. NO_{3-2} in the MRC database has a median value of 0.28 mg l^{-1} and a maximum value of 3.31 mg l^{-1}, which is quite low. The suggested value for the Mekong is 0.7 mg l^{-1}, which is about one standard deviation above the mean value and less than the inferred threshold of 1.69 for Total-N.

4.6. Ammonia (Aquatic Life)

MRC reports total ammonia as NH_4-N (ammonium) and does not calculate NH_3-N. Unionized (free) ammonia (NH_3) is much more toxic to aquatic life than NH_4. Guidelines for NH_3 values are highly variable due to the fact that that the percentage of total ammonia that is NH_3 depends on pH and temperature. Tests on salmonids indicate that half the test fish died at NH_3 levels of 0.03 mg l^{-1}. There is no research on levels of ammonia that are toxic to Mekong fish species; however, salmonids are known to be very sensitive to ammonia in contrast to carp (*Cyprinus carpio*), which are more tolerant. At mean temperature and pH of the Mekong, for NH_3 to reach the critical threshold value of 0.03, the value of NH_4 must be at least 3.0 mg l^{-1}, which is higher than any recorded value in the WQMN database. In the data sets used here, the highest recorded value of NH_4 is 0.037 mg l^{-1}. Ammonia is normally associated with municipal waste loads; therefore, it is not surprising that ammonia is much less than any threshold value.

Ammonia in the Thai classification is set at 0.5 mg l^{-1} for all classes and may be more realistic than the 0.03 established for salmonids; however, there is no detailed justification of this value in Thai waters. The comparable value of NH_4 at mean pH of 8.2 and temperature of 27.2 is 50 mg l^{-1} and seems much too high. The Viet Nam standard for NH_3 is provided for a temperature of 20 °C and pH = 8. The temperature is the highest provided by the Vietnamese standard for protection of aquatic life, but is very low relative to actual temperatures in the delta. The Viet Nam NH_3 value converts to a total ammonia concentration of 25 mg l^{-1}, which is much higher than western standards. Taiwan has two classes suitable for fish, class B = trout, sweetfish, perch and class C = carp, grass carp and mollusks. Both have an NH_3 value of <0.3 mg l^{-1} which, for the Mekong, is approximately NH_4 = 30 mg l^{-1}. The US-EPA (1999) uses a combined NH_3 + NH_4 approach expressed as total ammoniacal nitrogen (TAN). This requires that each sample value for TAN be evaluated against the chronic concentration criterion (CCC) that is calculated for each sample date from an equation that integrates the effects of pH and temperature. The CCC approach also is linked to sampling over a

30-day period, which is not satisfied by the WQMN data and must be assessed against a 3-year period for CCC exceedance. Therefore, it is recommended that the technical value of 0.1 mg l^{-1} for NH_3-N be used for this assessment. This is equivalent to 1.0 mg l^{-1} NH_4 at mean temperature and pH of the entire database.

4.7. Ammonia (Human Impact)

For human impact assessment, the value of ammonia will be different than that described above for protection of aquatic life as the intent is not to "protect," but rather to describe deviation from some "average" or reference condition due to human influence. Based on the WQMN database of 2005, the mean and median values for NH_4-N are 0.117 and 0.034, respectively. These values are biased upward due to the number of secondary stations outside of the main stem and of major tributaries and which can be heavily impacted from local waste sources. A value of 0.04 mg l^{-1} for NH_4 is suggested as the reference value as it is more reflective of the main-stem situation and is close to the Viet Nam national standard of 0.05 mg l^{-1} for Class 1 waters.

4.8. COD_{mn} (Human Impact)

COD is a measure of organic pollution, mainly from human, industrial, and animal wastes. It is useful for measuring human impact on water quality, but has little direct relevance for aquatic life. MRC uses COD_{mn} because the values of COD are quite low, so COD_{cr} is not sufficiently sensitive. According to Chapman (1996), COD_{cr} in unpolluted waters is <20 mg l^{-1} (as COD_{mn} would be approximately <10 mg l^{-1}) and includes, for example, natural waters high in humic acids. Data for the Mekong system show the mean and median, respectively, as 3.4 and 2.8. The purpose of COD is to evaluate human effects on the river system; therefore, the main issue is how to set a guideline value that reflects impact. It seems appropriate to use the period of record mean value as an arbitrary ambient value that establishes the average level of human impact. Therefore, it is suggested that values above the mean reflect human impact above ambient. Therefore, the suggested guideline for assessing human impact is <4 mg l^{-1}, which falls between Thai and Viet Nam standards and equals the Chinese Class 2 standard.

5. METHODOLOGY

The full data set from the MRC water quality database was employed for the period of record used in this analysis. Data were initially analyzed using Excel. Box and whisker plots were prepared using the statistical package SYSTAT; in some cases, extreme values were eliminated in order not to distort the graph output, however the data were not otherwise censored for statistical purposes. The horizontal line within each box indicates the median value; the top and bottom margins of the boxes (hinges) indicate the 25th and 75th percentile values of the distribution. A useful feature of the box and whisker plot is that the whiskers extend to values that are considered, statistically, as the maximum and minimum values, but exclude extreme values as defined by values that exceed the 25th or 75th percentile value by ±1.5× the range between the 25th and 75th quartiles (called "outside values") and by ±3× the range (far outside values). These outside values are indicted with asterisks and circles, respectively, and may be anomalous or erroneous values.

Contaminant data reported here are taken from MRC (2007). These data were collected in two field campaigns, during low water periods in 2003 and 2004. As there are no other systematic contaminants data for much of the Mekong basin, the intent of that study was to provide a broad-brush examination that covered the entire lower river basin as a basis for determining potential hotspots and to establish a baseline for future contaminant studies. The sampling sites were primarily a subset of the sites used by the MRC in its routine water quality program and included all benchmark sites used here. Water and bottom sediments were sampled. For bottom sediments, grab samples were taken 100 m apart in a triangular pattern and mixed to provide an integrated sample. For volatile parameters (benzene, toluene, ethylbenzene, and xylene (BTEX), CVOC, etc.), analyses were performed on bulk sediment samples. For the other parameters, the bulk samples were lyophilized (freeze dried) and screened at 250 μm in order to homogenize the samples. Routine water samples were analyzed by a Thai government laboratory; all organic and metals analyses were carried out in certified laboratories in France. The 2003 data indicated that routine data collected by MRC is reliable; therefore, the 2004 campaign focused exclusively on contaminants.

A limited number of bioassays was available for the diagnostic study using the amphipod, *Hyalella azteca*. The measured end points according to the French standard were survival (acute toxicity) and length (growth inhibition). Full details on the diagnostic study are found in MRC (2007).

6. OBSERVATIONS AND DISCUSSION

6.1. Aquatic Life

The parameters pH, DO, and total ammonia (NH_4) are central to sustainability of aquatic life. These parameters are illustrated in Fig. 12.2. There is a gradual decline in median annual values of pH, DO, and total ammonia from upstream to downstream. Mean pH regime falls a half pH unit from upstream to downstream; stations exhibit some evidence of temporal variance, however this may be partially a result of measurement error rather than real differences. With minor exceptions, the

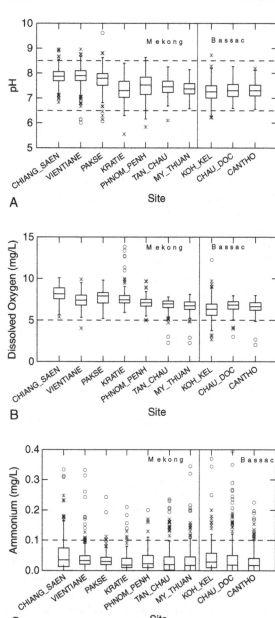

A

B

C

FIGURE 12.2 Box and whisker plots of results for parameters linked to aquatic life, 1986-2005 (1994-2005 in Cambodia). Sites are listed in order from upstream to downstream, with sites on the Bassac River indicated to the right of the vertical line. The horizontal line in each box indicates the median value and the upper and lower borders of the boxes indicate the 75th and 25th percentile values, respectively. Dotted lines indicate recommended threshold values.

data range for pH at all stations lies within the guideline values. For stations at and above Phnom Penh, the range of DO values exceeds the threshold of 5.0 mg l^{-1}. For stations below Phnom Penh, Koh Khel records values routinely less than the threshold. Other delta stations on the Bassac have outliers that are below the threshold. Ammonium (total ammonia) can exhibit substantial intra and interyear variation; however, this may be to some extent an artifact of the sampling and analysis. With the exception of the most upstream station, Chiang Saen, the Mekong stations record ammonia values generally lower than the threshold. Delta stations on the Bassac have a small part of their data distribution that exceed the threshold. In summary, stations for the most part conform to the requirements for aquatic life. Bassac River stations in Cambodia and Viet Nam tend to have higher values. The declining trend of total ammonia from Koh Khel downstream suggests that this may be influenced by Phnom Penh, which lies upstream.

6.2. Water Supply for Agricultural Use

Because of the lengthy dry season, irrigation is a prominent feature, especially for paddy rice. Agricultural land in some parts of the delta is flooded during the annual period of peak flow. Water quality for agricultural purposes is a complex issue reflecting crop type, irrigation practices, soil characteristics, and crop growth status. Water for irrigation is usually assessed for salinity, sodicity, and specific ion toxicity. As MRC has no data on sodicity nor on specific ion toxicity, both of which are site dependent, the evaluation can be done only for salinity.

Salinity guidelines for irrigation are provided by FAO and are shown in Table 12.4. According to FAO (1985), values above 70 mS m^{-1} are somewhat impaired, even for irrigation. Figure 12.3 illustrates that, for all mainstream stations, the entire data set is well below any level of concern. There is a general downward trend in salinity from Chiang Saen to the delta.

TABLE 12.4 Salinity guidelines for general irrigation and for paddy rice

Irrigation raw water	Units	Degree of restriction[a]		
		None (good)	Slight to moderate (fair)	Severe (poor)
Salinity—conductivity				
General	mS m^{-1}	<70	70-300	>300
Paddy rice	mS m^{-1}	<200	200-480	>480

Adapted from FAO (1985).
[a]None: 100% of yield; slight to moderate: 50-90% of yield; severe: ≤50% of yield.

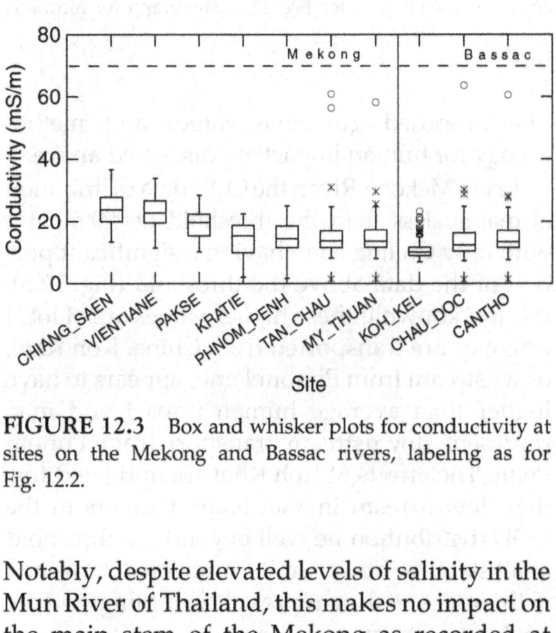

FIGURE 12.3 Box and whisker plots for conductivity at sites on the Mekong and Bassac rivers, labeling as for Fig. 12.2.

Notably, despite elevated levels of salinity in the Mun River of Thailand, this makes no impact on the main stem of the Mekong as recorded at Pakse, which is downstream of the confluence of the Mun River. Riparian farmers also use Mekong water for livestock and poultry; as the recommended threshold value is 500 mS m^{-1} (FAO, 1985), there is no limitation on main stem water for this purpose.

6.3. Eutrophication

As noted in the above discussion on threshold values, the N and P status of the Mekong relative to potential for eutrophication has received almost no research in the Mekong system. Therefore, the threshold values and their interpretation for the Mekong are uncertain. Figure 12.4 shows the distribution of P and N values for the benchmark stations. In preparing these graphs, a few extreme values were eliminated in order to better portray the data.

6.3.1. Phosphorus

Figure 12.4 illustrates that for stations in Thailand, Laos, and Cambodia, most of the data for Total-P is less than the threshold concentration. There is also a general decline in Total-P concentration from upstream to downstream, presumably reflecting dilution and relatively small sources of P between China and the Cambodian-Viet Nam border. The higher levels of Total-P at delta stations are apparent in Fig. 12.4 and reflect the large centers of population in this part of the basin.

6.3.2. Nitrogen

Values of $NO_3 + NO_2$ reported by the MRC are almost all nitrate nitrogen at prevailing pH conditions. Figure 12.4 shows that nitrate concentrations, with exception of some outliers, are much less than the threshold value of 0.7 mg l^{-1}. As for P, N also shows a downstream decrease from Chiang Saen to the Cambodia-Viet Nam border. Higher values at delta stations in Viet Nam reflect large populations in this location.

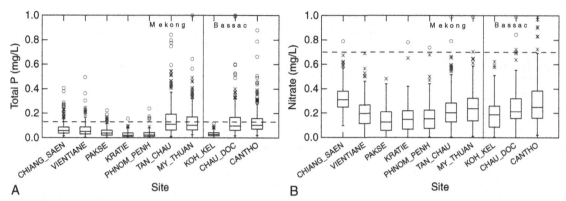

FIGURE 12.4 Nutrient conditions in the Mekong and Bassac rivers. Labeling as for Fig. 12.2. The graph for nitrate is actually nitrate + nitrite.

6.3.3. Summary

While this river system may be characterized as mesotrophic, it is unlikely that the main stems of the Mekong or Bassac rivers will produce significant levels of algae at prevailing levels of nutrients, and especially in view of light limitation from suspended sediment. Eutrophic conditions in canals of the delta may be a problem where nutrient levels can be much higher than in the main stems.

6.4. Human Impact

The objective of the four riparian countries is to maintain the good condition of water quality in the Mekong and Bassac rivers. The "human impact" indicator measures the status of water quality relative to an average condition of the river and is used to assess how individual sites perform relative to "average" conditions, and as a benchmark with which to compare future changes due to human impact. This is "NOT" a measure of water pollution per se and this assessment makes no judgment about pollution status.

Following an example from Hong Kong, an indicator of organic stress uses DO, COD (to replace BOD_5 used by Hong Kong, but which is not measured by MRC), and ammonia-N (NH_4).

The proposed guideline values and methodology for human impact are discussed above.

In the Mekong River, the COD data distribution is mainly less than the threshold of 4.0 mg l^{-1} with only Chiang Saen having a significant portion of the data above the threshold (Fig. 12.5). It is not known if these higher values are of local origin or are transported from China. Koh Khel, downstream from Phnom Penh, appears to have higher than average human impact and may represent downstream transport from Phnom Penh. The effects at Koh Khel are mitigated further downstream in Viet Nam. Outliers in the COD distribution lie well beyond the threshold at all stations; however, it cannot be determined if these are real or represent sampling and/or analytical errors.

Total ammonia has a similar downstream trend as COD with elevated levels at Chiang Saen, which decline downstream. Like COD, ammonia at Koh Khel is elevated, suggesting downstream impact from Phnom Penh.

In summary, human impact is currently not a current problem relative to the thresholds suggested. Values of both COD and ammonia at Koh Khel suggest that Phnom Penh may be a factor in water quality downstream. Notably, however, both COD and ammonia decline further downstream in the Viet Nam portion of

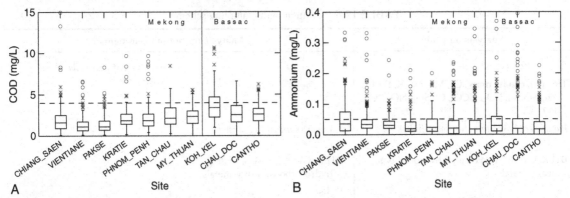

FIGURE 12.5 Parameters indicating human impacts on water quality in the Mekong and Bassac rivers. Labeling as for Fig. 12.2.

the delta. The fact that Tan Chau and Chau Doc on the Mekong and Bassac rivers, respectively, have the similar data distributions and similar medians, suggests that values observed at these two Viet Nam sites may be locally produced impacts. This is supported by the fact that relative to the rest of the lower Mekong basin, the population density in the delta region of Viet Nam is large. Anqiang Province, located on the Mekong-Bassac Rivers and containing Tan Chau on the Mekong and Chau Doc on the Bassac had a population of more than two million in 2004. Both rivers also support large fish farming operations in Viet Nam.

6.5. Contaminants

The contaminants sampling scheme of the diagnostic study (MRC, 2007) is noted in Table 12.5. For some contaminants, due to the cost of analysis, only selected sites were investigated, sufficient to permit an adequate portrayal of contaminants in the lower Mekong basin. Additional emphasis was placed on sediments due to their well-known affinity to many inorganic and organic contaminants.

6.5.1. Metals (Sediment)

Sediments were analyzed in 2003 and 2004 for arsenic, cadmium, chromium, copper, nickel,

lead, zinc, and mercury. The threshold effect concentration (TEC) and probably effect concentration (PEC) for arsenic and metals are taken from MacDonald et al. (2000) and provide an indication of the environmental significance of metals on benthic organisms. It should be noted that Mekong River bottom sediments are low in total organic carbon (TOC), barely reaching 0.1% at Tan Chau and elsewhere <0.5%.

The metal noted in Table 12.6 denotes that it exceeds an effects level in at least one of the two sampled years. It is worth noting that the diagnostic study found that a site at the Lao-China border displayed elevated levels of some metals. Also, metals levels above the TEC level were found at some other sites. More generally, however, there was an increase in metals concentrations at a station on the Tonle Sap River at Phnom Penh and at delta stations in Viet Nam, which are associated with large urban populations. Nevertheless, that study concluded that metals levels on sediments are low (indeed, many are less than detection at most sites) and are not of environmental significance at this time.

6.5.2. Metals (Water)

Metals in water were analyzed at some stations based on perceived probability of metals sources. Samples were, with three exceptions, always less than detection (zinc was detectable

TABLE 12.5 Contaminants sampling scheme

Analysis on water	Analysis on bottom sediments
Routine Parameters[a]	Bioassays with *Hyalella azteca* sediment
TSS, pH, conductivity, Ca, Mg, Na, K, Fe, Cu, Zn, Mn, Al, CO_3, HCO_3, Cl, SO_4, NO_3, NO_2, PO_4, COD	
Parameters linked to industrial pollution	*Parameters linked to industrial pollution*
BTEX, CVOC[c], PAHs, total hydrocarbons, heavy metals	BTEX, CVOC[c], PAHs, total hydrocarbons, heavy metals
	PCBs
	Dioxins and furans
Parameters linked to agriculture	*Parameters linked to agriculture*
Pesticides: organochloride, organophosphorous, triazines	Pesticides: organochloride, organophosphorous and triazines
Parameter possibly linked to Hyallella toxicity and mining activities[b]	*Parameter possibly linked to Hyallella toxicity and mining activities[b]*
Cyanide	Cyanide

Adopted from MRC (2007).
[a]Conducted only in 2003.
[b]Conducted only in 2004.
[c]CVOC in water and sediments was not detectable for 19 compounds and is not further reported in this chapter.

at two sites). Mercury is problematic as the detection limit of $0.5\,\mu g\,l^{-1}$ is far above the Canadian guideline of $0.026\,\mu g\,l^{-1}$ for inorganic mercury in water (CCME, 2002). Only at one site was the detection limit exceeded—at the most upstream site of Chiang Saen, where the observed concentration was $0.86\,\mu g\,l^{-1}$ and which is far higher than the Canadian guidelines. None of the stations in Viet Nam, for which water samples were analyzed, had detectable levels of metals.

6.5.3. Total Cyanides (Sediment)

Cyanide exists in many geochemical forms and total cyanide is not a particularly useful measure for the purpose of assessing environmental contamination. However, total cyanides were analyzed in 2004 on sediment samples at four stations (but not at the stations used in this

chapter) to provide some indication of presence/absence of cyanide in Mekong sediments. Persaud *et al.* (1992) suggest that concentrations below $0.1\,mg\,kg^{-1}$ are acceptable; US-EPA (1977) used a benchmark of $0.25\,mg\,kg^{-1}$ for classifying heavily polluted sediments. Results were all below the detection limit of $0.5\,mg\,kg^{-1}$ and are not likely to be significant for Mekong sediments. Cyanide data for water samples were not reported.

6.5.4. PCBs

Polychlorinated biphenyls (PCBs) were only analyzed on sediments as these are primarily transported in association with fine-grained mineral and organic solids. Concentrations of the seven toxic PCBs congeners on sediment were below the detection limit of $10\,\mu g\,kg^{-1}$. The probable effect level (PEL) value for total

TABLE 12.6 Summary of environmental significance of contaminants associated with sediments in main stem of Mekong and Bassac Rivers

| Site | Metals | | PCB[a] | | PAHs | Dioxins furans | Pesticides | |
	$\geq$TEC	$\geq$PEC[c]	>PEL[c]	BTEX[b]	>ISQG[d]	>ISQG[d]	Water	Sediment
Chiang Saen								?
Vientiane	As							?
Pakse								
Kratie	Hg				Yes			?
Phnom Penh								
Tan Chau	As							?
My Thuan								
Koh Khel								?
Chau Doc	As							?
Can Tho								

Empty cells mean less than a known effects level, or less than detection. Shaded rows were not included in the diagnostic study (MRC, 2007).
[a]Seven congeners were measured and are less than detection (refer to text).
[b]Relative to soil guidelines (CCME, 2002).
[c]Probable Effect Concentration (PEC) (MacDonald et al. 2000) and Probable Effect Level (PEL) (CCME, 2002) are derived differently, but have the same significance insofar as values above these levels are likely to cause biological impact.
[d]ISQG: Interim Sediment Quality Guideline (CCME, 2002).

PCBs (CCME, 2002) is 277 µg kg^{-1}. As the seven PCBs congeners analyzed are only 20-25% of total PCB, no conclusion can be drawn on their ecotoxicological significance; however as all measured values are less than the detection limit, the risk from PCB in sediments in the Mekong and Bassac rivers is probably very low.

6.5.5. BTEX (Benzene, Toluene, Ethylbenzene, and Xylene)

In both years, analyses on sediments and water were below detection (<1 µg l^{-1} for water and <0.05 mg kg^{-1} for sediment). There are no BTEX guidelines for sediments; however in Canada (CCME, 2004), the threshold concentrations in fine-grained agricultural soils are 0.03 mg kg^{-1} for benzene, 0.08 mg kg^{-1} for toluene, 0.018 mg kg^{-1} for ethylbenzene, and 2.4 mg kg^{-1} for xylene. It is concluded that

BTEX is not likely to be of significance in main stem of the Mekong.

6.5.6. PAHs (Sediment)

Total PAH (polycyclic aromatic hydrocarbon) concentrations were obtained by summing all the results of the 16 PAH congeners; nondetects were assigned a zero value. PAH congeners were less than detection at all sites with the exception of three stations. Therefore, total PAH could only be calculated for Kratie, Koh Khel, and Prek Kdam (on Tonle Sap River) with total PAHs concentrations of 55, 50, and 190 µg kg^{-1}, respectively. At Kratie, naphthalene, phenanthrene, and benzo(b)fluoranthene were measured at levels of 80, 60, and 50 µg kg^{-1}, respectively. Of these, naphthalene and phenanthrene exceed the Canadian Interim Sediment Quality Guidelines of 34.6 and 41.9 µg kg^{-1} (CCME, 2002).

6.5.7. PAHs (Water)

One or more of the congeners naphthalene, acenaphthene, and phenanthrene are found at barely detectable levels at several sites in the lower portion of the basin (Kratie, Prek Kdam on the Tonle Sap, Kho Khel, Tan Chau, and Chau Doc). None were detected at the upstream sites. As these are mainly associated with industrial production, sampling at low flow would maximize the probability of detecting these compounds.

6.5.8. Volatile Organohalogen Compounds (VOHCs)

Nineteen VOHC compounds were analyzed mainly on sediments and at several sites for water. No measurements were above detection.

6.5.9. Dioxins and Furans (Sediment)

Six dioxins (+6 toxic dioxin congeners) and six furans (+9 toxic furan congeners) were analyzed on sediment in view of their persistence in the environment and their affinity to particulate matter. Concern over dioxins is, in part, due to the widespread use of Agent Orange during the Viet Nam (American) war. Of the 210 dioxins and furans, only 17 are considered to be toxic. Toxicity of each of these 17 congeners is expressed as a toxicity equivalent factor (TEF). The concentrations of the individual compounds are multiplied by their respective TEF and are then summed to obtain a Toxicity Equivalent Index (I-TEQ, Safe and Phil, 1990). A probable effects level (similar in meaning PEC for metals and PAH) of 21.5 pg g^{-1} is proposed in the Canadian Environmental Quality Guidelines (CCME, 2002). Of the seven stations where dioxin was analyzed in the diagnostic study, except for Prek Kdam station on Tonle Sap River, all stations fluctuated around 0.1 pg g^{-1}. This is interpreted as the current background level of the Mekong and Bassac rivers for dioxins and furans and is well below the Canadian Interim Sediment Guideline of

0.85 pg g^{1}, and indicates that dioxin is not of concern in the main stem of the Mekong and Bassac Rivers. For Prek Kdam on the Tonle Sap, the I-TEQ is 0.91 and exceeds the Canadian sediment guideline of 0.85.

6.5.10. Pesticides

Twenty-four organochlorine, 9 organophosphorus, and 13 triazine pesticides were analyzed at most locations in water and sediment. Among the organochlorine pesticides were DDT and its metabolites and other banned pesticides such as aldrin and dieldrin. In the MRC study, the available detection limits for pesticides in water was adequate; however, they were too high for sediments to make meaningful conclusions about sediment-associated pesticides. Environmental guidelines are not available for all pesticides, therefore Table 12.6 notes only those for which guidelines are available. In every case, pesticides were less than detection. For water, this is significant as observed values are less than environmental effects guidelines.

6.5.11. Toxicity Assessment (Bioassay)

Only a limited number of bioassay results on one crustacean amphipod (H. azteca), was available for the diagnostic study (MRC, 2007). The choice of sites is noted in Table 12.7 and was made on the basis of perceived probability of human impacts. The bioassay was performed only on bottom sediment samples. The Lao/China border site was used in the diagnostic study to benchmark water quality as it entered the lower basin from China. Use of a single bioassay does not provide definitive information about toxicity due to the difference in response of various test organisms to different types of toxicants. Nevertheless, this bioassay at least flags potential toxicity issues that may merit further investigation.

Table 12.7 illustrates the test results for acute toxicity. Notably, two of the three samples from the Lao/China border were identified as toxic.

TABLE 12.7 Survival rates (Mean and SD) of *Hyalella azteca* after 14 days of exposure to sediment samples from seven stations sampled in 2003 and ten stations sampled in 2004

Code	2003		2004	
	Mean survival (%)	SD	Mean survival (%)	SD
Control #1	96	5.5	90	7.1
Lao/China border	30[a]	41.2		
Lao China border (upstream)			80	7.1
Lao China border (downstream)			64[a]	26.1
Chiang Saen	90	14.1	90	12.2
Luang Prabang			54[a]	11.4
Khong Chiam	98	4.5	72	16.4
Pakse	98	4.5	95	5.8
Control #2: 2003; Control #3: 2004	96	8.9	78	8.4
Lao/China border	0[a]			
Kratie	82[a]	16.4		
Neak Loeung			45[a]	12.9
Tan Chau	78[a]	1.1	52.5[a]	12.6
Koh Khel (Bassac River)			55[a]	5.8
Chau Doc (Bassac River)	96	5.5	55[a]	12.9
Prek Kdam (Tonle Sap River)			77.5	15.0

Adapted from MRC (2007).
Stations are on the Mekong mainstem unless otherwise noted, and are from upstream to downstream.
SD = Standard Deviation
[a]Significant difference with the control (Dunnett's test, $p < 0.05$ in 2003 or Bonferroni t-test, $p < 0.05$ in 2004).

This is not definitive evidence of toxicity emanating from China, but it does indicate that further work should address toxicity at this site in a more comprehensive way using both bottom and suspended sediments, and water samples. The one sample from Luang Prabang was toxic, however there are no identified sources of toxicity at or near this site. Of interest is the fact that all stations record significant toxicity downstream of Phnom Penh, both on the Mekong and the Bassac rivers. For the Viet Nam stations, there is no evidence from bottom sediment chemistry that might indicate what is causing the observed toxicity, although it is noted that the I-TEQ is slightly elevated downstream from Phnom Penh (but less than the CCME sediment guideline). This is, however, inconsistent with the nonlethality of sediments at Prek Kdam on the Tonle Sap River, but where the I-TEQ index is much higher. No definitive conclusions can be drawn from the bioassay results other than to flag certain areas that merit further investigation.

7. CONCLUSIONS

As one of the large rivers of the world, the Mekong River basin remains relatively undeveloped and without dams or diversions in the lower basin. Nevertheless, the four lower riparian countries have remained concerned about the status of water quality, but without the benefit of significant data on water quality status or trends. There is almost no reliable data on contaminants in the river. The MRC has addressed this problem though its water quality program carried out in concert with the four countries, and in 2003-2004 with the first comprehensive diagnostic study of water quality, including contaminants. That study concluded that the MRC water quality database for routine parameters is adequate.

This chapter has examined the MRC database for downstream trends in the main stems of the Mekong and Bassac rivers for a 20-year period, 1986-2005 (1994-2005 for Cambodia), and has focused on water quality for aquatic life, potential for eutrophication, water for agriculture, and human impact. As there are no water quality guidelines for the lower basin that are accepted by all four countries, recommendations are made here for threshold values as a basis for evaluating water quality. The principal observations of the diagnostic study (MRC, 2007) for contaminants are also summarized.

Median values at all stations conform to the requirements for aquatic life (pH, DO, and total ammonia), although stations in the lower part of the basin (downstream from Phnom Penh) have distributions that contain values that exceed threshold criteria. Conductivity at all main stem stations is well below the threshold limit for irrigation and livestock watering, indicating that there are no restrictions on agricultural use of Mekong River water. Nutrient parameters decline from upstream to downstream, reflecting lack of significant inputs between Chiang Saen and Phnom Penh. Values rise in the area below Phnom Penh, but it cannot be determined to what extent this is a transboundary issue, or is generated locally in the delta due to much higher amounts of anthropogenic activity in and adjacent to the Mekong and Bassac. The river system is characterized as mesotrophic using published trophic information for lakes;

it is unlikely that the main stems of the Mekong or Bassac rivers are likely to produce significant levels of algae at prevailing levels of nutrients, and especially in view of light limitation from suspended sediment. Eutrophic conditions in canals of the delta may be a problem where nutrient levels can be much higher than in the main stems. Human impact as measured by COD and total ammonia is currently not a current problem relative to the thresholds suggested. Values both of COD and ammonia at Koh Khel suggest that Phnom Penh may be a factor in water quality downstream.

The contaminants study has concluded that the main stems of the Mekong and Bassac are relatively free of environmentally significant contaminants, with many contaminants being less than detection. No sites have measurable contaminant levels that exceed international benchmarks for water or sediment contamination. Nevertheless, the evidence suggests that areas with large populations are associated with elevated levels of some metals. One site on the Tonle Sap River (Prek Kdam) that is close to wastewater sources from Phnom Penh has elevated levels of metals and some dioxins, and has an I-TEQ which slightly exceeds the Canadian sediment guideline of 0.85. Bioassay of a selection of bottom sediments using the amphipod *H. azteca* is not definitive, but does suggest that toxicity observed at the Lao/China border and at sites downstream of Phnom Penh merit additional investigation.

In summary, according to the suggested water quality threshold values, the water quality of the lower Mekong River and its distributary, the Bassac River, as measured with physico-chemical parameters, is suitable for aquatic life and for agricultural use. Because there are little or no data for microbiological parameters such as coliforms or other pathogens, this assessment can draw no conclusions on the suitability of Mekong and Bassac waters for direct human consumption or for water contact by riparian populations that use raw river water for washing.

ACKNOWLEDGMENTS AND DISCLAIMER

Water quality data, the contaminants information, and certain technical parts of this chapter, such as the discussion on water quality criteria, are used with the kind permission of the Mekong River Commission. However, the discussions and views expressed here are those of the author and do not necessarily reflect the views of the Commission or of the four riparian countries that comprise the Commission. Thanks to Dr Ian Campbell for producing the box and whisker plots and drafting of Fig. 12.1. The work of Christophe Pateron of the French company BURGEAP, which carried out the diagnostic study, is gratefully acknowledged. Dr Peter Hawkins, consultant to MRC, and Mr Pham Hien, water quality program officer of MRC, participated in a peer review process of the technical criteria that led to the suggested water quality thresholds used in this chapter. The work in the section on pollution sources is mainly taken from material prepared by the author for MRC (2007).

References

Agusa, T., Kunito, T., Iwata, H., Monirith, I., Tana, T. S., Subramanian, A., and Tanabe, S. (2005). Mercury contamination in human hair and fish from Cambodia: Levels, specific accumulation and risk assessment. *Environmental Pollution* **134**, 79-86.

ANZECC and ARMCANZ. (2000). "Australian and New Zealand Guidelines for Fresh and Marine Water Quality," National Water Quality Management Strategy Paper No. 4. Australian and New Zealand Environment and Conservation Council & Agriculture and Resource Management Council of Australia and New Zealand, Canberra, Australia, 939pp.

Campbell, I. C. (2007). Perceptions, data and river management: Lessons from the Mekong River. *Water Resources Research* **43**, doi:10.1029/2006WR005130.

CCME. (1995). "Canadian Water Quality Guidelines." Canadian Council of Resource and Environment Ministers, Ottawa, Canada.

CCME. (2002). "Canadian Environmental Quality Guidelines for the Protection of Aquatic Life." Canadian Council of Ministers of the Environment, Ottawa, Canada.

CCME. (2004). "Canadian Soil Quality Guidelines for the Protection of Environmental and Human Health." Canadian Council of Ministers of the Environment, Ottawa, Canada.

CERN (China Educational and Research Network). (undated). Work starts on Lancang River power station. http://www.edu.cn/20020125/3018726.shtml. Accessed 8 February 2008.

Chapman, D. (ed.). (1996). "Water Quality Assessments. A Guide to the Use of Biota, Sediments and Water in Environmental Monitoring," 2nd edition, UNESCO, WHO, UNEP. E. and FN Spon, London, 626pp.

China.Org. (n.d.). Xiaowan Dam: A reservoir for progress. http://www.china.org.cn/english/environment/42990.htm. Accessed 8 February 2008 (Attributed to China Daily, 16 September 2002).

CIIS (China Internet Information Services). (2002). No Pollution to Mekong River. http://us.tom.com/english/543.htm. Accessed 22 August 2006.

Dai, A and Trenberth, K. E. (2002). Estimates of freshwater discharge from continents: Latitudinal and seasonal variations. *Journal of Hydrometeorology* **3**(6), 660-687.

FAO. (1985). "Water Quality for Agriculture," FAO Irrigation & Drainage Paper No. 29 (Rev. 1). Food and Agriculture Organization of the United Nations, Rome, 174pp.

Guildford, S. J., Hecky, R. E., Taylor, W. D., Mugidde, R., and Bootsma, H. A. (2003). Nutrient enrichment experiments in tropical great lakes Malawi/Nyasa and Victoria. *Journal of Great Lakes Research* **29**(Suppl. 2), 89-106.

Hart, B. T., Bailey, P., Edwards, R., Hortle, K., James, K., McMahon, A., Meredith, C., and Swadling, K. (1991). A review of the salt sensitivity of the Australian Freshwater biota. *Hydrobiologia* **210**, 105-144. Kluwer Academic Publishers. Cited in http://www.dlwc.nsw.gov.au. Accessed 20 April 2006.

Hart, B. T., Jones, M. J., and Pistone, G. (2001). "Transboundary Water Quality Issues in the Mekong River Basin," Final Report (unpublished). Mekong River Commission, 63pp.

Lake Access. (2006). Dissolved oxygen in lakes. Reported in http://lakeaccess.org/russ/oxygen.htm. Accessed 13 April 2006.

Lehman, J. T. and Branstrator, D. K. (1993). Effects of nutrients and grazing on the phytoplankton of Lake Victoria. *Verhandlungen der Internationale Vereinigung für Theoretische und Angewandte Limnologie* **25**, 850-855.

Macdonald, D. D., Ingersoll, C. G., and Berger, T. A. (2000). Development and evaluation of consensus-based sediment quality guidelines for freshwater ecosystems. *Archives of Environmental Contamination and Toxicology* **39**, 20-31.

Moss, B. (1969). Limitation of algal growth in some central African waters. *Limnology and Oceanography* **14**, 591-601.

MRC. (1997). "Mekong River Basin Diagnostic Study," Mekong River Commission and United Nations Environment Programme, Bangkok, Thailand, 249pp.

MRC. (2003). "State of the Basin Report 2003," Mekong River Commission, Phnom Penh, 300 pp.

MRC. (2005). "Overview of the Hydrology of the Mekong Basin," Mekong River Commission, Vientiane, 73pp.

MRC. (2007). "Diagnostic Study of Water Quality in the Lower Mekong River Basin," MRC Technical Paper No. 15. Mekong River Commission, Vientiane, 57pp.

Mugidde, R., Hecky, R. E., Hendzel, L. L., and Taylor, W. D. (2003). Pelagic nitrogen fixation in Lake Victoria (East Africa). *Journal of Great Lakes Research* **29**(Suppl. 2), 76-88.

Ongley, E. D. (2006). "Water Quality Guidelines and Index: Recommendations for Mekong Water Quality Assessment," Unpublished report prepared for the Mekong River Commission, Vientiane, Laos, 28pp.

Ongley, E. D. and Wang, X. L. (2004). Transjurisdictional water pollution management in china: The legal and institutional framework. *Water International* **29**(3), 270-281.

Persaud, D., Jaagumaji, R., and Hayton, A. (1992). "Guidelines for the Protection and Management of Aquatic Sediment Quality in Ontario," ISBN 0-7729-9248-7. Ontario Ministry of the Environment, Water Resources Branch, Toronto, Canada, 23pp.

Safe, S. and Phil, D. (1990). Polychlorinated biphenyls (PCBs) dibenzo-*p*-dioxins (PCDDs), dibenzofurans (PCDFs), and related compounds: Environmental and mechanistic considerations which support the development of toxic equivalency factors (TEFs). *Critical Reviews in Toxicology* **21**, 51-87.

Tsadik, G. G. and Kutty, M. N. (1987). "Influence of Ambient Oxygen on Feeding and Growth of the Tilapia, *Oreochromis niloticus* (Linnaeus)." FAO Working Paper ARAC/87/WP/10, Port Harcourt (Nigeria).

UNEP. (1999). "Planning and Management of Lakes and Reservoirs: An Integrated Approach to Eutrophication," Technical Publication Series #11. International Environmental Technology Centre, United Nations Environment Programme, Japan, 375 pp. http://www.unep.or.jp/ietc/Publications/techpublications/TechPub-11/index.asp#1. Accessed 8 February 2008.

US-EPA. (1977). "Guidelines for the Pollutional Classification of Great Lakes Harbor Sediments." US-EPA, Region 5. As cited in BURGEAP. (2005). "Transboundary and Basin Wide Water Quality Issues in the Lower Mekong Basin," Final Report. BURGEAP Consultants. Mekong River Commission, Vientiane, 92pp.

US-EPA. (1999). "Update for Ambient Water Quality Criteria for Ammonia (Freshwater)," 147pp. EPA 822-R-99-014.

US-EPA. (2001). "Ambient Water Quality Criteria Recommendations. Information Supporting the Development of State and Tribal Nutrient Criteria. Rivers and Streams in Nutrient Ecoregion X," Publication EPA 822-B-01-016. United States Environmental Protection Agency, Washington DC, 116pp.

Xinhua New Agency. (2005). Water pollution worsening: SEPA. http://www.china.org.cn. Accessed 15 November 2007.

13

The Development and Application of Biomonitoring in the Lower Mekong River System

Ian C. Campbell,[1] *Bruce C. Chessman,*[2] *and Vincent H. Resh*[3]

[1]Principal Scientist, River Health, GHD, 180 Lonsdale Street, Melbourne 3000, Australia & Adjunct Research Associate, School of Biological Sciences, Monash University, 3800 Australia

[2]12 Patricia Place, Cherrybrook, NSW 2126, Australia

[3]Department of Environmental Science, Policy & Management, 137 Mulford Hall, University of California, Berkeley CA 94720-3114, USA

1. INTRODUCTION

The importance of the lower Mekong River to the people of the four riparian countries results from, to a large extent, the ecological products and services it supplies. People along the river depend on its fish, algae, and invertebrates as essential food and also as a source of income and trade goods (MRC, 2003). With the Mekong fishery valued at USD $1.5 billion per year (MRC, 2003, although van Zalinge *et al.* (2003) put the figure over USD $1.7 billion) and many of the fishers dependant on it having little in the way of skills or potential alternative sources of income, a loss of the ecological products of the Mekong would be catastrophic for the riparian countries. The river system also includes internationally significant biosphere reserves and endangered species (Campbell *et al.*, 2006). It therefore is vital that the ecology of the Mekong be monitored to ensure that the resource is being protected.

In the lower Mekong basin, an extensive program has monitored chemical water quality (Campbell, 2007; Ongley, this volume), but very little biological monitoring has been done. The chemical program was apparently established because the riparian countries were concerned about deteriorating water quality. Although people, both within and outside the basin, may perceive water quality as a major issue for the Mekong region (Barlow and Clarke, 2002), current evidence indicates that this is so only for restricted localities, and other far more pressing problems have emerged (Campbell, 2007; Campbell *et al.*, 2005). For example, over the next decade overfishing (MRC, 2003) and the impact of dams, including river regulation, flow modification, and barriers to fish migration (Adamson, 2001), are likely to be much greater threats to the river (Campbell, 2007; Campbell *et al.*, 2005).

Several hydroelectric dams have already been built on tributaries and even the main stream of the Mekong, several more are under construction, and others are planned. Of those on the tributaries, only that on Nam Ngum in Lao PDR has had a demonstrable effect on Mekong River flows (Interim Committee, 1988). However, in other cases including Pak Mun Dam in Thailand and Yali Dam in Viet Nam, adverse effects on human communities living along the rivers or on fish populations have been reported (Baird *et al.*, 2002; Fisheries Office, 2000; Roberts, 1993). Presently under construction are Xiaowan Dam in China, which will be the highest dam in the world on completion and the first large mainstream dam (Blake, 2001), and several more dams on the Se San River in Viet Nam (AMRC, 2002). Construction of Nam Theun II in Lao PDR appears likely in the near future, and it will also have significant effects on the mainstream. The changes in

flow that will be caused by the operation of these dams might impact the ecology of the downstream water bodies, including Tonle Sap Lake, and the fish diversity and fish catch. More detailed discussions on the potential impacts of dams appear elsewhere in this volume.

Chemical monitoring is very useful, but is not an effective surrogate for biological or ecological monitoring. The biota of rivers responds to a range of physical, chemical, and biological factors, not just chemical water quality. The earliest attempts at biological assessment of streams did focus on the responses of the biota to water chemistry, especially sewage pollution (Hynes, 1960). However, it is now understood that the biota responds to the composition of the stream bed, changes in the flow regime, the condition of the riparian vegetation and stream banks, the introduction of exotic species, the harvesting of native species, habitat modifications such as removal of woody debris, and many other stressors. Where the aim of biological monitoring is to infer chemical water quality, these other factors can confound the analysis (Resh, 1995). But when concerns are held for the maintenance of biological resources, it is desirable to monitor the overall condition of the river and its biota, not only the quality of the water in the narrow sense.

Biological monitoring of rivers has proved valuable in water resource management in both developed and developing countries (Bonada *et al.*, 2006; Hynes, 1960; Resh, 2007). It does not require complex laboratory equipment, which can be expensive to calibrate, recalibrate, and operate, and biological samples are relatively easy to preserve, or even process in the field. In some developing regions, biological monitoring can be difficult because of a lack of background information and identification keys for the biota, and a lack of capacity in local institutions for field and laboratory work (Gallacher, 2001). These problems apply less to the lower Mekong because the Mekong River Commission (MRC) has supported the development of a key for the identification of freshwater invertebrates (Sangpradub and Boonsoong, 2006), and several regional universities and research institutions employ specialists in freshwater biology.

Biological river assessment has been conducted within the greater Mekong region on a short-term, one-time basis in several projects. Thorne and Williams (1997), Mustow (1999, 2002), and Mustow *et al.* (1997) successfully used benthic invertebrates to assess the condition of the Ping River, a tributary of the Chao Phraya River, which flows through Chiang Mai in northern Thailand (outside of the Mekong River basin). An early attempt to establish a longer-term biological monitoring program in the Mekong system was made by the Mekong Committee, the precursor to the present MRC, in the 1980s. Then, monitoring sites were selected in Thailand, Lao PDR, and Viet Nam, staff from regional government agencies were sent to Sweden for training, and at least two rounds of sampling and sample analysis were conducted (Eriksson and Smith, undated; Grimås, 1988; Smith, undated). However, the program was not maintained.

The MRC was established in 1995 under an international agreement that included a provision "to make every effort to avoid, minimize and mitigate harmful effects that might occur to the environment, especially the water quantity and quality, the aquatic (ecosystem) conditions, and ecological balance of the river system, from the development and use of the Mekong River Basin water resources or discharge of wastes and return flows" (MRC, 2002). In order to demonstrate that it was fulfilling this mandate, it was necessary for the Commission to establish a program to monitor the ecological condition or health of the river directly. As a result, in 2002 the first steps were taken to develop and implement a long-term biological monitoring program.

2. THE MRC BIOLOGICAL MONITORING PROGRAM

From the outset, the MRC biological monitoring program had three objectives:

- It was intended to allow the Commission to demonstrate efficiently whether or not it was fulfilling its mandate to protect the aquatic life and ecological condition of the river.
- It was intended to operate as a basin-wide program applying, as far as technically possible, a uniform suite of indicators, sampling methods, and techniques for data analysis and interpretation to sites in all four member countries.
- It was intended to operate as a long-term program, conducted by nationals from the four countries.

2.1. Selection of Target Taxa

The first objective required the program to incorporate a broad range of organisms. People use a variety of aquatic products as food in different parts of the lower Mekong. Apart from fish, people harvest and consume a variety of aquatic organisms including turtles, algae, and invertebrates. In addition, the river system to be monitored is very large and physically variable, ranging from shallow stony riffles to deep pools and estuarine reaches. There was a concern from the outset that selecting just one narrow taxonomic group might not adequately represent the entire system of interest. A range of aquatic organisms has been used in river biomonitoring, each with its own set of advantages and disadvantages (Table 13.1), and several of these were selected for evaluation and use in the Mekong.

Of all the groups, benthic macroinvertebrates have been the organisms most frequently used in biological monitoring of rivers and streams for a number of reasons (Table 13.1) (Resh, 2008). Two habitat-based subsets of macroinvertebrates were chosen for the Mekong program: those from the littoral zone and those associated with fine sediments in the sublittoral benthos. The latter were included because while invertebrates on fine sediments might not be common in high-gradient or small tributary streams, where such

TABLE 13.1 Attributes that are listed as an advantage in at least 50% of literature sources examined for that group (modified from Resh, 2008).

Attribute for use in biomonitoring	Benthic macroinvertebrates (42 sources)	Algae (22 sources)	Fish (19 sources)	Zooplankton (9 sources)
Widespread: abundant, common, ubiquitous, etc.	X	–	–	–
Diverse: richness of species	X	–	–	X
Important to ecosystem: important trophic position or ecological role	–	–	X	X
Limited mobility: indicates local conditions, more or less sedentary	X	–	–	–
Long generation time: tracks changes over time, long-term integrators, bioaccumulate toxins	X	–	X	–
Short generation time: rapid responses to change, quick recovery from disturbance	–	–	–	–
Recreation value/consumed as human food: economic value	–	–	X	–
Easy taxonomy: easy to identify, good taxonomic keys available	–	–	X	–
Easy sampling: low field effort	X	X	–	–
Pre-existing information: good information on pollution tolerance, response to disturbance	–	–	X	–
Field examination possible: at least partially processed or identified in the field/useful for volunteers	–	–	–	–
Historical conditions can be indicated: fossils can reflect past conditions	–	–	–	–
Sensitivity to nutrients and herbicides: show sensitivity to specific plant stressors	–	X	–	–

habitats are less abundant, they might be a critical component of the ecosystems in downstream sites where fine sediments are the major benthic habitat. Conversely, the littoral habitats tend to be diverse in the upstream regions, but are often modified and less diverse in downstream reaches of a large river like the Mekong.

Filamentous macroalgae are harvested for human consumption in some parts of the Mekong River system, especially in Lao PDR around Luang Prabang. Consideration was given to assessing macroalgae, but quantification was

difficult at all sites, and often macroalgae were absent. Benthic diatoms were included in the program as an alternative representative of the primary producers. Diatoms have been used elsewhere in river biomonitoring (Whitton and Kelly, 1995), although not as frequently as macroinvertebrates.

The zooplankton has been used often in lake biomonitoring, especially via indices that indicate nutrient enrichment (Gannon and Stemberger, 1978; Whittier and Paulsen, 1992), but seldom in biological assessment of rivers.

However, biological assessment has rarely been attempted in rivers as large as the Mekong. Most riverine biomonitoring programs have involved small streams where the potamoplankton is not well developed. Given the indicative value of rotifers, in particular in lentic waters, and their ubiquity in the potamoplankton (Hynes, 1970), the zooplankton was seen as a potentially valuable component of the Mekong program.

Several other potential components were tested in a pilot study in 2003, but then abandoned, mainly because the time they required would have severely limited the number of sites that could be sampled. For example, primary production was tested during the trial year, but incubations required at least 4 h, and the additional information gained was not judged sufficiently useful. Fish could have been an important component, being a diverse assemblage of great social and ecological significance. However, experience in the pilot showed that setting nets and traps overnight would be needed to obtain an adequate sample. Rapid methods such as electric fishing could not be employed for reasons of public safety, logistics, and cost, and because locals might not have accepted that the fishing was for scientific purposes. It was hoped that a separate program might be established to assess changes in fish populations by other means such as monitoring of subsistence and commercial catches. However, this has not yet happened.

To ensure that the program operated as a basin-wide activity, a single multinational team conducted all sampling and analysis. This was in part a necessity, because there were insufficient experts in a single country to assemble national teams capable of conducting the full program. However, it was also seen as desirable both to promote international collaboration and to ensure that uniform sampling and analysis procedures were employed across the entire basin.

2.2. Design of the Sampling Program

Sites were selected on several criteria:

- It was necessary to evaluate sites that might be subject to particular stress, and so sites were selected downstream of major cities such as Vientiane and Phnom Penh, and on dammed tributaries such as the Se San and Nam Ngum.
- It was necessary to establish some reference sites on streams expected to be in a near-natural ecological condition and likely to remain in that condition, such as the Nam Ou in Lao PDR.
- Sites were selected in places that may be impacted by future development, such as the Sre Pok and the main stream near the border between Lao PDR and China.

The sampling program was intended to evaluate the current overall condition of the river system and provide a baseline for the future assessment of long-term changes; it was not expected that most sites would demonstrate great change in the first few years. However, four or five of the sites sampled in each year were also sampled in the following year, to allow for interyear comparisons. Because of budgetary constraints and limits on the time available to those involved, sampling was restricted to a period of about 3 weeks toward the end of the dry season in March each year. This allowed sampling of 16-21 sites per year.

Sampling only in the dry season meant that the data would reveal nothing about within-year variation at any site, but that was not an objective of the program. Sampling in the wet season would have been technically very difficult and even dangerous because of fast currents. The dry season was also arguably the time when the biota were likely to be under the greatest pressure from poor water quality, given the higher water temperatures and reduced dilution.

Routine sampling began in the dry season of 2004, and to the end of 2007 a total 51 sites had been sampled, 20 of them more than once. At each site, the sampling protocol was as follows:

- Littoral invertebrates were collected with a D-frame sweep-net, mesh size 475 μm, at up to 10 locations separated by about 20 m along one side of the river. Samples were sorted in the field, with all the invertebrates being picked out. Invertebrates were preserved for return to the laboratory.
- Invertebrates on soft substrata (mud, sand, and detritus) were collected with a Petersen grab, which samples 0.025 m² of the substratum. At each sampling site, 5 samples, each consisting of 4 grabs, were taken in midstream (if possible) and near each bank. If the substratum proved unsuitable for the grab at any point, a new one was selected. The samples were passed through a sieve, mesh size 300 μm, and all animals were picked out in the field where they were preserved for return to the laboratory.
- Zooplankton was collected in a 10 L sample of water from near the surface in midstream, and one near each bank of the river. Samples were filtered through a plankton net, mesh size 20 μm, and retained material was preserved for return to the laboratory.
- Benthic diatoms were sampled from 10 points at intervals of about 10 m along one shore. At each point, a stone or other solid object that had abundant algal cover was selected, and the algae from an area of 10 cm², defined with a plastic sheet mask, were brushed into a collecting bowl. The contents of the bowl were poured into a sample jar and preserved. In the laboratory, each sample was cleaned by digestion and a subsample of the diatom frustules was mounted on a microscope slide. The frustules were counted and the count was used to estimate density on the stone.
- All specimens were identified to the lowest practical level, normally genus or species.

Data were processed in a number of ways. A variety of indices of abundance, species diversity, species dominance, species richness, and tolerance were tested, but only three of these proved satisfactory indicators: number of individuals per sample (abundance), number of taxa per sample (richness), and average tolerance score per taxon (ATSPT). Tolerance scores were calculated for each taxon based on the average of a visually based site disturbance value determined at each site at which the taxon occurred.

Sites were evaluated through comparison with data from reference sites. In a populous region like the Mekong basin that has been inhabited by humans for millennia, it is not possible to find an adequate range of reference sites, on different types and sizes of streams, that are in completely pristine surroundings. So criteria were developed for designation of reference sites based on water chemistry (pH between 6.5 and 8.5; EC < 70 mS m^{-1}; DO > 5 mg l^{-1}), absence of evident major human local site disturbance (e.g., sand and gravel extraction, in-stream aquaculture, waste disposal, heavy boat traffic), and an absence of major upstream disturbance (no major dams or towns within 20 km and an absence of interbasin water transfers). Hence, data collected from these sites exemplified the biota expected under conditions of low development. Of 51 sites sampled in total, 14 met the criteria for reference sites (Fig. 13.1).

Index data for the reference sites were then analyzed and guideline values determined (Table 13.2). The guideline values were set at the 10th percentile values for the reference sites for the number of taxa and abundance and the 90th percentile value for ATSPT. ATSPT values

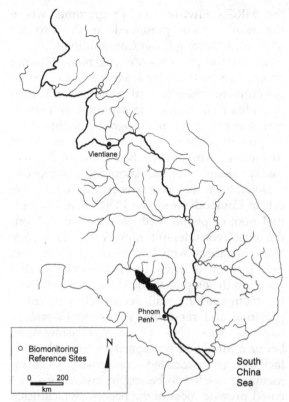

FIGURE 13.1 A map showing the localities of biomonitoring reference sites in the lower Mekong basin (after MRC, 2007).

TABLE 13.2 Indicator metric guidelines for four monitoring groups and their values at reference sites

Indicator metric	Median value at reference sites	Guideline value
Diatom richness (taxa/ subsample)	9.30	>6.54
Diatom abundance (number/subsample)	257.30	>136.22
Diatom ATSPT	35.58	<38.38
Zooplankton richness (taxa/sample)	12.67	>9.80
Zooplankton abundance (number/sample)	52.33	>22.33
Zooplankton ATSPT	38.58	<41.80
Littoral macroinvertebrate richness (taxa/sweep sample)	11.4	>5.37
Littoral macroinvertebrate abundance (number/ sweep sample)	124.8	>46.68
Littoral macroinvertebrate ATSPT	30.72	<33.58
Benthic macroinvertebrate richness (taxa/sample)	3.87	>1.84
Benthic macroinvertebrate abundance (number/ sample)	18.33	>4.13
Benthic macroinvertebrate ATSPT	35.36	<37.74

Low values were considered indicative of harm for all metrics except the average tolerance scores (ATSPT) for which high values were considered indicative of harm.

exceeding the threshold would indicate harm while, for the other two indices, values below the threshold indicate harm.

All sites were evaluated by comparison with the guideline values. Each of the three metrics for each of the four taxonomic groups was compared with guideline values, so for each site there were 12 guideline comparisons. Sites were rated based on the number of metrics that met or exceeded the guideline values. A four-level classification system was adopted. Class A, sites with 10 or more metrics meeting the guidelines, could be considered excellent, near pristine with unimpaired biodiversity. Class B, sites with 7-9 metrics meeting the guidelines, could be considered satisfactory, with a reduced level of biodiversity, but with most biological attributes in good condition. Class C, sites with 4-6 metrics meeting the guidelines, could be judged moderately impacted, with biodiversity notably reduced, most of the species present pollution tolerant and some biological attributes in poor condition. Class D, sites with 3 or fewer metrics meeting the guidelines, could be considered severely impacted, with a much reduced biodiversity, few pollution-sensitive species, and many biological attributes in poor condition. Where sites had been sampled in more than 1 year, the mean number of metrics meeting the guidelines was used to allocate the site to a category.

2.3. Mentoring and Project Development

A common approach to capacity building, in the Mekong region and elsewhere, has been intensive short-term training, either in workshops or through short courses in developed countries. Because of doubts about the efficacy of that approach, an alternative model was sought, that focused on developing high-level skills rather than acquiring content, and involved learning by doing.

Mentors were used as part of an alternative strategy to try to build the capacity to make the monitoring sustainable. The sampling, field measurements, and specimen identification were done by specialists recruited from the region, who already had field skills and taxonomic expertise, and were based in specialist research institutions or universities. Two international mentors, both with extensive experience in biological assessment of rivers, provided additional expertise. The regional experts, mentors, and specialists from the MRC secretariat jointly developed the overall program, in consultation with the National Mekong Committees in each country. In each year, at least one expert from each country and one mentor were present throughout the field component. The regional experts then took their samples back to their home laboratories for identification, maintaining contact with the mentors and MRC secretariat by email. The whole team reconvened in July or August each year for a 5-7-day workshop to jointly analyze and write up the data, and convened again at the annual technical meeting of

the MRC's Environment Programme, where the results were presented to the broader MRC and Mekong Basin Community.

Unlike the previous Mekong biomonitoring program, which selected team members from government chemical laboratories and then provided basic training in sampling and specimen identification, this program sought, as far as possible, team members from specialized institutions who already had many of the necessary skills. The selection of appropriate regional team members was critical to the project. In Cambodia and Lao PDR, there is a limited pool of people with relevant skills. From the outset, considerable effort was made to seek team members who had skills in taxonomy, field biology, or environmental science so that program development could focus on developing biomonitoring skills, especially data interpretation and reporting. Access to laboratory facilities was also a major consideration, because MRC had no capacity to provide such facilities or associated equipment, so team members needed to belong to institutions that could provide most of the necessary facilities.

Potential team members who already had higher degrees were sought in the four countries, and a decision was made to select team members from Cambodia and Lao PDR first, and then select team members from the countries with stronger capacity to "fill the gaps." It probably would have been possible to assemble the entire range of skills from either Thailand or Viet Nam, but that was not true for Cambodia or Lao PDR.

The mentors were a key element of the program. They contributed teaching and training, based on their experience in biomonitoring, which was important because the regional team members had little prior exposure to large-scale biomonitoring programs. The mentors were also aware of, and had access to, international literature and practice in biomonitoring, which was thus passed on to the regional experts. The mentors also assisted with quality assurance and the timing of their visits, set well in advance, helped to ensure the project remained on schedule.

For a mentoring approach to be successful, it needs to be long term. The team members need time to absorb a range of new knowledge and to develop new skills. These goals require time for practice and feedback. With team members who are not full-time participants in the program, practice requires additional time.

The project was never explicitly based on "training of trainers." However, many of the team members contracted were university academics active in teaching, and some regional members are training graduate students in biomonitoring. In general, the training of trainers probably works best when those initially trained are professional teachers to begin with.

The project can be evaluated against several criteria. The first is whether it succeeded in conducting a basin-wide evaluation and produced reliable data. The second is what the project tells us about the condition of the Mekong River, and whether the results can inform the management process. The third is whether the manner in which the project was structured and conducted proved to be an effective means of implementation.

3. RESULTS

The samples collected from the 51 sites sampled over the 4 years (77 sampling events) included 177 taxa of diatoms, 207 zooplankton taxa (mostly species level), 361 taxa of littoral macroinvertebrates (mostly genera), and 177 taxa of sublittoral benthic macroinvertebrates (mostly genera), some of which were the same as some of the littoral taxa. These numbers attest to the high aquatic biodiversity of the Mekong River system. Altogether, over 400,000 individual organisms were collected and identified. Full technical reports on the

"ecological health" of the lower Mekong have been prepared (including lists of taxa collected, details of analysis methods, and ranking of sites) and were published by the MRC in 2009. This analysis differs from those of the MRC in that we have used the average results for sites which have been assessed more than once, while they used the most recent result.

In most cases, few indicators lay outside of the guidelines and none of the 51 sites was rated in the worst category, category D, that signifies severe harm to the ecosystem. 15 sites were identified as Class A, excellent (Fig. 13.2), and 25 sites were categorized as Class B, satisfactory condition. 11 sites were categorized as Class C, moderately impacted. 12 of the

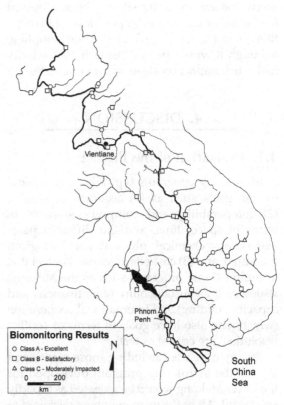

FIGURE 13.2 A map showing locations and ratings of the biomonitoring sites evaluated between 2004 and 2007 (after MRC, 2007).

reference sites were identified as Class A and 2 as Class B. Reference sites can occasionally fall in Class B because of inherent natural variability or human impacts that were not recognised when reference sites were selected.

Four of the Class C sites were located in the delta in Viet Nam, and two were located near Phnom Penh in Cambodia. These were all affected by urbanization and/or intensive agriculture and aquaculture, as was a site on the Stoeng Sangke River in Cambodia, one on the upper Se San River in Viet Nam and another on the Songkrahm River in Thailand. One site on the Nam Kam River in Thailand was impacted by a series of upstream weirs. It is not clear why the Mekong at Xieng Kok should rank so poorly. The river has extensive bedrock through that reach, and areas near the site had been subjected to blasting as part of the Upper Mekong Navigation project a few weeks before the sampling, although it would be surprising if that activity had much impact on algae and invertebrates.

4. DISCUSSION

4.1. Evaluation of the Project

Monitoring programs typically have several overall goals such as (1) accuracy of results, (2) comparability and comprehensiveness in terms of space, time, and the different parameters (e.g., chemical, physical, and biological) measured, and (3) cost-effectiveness. We feel that in multinational programs such as the Mekong, issues such as sustainability (e.g., financial and capacity building) and international cooperation (which may also have goals in terms of conflict resolution) are critical as well.

By any measure (including those mentioned above), the monitoring program developed for the lower Mekong would be assessed as broadly successful. All of the team members recruited to conduct the routine sampling have continued to participate in the project. Many sites have been surveyed, almost half on more than one occasion. The samples collected have been processed in a timely manner each year, and several technical reports have been published and are available through the MRC website.

Not all of the original goals of the project have yet been achieved. Some technical reports from particular years of the program are still in the process of publication. Other limitations include the absence to date of formal quality control processes in the taxonomic work, and the absence of any formal independent review of the project itself.

4.2. Evaluation of the River

The findings suggest that generally the aquatic ecosystems of the Mekong River and its major tributaries are not suffering major impacts from development. This can be attributed to the high volumes of flow in the rivers and consequent large capacity for dilution of wastes, the scarcity of major discharges of sewage and industrial wastes (sewage often being passed through wetland complexes before it reaches the river), and the fact that the construction of major dams is only now becoming widespread.

Estuaries and the lower reaches of watercourses are often the most stressed, because they are downstream of all the major stressors, including dams and pollution sources. They are also often the most intensively settled and utilized being flat, with water and fertile soils and close to the sea and ports.

The lower Mekong is no exception to this pattern. It has the highest population density in the basin and a wealthier agricultural sector that can afford fertilizers and pesticides. The acid sulfate soils of the delta require heavy fertilizer use to generate high productivity, with the consequent risks of fertilizer runoff to the river. Few of the towns and villages have sewage treatment systems, and there is intensive aquaculture based on cages in the river, which would be an added source of nutrients. Campbell (2007) identified nutrient enrichment of the delta as the only substantial water quality issue in the lower Mekong, based on an analysis of the MRC water quality database, so it is not surprising that the biological assessment should also detect moderate impact in this part of the river.

The impacts in Cambodia are more localized. Phnom Penh discharges most of its runoff and sewage effluent into wetlands north and south of the city. However, Phnom Penh Port does receive some local urban runoff, and presumably spillages during the loading and unloading processes for the ships. There is also seasonal, but substantial, fish processing carried out on the river banks associated with the dai fishery from December through to February each year, which may impact the river.

The data produced through the bioassessment program provide a solid basis against which future change in the basin can be evaluated. The Mekong basin will undergo substantial environmental change over the next 10-20 years. There will be rapid population growth, and rapid growth of cities such as Phnom Penh, Siem Reap, Vientiane, and Mukdahan as population, trade, and tourism grow. There are a number of large dams under construction in China, Viet Nam, and Laos and many more under discussion. These changes are all likely to impact the biota of the Mekong River, and the data already generated by the bioassessment program will allow the extent of change to be robustly assessed.

4.3. Evaluation of the Approach

The engagement of an international team of experts from the four countries and beyond, rather than individual national teams or a team consisting solely of foreign consultants, was extremely successful. Within the Mekong region, there is still residual mistrust among countries arising from both their recent and longer term political histories. One useful role of the MRC within the region is seen as building understanding between countries, and this project, by using a multinational approach rather than only four national teams, made a small contribution to that process.

The project also built on existing expertise in the region and ensured uniformity of methods so that results from different countries were truly comparable. The presence of citizens of each country throughout the field program each year ensured smooth logistics and effective communication with local officials and communities. The process of data analysis and interpretation also benefited greatly from the breadth of local knowledge brought by the national experts. In addition, their participation built further capacity within the region, helped greatly in communication of the results to others within their respective countries, and lent additional credibility to the findings.

The mentoring approach has been well received by participants in this and other similar exercises conducted by the MRC (Hart, 2004). Unfortunately, the approach has not been formally evaluated in this project, although Hart (2004) did so, but written feedback from several participants identified the opportunity to learn and gain confidence from the project as being far more valuable than could be achieved through workshops. They also believed that achieving the right combination of mentors and participants was important to achieving a successful outcome with this approach. However, mentoring is a long-term approach to training, and it may have limited appeal to donor agencies that often operate on shorter funding cycles than the 5 years of this project. The approach is also more susceptible to disruption if either the mentors or other members of the project team become unavailable over the longer period that the project operates.

4.4. Evaluation of the Program

A variety of evaluation techniques has been proposed to examine the technical aspects of biological monitoring programs, and Yoder and Barbour (2009) recently developed a list of critical technical elements that can be used in such evaluations. Using their principles, we examined 13 elements for the lower Mekong program and produced the following appraisal of each component:

(1) *Was the period of data collection consistent and appropriate to provide useful information?* All collections were made in March, at the low-flow period of the rivers in the region. This was an advantage for logistical considerations, but also provided consistency of results.

(2) *Was the spatial coverage adequate?* Collections were made from just below the Lao PDR-Chinese border to the most downstream freshwater areas before the river enters the South China Sea. Collections were not made annually at all sites, but were repeated at several sites to examine the possibilities of short-term annual variability.

(3) *Was a natural classification of habitats done?* Sites were selected in each of the subcatchments of the river, representing a variety of river types (except small streams). However, analysis showed that biological metrics were weakly related to natural gradients, so a classification was not attempted.

(4) *Were reference conditions established?* Reference conditions were established for all types of streams except the Mekong Delta, an area for which no reference sites were available for examination. However, because gradients of conditions were not evident, it was felt that the reference conditions developed in more upstream areas could suffice for the delta.

(5) *Were criteria for reference sites established?* These were established and involved assessments of human disturbance and water quality.

(6) *Was the taxonomic resolution adequate?* The program used the finest levels of taxonomic resolution possible with available keys, usually genus or species, for all four biological assemblages. The same national experts did the identifications throughout the program. Unfortunately, taxonomic quality assurance was not done to check the accuracy of the identifications.

(7) *Were sample collections done consistently?* Written protocols, training and overview by the mentors were used to maintain consistency throughout the program.

(8) *Was sample processing done consistently?* As above, protocols, training and overview of on-site processing were used.

(9) *Was data management accurate and consistent?* All data were stored in spreadsheets. Data checking was limited because of logistical constraints, however, each year, an August workshop was held to check and analyse data.

(10) *Were the correct ecological attributes measured?* A variety of metrics (richness, abundance, diversity and dominance) was calculated. Moreover, specific tolerance metrics were developed for the lower Mekong biota. The final choice of attributes was based on testing against independent data.

(11) *Were biologically significant endpoints or classifications developed?* In cooperation with the National Mekong Committees, a ranking system that led to an "Ecological Health Report Card" was developed.

(12) *Were diagnostic capabilities developed and used to evaluate the cause of impacts detected?* Although data were collected that could be used for this purpose, it was not done (but could be in the future).

(13) *Were there professional reviews?* Reviews were done by the National Mekong Committees and their professional staff, but there were no external reviews.

The above analysis suggests that the program included most of the critical technical elements. One of the elements that was not fully implemented, the verification of taxonomic identification, can be done in the future. However, the lack of reference conditions for the Mekong Delta is problematic because riverine sites that could qualify as reference conditions do not exist for this most downstream area of the lower Mekong.

5. CONCLUSIONS

Referring back to the objectives of the lower Mekong bioassessment program, the program has succeeded in establishing and operating as a basin-wide program applying a uniform suite of indicators, sampling methods, and techniques for data analysis and interpretation to sites in all four member countries. The results of the program demonstrate that, to a large extent, the aquatic life and ecological condition of the river have been protected throughout much of the basin. However, there have been serious declines in several high-profile species such as dolphins and giant fish (the Mekong catfish, giant barb, Jullien's golden carp, and others) (Poulsen *et al.*, 2002), and the delta is clearly suffering from an appreciable level of ecological stress.

It is too soon to judge whether bioassessment will successfully operate as a long-term program, conducted by nationals from the four countries. At the time of writing, the MRC bioassessment is being transferred to the individual countries, where training will be provided by the national experts, and the role of the mentors is not yet clear. The intention is that in future it will be run by separate national assessment teams in a similar fashion to the ongoing chemical monitoring program. Our hope is that the national teams will be able to maintain the program and continue to produce high quality data that are made publicly accessible.

ACKNOWLEDGMENTS

The bioassessment work described in this chapter was a project of the MRC. Many people assisted with the field work and the collection and analysis of samples and results including Ms Jane Lee, Mr Meng Monyrak, and Ms Arounna Vongskhamphouy from the MRC secretariat; Dr Supatra Parnrong Davison (Prince of Songkhla University, Thailand), Dr Tatporn Kunpradid, Ms Sutthawan Suphan, and Dr Yuwadee Peerapornpisal (Chiang Mai University, Thailand); Ms Nguyen Thi Mai Linh and Mr Pham Anh Duc (Institute of Tropical Biology, HCMC, Viet Nam); the late Dr Bounnam Pathoumthong and Mr Chanda Vongsambath (National University of Laos); and Mr Sok Khom (Cambodian National Mekong Committee).

References

Adamson, P. (2001). Hydrological perspectives of the Lower Mekong. *International Water Power and Dam Construction*, 16-21.

AMRC. (2002). Wider plans in the Se San River basin. Australian Mekong Resource Centre. http://www.mekong.es.usyd.edu.au/case_studies/sesan/widerplans/sesan3.htm. Accessed 30 November 2007.

Baird, I., Baird, M., Cheath, C. M., Sangha, K., Mckradee, N., Sounith, P., Nyok, P. B., Sarim, P., Savdee, R., Rushton, H., and Phen, S. (2002). "A Community-Based Study of the Downstream Impacts of the Yali Falls Dam Along the Se San, Sre Pok and Se Kong Rivers in Stung Treng Province, northeast Cambodia," Se San Protection Network Project and Partners for Development, Stung Treng, Cambodia, 89pp.

Barlow, M. and Clarke, T. (2002). "Blue Gold. The Battle Against Corporate Theft of the World's Water," Earthscan Publications, London, 278pp.

Blake, D. (2001). Proposed Mekong dam scheme in China threatens millions in downstream countries. *World Rivers Review*, 4-5.

Bonada, N., Prat, N., Resh, V. H., and Statzner, B. (2006). Developments in aquatic insect biomonitoring: A comparative analysis of recent approaches. *Annual Review of Entomology* **51**, 495-523.

Campbell, I. C. (2007). Perceptions, data and river management—Lessons from the Mekong River. *Water Resources Research* **43**, doi:10.1029/2006WR005130.

Campbell, I. C., Barlow, C. G., and Pham, G. H. (2005). Managing the ecological health of the Mekong River:

Evaluating threats and formulating responses. *Internationale Vereinigung für Theoretische und Angewandte Limnologie Verhandlungen* **29**, 497-500.

Campbell, I. C., Poole, C., Giesen, W., and Valbo-Jorgensen, J. (2006). Species diversity and ecology of Tonle Sap Great Lake, Cambodia. *Aquatic Sciences* **68**, 355-373.

Eriksson, L. and Smith, S. (undated). "The Chironomid Fauna in the Mekong River and its Tributaries in the Plain of Reeds," Report to the Interim Mekong Committee, 27pp.

Fisheries Office. (2000). "A Study of the Downstream Impacts of the Yali Falls Dam in the Se San River Basin in Ratanakiri Province, Northeast Cambodia," The Fisheries Office, Ratanakiri Province in cooperation with the Non-Timber Forest Products Project, Ratanakiri Province, Cambodia, 41pp.

Gallacher, D. (2001). The application of rapid bioassessment techniques based on benthic macroinvertebrates in East Asian rivers (a review). *Internationale Vereinigung für Theoretische und Angewandte Limnologie Verhandlungen* **27**, 3503-3509.

Gannon, J. E. and Stemberger, R. S. (1978). Zooplankton (especially crustaceans and rotifers) as indicators of water quality. *Transactions of the American Microscopical Society* **117**, 16-35.

Grimås, U. (1988). "Water Quality Investigations in the Lower Mekong Basin. Biological Monitoring: An Evaluation," Report to the Interim Mekong Committee, 37pp.

Hart, B. T. (2004). "Ecological Risk Assessment Training Program. MRC Environment Program Activity B-0009," Mentor Final Report. Water Studies Centre, Monash University, Clayton, Australia. 41pp.

Hynes, H. B. N. (1960). "The Biology of Polluted Waters," Liverpool University Press, Liverpool, UK, 202pp.

Hynes, H. B. N. (1970). "The Ecology of Running Waters," Liverpool University Press, Liverpool, UK, 555pp.

Interim Committee. (1988). "Water Balance study Phase 3 Report and Appendices, Investigation of Dry Season Flows," Report by the Institute of Hydrology, Wallingford, UK to the Interim Committee for Coordination of Investigations of the Lower Mekong Basin (ICCLMB), Bangkok, Thailand, 235pp.

MRC. (2002). "Agreement on the Cooperation for the Sustainable Development of the Mekong River Basin, 5 April 1995," Fourth Reprint, Mekong River Commission Secretariat, Phnom Penh, 19pp.

MRC. (2003). "The State of the Basin Report 2003," Mekong River Commission, Phnom Penh, 300pp.

MRC. (2007). "Report Card on the Aquatic Ecological Health of the Mekong River System; Principles and Application," Draft Report, September 2007. Mekong River Commission, Vientiane, 22pp.

Mustow, S. E. (1999). Lotic macroinvertebrate assemblages in northern Thailand: altitudinal and longitudinal distribution and the effects of pollution. *Natural History Bulletin of the Siam Society* **47**, 225-252.

Mustow, S. E. (2002). Biological monitoring of rivers in Thailand: Use and adaptation of the BMWP score. *Hydrobiologia* **479**, 191-229.

Mustow, S. E., Wilson, R. S., and Sannarm, P. (1997). Chironomid assemblages in two Thai water courses in relation to water quality. *Natural History Bulletin of the Siam Society* **45**, 53-64.

Poulsen, A. F., Ouch, P., Sintavong, V., Ubolratana, S., and Nguyen, T. T. (2002). "Fish Migrations of the Lower Mekong River Basin: Implications for Development Planning and Environmental Management," MRC Technical Paper No. 8. Mekong River Commission, Phnom Penh, 62pp.

Resh, V. H. (1995). Freshwater benthic macroinvertebrates and rapid assessment procedures for water quality monitoring in developing and newly industrialized countries. *In* "Biological Assessment and Criteria," Davis, W. S. and Simon, T. (eds.), pp 165-175. Lewis Publishers, Chelsea, Michigan.

Resh, V. H. (2007). Multinational, freshwater biomonitoring programs in the developing world: Lessons learned from African and southeast Asian river surveys. *Environmental Management* **39**, 737-748.

Resh, V. H. (2008). Which group is best? Attributes different biological assemblages used in freshwater biomonitoring programs. *Environmental Monitoring and Assessment* **138**, 131-138.

Roberts, T. R. (1993). Just another dammed river? Negative impacts of Pak Mun dam on fishes of the Mekong basin. *Natural History Bulletin of the Siam Society* **41**, 105-133.

Sangpradub, N. and Boonsoong, B. (2006). "Identification of Freshwater Invertebrates of the Lower Mekong River and Its Tributaries." Mekong River Commission, Vientiane, Lao PDR.

Smith, S. H. (undated). "Bottomfauna Monitoring in the Lower Mekong Basin. January-March 1987," Mission Report to the Interim Mekong Committee, 32pp.

Thorne, R. St. J. and Williams, W. P. (1997). The response of benthic macroinvertebrates to pollution in developing countries: A multimetric system of bioassessment. *Freshwater Biology* **37**, 671-686.

Van Zalinge, N., Degen, P., Nuov, S., Jensen, J. G., Nguyen, V. H., and Choulamany, X. (2003). The Mekong River system. *In* "Proceedings of the Second International Symposium on the Management of Large Rivers for Fisheries," Volume 1, Welcomme, R. L. and Petr, T. (eds.), pp 335-357. FAO and Mekong River Commission, Phnom Penh.

Whittier, T. R. and Paulsen, S. G. (1992). The surface waters component of the environmental monitoring and assessment program (EMAP): An overview. *Journal of Aquatic Ecosystem Stress and Recovery* **1**, 119-126.

Whitton, B. A. and Kelly, M. G. (1995). Use of algae and other plants for monitoring rivers. *Australian Journal of Ecology* **20**, 45-56.

Yoder, C. O. and Barbour, M. T. (2009). Critical technical elements of state bioassessment programs: A process to evaluate program rigor and comparability. *Environmental Monitoring and Assessment* **150**, 31-42.

CHAPTER

14

Watercourse Environmental Change in Upper Mekong

Daming He, Ying Lu, Zhiguo Li, and Shaojuan Li

Asian International Rivers Center (AIRC), 6th Floor of Wenjin Building, Yunnan University,
2 North Road of Green Lake, Kunming, Yunnan 650091, P.R. China

OUTLINE

1. GENERAL

1.1. Overview

The Lancang-Mekong River flows through six countries including China, Myanmar, Laos, Thailand, Cambodia, and Viet Nam. Its mainstream length is approximately 4880 km and the watershed area is 810,000 km². Within China, the river flows through three districts of Qinghai Province, Tibet Autonomous region, and Yunnan Province, and then flows out of China at the Nanla River outlet in Xishuangbanna prefecture in southern Yunnan.

The river originates in China on the north side of The Tang-gu-la Mountains, from the base of a small glacier on the Guo-zong-mu-cha Mountain in Zaduo County, Qinghai Province in China. The altitude of the source is 5244 m and the location is 94°41′E and 33°42′N

TABLE 14.1 Regional differences of the upper Mekong watershed in China

	Qinghai	Tibet	Yunnan	Total
Area (km^2)	38,700	38,500	90,200	167,400
Length of mainstream (km)	448	465	1217	2130
Watershed area (%)	21	22	57	100

(Zhou *et al.*, 2000). The length of the river's mainstream in China is 2130 km, of which about 57% is in Yunnan Province (Table 14.1).

In China, the river is called the Lancang, but internationally it is usually called the upper Mekong. This chapter provides an introduction to the biophysical environment and an analysis of the environmental changes in the upper Mekong basin, focusing mainly on that part within Yunnan Province, China.

The Lancang-Mekong River provides a very important ecological and economic corridor linking China and Southeast Asian countries. The upper Mekong River (Lancang River) is the key national base for the maintenance of biodiversity, hydropower, and mineral resource development, and also a key area for ecological conservation of the upstream regions of the great rivers in China. Rapid and dramatic changes in the watercourse are occurring as a result of a series of development programs including the China Great Western Development Program,

the Greater Mekong Subregional Economic Cooperation Program (GMS), and the development of the China-ASEAN free trade zone (10 + 1).

1.2. Topography

Approximately 95% of the upper Mekong watershed is classified as mountainous, and most of this belongs to the Hengduanshan Mountains system. The catchments in the upstream area are narrow, with many very high mountains (3500-6740 m above sea level), including the Nushan ranges in the west and the Yunling ranges in the east. Downstream, the mountains on each side of the main stream are relatively lower, about 1000-3000 m before the river reaches the lower Mekong (Table 14.2).

Because most of the basin is mountainous, its natural environment is characterized by steep slopes, unstable geology, and eroded soil. Soil erosion has been a serious problem in the basin because of montane agricultural development. The population, cultivated land, and villages and towns are concentrated into plains areas in the mountains, which totals about 4300 km^2 or 4% of the total area of the upper Mekong (Table 14.3).

1.3. Climate

The climate in the upper Mekong has significant spatial variations from north to south, ranging from a high-altitude northern plateau

TABLE 14.2 Characteristics of mainstream channels of the upper Mekong

River reach	Length (km)	Elevation of river bed (m)	Slope of river (%)		Depth of valley (m)	Valley shape
			Average	Maximum		
Changdu-Gongguoqiao	821	3210-1230	2.4	3.7	800-1500	V shape
Gongguoqiao-Jingyunqiao	213	1230-914	1.5	1.9	600-800	V shape
Jingyunqiao-Mengsong	495	914-465	0.9	2.7	400-800	V shape U shape

Source: He (1995).

TABLE 14.3 Land area distributions with different slopes of upper Mekong in Yunnan (km^2)

Counties/cities			Slope				
			0-8°	8-15°	15-25°	25-35°	>35°
Deqin county	Area	7296.75	17.84	135.21	1167.50	3383.95	2592.95
	%	100	0.24	1.85	16.00	46.38	35.53
Weixi county	Area	4461.13	66.87	245.65	1155.79	2023.25	969.57
	%	100	1.50	5.50	25.91	45.35	21.74
Lanping county	Area	4375.52	65.96	250.47	1279.79	2200.56	578.74
	%	100	1.51	5.72	29.25	50.29	13.23
Dali city	Area	1202.45	270.51	96.07	426.39	255.45	118.03
	%	100	22.50	7.99	38.45	21.24	9.82
Baoshan county	Area	4805.52	470.52	488.18	1314.80	1385.97	1146.05
	%	100	9.79	10.16	27.36	28.84	23.85
Changyuan county	Area	3784.52	164.92	331.85	1694.70	1115.32	477.64
	%	100	4.36	8.77	44.78	29.47	12.62
Pu'er[a] city	Area	3881.21	112.09	385.23	2333.26	934.89	115.74
	%	100	2.89	9.92	60.12	24.09	2.98
Menglian county	Area	1889.55	88.09	285.18	1055.76	435.14	25.37
	%	100	4.66	15.09	55.88	23.03	1.34
Menghai county	Area	5312.15	399.56	398.19	2067.04	1662.25	785.11
	%	100	7.52	7.50	38.91	31.29	14.78
Total basin	Area	10,7970	4380	8050	47,730	35,850	11,960
	%	100	4.0	7.5	44.2	33.2	11.1

Source: Plinston and He (1999).
[a]Simao city changed its name to Pu'er city on August 4, 2007.

climatic area through temperate, subtropical, and tropical by the border with Lao PDR (Fig. 14.1). It is mainly influenced by two monsoons: the southwest monsoon begins in May and lasts till late September with abundant water vapor from the Bangladesh Bay; the northeast monsoon has a reverse air stream from the mainland, relatively dry with limited rainfall.

Precipitation between the upper Mekong and the lower Mekong is quite different (Table 14.4) (Liu and Tang, 2001). The mean annual rainfall is more than 1000 mm in most areas of the upper Mekong basin, with the maximum of 2772 mm and a minimum of

664 mm during 1955–1990. But in the lower Mekong River basin, the mean annual rainfall ranges from 4000 mm in the mountains of Viet Nam, Cambodia, and Laos to 1000 mm in the northeast area of Thailand.

1.4. Water Resource

The upper Mekong basin is rich in water resources. The total annual water runoff of the basin in Yunnan Province is 500.22×10^8 m^3, and the average annual per capita consumption of water in the basin area was 8720 m^3 in 2004, which is much higher than the provincial

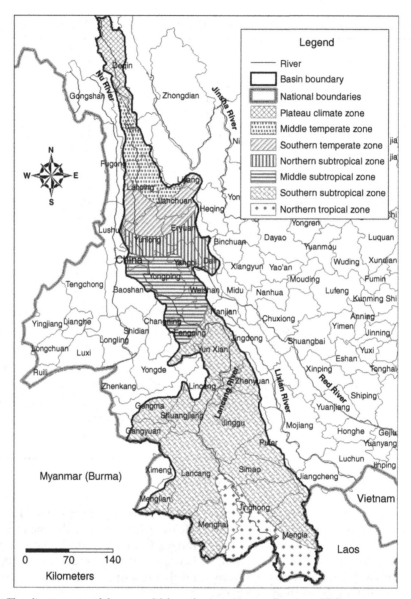

FIGURE 14.1 The climate zones of the upper Mekong basin in Yunnan Province, PRC.

average of 4771 m^3. The volume of water usage is relatively small, compared with the total water resource. In 2004, the total quantity of extractive water use was 25.53 × 10^8 m^3, only 5.10% of the basin's water runoff. The rate of water use in different parts of the basin is quite variable and often low (Table 14.5). The interannual variation in water utilization within the basin is also substantial; the average annual quantity of water used from 1999 to 2004 was 26.28 × 10^8 m^3.

The main purpose of water resources use in the upper Mekong River basin is to facilitate

TABLE 14.4 Monthly average precipitations in the Mekong river basin (in mm)

Basin	Month	January	February	March	April	May	June	July	August	September	October	November	December	Annual
Upper	Rainfall	17	19	27	41	98	174	224	228	132	102	43	19	1125
Mekong	%	1.52	1.65	2.43	3.67	8.70	15.5	19.9	20.3	11.7	9.10	3.81	1.72	100
Lower	Rainfall	8	15	40	77	198	241	269	292	299	165	54	14	1672
Mekong	%	0.5	0.9	2.4	4.6	11.8	14.4	16.1	17.5	17.9	9.9	3.2	0.8	100

Source: He (1995).

TABLE 14.5 Distribution and consumption of water resource in upper Mekong Yunnan part, 2004

Description	Water resource ($\times 10^8$ m^3 s^{-1})	Water use ($\times 10^8$ m^3 s^{-1})	Water resource utilization ratio (%)
Dali prefecture	104.10	12.52	12.0
Baoshan city	179.70	9.93	5.5
Lincang county	180.50	8.56	4.7
Pu'er city	251.61	10.02	4.0
Xishuanbanna county	91.14	5.56	6.1
Upper Mekong in Yunnan	500.22	25.53	5.1

Source: Yunnan Bureau of Hydrology and Water Resources (2004).

regional sustainability (Plinston and He, 1999). Objectives related to this purpose can be summarized under three major headings: the first, meeting the needs of social and economic development in the basin area, which includes water supply for the development of hydropower, navigation, and irrigation, and water supply for cities, towns, and communities; the second, reducing water-related disasters, both natural and artificial, including flood control, water pollution prevention, and soil erosion control; and the last, maintaining the balance of the river ecosystem, including the conservation of river wetlands and biodiversity conservation, especially the conservation of aquatic animals and tropical and subtropical forest systems.

In 2003, of the total water usage was 25.24×10^8 m^3; 84.62% was water supplied for agriculture, 5.80% for industries, and 9.58% for domestic uses (Yunnan Bureau of Hydrology and Water Resources, 2005). Over the long term, the agricultural sector will be the biggest water user in the upper Mekong basin. The construction projects for water utilization on the upper Mekong mainly focus on irrigation and can be divided into two categories, storage projects and pumping projects with a range of sizes (Table 14.6).

1.5. Hydropower Development

The Lancang-Mekong River has plentiful hydropower resources mainly concentrated in China, Laos, and Cambodia. The upper Mekong hydropower cascade is a project designed to take advantage of an 810-m drop over a 750-km river section in the middle and lower sections of the Yunnan stretch (Plinston and He, 1999).

Hydropower development in the upper Mekong watershed dates back to 1946 with the construction of the Tianshengqiao hydropower station near Dali, with a total installed capacity was 400 kW (2 × 200 kW). By the 1980s, the focus of hydropower development plans in the watershed had moved toward the main channel of the upper Mekong, and a plan for an eight-dam hydropower cascade had

TABLE 14.6   Water supplement projects summary in upper Mekong, Yunnan Province, 2000

		Project scale		
		Middle	Small	Pond and small dam
Storage project	Number	16	627	3449
	Storage capacity ($\times 10^4$ m^3)	43,772	45,597	4010
	Supply capacity ($\times 10^4$ m^3)	31,656	39,308	3320
	Current water supply ($\times 10^4$ m^3)	28,461	36,387	4491
	Designed water supply ($\times 10^4$ m^3)	36,876	41,891	5023
Pumping project	Number		884	
	Pumping capacity (m^3 s^{-1})		91	
	Current water supply ($\times 10^4$ m^3)		26,990	
	Designed water supply ($\times 10^4$ m^3)		32,576	

Source: Yunnan Institute of Water Resources and Hydropower Engineering, Investigation, Design and Research, and Yunnan Bureau of Hydrology and Water Resources (2001).

been developed (Fig. 14.2 Table 14.7). The first dam constructed on the upper Mekong mainstream was Manwan, finished in 1996. As of October 2003, the second dam, Dachaoshan, is in full operation, with each of its six, 225 MW generators now installed. The third dam being constructed is Xiaowan and is seen as an iconic project for the Great Western China Development Strategy. The power production from Xiaowan is considered an essential element of the "West to East" energy transfer. In addition, preliminary work has begun on Nuozhadu Dam (Dore and Yu, 2004; He and Chen, 2002).

Proponents argue that the dams have the potential to offer limited flood control, more assured dry-season flows, increased navigation options, reduced saline intrusion in the delta, and extra irrigation opportunities for downstream countries like Thailand. In addition to the rapidly expanding grid system within China, the electricity produced will be able to enter the Mekong 13-Region electricity grid. A particularly sanguine view is that "upstream development of hydropower will not sharpen the conflict of multi-objective competitive uses and will give benefits to downstream for the development of irrigation, navigation, and

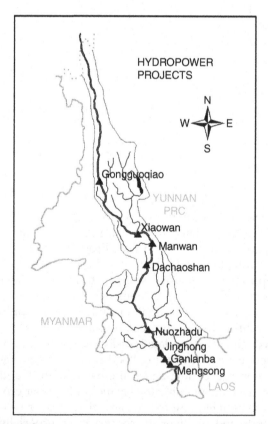

FIGURE 14.2   The locations of the eight proposed dams on the upper Mekong River.

TABLE 14.7 Main features of cascade hydropower projects on middle and lower reaches of upper Mekong river

Upper Mekong	Watershed (km²)	Average inflow (mcm)	Normal storage level (m)	Total storage (mcm)	Installed capacity (MW)	Annual energy (GWh)	Inundated area (ha)	Locally displaced people	Wall height (m)	Status
Gonguoqiao	97,200	31,060	1319	510	750	4060	343	4596	130	Designed
Xiaowan	113,300	38,470	1240	14,560	4200	18,990	3712	32,737	300	Building 2001-2010
Manwan	114,500	38,790	994	920	1500	7805	415	3513	126	Built 1986-1996
Dachaoshan	121,000	42,260	899	890	1350	7021	826	6100	118	Built 1996-2003
Nuozhadu	144,700	55,190	812	22,400	5500	23,777	4508	23,826	254	Designed
Jinghong	149,100	58,030	602	1233	1500	8059	510	2264	118	Building 2003-2010
Ganlanba	151,800	59,290	533	n/a	250	780	12	58	n/a	Designed
Mengsong	160,000	63,700	519	n/a	600	3380	58	230	n/a	Designed

Source: Dore and Yu (2004).

hydropower, and for flooding control" (Plinston and He, 1999).

Over the past 10 years, concerns about the influences of large hydropower dams on the upper Mekong River and their related influences on ecological and social systems along the river downstream have been the focus of much heated debate. Many have suggested that the dams will improve downstream flood control, irrigation, navigation, pollution control, and aquaculture. Others have argued that the dams will obstruct the path of migratory fish, threatening biodiversity and reducing the catches upon which millions of human lives depend. Still others have raised concerns that sediment trapping behind the dams and flow pulse alterations may increase downstream bank erosion and reduce the quality of fish habitat as far downstream as the delta area in Viet Nam (Dore and Yu, 2004; Quang and Nguyen, 2003; Reuters News Service, 2001; Roberts, 2001). There is also some literature indicating that environmental changes (e.g., climatic factors and land use) are the main cause of the hydrometeorological variations in the Mekong River basin (Fu et al., 2006; He et al., 2006; Li et al., 2006; MRC, 2004).

Downstream effects of the construction and operation of dams are extremely complex because they are strongly dependent on the condition of the socioeconomic and environmental systems in the downstream area. That is why the precise nature of the impact of construction and operation of such projects on downstream is frequently overlooked, avoided, and remains a subject of much uncertainty (Adams, 2000).

Due to the interconnected nature of hydrologic systems and the dynamic nature of river flow, transboundary rivers present a formidable challenge to the fixed and arbitrary political boundaries that make up today's world. This is especially true regarding how much of an impact the upstream development may have on downstream countries, communities, and

ecosystems. The transboundary issues from watercourse environmental change, for example, caused by the upstream hydropower development in the Mekong region have resulted in much misunderstanding, even if some conclusions of scientific researches are inconsistent (He et al., 2006; Li et al., 2006). The major factors contributing to misunderstandings are:

1. Lack of sufficient and quantitative data and analysis dealing with the whole basin, especially data on environmental changes;
2. Lack of standardization of data sources between riparian countries because of the use of different monitoring standards for hydrology, meteorology, and water quality in different regions;
3. Asymmetric information transfer between upper and lower Mekong, and between the riparian countries. Most information, for example, is produced as research reports in the gray literature which is not generally accessible, rather than being published as academic articles that are freely available. Most of the information in riparian countries is circulated in the local language and so cannot easily be shared internationally;
4. Absence of multilevel upstream-downstream dialog mechanisms. Because of sovereignty and international relationship issues, transboundary ecosecurity issues in international rivers are often sensitive, and it is difficult to establish joint studies among the riparian countries, which limits participant cooperation and dialog basin-wide.

1.6. Data and Methods

Economic globalization and regionalization in recent years have led to an increasing number of questions and conflicts regarding management and use of transboundary natural resources. Natural resource considerations play an increasingly important role in national security, geopolitics, economic cooperation, and regional stability,

TABLE 14.8 The list of database and methods used in III, IV, and V research

Item	Data	Method
Runoff	Monthly flow record at Jiuzhou, Gajiu, and Yunjinghong stations in upper Mekong (1956-2001)	Comparative analysis method
Water level	Daily water level at Jiuzhou (1960-2003), Gajiu (1960-2001) and Yunjinghong stations in upper Mekong (1960-2003)	Binomial coefficient weighted average method and 5-year moving average method
		Correlation analysis
Sediment	Daily sediment data at Yunjinghong station in upper Mekong and Chiang Saen station in lower Mekong (1987-2003)	Correlation analysis
		Regression analysis
		Granger causality test

and are the focus of an everexpanding amount of academic and policy-oriented research (Dalby, 2002; He *et al.*, 2001; Michel, 2002; United Nations, 2003). Under the impetus of upstream-downstream linkages promoted by large-scale regional economic cooperation, transboundary ecosystem change and its related ecological security issues are currently the focus of numerous international cooperative research projects, especially in the region of the Lancang-Mekong river basin (Badenoch, 2002; Chen, 2003; Goh, 2004; Ratner, 2003; United Nations Development Programme *et al.*, 2000; Wolf, 2001).

Among all these transboundary issues in Mekong region, watercourse changes related to hydropower development have highest priority and are most critical. So, this has been selected as the major topic for analysis in this chapter with analyses based on the database and methods identified in Table 14.8.

2. GEOHYDROLOGICAL CHARACTERISTICS

2.1. River System and Runoff Supplement

Within Yunnan Province, the upper Mekong River mainstream is 1220 km long with a catchment area of 90,200 km^2. The average gradient

of the mainstream is 0.145%; the average annual runoff depth is 584 mm; the annual runoff is 518×10^8 m^3; and the average annual water output per unit area is 57.4×10^4 m^3 km^{-2}.

The pattern of river system development is mainly controlled by geology, with valleys formed and developed along faults. The catchment is narrow with many tributaries. There are 96 branches belonging to the upper Mekong River system in Yunnan Province, China (Fig. 14.3 Table 14.9).

The runoff in the upper Mekong River is mainly from direct precipitation, groundwater, and melt-water from ice and snow. The upper stream region above Changdu is part of Qinghai-Tibet plateau, where the climate is cold and dry. The runoff is supplied by groundwater and melt-water in spring; in other seasons, it comes from rainfall and groundwater, each accounting for 50% of the total. The variation in runoff within the year is moderate. A correlation between runoff and temperature is evident. The high water level from June to August is the period when concentrated rainfall and the warmest weather occur.

In the lower stream, below Changdu, the runoff is dominantly influenced by the subtropical and tropical climates. The monsoon rainfall is the main component. About 60% of the total runoff is from precipitation, the rest is from groundwater. Toward the downstream, the annual runoff is concentrated more in particular months.

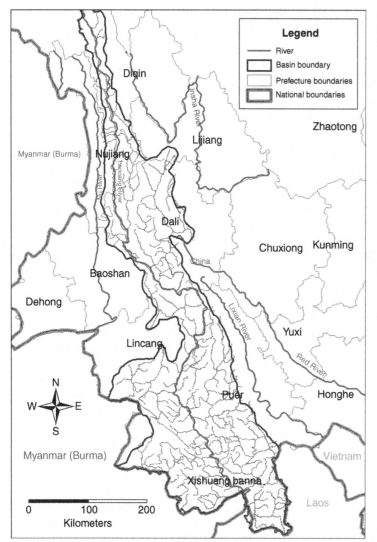

FIGURE 14.3 The tributary network of the upper Mekong River basin in Yunnan Province, PRC.

2.2. Runoff Distribution

2.2.1. *Spatial Variation*

River discharge in the Lancang-Mekong watershed increases dramatically from north to south, with large regional differences (Fig. 14.4; Table 14.10). The left bank (windward slope) has a much larger inflow than the right bank (leeward slope). On the left bank, for example, there are 15 tributaries with a catchment area above 5000 km^2, but only 7 tributaries with similar catchment area on the right bank (He, 1995). In the gorge area, toward the bottom of the valley, the surface runoff and precipitation are lower because of the high mountains.

2.2.2. *Temporal Variation*

There is a substantial difference in the temporal distribution of flow between upstream and downstream, which can be divided into two

TABLE 14.9 Hydrology characteristic values of main branches in the upper Mekong river, Yunnan, China

Name of branches	Location in counties/cities	Length of river (km)	Average slope of river (%)	Catchments area (km^2)	Annual runoff ($\times 10^8$ m^3)		
					Normal year	Wet year	Dry year
Yongchun	Weixi	57.0	2.89	691	3.39	4.20	2.54
Gongjiang	Lanping	96.2	1.49	1447	7.24	8.69	5.72
Bijiang	Lanping-yunlong	180.0	0.78	2646	11.90	15.20	8.33
Yongshun	Yongping	105.0	0.88	1244	3.48	4.32	2.61
Heihui	Lijiang-Fengqing	349.0	0.33	12,075	49.50	63.40	34.70
Misha	Lanping-Eryuan	73.0	2.06	994	4.97	5.96	3.93
Xier	Jianchuan-Eryuan	56.7	1.00	1279	5.12	6.55	3.58
	Xiaguan-Yangbi	23.0	2.67	2714	8.68	10.80	6.51
Shengbei	Yunlong-Yangbi	128.0	1.02	1690	8.46	11.10	5.58
Shundian	Fengqing-Yunxian	200.0	0.49	2604	20.60	23.90	17.10
Shunning	Fengqing-Yunxian	51.0	1.73	463	3.24	4.02	2.43
Menluoba	Jinggu	96.0	0.69	1516	10.60	13.10	7.95
Xiaohe	Gengma-Shuanjiang	181.0	0.51	5758	30.50	39.00	21.40
Mengmeng	Lincang-shuanjiang	88.0	0.89	1350	7.83	9.71	5.87
Manpa	Lincang	51.0	2.35	564	3.27	4.19	2.29
Lama	Jinggu	50.0	1.72	436	2.27	2.81	1.70
Weiyuan	Zengyuan-Pu'er	283.0	1.50	8461	59.20	75.80	41.70
Jinggu	Zengyuan-Jinggu	83.0	0.92	634	4.44	5.51	3.33
Xiaohei	Jinggu-Pu'er	114.0	1.30	1978	13.30	16.50	9.98
Pu'er	Pu'er	100.0	0.62	1223	5.75	6.90	4.54
Xima	Pu'er	28.0	1.46	296	1.18	1.42	0.93
Heihe	Lancang	132.0	0.92	2089	15.20	18.80	11.40
Nanna	Menghai	86.7	1.79	836	5.02	6.02	3.97
Liusha	Menghai-Jinghong	91.2	0.69	1992	9.36	10.90	7.77
Buyuan	Pu'er-Jinghong	307.0	0.33	7184	57.50	66.70	47.70
Nanpuding	Pu'er-Jinghong	110.0	0.43	1175	7.52	8.72	6.24
Molao	Jinghong	58.0	2.13	503	4.53	5.25	3.76
Nana	Menghai-Jinghong	130.0	0.65	1848	9.24	11.50	6.93
Nanla	Mengla	183.0	0.41	4934	19.20	22.90	16.40
Nanmuwo	Mengla	67.0	0.73	672	3.02	3.50	2.51
Nanlei	Menglian	85.0	0.58	1961	11.80	14.20	9.32
Nanla	Lancang-Menghai	216.0	0.55	3404	25.00	31.10	20.50
Nanzhe	Menghai	44.2	2.62	413	3.22	3.99	2.42

Source: Mekong Development Research Network (1993).

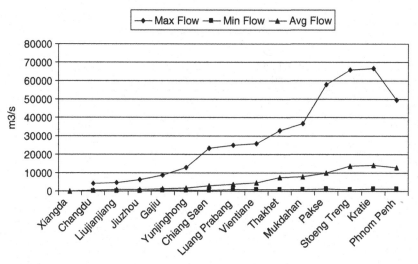

FIGURE 14.4   Maximum, minimum and mean annual flows at sites along the upper (Lancang) and lower Mekong River.

TABLE 14.10   Seasonal distribution of runoff in mainstreams and tributaries of the upper Mekong river

Coefficient of variation (Cv)	River	Hydrological station	Catchment area (km²)	Total runoff (×10⁸ m³)	Wet season (May-October)		Dry season (November-April)	
					Runoff (×10⁸ m³)	Percentage in whole year (%)	Runoff (×10⁸ m³)	Percentage in whole year (%)
–	Mainstream	Xiangda	17,795	41.55	33.78	81.3	7.77	18.7
–	Mainstream	Changdu	53,800	151.90	123.34	81.2	28.56	18.8
0.32	Mainstream	Liutong river	76,690	252.90	210.16	81.3	42.74	16.9
0.31	Mainstream	Jiuzhou	87,205	294.90	231.90	78.6	63.00	21.4
0.30	Mainstream	Gajiu	107,681	399.20	316.36	79.3	82.84	20.7
0.30	Mainstream	Yunjinghong	140,933	566.10	432.00	76.3	134.10	23.7
0.19	Yongchun river	Tangshang	202	0.83	0.53	63.7	0.30	36.3
0.37	Heihui river	Yangzhuangping	4330	20.50	15.72	76.7	4.78	23.3
0.37	Xier river	Xiaguan	2565	8.11	4.68	57.7	3.43	42.3
0.37	Shundian river	Taipingzhuang	2303	20.72	15.62	75.4	5.10	24.6
0.37	Heijiang river	Daxinshan	8454	52.03	40.74	78.3	11.29	21.7
0.37	Buyuan river	Yanan	6206	47.30	36.92	78.1	10.38	21.9

Source: He (1995).

parts, based on the different hydrogeographic conditions. Above Jiuzhou, wet season usually starts in April or May; the interannual and annual flow differences are relatively small and the runoff in dry season is less than 20% of the annual total runoff. From Jiuzhou to Chiang Saen, high flows usually arrive in May or June and the peak in August or September (Table 14.10).

The variation of runoff in the basin is relatively stable from year to year under natural conditions; the coefficient of variation of annual runoff (Cv) is between 0.14 and 0.24 at the river basin level, and between 0.15 and 0.24 in the mainstream (He, 1996; Mekong Development Research Network, 1993).

2.3. Runoff and Water Level

The runoff variation is affected by multiple factors, such as the monsoon climate, the longitudinal range-gorge topography, and human activities in the mountains. The effects of these factors are quite different in time and space. The longer term changes in the runoff and water level are usually controlled by natural factors, and only those at daily scales can be significantly related to the disturbance from the cascade dam building and operations in the mainstream.

Transboundary runoff variation in the dry season has received international attention especially in lower Mekong mainstream since 1993 (Dore and Yu, 2004; MRC, 2004; Quang and Nguyen, 2003; Roberts, 2001). Following four continuous years of abundant precipitation from 1988 to 1991, annual runoff at the three mainstream hydrological sites of Jiuzhou, Gajiu, and Yunjinghong in upper Mekong in China dropped appreciably during the dry years 1992 and 1994 (Fig. 14.5). Climate variability was the major influence on flow variation. Possible influences from cascade dams already constructed only become evident at timescales of one to several days. Further analysis of the linkage between runoff and precipitation data in upper Mekong basin through wavelet transforms confirms that the principal factor influencing annual runoff variation on upper Mekong is precipitation variation (You et al., 2005).

The average percentage of water resources utilization within the upper Mekong area in Yunnan was 3.8% from 1990 to 2000, and was only 4.3% in 2000. So, the water uses during the dry season in upper Mekong could not have a noticeable influence on runoff variation. This is unlikely to change in the future because of the very limited available cultivatable land resources.

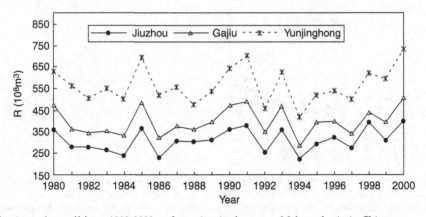

FIGURE 14.5   Annual runoff from 1980-2000 at three sites in the upper Mekong basin in China.

2.4. Water Level in the Mainstream

2.4.1. Long-Term Variations in Water Levels

Water level data at mainstream gauging stations can be compared before and after dam construction to test to the impact of dam operations. We compared the interannual variation of annual mean, yearly maximum, yearly minimum, wet season (June-October) average and extreme dry period (March-May) average, and average water level at three sites, Jiuzhou, Gajiu, and Yunjinghong, based on the daily water level records from 1960 to 2003 (Fig. 14.6). The three sites exhibited similar features from 1960 to 2002, high at the same time in the wet years (e.g., 1966, 1991) and low at the same time in the dry years (e.g., 1992),

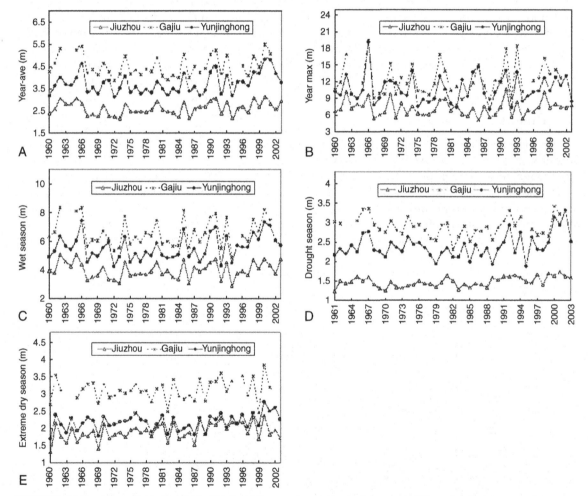

FIGURE 14.6 A comparison analysis for the interannual variations of a range of flow characteristics at three sites in the mainstream of upper Mekong in China: (A) annual mean water levels; (B) yearly maximum water levels; (C) wet season average water levels; (D) drought season average water levels; (E) extreme dry period average water levels.

except for the yearly minimum. After the construction of dams in the main stream in 1992, their variations keep the same patterns, indicating that the construction of the dams mainly influenced short-term water levels, such as daily variations.

2.4.2. Multi-Timescale Water Level Variations

The binomial coefficient weighted average and the 5-year moving average was used to analyze multi-timescale water level variation characteristics from 1960 through 2003 in the upper and lower reaches of the upper Mekong in China (Table 14.11; Fig. 14.7) (Li et al., 2006). Three patterns were obvious:

1. The average annual, wet season and dry season water levels increase at Jiuzhou and Yunjinghong sites.
2. A stable 9- or 10-year cycle of annual mean water levels occurred at the three sites during the past 40 years, which is exactly consistent with the periodic oscillation pattern of solar spot activity. It suggests that annual mean of transboundary runoff in the upper Mekong River may be directly related to the supply of solar energy.
3. The water level variations are usually more different on smaller timescales and more similar on larger timescales. Variation appears

to have been reduced in the most recent 14–15 years, but this will be confirmed in future work.

2.5. Influences of Cascade Hydropower Development

2.5.1. Influence of Existing Dams on Flow Variation

Since 1987, when construction of the first mainstream dam, Manwan, was commenced, disturbance from dam development has been another potential factor influencing runoff and water level in the upper Mekong mainstream (Li et al., 2006).

Before the construction of Manwan, variations of monthly runoff allocation at the three sites were consistent (Fig. 14.7A). During construction of the dam, however, differences in monthly runoff allocation became evident (Fig. 14.7B). Clearly, construction of the dam had an influence on the monthly distribution of flow volume allocation at the site below the Manwan dam. After the reservoir behind the dam began filling, the allocation at the three sites returned to more consistent levels (Fig. 14.7C). Yet in 2000, when Manwan dam was under operation and Dachaoshan dam was close to completion, the flow allocation again appeared more or less consistent at the three sites (Fig. 14.7D).

TABLE 14.11 The characteristics of multi-timescale water level variations

		Annual mean (year)	Yearly maximum (year)	Yearly minimum (year)	Wet season average (year)	Extreme dry season average (year)
Interannual oscillation	Jiuzhou	7, 4-5, 2	7, 4-5, 2	8, 4-5, 2	6-7, 5, 2-3	5-6, 3
	Gajiu	7, 8-9	7-8, 5-6, 2	6, 2-3	8, 6, 4, 2-3	7, 2-3
	Yunjinghong	4-5, 2	7, 4-5, 2-3	7, 5-6, 3-4	6, 4-5, 2	7, 4-5, 2
Interdecadal cycle	Jiuzhou	15, 12, 10-11	14, 11-12, 9	14, 9	11-12, 9	14, 11-12, 9
	Gajiu	15, 12-13, 9	11-12, 9	–	12-13, 9-10	9-10
	Yunjinghong	12-13, 8-9	11-12	11, 9	14, 12, 9,	12, 9-10

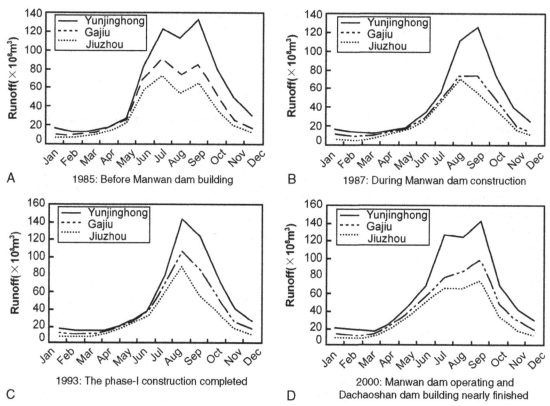

FIGURE 14.7 The influence of the construction of Manwan and Dachaoshan dams on monthly runoff in the upper Mekong River: (A) 1985, prior to building Manwan Dam; (B) 1987, during construction of Manwan dam; (C) 1993, after phase-I of construction was completed; (D) 2000, with Manwan Dam operating and Dachaoshan Dam construction almost finished.

TABLE 14.12 Correlation coefficients (Cc) of multi-timescale water level between three sites (most of them passed 95% confidence verification)

Cc	Year average I_b	Year average I_a	Year maximum I_b	Year maximum I_a	Year minimum I_b	Year minimum I_a	Wet season average I_b	Wet season average I_a	Extreme dry average I_b	Extreme dry average I_a
Jiuzhou-Gajiu	0.882	0.955	0.807	0.808	0.571	-0.265	0.930	0.983	0.864	0.917
Gajiu-Yunjinghong	0.903	0.926	0.780	0.926	0.716	0.578	0.917	0.960	0.901	0.823
Jiuzhou-Yunjinghong	0.732	0.845	0.530	0.823	0.531	0.238	0.760	0.932	0.858	0.673

Note: I_b, influence before dams' construction (1960–1992); I_a, influence after dams' construction (1993-2003).

Currently, when the two dams are under normal operating conditions, the influence on downstream monthly runoff allocation in year is not obvious. Yet at timescales shorter than a month, especially daily and hourly, influences of the dams on flow allocation are more apparent. A correlation analysis of the data for the three mainstream sites (Table 14.12) led to three conclusions (they all are greater than 95% confidence except for that of the yearly minimum water level in the second phase):

1. Before the disturbance due to dam construction (the first phase), all of the correlation coefficients of the yearly minimum among three sites were significant at the 90% confidence level, but following dam construction (the second phase), their correlation coefficients were no longer significant, there was no longer a positive correlation between the three sites.
2. During the extreme dry period, the correlation coefficients of the mean water level between Gajiu and Yunjinhong and Jiuzhou and Yunjinhong reduced after dam construction, although the correlation between Jiuzhou and Gajiu increased.
3. The correlation coefficient for annual maximum water level between Jiuzhou and Yunjinghong sites was appreciably higher after the dam construction. This probably reflects the dams' regulation for flood control during wet periods, which needs to be assessed further based on the daily records between the upper Mekong and the lower Mekong sites.
4. Correlation coefficients for annual mean and wet-period mean water levels among these three sites show no obvious changes between the two phases. For example, the correlation coefficient of annual means between Gajiu and Yunjinghong sites after the dams is 0.023 higher than before the dams. This means that the dam operations have little regulating effect on annual mean and wet-period mean water levels in downstream.

2.5.2. Simulating the Influences of Different Cascade Dams

Computer simulations have been used to make predictions about impact of the upper Mekong dams on flows to the lower Mekong

(He, 1996; He and Chen, 2002). Designating the annual average discharge of natural flow as Qn, the discharge regulated by Xiaowan dam as Qs, and that regulated by Xiaowan and Nuozhadu as Qs + l, a simulation was run based on 30 years of dry-season flow records from 1953 to 1982 (Fig. 14.7A). The system model frequency was set at 50%, 75%, and 97% for upper Mekong outflow into lower Mekong (He, 1996). A second simulation was run based on 40 years (1953-1993) of annual flow data analyzing the flow regulation effects of the construction periods of Manwan, Dachaoshan, Jinghong, and Xiaowan on monthly flow allocation for upper Mekong outflow into the lower Mekong (Fig. 14.7B) (He and Chen, 2002). These relational analyses and simulations reveal that:

1. Over the course of the year, flow volumes on the lower Mekong increase during the wet season and decrease during the dry season. The two dams already completed on the upper Mekong, Manwan and Dachaoshan, can generally be considered as run-of-river dams with limited flow regulation capacity; at yearly, monthly, and semimonthly timescales, the influence of these dams on lower Mekong flow volumes will not be apparent. However, during the dry season, the influences of these two dams on downstream main channel flows in areas nearest Yunnan (upstream from Luang Prabang) on daily volumes are obvious, especially as a result of sudden stoppages or releases of flow of the two dams.
2. Once the Xiaowan dam is completed, dry season average flows at the border section of the main channel from upper Mekong into lower Mekong River during a normal year (P = 50%) will be equal to 1.53 times the natural flow volumes; during a dry year (P = 75%), it will be equal to 1.76 times the natural flow; and during an especially dry year (P = 97%), it will be equal to 2 times the

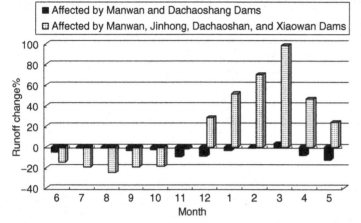

FIGURE 14.8 The percentage change in monthly runoff resulting from cascade dam development on the upper Mekong River.

natural flow. Once Xiaowan and Nuozhadu are both completed, these flow volumes will be 1.8, 2.13, and 2.71 times the natural flow, respectively (see Fig. 14.8).

3. Following completion of Xiaowan and Nuozhadu, the upper Mekong cascade will be able to seasonally regulate flow volumes as much as 100%, but yearly regulation will only reach 23%. This will have significant effects on distribution of flows within a given year; flows during the rainy season months of July, August, and September will be reduced on average by 750, 1200, and 750 m^3 s^{-1}, respectively. Flows during the dry season months of January, February, March, and April will be increased on average by 499, 500, 600, and 350 m^3 s^{-1}, respectively.

The early combined flow regulation capacity of Xiaowan and Nuozhadu, estimated by the MRC (China Daily, 2002), would increase dry season flows by 320 m^3 s^{-1}, which is clearly low. Ma Hongqi, chief engineer of the Yunnan upper Mekong Hydropower Development Company, estimated that the regulatory capacity of Xiaowan alone will increase average dry season (November-May) flows into the lower Mekong by 39.7% (He and Chen, 2002).

Despite the fact that different people arrive at different estimates of the actual regulating capacity of the dams, many have agreed (Campbell, 2007; Campbell and Manusthiparom, 2004; Cogels, 2007; Fu *et al.*, 2006; He *et al.*, 2005, 2006; Li *et al.*, 2006; MRC, 2004) that present environmental changes (e.g., climatic factors and land use) are the main cause of the hydrological variations in the Mekong River basin. This regulating capacity will have the positive influences of reducing flooding downstream, increasing irrigation potential during the dry season, improving river navigation conditions, and reducing saltwater intrusion at the delta. In 2001, David Jezeph, water resources consultant to the United Nations Economic and Social Commission for Asia and the Pacific (UNESCAP), noted that "the upper Mekong dams already built are too small and will not have an obvious influence on the river stage downstream, and the whole cascade indeed may also have several positive influences in future" (Reuters News Service, 2001).

There is still abundant local inflow in the upper reaches of the lower Mekong not influenced by dam construction. The distance along upper Mekong main channel from the Yunjinghong hydrologic site to the mouth of the Nanla River (where the borders of China, Laos, and Myanmar meet) is 102 km, over which significant tributaries enter the main channel. On the right bank in lower-upper Mekong mainstream, are the 92-km-long Liusha River and 140-km-long Nan'a River. On the left bank, are the 262-km-long Buyuan River and the 171-km-long Nanla River. The total area drained by this section of upper Mekong basin is 17,057 km^2, with annual runoff reaching some 113.7 × 10^8 m^3, or 15.44% of the total annual average flow (736 × 10^8 m^3) of upper Mekong watershed (He, 2005). Since this portion of the runoff is below the Jinghong dam, it is not affected by construction or operation of the cascade dams above Jinghong dam. Thus even if all dams upstream of Jinghong hydrological site are built in future, the regulatory capacity of the cascade on discharge into the lower Mekong (outside Yunnan) will be 84.56%.

3. SOIL EROSION AND SEDIMENT

3.1. Soil Erosion

Historically, soil erosion and sedimentation in the upper Mekong have been low due to the natural conditions and high rate of forest cover. However, with increasing development in the watershed area, soil conditions have been threatened. Soil conservation in the watershed area is influenced not only by natural factors (such as climate, precipitation volume, precipitation intensity, and surface coverage), but also by human factors such as land use, cultivation systems, and construction of large-scale infrastructure projects, such as dams and highways.

Compared with other major water systems in Yunnan Province (Table 14.13), although the total volume of sediment transported is large, soil erosion rate is relatively low in upper Mekong.

Seasonal variations of sediment levels in upper Mekong River are closely related to seasonal precipitation and runoff volumes. Sediment transportation is mainly concentrated in the wet season. June and August, for example, account for 80-95% of the annual total sediment

TABLE 14.13 Comparison of the soil erosion in the water system of Yunnan Province

Basins	Watershed area (km^2)	Sediment transport volume (×10^4 t year^{-1})	Erosion speed (mm year^{-1})
Upper Mekong	88,536	9571	0.80
Irrawaddy River	19,023	1877	0.73
Nujiang River (Salween River)	33,423	4180	0.93
Jinsha River (Yangtze River system)	109,679	17,284	1.17
Yuanjiang River (Red River)	74,829	10,692	1.06
Yunnan (total)	383,210	50,813	0.98

Source: Yunnan Provincial Development and Reform Commission, Yunnan Provincial Bureau of Land and Resources (2004).

transfer. In the upper Mekong river basin in Yunnan Province, the annual soil erosion volume is 28 million tons per year, the average annual amount of sediment transported by the river is 61 million tons, and the average silt content is 0.57-1.35 kg m^{-3} (Huang, 1996).

In the upstream area of Gongguoqiao, the impact of human activity is limited due to the turbulent stream, deep water, and sparse population. The annual average silt content is 0.73 kg m^{-3} and the annual silt transport amount accounts for 214 million tons. Downstream of Gongguoqiao, the annual silt transport amount is 632 million tons, accounting for 75% of the total annual silt transport in the whole watershed of upper Mekong.

3.2. Sediment Responses to Mainstream Hydropower Development

By the 1980s, the focus of hydropower development plans in the watershed had moved toward the main channel of the upper Mekong.

In recent years, assessment reports by some independent researchers suggested that sediment in the lower Mekong carried from upper Mekong comprised almost half of the total annual load of 15-17 million tons, though the upper Mekong contributes only about 16% of the flow. They raised concerns that the cascade of dams in the upper Mekong could cause a substantial decrease in the sediment input to downstream, threatening the fishery production of riparian countries, accelerating bank erosion, and imperiling the ecological security and sustainable development of downstream communities (Gupta *et al.*, 2002; International Rivers Network, 2002a,b; Kummu *et al.*, 2004). Based on the basin sedimentary data analyses, Fu *et al.* (2006) suggest that the sediment in the lower Mekong basin originated mostly from northern Laos region rather than the upper Mekong.

To determine sediment variation and its response to the construction and operation of Lancang cascade dams, we used 17 years (1987-2003) of sediment data from Yunjinghong hydrological station in Yunnan Province and Chiang Saen hydrological station in northern Thailand (Fig. 14.9). The furthest downstream information on sediment in China is recorded by Yunjinghong station, while Chiang Saen hydrological station represents a locality recording the output of sediment to the lower Mekong basin after the river flows through northern Laos, which has serious soil erosion problems (Roder *et al.*, 1997; Rumpel *et al.*, 2006).

Two dams and power stations built between 1987 and 2003 had the potential to influence sediment loads at the two hydrological stations. Manwan dam, about 400 km upstream from Yunjinghong and 750 km upstream of Chiang Saen hydrological station, was closed in 1987 and commenced power generation in 1996; and Dachaoshan dam, about 80 km upstream of Yunjinghong and 430 km upstream from Chiang Saen hydrological station, was finished

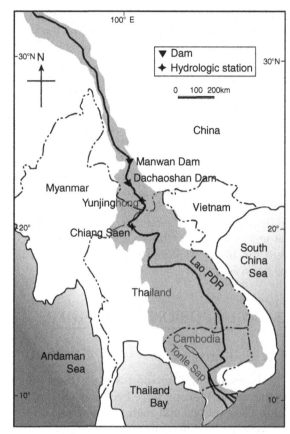

FIGURE 14.9 The locations of Yunjinghong and Chiang Saen hydrological stations.

in 2003. Construction of dams can disturb sediment and water storages and trap sediment from upper reaches; both effects have been recorded in the downstream sediment data of Yunjinghong and Chiang Saen. The data series include monthly and annual mean sediment concentrations, and the maximum and minimum of mean monthly sediment concentration in each year. Synchronous monthly sediment concentrations from two stations were contrasted, while annual sediment concentrations were used to test correlation and causality.

The river reaches below Manwan dam and above Yunjinghong hydrological station flow through mountainous areas with rocky topography, deep loose surface soil, and little vegetation cover. The river reach has a high channel slope, high-velocity flow, and strong transporting capacity, suggesting that coarse sediments should easily be carried away. But the river reach near Jinghong is characterized by a wide valley, small channel slope, and better vegetation cover, so the reach around Jinghong acts as a sink for coarse sediment from upstream. When the river flows out of China and into the reach in the upper lower Mekong basin, the watercourse is characterized by a high channel slope and the catchment has problems such as slash-and-burn cultivation, unsustainable development, and inappropriate land use resulting in decreasing vegetation cover (Chaplot et al., 2005). The watershed of this reach, especially northern Laos region, acts as the most serious zone of water and soil erosion and the main source of sediment in the mainstream of lower Mekong (Huang 1996; Li et al., 2006; You 1999, 2001).

Plots of monthly sediment concentrations from Yunjinghong and Chiang Saen hydrologic stations indicate that their sediment concentrations at the period from the end of 1987 to 1988 were strongly influenced by impacts caused by the construction of Manwan dam (Fig. 14.10), though it was a normal runoff period (Li et al., 2006). The sediment concentrations in 1991 are also very large because it was a high flow year. However, during the period 1997-2003, the pattern of sediment concentrations from two stations is different, suggesting that the closure of Dachaoshan dam might have affected the downstream sediment levels because Manwan dam in the upper reach had been constructed and stored water and trapped sediment and regulated the downstream runoff and water level.

The curves also show that the sediment concentration series from Yujinghong station decreased throughout the period, while that from Chiang Saen decreased from 1987 to

1997, but increased from 1997 to 2003. We argue that these different trends were probably caused by different degrees of land development in the catchments of Yunjinghong and Chiang Saen, leading to different sediment supply patterns in the reaches with rapid development through economic cooperation along the frontier areas in recent years.

Linear regression was used to assess the relationship between the one-to-one sediment series from Yunjinghong and Chiang Saen hydrologic

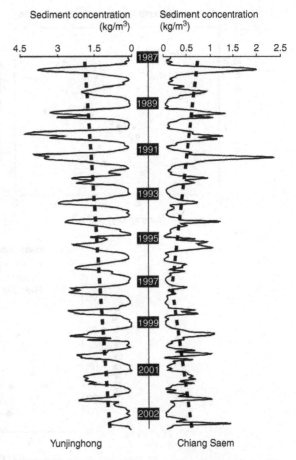

FIGURE 14.10 Parallel plots illustrating the correlations between monthly mean sediment concentrations from Yunjinghong and Chiang Saen hydrological stations, Lancang-Mekong river basin.

stations. There was no relationship between the maximum mean monthly sediment concentrations at the two stations ($R^2 = 0.158$, $p > 0.05$), but significant ($p < 0.05$) relationships were found between the annual and minimum mean monthly sediment concentrations in each year ($R^2 = 0.474$ and 0.260, respectively), although the relationship explained only a small amount of the variability (Fig. 14.11). There was no significant relationship between the sediment concentrations from the two stations.

Granger causality tests (Fuller, 1976; Fu et al., 2006) on the time series indicate that there were no significant ($p > 0.1$) Granger causalities between the annual mean sediment concentration series or the minimum mean monthly sediment concentration series pairs. However, there is a significant Granger causality between the maximum mean monthly sediment concentrations of the two sites ($p < 0.1$). These results are reasonable because the sediment from China could be probably only be transported

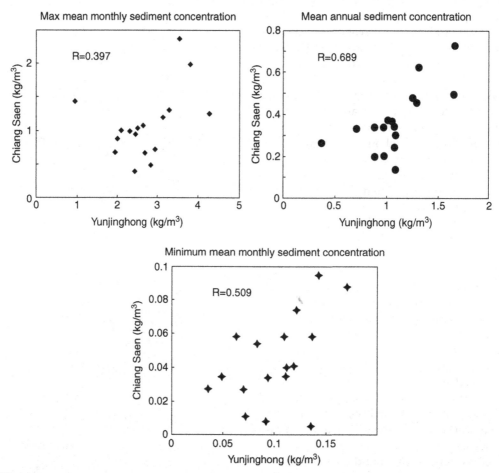

FIGURE 14.11   Regression scatter plots of the annual series of maximum, minimum and mean monthly sediment concentrations from Yunjinghong and Chiang Saen hydrological stations.

over long distances during periods of high flow when there are high sediment concentrations and turbulence along the reaches. In summary,

1. The variation of the maximum mean monthly sediment concentration each year from Yunjinghong station is the cause of variation of the maximum mean monthly sediment concentration each year at Chiang Saen.
2. The variations of annual mean sediment concentration and minimum mean monthly sediment concentration each year at Yunjinghong hydrologic station are not the cause of variations of the annual mean sediment concentration and minimum mean monthly sediment concentration each year at Chiang Saen.

4. WATER QUALITY

4.1. Trends

The water quality of the mainstream in the upper reach of the upper Mekong River has largely remained in a natural status due to its small population and relatively abundant water (Li, 2001). Water quality meets level II of the PRC national water quality standards (Table 14.14). The river is seriously polluted at Manwan below the Yongbaoqiao station of the middle reach. In Jinglinqiao station (near Manwan), the water was polluted and the quality met level V in 1992 (Table 14.15). The main pollutants are wastewater from paper plants and domestic wastewater in Dali city; the main contaminants are Hg and P. The water quality recovers to level II at Yakou station by adsorption to suspended sediment. Water pollution at Xishuangbanna section is more serious and water quality there was worse than level III except in 1993 and 1994, mainly due to high levels of chemical oxygen demand (COD), P, and N.

TABLE 14.14 Environmental quality standards for surface water (GB 3838-2002)

Level type	Description
Level I	Mainly applicable to the water from springs, and national nature reserves
Level II	Mainly applicable to first class protected areas for centralized sources of drinking
Level III	Mainly applicable to second class protected areas for centralized sources of drinking water, protected areas for common fishes and swimming
Level IV	Mainly applicable to water for industrial use and non-contact recreation.
Level V	Mainly applicable to the water bodies for agricultural use and landscape aesthetics

54×10^6 t of wastewater was discharged into the upper Mekong in 1990, of which industrial wastewater accounted for 85%. There were 47×10^3 t pollutants in the wastewater, of which COD comprised 44×10^3 t and accounted for 95% of the total pollutant load. The industrial wastewater discharge of the basin is 1657 t per 10,000 yuan output, which is 9.2 times the national average (Li, 2001). The wastewater treatment rate is only 12%, which is far behind the China national average of 60%.

By 1997, the fresh water consumption was 152×10^6 t in the major prefectures and counties of the basin, with a recycle rate of 33%. Industrial wastewater discharge had increased to 141×10^6 t and industrial solid waste discharge was 210×10^3 t. The rate of industrial wastewater treatment in 1997 was higher than that of 1990, but at 35% it was still far behind the China national level.

Since there are not enough city wastewater treatment plants along the upper Mekong River, domestic wastewater is becoming the major source of organic pollution in the basin,

TABLE 14.15 Results of upper Mekong water quality and assessment during 1990-1997

Station	1990	1991	1992	1993	1994	1995	1996	1997	Main pollutants
Yongbaoqiao	I	II	II	III	II	II	II	II	
Manwan			>V	II	II	III	III	IV	Hg
Jinglinqiao	II	II	V	IV	IV	III	II	IV	P
Yakou			II	II	I	II	II	II	
Xishuangbanna	IV		IV	II	II	IV	>V	>V	P, COD
Menghan	V	V	IV	III	IV	>V	IV	>V	P, N, COD

Source: Li (2001).

especially in Dali city. Jinghong city is located at the lower reaches of the upper Mekong River with a smaller population, but has a quite significant effect on the river water quality since it is very close to the mainstream of the river.

There were 257×10^3 t of agricultural fertilizer consumed in the basin in 1997 (Table 14.16), of which nitrogenous fertilizer amounted to 139×10^3 t, phosphorus fertilizer 40×10^3 t, potash fertilizer 28×10^3 t, mixed fertilizer 49×10^3 t, and the average consumption was 150 kg hm^{-2}. Assuming that 30% of the fertilizers are absorbed, 77×10^3 t of the fertilizer would be discharged into the upper Mekong,

including 42×10^3 t of nitrogen, 12×10^3 t of phosphorus, and 15×10^3 t of mixed chemicals (Statistical Bureau of Yunnan Province, 1998). These are the major nitrogen and phosphorus pollution sources of the upper Mekong.

Soil erosion is another source of water pollution of the upper Mekong. Deforestation and unstable slopes have caused serious soil erosion in the middle and lower stream of the basin causing sharp increases of suspended sediments. Meanwhile, the nitrogen and phosphorus also increase as silt load increases, worsening the water quality, especially in the Xishuangbanna section.

TABLE 14.16 Fertilizers used in agriculture of upper Mekong in 1997 (Unit: t)

	Total amount	N	P	K	Mixed	Arable land (hm^2)	Amount (kg hm^{-2})
Diqin	6438	3689	2264	105	380	3.8	105
Dali	88,593	40,143	13,895	10,876	23,679	19.87	240
Baoshan	77,932	42,206	13,333	10,987	11,406	16.35	240
Lincang	45,456	26,978	5994	2110	10,374	24.53	120
Pu'er	34,264	22,326	6106	2886	2946	34.01	90
Xishuangbanna	10,672	7402	982	1542	746	11.26	75
Total	256,917	139,055	40,310	28,401	49,151	106.02	150

Source: Statistical Bureau of Yunnan Province (1998).

Generally, the main pollutants of the main-stream of the upper Mekong are organic material, nitrogen, and phosphorus. These mainly come from the tributaries. Water pollution is becoming more serious associated with population increases and economic development. Soil erosion also exacerbates water pollution.

4.2. Influences of Dam Construction

In addition to water levels and discharge volumes, lower Mekong countries and international organizations are also paying increasing attention to the impact of upper Mekong dams on water quality. Analysis of more than a decade (1988-2002) of water quality data for the upper Mekong watershed shows that over the past 10 years, organic pollutant indicators in the watershed within Yunnan have decreased, while toxics and heavy metals have remained relatively stable, indicating that overall water quality on major rivers and tributaries within the upper Mekong has improved (see Fig. 14.12).

Using data on water quality indicators measured at six mainstream upper Mekong monitoring stations (Gajiu, Jinglinqiao, Yakou, Yunjinghong, Menghan, and Guanlei) over the period 1986-2002, an analysis was conducted over three 5-year impact periods: 1988-1992, before the Manwan reservoir began to fill; 1993-1997, the first 5 years after the reservoir began to fill; and 1998-2002, the second 5 years after the reservoir began to fill, and during the period when Dachaoshan reservoir also began to fill (He *et al.*, 2006). During the first period, water quality over all six river segments met the water environmental services demand of 66.7% of the time (Table 14.17); during the second period, 59.2%; and during the third period, 73.3%.

Water quality impacts on the lower reaches of the upper Mekong were minor, and overall water quality tended to remain stable. Construction and operation of Manwan and Dachaoshan together had no influence on downstream water quality at the Yunnan Province border, but water quality in the Manwan reservoir declined due to the accumulation of heavy metals and other pollutants. Presumably as a consequence, toxicants and heavy metals decreased in the river. Water quality at the border station of Guanlei was good, generally meeting water environmental services demand for that section of the river.

FIGURE 14.12 Temporal trends in organic (dashed line) and toxic and heavy metal (solid line) pollution indicators in the upper Mekong in Yunnan, plotted on an arbitrary scale.

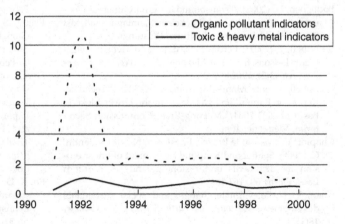

TABLE 14.17 Water quality assessment results on mainstream of upper Mekong (1988-2002)

		Gajiu	Jinglinqiao	Yakou	Yunjinghong	Menghan	Guanlei
Period 1	1988	IV	V	V	III	III	
	1989	III	III	III	II	III	
	1990	III	II	II	III	IV	
	1991	IV	II	II	IV	>V	
	1992	III	V	II		III	
Period 2	1993	III	IV	II	III	III	
	1994	II	IV	I	II	IV	
	1995	III	V	II	III	IV	
	1996	III	III	I	V	III	III
	1997	IV	IV	II	IV	IV	IV
Period 3	1998	III	III	II	IV	IV	III
	1999	III	III	I	IV	III	II
	2000	II	III	II	>V	V	II
	2001	III	III	II	III	III	III
	2002	II	II	II	IV	IV	IV

References

Adams, W. (2000). "The Downstream Impact of Large Dams," Third Draft prepared for the World Commission on Dams Secretariat, 31pp. http://www.dams.org/docs/kbase/thematic/drafts/tr11_reviewdraft_annexa.pdf. Accessed 24 February 2008.

Badenoch, N. (2002). "Transboundary Environmental Governance—Principles and Practice in Mainland Southeast Asia," World Resources Institute, Washington, 44pp.

Campbell, I. (2007). Perceptions, data, and river management: Lessons from the Mekong River. *Water Resources Research* 43(2), W02407.

Campbell, I. and Manusthiparom, C. (2004). "Technical Report on Rainfall and Discharge in the Lower Mekong Basin in 2003-2004," Mekong River Commission Secretariat, Vientiane, 10pp.

Chaplot, V., Coadou le Brozec, E., Silvera, N., and Valentin, C. (2005). Spatial and temporal assessment of linear erosion in catchments under sloping lands of northern Laos. *Catena* 63, 167-184.

Chen, Y. (2003). Climate changes and social sustainable development. *Advances in Earth Sciences (in Chinese)* 18(1), 1-3.

China Daily. (2002). Xiaowan dam, a reservoir in progress. September 16, 2002. http://www.china.org.cn/english/environment/42990.htm. Accessed 24 February 2008.

Cogels, O. (2007). Mekong hydropower development is good. Bangkok Post, January 9, 2007. http://www.greengrants.org.cn/read.php?id=1353. Accessed 24 February 2008.

Dalby, S. (2002). "Security and Ecology in the Age of Globalization," ECSP Report, 14pp. http://www.wilsoncenter.org/topics/pubs/Report_8_Dalby.pdf. Accessed 24 February 2008.

Dore, J. and Yu, X. G. (2004). "Yunnan Hydropower Expansion: Update on China's Industry Reforms and the Nu, Lancang, and Jinsha Hydropower Dams," Working Paper from Chiang Mai University's Unit for Social and Environmental Research, and Green Watershed. Thailand, Chiang Mai University, Chiang Mai, 38pp. http://www.mpowernet.org/download_pubdoc.php?doc=2586. Accessed 24 February 2008.

Fu, K. D., He, D. M., and Li, S. J. (2006). Response of downstream sediment to water resource development in mainstream of the Lancang River. *Chinese Science Bulletin* 51(Suppl.), 119-126.

Fuller, A. (1976). "Introduction to Statistical Time Series," John Wiley Press, New York, 726pp.

Goh, E. (2004). "China in the Mekong River Basin: The Regional Security Implications of Resource Development on the Lancang Jiang," Institute of Defense and Strategic Studies, Singapore, 26pp.

Gupta, A., Hock, L., Xiaojing, H., and Ping, C. (2002). Evaluation of part of the Mekong river using satellite imagery. *Geomorphology* **44**(3-4), 221-239.

He, D. M. (1995). Analysis of hydrological characteristic in Lancang-Mekong River. *Yunnan Geographic Environment Research* **7**(1), 58-74.

He, D. M. (1996). Facilitating sustainable development through integrated multi-objective utilization and management of water resources environment in the Lancang-Mekong watershed. *Yunnan Geographic Environment Research* **8**(1), 25-36.

He, D. M. (2005). Transboundary effects and countermeasures of hydropower cascade development on Lancang river mainstream (a research report for the GMS Summit in 2005), Kunming, China, 9pp.

He, D. M. and Chen, L. H. (2002). The impact of hydropower cascade development in the Lancang-Mekong Basin. *Yunnan Mekong Update & Dialogue* **5**(3), 2-4.

He, D. M., Kung, H., and Chapman, E. C. (2001). The Lancang-Mekong River subregion: Going towards sustainable development through international cooperation. *In* "Towards Cooperative Utilization and Coordinated Management of International Rivers," He, D. M., Zhang, G. Y., and Kung, H. T. (eds.), pp 223-232. Science Press, New York.

He, D. M., You, W. H., and Magee, D. (2005). Correlation and multi-timescale characteristics of the Lancang River's monthly flow variations in Yunnan (China). *In* "Proceedings of International Symposium on Role of Water Sciences in Transboundary River Basin Management," Herath, S., Dutta, D., Weesakul, U., and Das Gupta, A. (eds.), pp 33-41. United Nations University, Asian Institute of Technology and Thammasat University, Bangkok.

He, D. M., Feng, Y., Gan, S, Magee, D., and You, W. H. (2006). Transboundary hydrological effects of hydropower dam construction on the Lancang River. *Chinese Science Bulletin* **51**(Suppl.), 16-24.

Huang, Y. (1996). Transported sediment variation of Lancang River in Yunnan section. *Yangtze River (in Chinese)* **27**(1), 33-35.

International Rivers Network. (2002a). China's upper Mekong dams endanger millions downstream. Briefing Paper 3, Berkeley. http://www.internationalrivers.org/files/03.uppermekongfac.pdf. Accessed 18 June 2008.

International Rivers Network. (2002b). Navigation project threatens livelihoods, ecosystem. Briefing Paper 2, Berkeley. http://www.internationalrivers.org/files/02.navfactshet.pdf. Accessed 18 June 2008.

Kummu, M., Koponen, J., and Sarkkula, J. (2004). Upstream effects on lower Mekong floodplains: Tonle Sap case study. *In* "Advances in Integrated Mekong River Management", Proceedings of an International Conference," Vientiane, Lao PDR, 6pp. http://users.tkk.fi/~mkummu/publications/kummu&al_RR2002_Vientiane_2004.pdf. Accessed 24 February 2008.

Li, L. J. (2001). Study on water quality of the Lancang River. *In* "Towards Cooperative Utilization and Coordinated Management of International Rivers," He, D. M., Zhang, G. Y., and Kung, H. T. (eds.), pp 136-141. Science Press, New York.

Li, S. J., He, D. M., and Fu, K. D. (2006). The correlation of multi-timescale characteristics of water level processes in Lancang-Mekong River. *Chinese Science Bulletin* **51**(Suppl.), 50-58.

Liu, H. and Tang, H. X. (2001). The comparative analysis of the hydrological characteristics in the Lancang-Mekong River basin. *In* "Towards Cooperative Utilization and Coordinated Management of International Rivers," He, D. M., Zhang, G. Y., and Kung, H. T. (eds.), pp 130-135. Science Press, New York.

Mekong Development Research Network. (1993). "Investigation and Study of the Current Status of the Lancang River—Mekong River in Yunnan," pp 3–15. Kunming, PRC, 106pp.

Michel, T. K. (2002). "Resource Wars: The New Landscape of Global Conflict," Henry Holt & Company, New York, 304pp.

MRC. (2004). Mekong's low flows linked to drought. http://www.mrcmekong.org/MRC_news/press04/26-mar-04.htm. Accessed 18 June 2008.

Plinston, D. and He, D. M. (1999). "Policies and Strategies for the Sustainable Development of the Lancang River Basin," Water Resources and Hydropower Report of TA 3139-PRC. Asian Development Bank, Manila, 36pp.

Quang, M. and Nguyen, P. E. (2003). Hydrological impacts of China's Upper Mekong dams on the lower Mekong river. http://mekongriver.org/publish/qghydrochdam.htm. Accessed in May 2005.

Ratner, B. D. (2003). The politics of regional governance in the Mekong River Basin. *Global Change* **15**(1), 59-76.

Reuters News Service. (2001). China dams could help Laos fight floods. August 7, 2001, Bangkok, Thailand. http://www.planetark.org/dailynewsstory.cfm?newsid=11920. Accessed 18 June 2008.

Roberts, T. (2001). Downstream ecological implications of China's Lancang Hydropower and Mekong Navigation Project. International River Network (IRN). http://www.irn.org. Accessed in May 2005.

Roder, W., Phengchanh, S., and Maniphone, S. (1997). Dynamics of soil and vegetation during crop and fallow period in slash-and-burn fields of northern Laos. *Geoderma* **76**, 131-144.

Rumpel, C., Alexis, M., Chabbi, A., Chaplot, V., Rasse, D. P., Valentin, C., and Mariotti, A. (2006). Black carbon contribution to soil organic matter composition in tropical sloping land under slash and burn agriculture. *Geoderma* **130**, 35-46.

Statistical Bureau of Yunnan Province. (1998). "Statistics Yearbook of Yunnan Province in 1998," China Statistics Press, Beijing, 591pp.

United Nations. (2003). Joint declaration on the promotion of tripartite cooperation among the People's Republic of China, Japan and the Republic of Korea, Bali, Indonesia. http://www.aseansec.org/15284.htm. Accessed 24 February 2008.

United Nations Development Programme, World Bank, and World Resources Institute. (2000). "World Resources 2000–2001: People and Ecosystems: The Fraying Web of Life," WRI Press, Washington, DC, http://archive.wri.org/publication_detail.cfm?pubid=3027#1. Accessed 24 February 2008, 400pp.

Wolf, T. A. (2001). Dehydrating conflict. *Foreign Policy* **9**, 2-9.

You, L. Y. (1999). A study on temporal changes of river sedimentation in Lancang River Basin. Scientia Geographica Sinica **54**(Suppl.), 93-100.

You, L. Y. (2001). Scouring and silting changes of Lancang River (Mekong River) and its development tendency. *Geographical Research (in Chinese)* **20**(2), 178-183.

You, W. H., He, D. M., and Duan, C. Q. (2005). Climate change of the Longitudinal Range-Gorge in Yunnan and its influence on the river flow. *Acta Geographica Sinica* **60**(1), 95-105.

Yunnan Bureau of Hydrology and Water Resources. (2001). "Integrated Report on Sustainable Utilization of Water Resource in Yunnan Province," Kunming, PRC, 243pp.

Yunnan Bureau of Hydrology and Water Resources. (2001-2005). Yunnan Water Resource Bulletin 2001, 2004, 2005. Kunming, Yunnan, PRC. <http://www.ynswj.gov.cn/article_list.asp?SortId=5&NsortId=79>. Accessed 3 July 2009.

Yunnan Provincial Development and Reform Commission, Yunnan Provincial Bureau of Land and Resources. (2004). "Integrated Remote Sensing Report on Land Resources of Yunnan Province," Yunnan Science and Technology Press, Kunming, 257pp.

Zhou, C. J., Guan, Z. H., and Len, Y. F. (2000). The source of Mekong River. *Advances in Science and Technology of Water Resources* **20**(03), 15-17.

Conserving Dolphins in the Mekong River: The Complex Challenge of Competing Interests

Isabel Beasley,[1] Helene Marsh,[1] Thomas A. Jefferson,[2] and Peter Arnold[3]

[1]School of Earth and Environmental Sciences, James Cook University, Townsville, QLD, 4811, Australia
[2]Southwest Fisheries Science Center, NOAA Fisheries, 3333 North Torrey Pines Court, La Jolla, CA 92037-1508, USA
[3]Museum of Tropical Queensland, 70-102 Flinders Street, Townsville, QLD 4810, Australia

1. INTRODUCTON

Attempts to conserve the critically endangered population of Irrawaddy dolphins (*Orcaella brevirostris*) (Owen in Gray, 1866) inhabiting the lower Mekong River (Fig. 15.1) are an example of the challenge of conserving endangered species in complex economic, political, and social situations. This Irrawaddy dolphin population is small, declining, and facing numerous threats to its survival. The subpopulation inhabiting the Mekong River was classified as critically endangered by the World Conservation Union (IUCN) in 2004 (Smith and Beasley, 2004).

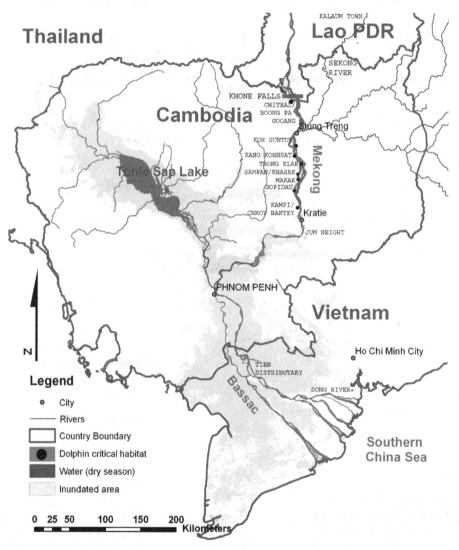

FIGURE 15.1   Map showing the study area of the lower Mekong River. The area ranged from the Laos/Cambodian border (Muang Khong) south to the Vietnamese Delta (including Tonle Sap Lake). The Kratie to Khone Falls river section is the only area in the river where dolphins were sighted. The southern part of this river stretch is in Kratie Province, the upper part is in Stung Treng Province. Base map produced by Matti Kummu and reproduced with his permission.

The Irrawaddy dolphin was originally described as the short-snouted porpoise from a specimen found at the mouth of the Vishakhapatnam (Vizagapatam) River, along the east coast of India, in 1852 by the Englishman Sir Richard Owen (Owen, 1866). This small, unusual dolphin reaches a length of 2.75 m, is uniform gray in color with a white belly, has a rounded forehead, small dorsal fin, and disproportionately large paddle-like flippers (Fig. 15.2).

The Irrawaddy dolphin is a facultative freshwater cetacean (i.e., it inhabits both fresh and marine waters) and is subject to increasing human-induced threats as a result of its reliance on riverine and coastal habitats (Stacey and Leatherwood, 1997; Smith and Jefferson, 2002). The known global distribution of Irrawaddy dolphins is shown in Fig. 15.3.

FIGURE 15.2 Irrawaddy dolphin calf from Koh Pidau Pool, Kratie Province, Mekong River. (See Color Plate 23)

However, knowledge of the distribution of this species throughout its coastal range is still incomplete.

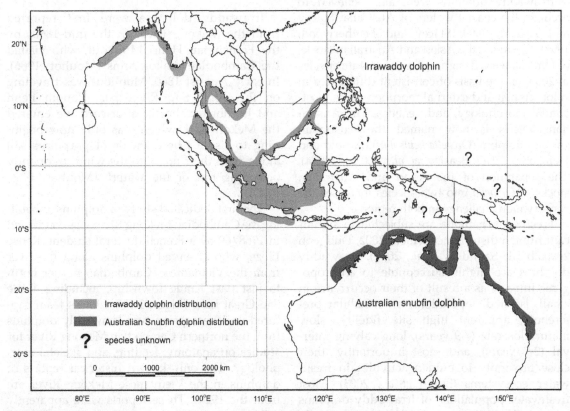

FIGURE 15.3 Distribution map of the Irrawaddy and Australian Snubfin dolphin.

There are five freshwater Irrawaddy dolphin populations. Three populations occur in major Asian river systems: (1) the Mahakam population of Kalimantan, Indonesia (estimated population size of 70 individuals; Kreb, 2004; Kreb, 2005; Kreb *et al.*, 2007); (2) the Ayeyarwady population of Myanmar (estimated population size 59-72 individuals; Smith *et al.*, 2007); and (3) the Mekong population of southern Lao PDR (hereafter referred to as Laos), Cambodia and Viet Nam (estimated population size 108-146 individuals; Beasley, 2007; Beasley *et al.*, 2007; see below). Two populations occur in brackish/freshwater lakes: (1) the Songkhla Lake population in Thailand (estimated population size of <20 individuals; Beasley *et al.*, 2002a; Kittiwattanawong *et al.*, 2007); and (2) the Chilka Lake population in India (estimated population size of at least 85 individuals, Pattnaik *et al.*, 2007).

Irrawaddy dolphins were also believed to occur in the coastal waters of Australia (Stacey and Arnold, 1999; Stacey and Leatherwood, 1997). However, the Asian and Australian stocks of *Orcaella* were designated as a separate species in 2005, on the basis of consistent differences in color, cranial and external morphometrics, postcranial morphology, and genetics. The Australian species is now named the Australian snubfin dolphin (*Orcaella heinsohni*) (Arnold and Heinsohn, 1996; Beasley *et al.*, 2002b, 2005a). The separation of the Asian and Australian stocks of *Orcaella* into two species increases the conservation challenge in both regions.

Freshwater habitats are subjected to significant human disturbance (Abell, 2002; Dudgeon 2000a,b,c,d; Saunders *et al.*, 2002). Irrawaddy dolphins are highly susceptible to anthropogenic impacts as a result of their occurrence in small, isolated populations, strict habitat preferences, apparent high site fidelity, slow maturation rate (7-9 years), long calving interval (2-3 years), and most importantly, their close proximity to human activities in freshwater ecosystems (Smith *et al.*, 2003). Most freshwater populations of Irrawaddy dolphins are small, declining, and listed as critically endangered by the IUCN. Nonetheless, there has been a notable lack of on-the-ground conservation measures to conserve most of these populations.

In this chapter, we present the results of recent research and conservation efforts focused on the Irrawaddy dolphin population inhabiting the lower Mekong River of southern Laos, Cambodia, and Viet Nam. We also discuss the potential for successful future conservation of the dolphins and the Mekong River ecosystem; in an environment in which significant economic, political, and social considerations are influencing management initiatives.

2. HISTORICAL ACCOUNTS OF IRRAWADY DOLPHINS IN THE MEKONG RIVER

Irrawaddy dolphins were first reported from the Mekong River in the mid-1860s by the Frenchman Henri Mouhout, who visited the Cambodian Ankor ruins (Mouhot, 1966). In early August 1860, Mouhout was traveling on the Tonle Sap River past Phnom Penh and he noted "shortly afterward we entered the Mekon [sic], which was only now beginning to rise ... here shoals of porpoises sail along with their noses to the wind, frequently bounding out of the water" (Mouhot, 1966, p. 173).

The first dedicated study of dolphins' inhabiting the Cambodian Mekong River was conducted in 1968/69 by a French doctoral student, Renee Lloze, who observed dolphins along the river from the Vietnamese/Cambodian border north to just past Kratie township, including Tonle Sap Great Lake (Lloze, 1973). Lloze's team captured and necropsied two Irrawaddy dolphins from the northern Cambodian Mekong River for studies on anatomy, feeding, and skeletal morphology. The only known historical reports of dolphins in the Vietnamese Mekong River are from the 1920s. These reports were apparently collected by Frenchmen Gruvel (1925) and Krempf (1924-1925) (cited by Lloze 1973).

These early records suggest that dolphins historically occurred throughout the lower Mekong River, from the bottom of Khone Falls (Fig. 15.1), south to the Vietnamese Delta (including Tonle Sap Great Lake), perhaps numbering at least a few thousand individuals. No historical or contemporary dolphin records are known from the mainstream Mekong River north of the Khone Falls. As a result of political instability, war, and internal conflict, little research had been conducted on the Mekong dolphin population before the early 2000s, as described below.

2.1. Country Status

2.1.1. Laos

In the early 1990s, field research by Canadian Ian Baird of the Lao Community Fisheries and Dolphin Protection Project (LCFDPP) confirmed the presence of Irrawaddy dolphins in southern Laos, and to a lesser extent in northeast Cambodia. Baird and his Laotian counterparts conducted studies on the dolphins' distribution and feeding and investigated mortality rates and causes at Chiteal Pool (known as "Vern Nyang," or "Boong Pa Gooang" in Laos) on the Laos/Cambodian border from 1991 to 1997 (Baird and Mounsouphom, 1994, 1997). Baird also conducted interviews with local fishers in the Sekong, Srepok, and Sesan rivers (which converge with the Mekong River at Stung Treng, Cambodia, 500 km from the river mouth) in Cambodia and Laos. These interviews confirmed that dolphins historically ascended all three rivers, to approximately 280 km up the Sekong River (to Kaleum District) in southern Laos. A German scientist investigating the use of nontimber products also confirmed through interviews that two dolphins had been shot near Sekong town in the Sekong River of southern Laos in 1990 (Bergmans, 1995).

2.1.2. Cambodia

Very little research had historically been conducted on dolphins inhabiting the Cambodian Mekong River. A Cambodian national, Touch Seang Tana, conducted the first studies from 1994 to 2000. Tana conducted observations, interviews, and opportunistically collected carcasses, concluding in a 1996 report that the species was rare in Mekong River waters as a result of human activities, including direct persecution for oil extraction in Tonle Sap Great Lake during the mid-1970s (Perrin et al., 1996). Baird combined boat surveys and interviews to assess the abundance and distribution of dolphins in the upper Cambodian Mekong River in 1996, and estimated that no more than 200 individuals remained in the entire river (Baird and Beasley, 2005). No further studies on the dolphins were conducted in Cambodia, until the Mekong Dolphin Conservation Project (MCDP) began in 2001 (see Section 3).

2.1.3. Viet Nam

There are very few historical dolphin records from Viet Nam, and no previous dedicated studies. There are three recently confirmed reports (with photographs) of Irrawaddy dolphins found dead in fishers' nets from the Vietnamese Mekong River near the Vietnamese/Cambodian border. These carcasses were discovered in 2000, 2002, and 2005 (Beasley et al., 2005b; Chung and Ho, 2002). Irrawaddy dolphin specimens have recently been discovered in various Vietnamese whale temples in Vung Tau and Binh Thang, near the Mekong River Delta (Beasley et al., 2002b; Smith et al., 1997). However, these specimens are probably from coastal populations, as there are local reports of Irrawaddy dolphins occurring along the Vietnamese coast (Beasley et al., 2005b).

2.2. Existing Dolphin-Watching Tourism

In the Mekong River, dolphin-watching tourism is facilitated by the reliable occurrence of dolphins in small deep-water pools throughout the year. There are two locations where tourists can view Irrawaddy dolphins in the Mekong River: (1) Chiteal Pool on the Laos/Cambodian border; and (2) Kampi Pool in Kratie Province, Cambodia.

Dolphin-watching tourism at Chiteal Pool was initiated by three Laotian villages in the late 1990s. This initiative initially used rowboats, expanding to boats with engines by the early 2000s. Cambodia nationals initiated small-scale tourism to observe the dolphins at Chiteal Pool in the early 2000s, using fast speedboats from Stung Treng township.

In the late 1990s, the international non-government agency Community Aid Abroad initiated small-scale dolphin-watching tourism using rowboats at Kampi Pool, 15 km north of Kratie township. This industry subsequently expanded to larger boats with engines in the early 2000s. Only seven families from the village were involved in this dolphin-watching tourism, an arrangement that subsequently created conflict and resentment among other village members who were not allowed to participate. Section 4 further discusses the biological, social, economic, and political ramifications of the dolphin-watching tourism.

3. THE MEKONG DOLPHIN CONSERVATION PROJECT

In January 2001, the MDCP was initiated as part of the first author's Ph.D. research at James Cook University, Australia (Beasley 2007). This project represented the first comprehensive attempt to research and conserve the dolphin population along the entire lower Mekong River. Research was the primary focus of activities from 2001 to 2002, and dedicated conservation activities began in 2003. All activities were conducted in cooperation with the Cambodian Department of Fisheries, which was extremely supportive of all aspects of the project. Beasley was the full-time project manager for 4.5 years. The MDCP also comprised one full-time senior level Department of Fisheries official, various part-time provincial fisheries officials, and two full-time local staff. All project activities were designed to contribute

toward a comprehensive understanding of the dolphin population on which to base initiatives to ensure the population's long-term survival.

3.1. The Critically Endangered Status of the Mekong Dolphin Population

The research results from the MDCP till April 2005, confirmed that the Irrawaddy dolphin population inhabiting the Mekong River is very small, declining, and facing continuing threats.

3.1.1. Dolphin Distribution and Ranging Patterns

MDCP conducted boat surveys from 2001 to 2005 along the entire lower Mekong River from the bottom of Khone Falls on the Laos/Cambodian border south to the Vietnamese Delta (including Tonle Sap Great Lake) (see Fig. 15.1). Over 14,000 km of survey effort were conducted along the river. These surveys confirmed that Irrawaddy dolphins are now primarily distributed along the Kratie township to Khone Falls river stretch (approximately 190 km) and rarely move south of Kratie township, even during the wet season. Although up to a hundred deep-water areas have been recorded between Kratie and Khone Falls, dolphins frequently occur in only 12 deep-water pools. Deep-water areas in the Mekong River have also been described as essential for fish and fisheries during the dry season (Coates et al., 2003; Poulsen and Valbo-Jorgensen, 2001; Vannaren and Sean, 2001).

Intensive photo-identification studies were conducted to investigate the ranging patterns of individual dolphins, which were identified using distinctive features on their dorsal fins (such as nicks and injuries). Individual Irrawaddy dolphins exhibited extremely high site fidelity and preferred particular areas of the river. Analysis of the ranging patterns for the 15 most frequently sighted identifiable individuals showed that on average, each individual ranged over only 16 km^2 in the dry season (range: 0.7-73.0 km^2), and used the same deep water pools each year. During the wet season, as a result of increased water levels and less obstruction to dolphin movements along the river, the area over which individual dolphins ranged expanded to 42 km^2 (range: 0.9-99.0 km^2).

Four largely discrete subpopulations of dolphins were evident: (1) Kampi, (2) Koh Pidau, (3) Stung Treng, and (4) Chiteal (Fig. 15.4). Although dolphins from the Kampi and Koh Pidau subpopulations interacted during the wet season at Phum Kreing (2 km upstream from Kampi Pool), dolphins from the Stung Treng and Chiteal subpopulations appeared isolated both from each other, and from the other subpopulations.

3.1.2. A Critically Small Population

The abundance of the Irrawaddy dolphin population in the Mekong River was estimated using three methods: (1) capture-recapture analysis of photo-identified individuals, (2) line-transect methodology, and (3) direct count survey methodologies. The three survey methodologies were compared to ascertain the most appropriate survey technique for accurate and precise long-term monitoring (Beasley, 2007).

Boat surveys using direct counts and line-transect methodologies were undertaken throughout the lower Mekong River south of Khone Falls. During these surveys, dolphins were sighted only in the Kratie to Khone Falls river section—no dolphins were sighted south of Kratie township.

Ninety-nine adult dolphins were individually photo-identified during the 4-year study period, with 83% of the population estimated to be photographically identifiable (Fig. 15.5). A closed capture-recapture model (incorporating known mortalities) was used to estimate the size of the total population using photo-identification. Based on the results obtained from photo-identification, the total Irrawaddy dolphin population in the Mekong River was 127 individuals (range: 108-146), as of April

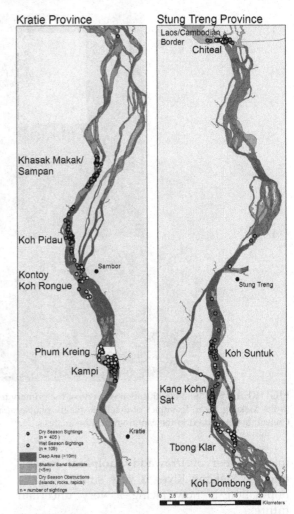

FIGURE 15.4 Distribution of Irrawaddy dolphins inhabiting the Mekong River, based on all dolphin sightings obtained between January 2001 and April 2005. Kratie province is shown on the left map and Stung Treng Province, which is further north (see Fig. 15.6) on the right map. Sightings are separated into dry season (dark dots) and wet season (light dots). Map created by Erin LaBreaque and reproduced with her permission.

2005. Comparisons of survey techniques indicate photo-identification is the preferred methodology for population monitoring because of its efficiency and precision. Irrespective of the differences between survey methodologies, the

FIGURE 15.5   Photo-identification was one of the primary techniques used to estimate abundance of Irrawaddy dolphins in the Mekong River. Examples of excellent quality photographs of two well-known dolphins; KA05 Rags (top) and CH01 Chiteal (bottom), used in the capture-recapture analysis.

total number of Irrawaddy dolphins inhabiting the Mekong River is very small, and the population is now facing a very uncertain future.

Very small populations are at risk, simply because of their size (Berger, 1990; Caughley and Gunn, 1996; Reed et al., 2003; Soule, 1986). Small populations are particularly susceptible to threats such as demographic stochasticity, environmental stochasticity (including natural catastrophes) and genetic stochasticity, (Caughley and Gunn, 1996). As a result of the small remaining Mekong dolphin population, it will be difficult to detect a statistically significant declining trend (Taylor and Gerrodette, 1993). By the time a trend is detected with a high level of statistical confidence, the population will be

approaching local extinction (Beasley, 2007). Scientists and managers have emphasized the need for a precautionary approach toward management of seemingly small and declining populations (Mayer and Simmonds, 1996; Pichler et al., 2003; Thompson et al., 2000). Such a precautionary approach should be taken for the Irrawaddy dolphin population inhabiting the Mekong River.

3.1.3. Social Structure

School dynamics and social structure were investigated using the photo-identified individuals. Average group size during the dry season was 6.8 dolphins + S.E. 0.2 (range: 1-19; $n = 405$); the corresponding figures for the wet season were 5.7 dolphins + S.E. 0.41

(range: 1-34; $n = 107$). Analysis of the association patterns of individual dolphins (the frequency that individuals associate with other individuals in the population), were conducted using the computer program SOCPROG (Whitehead 1997, 2005). Association values ranged from 1 (animals always seen together) to 0 (individuals never seen together). The resulting association values demonstrated that individual dolphins were seen with a particular companion significantly more often than would be expected by chance. This indicated that the population is highly structured, with most individuals having preferred, long-term associates.

From a management standpoint, there are two primary implications of social structure studies.

Firstly, it is critical that effective on-the-ground conservation efforts are focused on each of the four subpopulations and their associated critical habitats. Some subpopulations are small (e.g., Chiteal); however, conservation and management efforts should continue to be a high priority for each area.

Secondly, translocation programs to repopulate critical areas are probably not a viable conservation option. Translocation programs in which dolphins from one community (e.g., Kampi) are moved to a smaller community (e.g., Chiteal) are unlikely to be beneficial to the long-term survival of the population. In addition to the high probability of mortality during capture and transport (Fisher and Reeves, 2005), the removal of one or more individuals from a socially stable group may negatively affect the group that the individual(s) are taken from, the individual(s) that are translocated to a new group, and/or the new group that the individual(s) are translocated into.

3.1.4. Unsustainable Dolphin Mortalities

In 2001, MDCP initiated a carcass recovery program throughout the lower Mekong River to collect and conduct necropsy on all dead dolphins and attempt to determine the cause of mortalities. During the first few years of the stranding program few local people were aware of the reporting procedure (or MDCP), a situation that resulted in only a few carcasses being reported, often many months after a dolphin had died. However, at the end of 2002, MDCP conducted a large-scale awareness campaign about the importance of reporting dolphin carcasses. As a result of these efforts, dolphin carcasses were often reported within days of death from 2003 onwards. A total of 54 dolphin carcasses were recovered and/or confirmed between January 2001 and April 2005 (Gilbert and Beasley, 2006). Forty-three percent of all carcasses recovered were newborns (see Table 15.1). Interestingly, only one carcass of a juvenile dolphin has been recovered since 2001, potentially indicating that there is very little recruitment into the population as a result of the high number of newborn mortalities. Newborn mortalities after 2005 have continued to be high, with 16 in 2006, and nine as of April 2007 (WWF Cambodia Program, personal communication).

The cause of the high number of newborn deaths is unknown. If the Irrawaddy dolphin population inhabiting the Mekong River has any chance for survival, it is imperative that the cause(s) for newborn mortality are established, and subsequently managed. The population will not survive in the long term unless newborn survival increases.

Entanglement in gill nets and direct deaths through destructive fishing practices (e.g., dynamite fishing) are known causes of mortality of some adult dolphins. Other potential indirect causes of dolphin mortality include habitat degradation, contaminants, disease, boat harassment and noise, boat collision, reduced fish stocks, and inbreeding depression. As of April 2005, the Irrawaddy dolphin population in the Mekong River was estimated to be declining at a yearly rate of at least 4.8% on the basis of the mortality rate evident in the carcass

TABLE 15.1   Total confirmed and unconfirmed dolphin mortalities in the Mekong River from January 2001 to April 2005

Year	Confirmed deaths	Unconfirmed deaths	Confirmed adults	Confirmed newborns/ young calves	Confirmed unknown age
Pre-2001	8	–	4	–	4
2001	3	–	2	1	–
2002	5	1	4 [a]	1	–
2003	15	3	10	5	–
2004	16	–	5	11	–
(Jan-April) 2005	7	–	2	5	–
Total	54	4	27	23	4
Total (2003 to April 2005)	38 (31)	3 (3)	17 (15)	21 (16)	–
Average/year (Jan 2003 to April 2005)	13 (15.5)	1 (1.5)	6 (7.5)	7 (8)	–

Note: Data obtained from Jan 2003 onward is considered to be reliable and close to representative of the majority of dolphins that died in Mekong River. Totals and averages are presented for January 2003 to April 2005 in the last two rows in brackets. Confirmed totals include one carcass that was reported from the Vietnamese Mekong River (photographic evidence was available).
[a]One animal recorded as a juvenile (OBRE02-01/04) is listed with the adults to differentiate it from newborns/young calves.

recovery program and field evidence that few newborns survive for more than one month.

An example of the unsustainable mortalities and population decline is evidenced at Chiteal Pool, situated on the Laos/Cambodian border. In 1991, Baird (1991, p. 2) stated "Although it is difficult to estimate the dolphin population near Hangorn village [Chiteal Pool] at this time, our group did see 20-30 animals there at one time." Based on recent surveys, the Chiteal population has experienced a significant decline since 1991, and numbered only nine individuals, as of April 2005 (Beasley, 2007). As a result of high site fidelity exhibited by individual dolphins in the river and resightings through photo-identification, it is highly unlikely that this decline is associated with movements out of the area. Interestingly, one individual photo-identified in 1996 by Stacey (1996), has remained resident at Chiteal Pool through the duration of MDCP activities (2001-2005), and still remained in the pool in 2007, a period of 11 years.

Based on the estimated Mekong dolphin population size (Beasley, 2007) and typical growth rate of a cetacean population (4% per year, calculated from Wade, 1998), the most conservative level of anthropogenic mortality that the Mekong dolphin population can currently withstand (the Potential Biological Removal) is less than one individual per year (Beasley, 2007). If the Irrawaddy dolphin population inhabiting the Mekong River has any chance of survival, the primary management goals related to mortality reduction must be to (1) determine the cause(s) of newborn mortality and subsequently mitigate the causative factors, and (2) reduce anthropogenic mortality to zero (ideally in cooperation with local communities).

3.1.5. *Local Knowledge and Perceptions*

Interviews with local people can provide extensive information about flora, fauna, and ecosystem functioning that would take researchers years, if not decades, to obtain independently (Johannes *et al.*, 2000). Few historical accounts of dolphin occurrence in the Mekong River exist with which to compare current abundance and distribution records. However, as detailed in Section 2, the few historical published accounts indicate that dolphins were present throughout the lower Mekong River from the bottom of Khone Falls south to at least the Vietnamese/Cambodian border.

MDCP conducted extensive interviews with local communities throughout the lower Mekong River and associated tributaries (including Tonle Sap Great Lake) south to the Vietnamese Delta, to investigate the historical distribution of dolphins in the river and local perceptions toward dolphins and their conservation. A total of 497 local villagers were interviewed. The interviewees were mainly male fishers who spent much of each day on the river.

These interviews suggest a major decline in dolphin occurrence and abundance throughout most of the river. Reports confirm that dolphins previously occurred regularly from Kratie township south to the Vietnamese Delta, during both dry and wet seasons. Dolphins are now rarely sighted in this region. Interviewees identified the Kratie to Khone Falls river segment as the most important habitat remaining for dolphins in the lower Mekong River. These conclusions confirm the results of the MDCP dedicated boat surveys.

Local communities in both Cambodia and Laos hold very positive attitudes toward Irrawaddy dolphins, a situation which significantly assists with securing local cooperation for management strategies. Part of this local reverence results from local folklore about the Irrawaddy dolphin's human origins (see Box I and Box II).

Although dolphins are not currently deliberately caught in the Mekong River, dolphins reportedly suffered significant human-induced mortality during the 1950-1980s. The first major episode of direct catch reportedly occurred in the early 1950s, and was caused by a fishing lot owner in Tonle Sap. This man apparently did not want dolphins eating fish near his fishing lot, so he ordered his workers, at least once, to use a seine-net to catch and then kill as many dolphins as possible in the area around his fishing lot. The significant direct catch of dolphins before, during, and directly after the Vietnamese War/Pol Pot Regime periods, is probably responsible for a major decline in dolphin numbers. During the Pol Pot regime, many dolphins were allegedly captured in Tonle Sap Lake by the Khmer Rouge who used the oil from dolphins in lamps and motorbike and boat engines, and also ate dolphin meat. After the Pol Pot Regime when guns were abundant throughout the country, Vietnamese and Khmer soldiers reportedly shot at dolphins for target practice. Many interviewees from Stung Treng Province in Cambodia reported that they had observed groups of dead dolphins floating dead downstream after the Pol Pot Regime (Beasley, 2007).

Few of the respondents to the MDCP survey had recently sighted dolphins in the river south of Kratie township. Elderly respondents south of Kratie township reported that historically dolphins regularly occurred in the river in front of their village. However, now children in these same villagers have never seen dolphins and many children believe dolphins are mythical creatures, similar to dragons (Beasley, 2007).

3.2. MDCP Conservation Initiatives

MDCP research established that the Irrawaddy dolphin population inhabiting the Mekong River was small and facing unsustainable mortalities. In 2003, in parallel with

---

BOX I

CAMBODIAN DOLPHIN FOLKLORE

The legend of the dolphin in Cambodia refers to beliefs and gods from the Indian Brahamic civilization, which left a strong mark on the Khmer civilization (Lloze, 1973). According to the legend, as quoted from Lloze (1973):

There was once, near a Khmer village, a banian tree inhabited by a spirit. A young girl came to the tree one day to make an offering to the spirit, who, highly moved, recognised in her the woman that he had loved during one of his previous lives. In order to live once again with her, as he was still in love with her, he asked for the help of the powerful god Indra, who gave the spirit the power to change into a python so that he could go and see the young girl without being recognised by the people of the village. The spirit was therefore able, each night, to go and pay a loving visit to his beloved, to whom he had, of course, made it known who he was.

To complete the happiness of his lover, and to reward her parents for their co-operation, the spirit revealed to them the location of a treasure hidden in the forest that made the family very rich. This story spread and made the simple people very envious. In a neighbouring village, a peasant couple thought that it would be enough to marry their daughter off to a python in order to acquire a great fortune. The peasant therefore went to the forest and soon found an enormous python that was half dead of starvation. He brought the python home and the preparations for the big ceremony began immediately. That night, the young bride was delivered to her starving husband, who, famished, started to devour her from the feet upwards. The cries of the poor bride made no difference as the parents were determined that this marriage be consummated.

The calm that descended again on the married couple's room raised the suspicions of the mother, who went to investigate. She went into the room and immediately understood the cause and effect between the disappearance of her daughter and the distended stomach of the full husband, and raised the alarm in the household. The father immediately opened the stomach of the animal and freed the girl, who was still alive but covered in foul-smelling mucus. Try as they might, washing her in warm water had no effect and the smell remained. The young girl decided to take a bucket and to go and wash herself with the water of the Mekong River. No result. Confused, shamed and desperate, she decided to throw herself into the river, after putting the wooden bowl on her head. Touched by her beauty and her youth, the spirit of the river took pity on her and turned her into a dolphin. This is how the legend of the dolphin came about, this extraordinary animal with the body of a woman, and the rounded and bald head, as if covered by a receptacle with a rounded base.

---

continuing research, MDCP initiated a series of conservation initiatives that aimed to contribute to conservation of the remaining dolphin population. These activities consisted of (1) large-scale local and government notification of the dolphin carcass recovery program and importance of reporting dolphin carcasses; (2) public-awareness raising of dolphin conservation efforts through workshops, production of posters, information leaflets, and a Mekong dolphin coloring book; (3) increased enforcement of existing fisheries regulations through provision of a boat, engine and per diems to the local fisheries office; (4) initiation of an Integrated Conservation Development Project (ICDP) named "Dolphins for Development," which aimed to

BOX II

LAO DOLPHIN FOLKLORE

Baird commented (1991, pg. 4), "One of the main reasons why many Laos people believe that dead people are reincarnated as dolphins is because there is a widely known traditional Lao fairy-tale about dolphins that has helped to popularise this idea".

The following account of the fairy tale was provided by Ian Baird and is reproduced with his permission. It should be noted that there are various versions of this fairy tale, each somewhat different. This is one common version:

"Long ago, in the time of our ancestors, there was a young princess in Luang Phrabang named Nang Sida. She liked to trade, and asked her father, the king, if she could go to do commerce in China. He reluctantly agreed, but he insisted that a servant be assigned to protect her during the trip. His name was Thao Kha. Although it was closer to travel overland from Luang Phrabang to China, the journey over the mountains was hard and dangerous. Therefore, it was decided that it would be better to travel down the Mekong River to the South China Sea, and from there to travel along the coast to China. Therefore, a large raft was prepared, and all kinds of food and goods for trading were put on the raft. Finally Nang Sida and Thao Kha started downstream. A chicken, a duck, a frog, a peafowl and a drongo bird were also on the raft with them. During this age it was said that people and animals could talk to each other.

After travelling many days down the Mekong River by raft, they approached the Khone Falls just north of the present-day border between Laos and Cambodia. Both Nang Sida and Thao Kha were unfamiliar with this part of the Mekong River, and did not know about the Waterfalls along the river. As they approached the Li Phi Waterfalls (Somphamit Waterfalls) (not the Khone Phrapheng Waterfalls as some people have incorrectly

reported), the chicken was standing at the front of the boat, and was the first to see the Falls. He called out, "chote, chote, chote" (the sound of a chicken, which also means 'stop, stop, stop' in English). The duck then became aware of the impending disaster and called out, "vat, vat, vat" (the sound of a duck, which also means to paddle to shore in English). The frog decided to take a look and jumped into the water. He could see the Falls from underwater, and when he came up, he called out, "leuk, leuk, leuk" (the sound of a frog, which also means 'deep, deep, deep' in English). The drongo bird flew up in the air and could see that the raft was going to go over the high Waterfalls. He called out, "sia khong sia sen, sia khong sia sen" (the sound of a drongo, which also means that one will lose his life in English). Finally, the peafowl called out, "peo vong, peo vong, peo vong" (the sound of a peafowl, which also means to follow the main channel of water in English).

Nang Sida and Thao Kha heard all the animals, and consulted about the situation. Nang Sida said, "Most of the animals say not to go any further, but the peafowl says to go ahead. Who should we listen to?" Thao Kha answered that the peafowl was larger than the other animals, and was their leader, so they should follow its advice. Nang Sida agreed, and so they kept going straight downriver, and soon after the raft went over the Li Phi Waterfalls. Those birds that could fly were able to survive, but everyone else died. Nang Sida was reincarnated as 'Nok Sida' (a river tern bird), and Thao Kha was reincarnated as 'Pa Kha' (a dolphin).

This story can explain some of the behaviour of the dolphins to this day. The dolphins never travel upriver of the Khone Falls, because Thao Kha dares not enter Laos. He fears that the king of Luang Phrabang will harshly punish him for not being

Continued

BOX II *(cont'd)*

---

able to follow his duty and protect the life of Nang Sida.

Secondly, one can observe that wherever there are dolphins in the Mekong River, there are river terns flying near the surface of the river nearby. River terns eat fish, as do dolphins. When the dolphins chase small minnows, they often try to escape, swimming near the surface of the water. This provides the river terns with the perfect opportunity to dive down and catch minnows to

eat. The dolphins appear to provide them with good chances for feeding. This can, in fact, be observed in nature, and it is believed that this occurs because Thao Kha is still the servant of Nang Sida, and therefore is obliged to help her find food to eat.

This story illustrates how the dolphin and the river tern came to be, and why they are as they are today."

---

provide direct tangible benefits to the local community in return for their cooperation with conservation activities; and (5) development of a Mekong Dolphin Conservation and Management Plan (which was formally adopted as Cambodian national policy in January 2005).

Probably the most significant conservation activity to contribute toward the conservation of the remaining dolphin population was the Dolphins for Development ICDP. This project was conducted in collaboration with a local Cambodian NGO, the Cambodian Rural Development Team (CRDT).

3.2.1. Dolphins for Development ICDP

It is now widely acknowledged that community participation is crucial to the long-term success of conservation strategies (Alpert, 1996). To encourage local coexistence with wildlife, there is a need to estimate and offset the economic costs of wildlife conservation, and importantly, to make wildlife conservation beneficial to local people (Prins, 1992; Prins et al., 2000). The feasibility of community-based conservation initiatives (Berkes, 2004; Barrett et al., 2001), and the usefulness of establishing protected areas (Wilke et al., 2006) are still hotly debated. Recent attempts have been made to investigate strategies and incentives to increase local community cooperation with endangered

species' conservation, primarily in terrestrial protected areas (Ferraro, 2002). These conservation incentives lie on a spectrum from indirect (e.g., diversification of livelihood projects) to direct (e.g., land purchases), with respect to their conservation objectives (Ferraro, 2001; Ferraro and Kiss, 2002; Ferraro and Simpson, 2001; Main et al., 1999; McShane and Wells, 2004).

The Dolphins for Development ICDP aimed to provide tangible benefits to local communities in exchange for their cooperation with conservation efforts. Project components included (1) rural development and diversification of livelihoods; (2) management of the existing community-based ecotourism; (3) education and awareness raising; and (4) strengthening stakeholder relationships. The first Dolphins for Development project was initiated in Kampi Village (adjacent to Kampi Pool) in April 2004. A second project was initiated at Chiteal Village (on the Laos/Cambodian border), in December 2005 (Beasley, 2005).

CRDT was able to build on the existing relationships established by MDCP in each village to initiate the development projects. CRDT then regularly emphasized the close link between dolphin and habitat conservation and the development and livelihood diversification activities that the community was receiving. Although the project was limited by the seasonal flooding of the areas, the low capacity of villagers, and occasional political interference, examples of observable measures of short-term success included (1) increased infrastructure in the village; (2) diversification of livelihoods through provision of livestock and seeds; (3) increased ability of the villagers for infrastructure construction and livestock care; (4) community benefit from the dolphin-watching tourism through a community development fund; and (5) apparent (but unquantified) reduction of fishing activity in Kampi and Chiteal pools.

Interviews to gauge each community's knowledge of dolphins, perceptions of dolphins, and conservation and socioeconomic status were conducted prior to the implementation of the Dolphins for Development project. It will now be important to repeat the questionnaire, to assess the success/failure of various project components (see the importance of project evaluation in Margoluis and Salafsky, 1998).

4. THE ECONOMIC, POLITICAL, AND SOCIAL COMPLEXITIES OF MEKONG DOLPHIN CONSERVATION

The three countries of the lower Mekong region; Laos, Cambodia, and Viet Nam, are developing quickly and face burgeoning pressures from human overpopulation, excessive exploitation of resources, poverty, lack of basic services, and wide-scale corruption at all social levels. When developing strategies to conserve and manage the Irrawaddy dolphin population in the Mekong River (as well as other flora and fauna), consideration of economic, political, biological, and social factors is of major importance. It is now recognized that successful conservation biology requires the integration of all these considerations (McShane and Wells, 2004). As Stankey and Shindler (2006) state:

effective policies for management of rare and little-known species must not only be scientifically valid and cost-effective but also consistent with prevailing social beliefs and values. Failure to foster understanding and support will leave management dominated by conflict and continued species loss.

The conservation situation in the Mekong River is beginning to resemble the dire situation in the Yantze River, China, where efforts to manage the Yangtze River dolphin, or baiji (*Lipotes vexillifer*) have failed. The baiji has recently been proclaimed "effectively extinct" after a large-scale survey throughout the river failed to sight a single individual (Lovgren, 2006). This unfortunate situation is largely as a result of extreme anthropogenic pressures (e.g., dam construction, agricultural and industrial pollution, riverine development, boat traffic, and fishing) associated with an exploited habitat, where 5% of the world's total human populations lives (Dudgeon, 2005; Yang *et al.*, 2006). Although baiji conservation efforts have been evident since the 1980s, the commitment to baiji conservation efforts by international NGOs, the Chinese government, and various stakeholders has been debated (Reeves and Gales, 2006). A major impediment to baiji conservation is the severely degraded state of the Yangtze River. There are no prospects for improvement in the near future (Dudgeon, 2005; Reeves and Gales, 2006). Recent debates regarding baiji conservation efforts are evident in the literature and unfortunately it appears that the baiji will be the first modern day cetacean species to become extinct (Dudgeon, 2005; Reeves and Gales, 2006; Wang *et al.*, 2006; Yang *et al.*, 2006).

Regardless of the future directions of baiji conservation efforts (if any), it is imperative that lessons are learnt from the unsuccessful efforts to date. The problems of habitat destruction, high human population growth in the catchment, and lack of stakeholder involvement and commitment need to be considered and appropriate solutions applied to conservation of the Irrawaddy dolphin population inhabiting the Mekong River, and other freshwater dolphin populations.

4.1. Early Progress with Dolphin Conservation in the Mekong River

Although various issues associated with dolphin-watching tourism at Chiteal and Kampi pools were evident before MDCP began in the late 1990s, there was little economic and political interest in dolphin conservation in the region during the early 2000s. The Cambodian government (through the Department of Fisheries) was supportive of research and conservation activities, and MDCP was able to conduct most activities with little government intervention. The dolphin-watching tourism industry presented one of the greatest challenges to conservation (see Section 2.2), particularly at Chiteal Pool on the Laos/Cambodian border, where few restrictions were imposed on the boats involved with dolphin-watching tourism, or the revenue generated from the industry. The situation was initially somewhat different at Kampi Pool. An agreement for increased community benefit was reached between government and the Kampi community at the end of 2004 (discussed further in Section 4.4), and it seemed that the area showed promise for positive ecotourism.

4.2. Recent Developments

In January 2005, the World Wildlife Fund for Nature (WWF) Cambodia Program assumed formal responsibility for dolphin research and management in the Mekong River, as part of their Mekong Living Water's Initiative. In mid-2005, the Cambodian government developed the Commission for Dolphin Conservation, to direct future dolphin conservation efforts. The Commission initially had responsibilities for both dolphin conservation and the promotion of tourism development, a potential conflict of interest when dolphins are a major target of tourism efforts. Large-scale tourism development focusing on the dolphins is now planned for the Kampi to Khone Falls river section. Already the number of tourism boats operating at Kampi Pool has increased to 20 (compared to 9 in 2005).

In 2005, the Cambodian Government also decreed that the Kratie to Khone Falls river stretch would be a gill net free area, in a well-intentioned attempt to conserve the remaining dolphin population from a major source of anthropogenic mortality. Unfortunately, there was no prior consultation with the local communities along the river, who commonly use gill nets for subsistence fishing. Additionally, no alternative livelihoods were provided (except independently by CRDT at Kampi and Chiteal Villages in 2004), or gear modification trials conducted, before this legislation came into effect. These well-enforced initiatives have inevitably alienated the local people, despite the communities' positive perceptions toward dolphins and demonstrated willingness to participate in small-scale management efforts previously initiated by MDCP.

4.3. Large-Scale Conservation Concerns

Serious concerns for the survival of the dolphin population now exist resulting from various sources of habitat degradation.

4.3.1. Dam Construction

Plans for dam construction across the mainstream Mekong River in southern Laos and Cambodia are of paramount concern. The construction of a single large run-of-the-river dam in the mainstream Mekong River in southern Laos or Cambodia could quickly catalyze the extinction of the remaining Mekong dolphin population. The negative effects of large-scale dams on major river systems are well documented (Dudgeon, 2000a,b,c,d; McCully, 2001).

Two dams are currently in the planning stage. One situated just north of Sambor Kratie Province, is currently being investigated by a Chinese company. There are reportedly two options for this dam, one being a large-run-of-the-river dam that would block all traffic and fish migrations, and the other a smaller dam across part of the river, that would still allow boat traffic to pass (TERRA, 2007).

A second dam, the Don Sahong dam, is now planned for the mainstream Mekong River in the Khone Falls area, Khong District, Champasak Province, southern Laos. Recently, a Malaysian company, Mega First Corporation Berhad (MFCB), reportedly signed an agreement with the Laos government to conduct an 18-month feasibility study for the Don Sahong dam, with the goal of signing an agreement to build the dam if the study's results are favorable (TERRA, 2007). Construction of this dam would have significant negative ramifications for fisheries, the environment, tourism, and all communities along the Mekong whose livelihood depends on sustained fisheries, including communities in neighboring countries such as Cambodia, Viet Nam, and Thailand. As stated by Phil Hirsch, Director, Australian Mekong Resource Centre, in an open letter to governmental and international agencies responsible for managing and developing the Mekong River "the location of this proposed dam is probably the worst possible place to site a 240 MW project since it is the point of maximum concentration of fish migration in the river that supports to world's largest freshwater fishery" (TERRA, 2007).

4.3.2. Contaminants and Disease

As mentioned previously, the Irrawaddy dolphin population will not survive in the long term in the Mekong River unless the cause(s) of newborn mortality is established, and managed effectively. A major cause for concern, particularly for newborns, is the potential for contaminants to be released into the Mekong ecosystem, working up the trophic level to potentially lethal levels in the dolphin population. In many cetacean populations, contaminants are known to be passed from the mothers' milk onto newborns, which can lead to reproductive failure, immunosuppression, and congenital defects (De Guise et al., 1995). An example of the effects of contaminants is shown by a high incidence of tumors in the St Lawrence beluga (Delphinapterus leucas) population of Canada. This cetacean population has been recorded as having some of the highest concentrations of polychlorinated biphenyls (PCBs) in blubber of all cetaceans (De Guise et al., 1995; Gauthier et al., 1998).

Small-scale gold mining is prevalent in Cambodia, as documented in a recent report by Oxfam America (Sieng, 2004). Associated with these gold mines is the potential for mercury (used to separate gold from sediment), to enter waterways and accumulate in fish, which are subsequently eaten by dolphins (and humans). Other contaminants originating from agriculture uses and/or industry are additional cause for concern.

Disease and toxins in the water system may also affect the dolphins' health, or even cause death, particularly in the event that the dolphins' immune system may be already compromised by other factors, such as habitat degradation, stress from boats, lack of prey, and/or reduced genetic fitness.

4.4. Dolphin-Watching Tourism: A Case Study of Economic Interests Influencing Conservation

As explained above, there are two locations in the Mekong River where tourists can view Irrawaddy dolphins, Chiteal and Kampi pools. Tourism was initiated at Kampi Pool in 1997 by an international NGO, Community Aid Abroad, with a local committee of seven villagers from Kampi Village. From 1997 to 2000, viewing of dolphins was conducted sporadically from land, with no formal management. International tourists were also able to view dolphins by small rowboat, opportunistically, for a small fee (US$1). Only the seven families were allowed to participate in the dolphin-watching tourism. In 2001, the seven villagers changed the small rowboats to larger "stand-up" paddle-boats with motors and sunshades over the boats for the tourists. These arrangements ensured tourist comfort and enabled dolphin viewing all year round (previously, the small rowboat was unsafe when the current was strong during the wet season).

In 2002, the Kratie Tourism Department became formally responsible for dolphin-watching tourism at Kampi Pool and cooperated with the seven families. No other families were allowed to participate in the venture and the financial benefits (50% of revenue) were distributed only to the seven families, with Kratie Tourism Department receiving the remaining revenue. Most villagers were unable to participate in the tourism but had lost their rights to fish in the pool as a result of a Provincial Decree prohibiting fishing in the pool in the early 2000s. Conflict was rife and the seven families became segregated from the other villagers. No management plan existed for tourism development and the boats were unregulated. Local people were unaware that the sound from the boat motors and the boats' activities had the potential to interfere with the dolphins' daily activities. Additionally, villagers were unable to communicate with foreign tourists and no information (verbal or printed) was provided to the tourists regarding the dolphins, or their conservation status in the river. Thus, the situation was unmanaged and unregulated and unable to contribute to dolphin conservation or management.

In March 2004, MDCP initiated a project to promote the sharing of revenue to the local community from the dolphin-watching tourism industry, as part of the Dolphins for Development project (see Section 2.1). The aims of the community-based tourism project were to (1) promote community benefit from dolphin-watching tourism implemented prior to this project's inception; (2) encourage effective management of this industry to minimize threats to the dolphin group inhabiting this area; and (3) promote visitor satisfaction and awareness raising of dolphin conservation and status.

MDCP developed dolphin-watching guidelines in cooperation with the boat owners and Kratie Department of Tourism, and produced awareness materials for tourists (printed in both English and Khmer). To promote community benefit, discussions and workshops were undertaken with all relevant stakeholders throughout the remainder of 2004. In December 2004, a written agreement was finalized and signed by the Kratie Department of Tourism to ensure that an entrance fee (US$2 per international tourist, US$0.15 per national tourist) would be introduced and shared between the community (40% for development activities), Department of Tourism (30% to ensure maintenance of the tourism site), and Department of Fisheries (30% for dolphin conservation activities). However, all revenue from the boat hire (US$2-4 per hour for each boat) continued to go to the government and boat owners only. Critical to the success of this agreement was that the community had the capacity to manage these funds adequately and that all activities were accountable and transparent to preempt corruption. CRDT played an essential role in this process through the establishment and development of a Village Development Committee (VDC) that was able to initiate an effective process for management.

The newly formed Government Commission for Dolphin Conservation cancelled this agreement in January 2007—despite a national policy on poverty alleviation. The commission instead allowed the community to operate two of the 20 tourist boats cooperatively, and distribute the revenue gained from these two boats among the remaining 124 families. A flat entrance fee of US$7 per person was charged, as of April 2007; US$1.50 was distributed to the relevant boat owners, and all other proceeds went to the government—no other revenue from the entrance fee went back to the community. Allowing more tourist boats to operate in the pool has significantly reduced the benefits to each boat owner, exacerbated village hostilities, and increased the level of boat harassment dolphins are exposed to daily.

There is no information on the effects the 20 tourist boats currently operating at Kampi Pool are having on the resident dolphins. However, studies of other dolphin populations indicate that continuing dolphin-watching tourism may have detrimental impacts. Such reported impacts have included (1) changes in swim direction (Lemon et al., 2006; Nowacek et al., 2001); (2) lengthened interbreath intervals (Lusseau, 2003a); (3) reduction in inter-individual distances (Bejder et al., 1999); (4) changes in the types of surface behaviors exhibited (Lemon et al., 2006); reductions in resting behavior (Constantine et al., 2004; Lusseau, 2003b); (5) an increase in breathing synchronicity between individuals (Hastie et al., 2003); (6) and increased rates of whistle production (Buckstaff, 2004).

The cumulative short-term effects outlined above may result in serious long-term conservation concerns. For example, in Shark Bay, Western Australia, bottlenose dolphins moved out of their preferred habitat in response to increased dolphin-watching tourism and the reproductive potential of females exposed to dolphin-watching tourism appears to have decreased (Bejder et al., 2006).

As a result of the critically endangered status of the Mekong dolphin population, adequate studies on the effects of dolphin-watching tourism on dolphins in the Mekong River would be beneficial before the industry expands further at Kampi Pool, or to other areas of the river that dolphins are known to inhabit.

5. THE POTENTIAL FOR SUSTAINABLE CONSERVATION

The Irrawaddy dolphin population inhabiting the Mekong River now numbers fewer than 127 individuals, is declining, and facing continued threats. The survival of the dolphins is inextricable linked to the well-being of local communities along the river and requires their support for conservation.

To be effective, conservation initiatives must be ecologically, managerially, socially, and economically sustainable (Fig. 15.6). The present management arrangements for the dolphins in the Mekong River do not meet these criteria. While the complete ban on gill net fishing over 190 km river stretch may be designed to remove a significant threat to the dolphins, this arrangement is not sustainable from a social or managerial perspective because it has the potential to negatively affect thousands of local subsistence villagers along the river, and is likely to create significant resentment and hostility toward dolphins in the river. The present tourism arrangements at Kampi Pool are also unlikely to be locally accepted or sustainable, particularly since the effects of the dolphin-watching tourism on dolphin behavioral ecology remain unknown.

Establishing the cause(s) of newborn mortality is of critical importance to the populations' long-term survival. The retrieval of fresh dolphin carcasses to enable necropsy and adequate tissue examination by a qualified veterinarian is essential to this process. As evidenced during the MDCP, as a result of the remoteness of the river section that the dolphins inhabit, dolphin carcass retrieval fundamentally relies on the support and cooperation of the local community living along the river. Thus, the merits of a large-scale ban on gill-netting which may reduce dolphin entanglements must be carefully balanced against the negative effects of

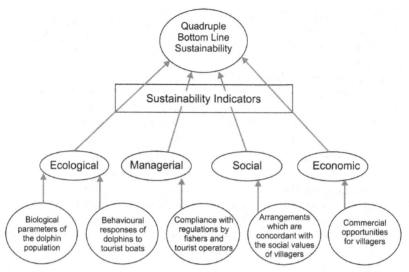

FIGURE 15.6 Conceptual framework for the arrangements required to ensure the sustainability of the Mekong River dolphin population. Modified from a diagram developed by Alastair Birtles and his research group at James Cook University.

alienating the local community and reducing their participation in conservation efforts, such as carcass recovery.

As with many conservation programs worldwide, it is a difficult management decision to gauge if short-term gains are worth the long-term loss of community support. For subsistence communities, the long-term loss may not just be a lack of support for the conservation efforts toward that species, but a loss of appreciation and support for any future conservation programs. As stated by Taylor and Gerrodette (1993), "endangered populations leave little margin for recovery from incorrect management decisions."

Habitat degradation is also a serious concern. Even with the most comprehensive management plan accepted by all stakeholders, dolphins will not survive in the river if adequate habitat is not available. Dolphins rely on deep-water areas during the dry season and annual fish migrations to replenish fish stocks. The construction of

dams along the mainstream lower Mekong River (particularly southern Laos or Cambodia), will no doubt substantially increase the risk of the Mekong dolphin population's extinction.

6. CONCLUSIONS

The freshwater Irrawaddy dolphin is a charismatic mega-vertebrate distributed along important river ecosystems. Such species have the potential to be effective flagship species for more generic freshwater biodiversity conservation initiatives that would benefit other riverine flora and fauna and local subsistence communities (Fig. 15.7).

The Irrawaddy dolphin population inhabiting the Mekong River is a remarkable natural asset. It will be a major loss to the people, government, culture, and environment of all lower Mekong countries if this dolphin population disappears forever.

FIGURE 15.7 An Irrawaddy dolphin from Kampi Pool, Kratie Province, Mekong River. The Irrawaddy dolphin is a charismatic mega-vertebrate species that is ideal as a flagship species in the Mekong River. Photograph: Yim Sak Sang. (See Color Plate 24)

7. RECOMMENDATIONS

Based on four and a half years' experience researching and conserving the Mekong dolphin population, we provide the following recommendations for ways forward to contribute to the effective conservation of dolphins in the river.

- Continue and expand the carcass recovery program, to ensure a qualified veterinarian examines all carcasses to (1) determine cause of death; (2) examine tissue samples for contaminants and disease; and (3) collect relevant samples for life history analyses. Establishing the cause of newborn death should be the highest priority.
- Initiate immediate discussions and cooperation with local communities and relevant stakeholders to find a sustainable resolution to the gill net entanglements and preservation of fish stocks in the river. A detailed socioeconomic study of the effects of the gill net ban, local perceptions and potential solutions would be very beneficial.
- Based on the results of the socioeconomic survey above, reassess the ban on gill-netting from the entire 190 km river stretch, and if appropriate, instead focus conservation and rural development efforts on the deep-pool habitats commonly frequented by dolphins during the dry season.
- Conduct immediate studies on the effects of dolphin-watching tourism on dolphins in the river and initiate management as required. These studies are essential before the industry expands further at Kampi Pool, or to other areas of the river dolphins are known to inhabit.
- Conduct an independent review and evaluation of dolphin conservation (WWF Cambodian Mekong Dolphin Conservation Project and Government Commission on Dolphin Conservation) and rural development/livelihood diversification (Cambodian Rural Development Team) activities as a matter of priority, with appropriate program adaptation if required.
- Continue dedicated long-term monitoring of the dolphin population, preferably through photo-identification.
- Encourage all stakeholders (particularly high-level government officials), to express significant concern over plans to construct any dams on the mainstream Mekong River.

ACKNOWLEDGMENTS

We would like to thank the various funding agencies for providing the financial assistance that ensured fieldwork could be completed. Thanks to the Mekong River Commission—Environment Program, Ocean Park Conservation Foundation, British Embassy—Phnom Penh, Rufford Foundation Small Grants

Program, Whale and Dolphin Conservation Society, the Society for Marine Mammalogy, IUCN Mekong Wetlands Biodiversity Project, Wildlife Conservation Society—Research Fellowship Program, School of Tropical Environment Studies and Geography, James Cook University, and Aruna Technologies.

Many thanks to the Cambodian Ministry of Forestry and Fisheries (MAFF) and the Cambodian Department of Fisheries for their full support of research and conservation activities in Cambodia. Our sincerest thanks to Excellency Chan Sarun (Minister of Agriculture, Forestry and Fisheries), Excellency Nao Thuok (Director General, Department of Fisheries, Cambodia), Mr Sam Kim Lun (Chief of Kratie Fisheries Office), and Mr Mao Chan Samon (Chief Stung Treng Fisheries office). Thanks also to all the Department of Fisheries counterparts for their efforts in the project, which has ensured its continued success, thanks in particular to Mr Phay Somany, Mr Kim Sokha and Mr Sean Kin.

Special thanks to Yim Saksang and Lor Kim San who were local MDCP local counterparts throughout the duration of the project. Additionally, we would sincerely like to thank the Cambodian Rural Development Team (CRDT) team members, particularly Brendan Boucher, Or Channy, Sun Mao, Hang Vong, Hean Pheap, and Ty Ratana. CRDT's dedication and professionalism throughout all aspects of the Dolphins for Development project were integral to its success, and could not have been achieved by any other group in Cambodia.

References

Abell, R. (2002). Conservation biology for the biodiversity crisis: A freshwater follow-up. *Conservation Biology* **16**(5), 1435-1437.

Alpert, P. (1996). Integrated conservation and development projects: Examples from Africa. *BioScience* **46**(11), 845-855.

Arnold, P. and Heinsohn, G. E. (1996). Phylogenetic status of the Irrawaddy dolphin *Orcaella brevirostris* (Owen in Gray): A cladistic analysis. *Memoirs of the Queensland Museum* **39**, 141-204.

Baird, I. G. (1991). "Preliminary Survey of the Irrawaddy Dolphin (*Orcaella brevirostris*) in Lao PDR," Unpublished report submitted to the Earth Island Institute and the Whale and Dolphin Conservation Society, 12pp.

Baird, I. G. and Beasley, I. L. (2005). Irrawaddy dolphin *Orcaella brevirostris* in the Cambodian Mekong River: An initial survey. *Oryx* **39**(3), 301-310.

Baird, I. G. and Mounsouphom, B. (1994). Irrawaddy dolphins (*Orcaella brevirostris*) in southern Lao PDR and northeastern Cambodia. *Natural History Bulletin of the Siam Society* **42**, 159-175.

Baird, I. G. and Mounsouphom, B. (1997). Distribution, mortality, diet and conservation of Irrawaddy dolphins (*Orcaella brevirostris*) in Lao PDR. *Asian Marine Biology* **14**, 41-48.

Barrett, C. B., Brandon, K., Gibson, C., and Gjertsen, H. (2001). Conserving tropical biodiversity amid weak institutions. *Bioscience* **51**(6), 497-502.

Beasley, I. L. (2005). "Promoting the Diversification of Livelihoods Through Eco-tourism—A Pilot on the Mekong River, Kratie Province" Unpublished Report. Mekong Dolphin Conservation Project, Phnom Penh, 42pp.

Beasley, I. L. (2007). "Conservation of Irrawaddy Dolphins (*Orcaella brevirostris*) in the Mekong River: Biological and Social Considerations Influencing Management," Ph.D. Thesis. School of Earth and Environment Studies, James Cook University, Townsville, Australia, 427pp.

Beasley, I. L., Chooruk, S., and Piwpong, N. (2002a). The status of the Irrawaddy dolphin, *Orcaella brevirostris* in Songkhla Lake, Southern Thailand. *The Raffles Bulletin of Zoology* **10**(Suppl.), 75-83.

Beasley, I. L., Arnold, P., and Heinsohn, G. E. (2002b). Geographic variation in skull morphology of the Irrawaddy dolphin, *Orcaella brevirostris* (Owen in Gray 1866). *The Raffles Bulletin of Zoology* **10**(Suppl.), 15-34.

Beasley, I. L., Robertson, K., and Arnold, P. (2005a). Description of a new dolphin, the Australian snubfin dolphin *Orcaella heinsohni* sp. N. (Cetacea, Delphinidae). *Marine Mammal Science* **21**(3), 365-400.

Beasley, I. L., Sinh, L. X., and Hung, H. H. (2005b). "Mekong River Irrawaddy Dolphin Status in the Vietnamese Delta (May 2005 surveys)," Unpublished report submitted to the Mekong Wetlands Biodiversity and Sustainable Use Program, IUCN/MRC/UNDP, 38pp.

Beasley, I. L., Phay, S., Gilbert, M., Chanthone, P., Yim, S., Lor Kim, S., and Kim, S. (2007). Review of the status and conservation of Irrawaddy dolphins *Orcaella brevirostris* in the Mekong River of Cambodia, Lao PDR and Vietnam. In "Status and Conservation of Freshwater Populations of Irrawaddy Dolphins," Smith, B. D., Shore, R. G., and Lopez, A. (eds.), pp 67-82. Wildlife Conservation Society Working Paper No. 31. Wildlife Conservation Society, New York, 115pp.

Bejder, L., Dawson, S. M., and Harraway, J. A. (1999). Responses by Hector's dolphins to boats and swimmers in Porpoise Bay, New Zealand. *Marine Mammal Science* **15**, 738-750.

Bejder, L., Samuels, A., Whitehead, H., Gales, N., Mann, J., Conner, R., Heithaus, M., Watson-Capp, J., Flaherty, C., and Krutzen, M. (2006). Decline in relative abundance of bottlenose dolphins exposed to long-term disturbance. *Conservation Biology* **20**, 1791-1798.

Berger, J. (1990). Persistence of different sized populations: An empirical assessment of rapid extinctions in bighorn sheep. *Conservation Biology* **4**, 91-98.

Bergmans, W. (1995). On mammals from the People's Democratic Republic of Laos, mainly from Sekong Province and Hongsa Special Zone. *International Journal of Mammalian Biology* **60**, 286-306.

Berkes, F. (2004). Rethinking community-based conservation. **18**(3), 621-630.

Buckstaff, K. C. (2004). Effects of watercraft noise on the acoustic behavior of bottlenose dolphins, *Tursiops truncatus*, in Sarasota Bay, Florida. *Marine Mammal Science* **20**, 709-725.

Caughley, G. and Gunn, A. (1996). "Conservation Biology in Theory and Practice," Blackwell Science, London, 459pp.

Chung, B. D. and Ho, D. T. (2002). A review of the results of the studies on marine mammals in Vietnamese waters. Working paper CMS/SEAMAMII/Doc.4 presented *In* "Second International Conference on the Marine Mammals of Southeast Asia," Dumaguete City, Philippines, 5pp.

Coates, D., Poeu, O., Suntornratana, D., Tung, N. T., and Viravong, S. (2003). "Biodiversity and Fisheries in the Lower Mekong Basin," Mekong Development Series No. 2. Mekong River Commission, Phnom Penh, Cambodia, 30pp.

Constantine, R., Brunton, D. H., and Dennis, T. (2004). Dolphin-watching tour boats change bottlenose dolphin (*Tursiops truncatus*) behaviour. *Biological Conservation* **117**, 299-307.

De Guise, S., Martineau, D., Béland, P., and Fournier, M. (1995). Possible mechanisms of action of environmental contaminants on St. Lawrence beluga whales (*Delphinopterus leucas*). *Environmental Health Perspectives*. **103**(4), 73-77.

Dudgeon, D. (2000a). Conservation of freshwater biodiversity in Oriental Asia: Constraints, conflicts and challenges to science and sustainability. *Limnology* **1**, 237-243.

Dudgeon, D. (2000b). The ecology of tropical Asian rivers and streams in relation to biodiversity conservation. *Annual Review of Ecology and Systematics* **31**, 239-263.

Dudgeon, D. (2000c). Large-scale hydrological changes in tropical Asia: Prospects for riverine biodiversity. *BioScience* **50**(9), 793-806.

Dudgeon, D. (2000d). Riverine biodiversity in Asia: A challenge for conservation biology. *Hydrobiologia* **418**, 1-13.

Dudgeon, D. (2005). Last chance to see: Ex situ conservation and the fate of the baiji. *Aquatic Conservation: Marine and Freshwater Ecosystems* **15**, 105-108.

Ferraro, P. J. (2001). Global habitat protection: Limitations of development interventions and a role for conservation performance payments. *Conservation Biology* **15**(4), 990-1000.

Ferraro, P. J. (2002). The local costs of establishing protected areas in low-income nations: Ranomafana National Park, Madagascar. *Ecological Economics* **43**, 261-275.

Ferraro, P. J. and Kiss, A. (2002). Direct payments to conserve biodiversity. *Science* **298**, 1718-1719.

Ferraro, P. J. and Simpson, R. D. (2001). Cost-effective conservation: A review of what works to preserve biodiversity. *Resources* **143**, 17-20.

Fisher, S. J. and Reeves, R. R. (2005). The global trade in live cetaceans: Implications for conservation. *Journal of International Wildlife Law and Policy* **8**, 315-340.

Gauthier, J. M., Pelletier, E., Brochu, C., Moore, S., Metcalfe, C. D., and Béland, P. (1998). Environmental contaminants in tissues of a neonate St Lawrence Beluga whale. (*Delphinapterus leucas*). *Marine Pollution Bulletin* **36**(1), 102-108.

Gilbert, M. and Beasley, I. L. (2006). "Mekong River Irrawaddy Dolphin Stranding and Mortality Summary: January 2001-December 2005," Wildlife Conservation Society—Cambodia Program, Phnom Penh, 39pp.

Gruvel, A. (1925). "L'Indochine, ses Richesses Marines et Fluviales," Society for the Education of Geological Maritime College, Paris, 215pp (not viewed).

Hastie, G. D., Wilson, B., Turfft, L. H., and Thompson, P. M. (2003). Bottlenose dolphins increase breathing synchrony in response to boat traffic. *Marine Mammal Science* **19**, 74-84.

Johannes, R. E., Freeman, M., and Hamilton, R. (2000). Ignore fishers knowledge and miss the boat. *Fish and Fisheries* **1**, 257-271.

Kittiwattanawong, K., Chantraporysl, S., Ninwat, S., and Chooruk, S. (2007). Review of the status and conservation of Irrawaddy dolphins *Orcaella brevirostris* in Songkhla Lake of Thailand. *In* "Status and Conservation of Freshwater Populations of Irrawaddy Dolphins," Smith, B. D., Shore, R. G., and Lopez, A. (eds.), pp 83-89. Wildlife Conservation Society Working Paper No. 31. Wildlife Conservation Society, New York, 115pp.

Kreb, D. (2004). "Conservation and Social Ecology of Freshwater and Coastal Irrawaddy Dolphins in Indonesia," Ph.D. Dissertation. Institute for Biodiversity and Ecosystem Dynamics, University of Amsterdam, Amsterdam, 230pp.

Kreb, D. (2005). Abundance of freshwater Irrawaddy dolphins in the Mahakam River in East Kalimantan, Indonesia, based on mark-recapture analysis of photo-identified individuals. *Journal of Cetacean Research and Management* 6(3), 269-277.

Kreb, D., Budiono, and Syahrani. (2007). Review of the status and conservation of Irrawaddy dolphins *Orcaella brevirostris* in the Mahakam River of Indonesia. *In* "Status and Conservation of Freshwater Populations of Irrawaddy Dolphins," Smith, B. D., Shore, R. G., and Lopez, A. (eds.), pp 53-66. Wildlife Conservation Society, Working Paper No. 31. Wildlife Conservation Society, New York, 115pp.

Krempf. (1924-1925). "Campagnes de Lanessan Rapp," Institute Oceanography Indochine, Hanoi, 255pp (not viewed).

Lemon, M., Lynch, T. P., Cato, D. H., and Harcourt, R. G. (2006). Response of travelling bottlenose dolphins (*Tursiops aduncas*) to experimental approaches by a powerboat in Jervis Bay, New South Wales, Australia. *Biological Conservation* 127, 363-372.

Lloze, R. (1973). "Contribution a L'etude Anatomique, Histologique et Biologique de l'*Orcaella brevirostris* (Gray, 1866) (Cetacea, Delphinidae) du Mekong," Ph.D. Dissertation. L'Universite Paul Sabatier de Toulouse, France, 598pp.

Lovgren, S. (2006). China's rare dolphin now extinct, experts announce. National Geographic News. http://news.nationalgeographic.com/news/2006/12/061214-dolphin-extinct.html. Accessed 01 February 2007.

Lusseau, D. (2003a). Male and female bottlenose dolphins *Tursiops* spp. have different strategies to avoid interactions with tour boats in Doubtful Sound, New Zealand. *Marine Ecology Progress Series* 257, 267-274.

Lusseau, D. (2003b). Effects of tour boats on the behavior of bottlenose dolphins: Using Markov chains to model anthropogenic impacts. *Conservation Biology* 17, 1785-1793.

Main, M., Roka, B. F., and Noss, R. F. (1999). Evaluating costs of conservation. *Conservation Biology* 13(6), 1262-1272.

Margoluis, R. and Salafsky, N. (1998). "Measures of Success: Designing, Managing and Monitoring Conservation and Development Projects," Island Press, Washington, DC, 382pp.

Mayer, S. and Simmonds, M. (1996). Science and precaution in cetacean conservation. *In* "The Conservation of Whales and Dolphins," Simmonds, M. and Hutchison, J. D. (eds.), pp 391-406. Chichester, John Wiley and Sons, New York.

McCully, P. (2001). "Silenced Rivers: The Ecology and Politics of Large Dams," Zen Books, London, 249pp.

McShane, T. O. and Wells, M. P. (2004). Integrated conservation and development. *In* "Getting Biodiversity Projects to Work: Towards More Effective Conservation and Development," McShane, T. O. and Wells, M. P. (eds.), pp 3-9. Columbia University Press, New York, 464pp.

Mouhot, H. (1966). "Henri Mouhot's Diary—Travels in the Central Parts of Siam, Cambodia and Laos During the years 1858-1861" Oxford University Press, London, 238pp.

Nowacek, S. M., Wells, R. S., and Solow, A. R. (2001). Short-term effects of boat traffic on bottlenose dolphins, *Tursiops truncatus*, in Sarasota Bay, Florida. *Marine Mammal Science* 17, 673-688.

Owen, R. (1866). On some Indian Cetacea collected by Walter Elliot, Esq. *Transactions of the Zoological Society of London* 6, 17-47.

Pattnaik, A. K., Sutaria, D., Khan, M., and Behera, B. P. (2007). Review of the status and conservation of Irrawaddy dolphins *Orcaella brevirostris* in Chilika Lake of India. *In* "Status and Conservation of Freshwater Populations of Irrawaddy Dolphins," Smith, B. D., Shore, R. G., and Lopez, A. (eds.), pp 41-52. Wildlife Conservation Society, Working Paper No. 31. Wildlife Conservation Society, New York, 115pp.

Perrin, W. F., Dolar, M. L. L., and Alava, M. N. R. (1996). Report of the workshop on the biology and conservation of small cetaceans and dugongs of Southeast Asia, UNEP(W)/EAS WG.1/2. UNEP, Bangkok, Thailand, 101pp.

Pichler, F. B., Slooten, E., and Dawson, S. M. (2003). Hector's dolphins and fisheries in New Zealand: A species at risk. *In* "Marine Mammals: Fisheries, Tourism and Management Issues,"Gales, N., Hindell, M., and Kirkwood, R. (eds.). CSIRO, Collingwood, 460pp.

Poulsen, A. F. and Valbo-Jorgensen, J. (2001). Deep pools in the Mekong River. *Mekong River Commission: Catch and Culture* 7, 10-11.

Prins, H. H. (1992). The pastoral road to extinction: Competition between wildlife and traditional pastoralism in East Africa. *Environmental Conservation* 19, 117-123.

Prins, H. H., Grootenhuis, J. J., and Dolan, T. (2000). "Wildlife Conservation by Sustainable Use," Kluwer Academic Publishers, Boston, 496pp.

Reed, D. H., O'Grady, J. J., Brook, B. W., Ballou, J. D., and Frankham, R. (2003). Estimates of minimum viable population sizes for vertebrates and factors influencing those estimates. *Biological Conservation* 113(1), 23-34.

Reeves, R. R. and Gales, N. J. (2006). Realities of baiji conservation. *Conservation Biology* 20(3), 626-628.

Saunders, D. L., Meeuwig, J. J., and Vincent, C. J. (2002). Freshwater protected areas: Strategies for conservation. *Conservation Biology* 16(1), 30-41.

Sieng, S. (2004). "Small-scale Gold-Mining in Cambodia: A Situation Assessment," Oxfam America, Phnom Penh, Cambodia, 37pp.

Smith, B. D. and Beasley, I. L. (2004). *Orcaella brevirostris* (Mekong River subpopulation). *In* "IUCN 2004. 2004

IUCN Red List of Threatened Species." www.redlist.org. Downloaded on 1 November 2004.

Smith, B. D. and Jefferson, T. A. (2002). Status and conservation of facultative freshwater cetaceans in Asia. *The Raffles Bulletin of Zoology* **10**(Suppl.), 173-187.

Smith, B. D., Jefferson, T. J., Leatherwood, S., Ho, D. T., Thuoc, C. V., and Quang, L. H. (1997). Investigations of marine mammals in Vietnam. *Asian Marine Biology* **14**, 145-172.

Smith, B. D., Beasley, I. L., and Kreb, D. (2003). Marked declines in populations of Irrawaddy dolphins. *Oryx* **37**(4), 401.

Smith, B. D., Tun, T., and Tun, M. T. (2007). Review of the status and conservation of Irrawaddy dolphins *Orcaella brevirostris* in the Ayeyarwady River of Myanmar. *In* "Status and Conservation of Freshwater Populations of Irrawaddy Dolphins," Smith, B. D., Shore, R. G., and Lopez, A. (eds.), pp 21-39. Wildlife Conservation Society, Working Paper No. 31, Wildlife Conservation Society, New York, 115pp.

Soule, M. E. (1986). "Conservation Biology: The Science of Scarcity and Diversity," Sinauer Associates, Inc., Sunderland, MA, 584pp.

Stacey, P. J. (1996). "Natural History and Conservation of Irrawaddy Dolphins, *Orcaella brevirostris*, with Special Reference to the Mekong River, Laos PDR."M.Sc Thesis. University of Victoria.

Stacey, P. J. and Arnold, P. A. (1999). *Orcaella brevirostris*. *Mammalian Species* **616**, 1-8.

Stacey, P. J. and Leatherwood, S. (1997). The Irrawaddy dolphin *Orcaella brevirostris*: A summary of current knowledge and recommendations for conservation action. *Asian Marine Biology* **14**, 195-214.

Stankey, G. H. and Shindler, B. (2006). Formation of social acceptability judgments and their implications for management of rare and little-known species. *Conservation Biology* **20**(1), 28-37.

Taylor, B. L. and Gerrodette, T. (1993). The uses of statistical power in conservation biology: The vaquita and northern spotted owl. *Conservation Biology* **7**(3), 489-500.

TERRA. (2007). MRC silent as mainstream dams move forward (Press Briefing). www.wrm.org.uy/countries/Thailand/MRC/Mekong_mainstream_dams_media.pdf. Downloaded on 1 January 2008.

Thompson, P. M., Wilson, B., Grellier, K., and Hammond, P. S. (2000). Combining power analysis and population viability analysis to compare traditional and precautionary approaches to conservation of coastal cetaceans. *Conservation Biology* **14**(5), 1253-1263.

Vannaren, C. and Sean, K. (2001). "Fisheries Preservation in the Mekong River Pools in Stung Treng and Kratie Provinces, Cambodia,"Unpublished report submitted to Cambodian Department of Fisheries, 4pp.

Wade, P. R. (1998). Calculating limits to the allowable human-caused mortality of cetaceans and pinnipeds. *Marine Mammal Science* **14**, 1-37.

Wang, D., Zhang, X., Wang, K., Wei, Z., Wursig, B., Braulik, G., and Ellis, S. (2006). Conservation of the baiji: No simple solution. *Conservation Biology* **20**(3), 623-625.

Whitehead, H. (1997). Analysing animal social structure. *Animal Behaviour* **53**(5), 1053-1067.

Whitehead, H. (2005). "Programs for Analysing Social Structure: SOCPROG 2.2 (for MATLAB 6.5, Release 13)." Department of Biology, Dalhousie University, Halifax, Nova Scotia, Canada.

Wilke, D. S., Morelli, G., Demmer, A. J., Starkey, M., Telfer, P., and Steil, M. (2006). Parks and people: Assessing the human welfare effects of establishing protected areas for biodiversity conservation. *Conservation Biology* **20**(1), 247-249.

Yang, G., Bruford, M. W., Wei, F., and Zhou, K. (2006). Conservation options for the baiji: Time for realism? *Conservation Biology* **20**(3), 620-622.

16

Development Scenarios and Mekong River Flows

Ian Campbell

Principal Scientist, River Health, GHD, 180 Lonsdale Street, Melbourne 3000, Australia & Adjunct Research Associate, School of Biological Sciences, Monash University, 3800 Australia

1. INTRODUCTION

The Mekong is a large predictable flood pulse river (Fig. 16.1). The flood pulse concept was first enunciated by Junk et al. (1989) who pointed out that many large river floodplain systems, especially in tropical regions, showed regular annual flood pulses during which large areas of the flood plain are inundated. As a consequence, such systems have tight ecological linkages between the floodplain and the riverine ecosystems, and often very high productivity and biodiversity (Junk and Wantzen, 2004). The Mekong is one such system, as

are the Amazon and Pantanal in Brazil (Junk et al., 1989; Sioli, 1984), the Mississippi in the USA, and many rivers in the wet-dry tropics of Australia (Douglas et al., 2005).

By comparison with many other rivers, the size of the flood pulse in the Mekong is extremely predictable. One measure of the variability of river flow is the coefficient of variation (Cv), a measure of the variability of the average annual discharge from year to year, standardized to allow for differences in river size. MRC (2003a) noted that the Cv of the annual flow of the Mekong at Chiang Saen is only 0.2, while the worldwide average for rivers with

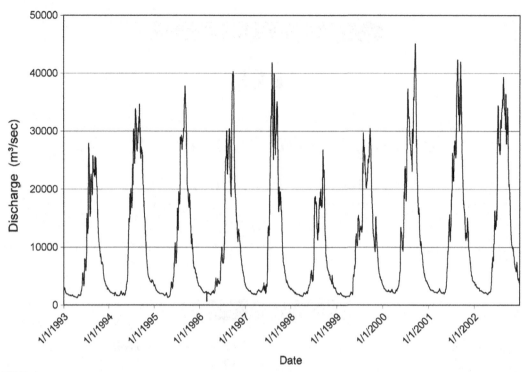

FIGURE 16.1   Daily discharge for the Mekong River at Pakse in southern Lao PDR from 1993 to 2003, illustrating the very strong annual flood pulse.

catchments larger than 10^5 km^2 is 0.33 (McMahon *et al.*, 1992). Further downstream, at Pakse, the site closest to the Cambodian floodplain for which there is a reliable long-term discharge data set, the Cv has dropped to 0.16, less than half the global average.

The timing of the Mekong flood is also fairly constant. Over the 10-year period between 1992 and 2003, the earliest the flood peak at Pakse occurred on July 20 and the latest on September 29. On average, it arrived on September 1 with a standard deviation of 23 days.

Like other flood pulse rivers, the Mekong is highly productive and highly diverse. The annual fish catch in the Mekong, estimated at about 2% of the world's total (see Hortle, Chapter 9), is one indicator of the productivity of the system. The Mekong is also notable for its high fish diversity (see Valbo-Jorgensen

et al., Chapter 8) and is a global diversity hot spot for freshwater mollusks (WCMC, 2007, Attwood, Chapter 11). The species diversity of the Tonle Sap Great Lake, within the Mekong system, is also high (Campbell *et al.*, 2006).

The dai fish catch in the Tonle Sap River, the only reliable fish catch data for the Mekong basin, correlates with the depth of the annual flood. The peak of the annual flood height explained 43% of the variation in the catch between 1995 and 2004 (data from Hortle *et al.*, 2004), while the maximum depth of the flood persisting for 31 days explained 46% of the variation. Whether the depth of the flood influences fish catch by increasing production, or whether with greater water depth illegal fishing techniques such as brush traps and floodplain fences are less effective, is not clear. But the former is widely believed.

Both government officials and civil society in the lower Mekong countries are very concerned about the potential for upstream developments to alter downstream river flows. Campbell (2007) documented results from workshops at which participants, mainly government officials from the four lower Mekong countries, were asked to identify the most significant transboundary environmental issues. "Dams and reduced dry season flows" was the second most serious issue identified after "water quality." It was identified as a serious issue in all five workshops from which data were available.

This concern has been reflected in the popular media with low flows during recent droughts often being blamed on dams in China. For example, during the drought in 2004 and earlier, articles implicating Chinese dams as a causative agent behind changes to Mekong River flows appeared in a range of regional papers and online media (Asia Times Online, 2002; Cambodianonline, 2004; Samabuddhi, 2004,) as well as in newspapers in Australia and Britain (Vidal, 2004), New Scientist (Pearce, 2004), and elsewhere. Subsequently, follow-up articles have appeared frequently in regional media (e.g., Bangkok Post, 2005, 2006). An analysis conducted by the MRC (MRC, 2004a) concluded that, while there was evidence of hydrological impacts of Manwan dam in China on flow variability at Chiang Saen in Thailand, there was no evidence that the existing Chinese dams played any role in contributing to the 2004 low flows. The low flows became more extreme downstream, and were evidently caused by reduced wet season rainfall throughout the basin.

Historically, Thailand has been the country most interested in diverting Mekong waters. Thailand sees the Mekong as a potential source of water that could be used to supplement water supplies in the central rice growing area in Bangkok as well as in the relatively dry north-east region. A number of schemes have been proposed at various times to divert water from the Mekong, or its tributaries, for use in Thailand. One well-known proposal is the Kok-Ing-Nan water diversion scheme proposed in 1994, which would divert water from Kok and Ing rivers, tributaries of the Mekong, into the Nan, a tributary of the Chao Phraya, which flows through central Thailand and Bangkok (Hori, 2000; Osborne, 2004). Another is the Kong-Chi-Mun plan, which would divert Mekong water from a location in the vicinity of Nong Khai into the Chi and Mun rivers, whence it could be used for dry season irrigation (Hori, 2000; Osborne, 2004). The scheme was first enunciated in 1992 and has recently been more circumspectly hinted at as part of the Green Isaan proposals vaguely espoused by the Thaksin government before its demise (Prateechaikul, 2003).

The Thai diversion plans have been controversial within Thailand, and have caused consternation in other Mekong riparian countries. Viet Nam is reported to have expressed strong opposition to these projects on the grounds that they would exacerbate salinity intrusion problems in the Mekong delta (Hori, 2000). The Lao government is also said to have expressed concerns about the impact the water extractions would have on dry season navigation in the river.

More recently, a large number of dam proposals in Laos and Thailand have been officially announced, or leaked to the press by NGOs (Nation, 2007). These include six new hydropower dams in Laos, including Don Sahong dam at Khone Falls, a mainstream dam near Ubon Ratchathani, where the river forms the Thai-Lao border and the Sambor Rapids dam in Cambodia. In an unusual step for the Mekong region, government officials from Cambodia have publicly criticized the Lao government for a lack of transparency, and suggested they would work to stop the construction of the Lao dams (Bangkok Post, 2007), indicating the level of concern they feel.

2. MODELING DEVELOPMENT IMPACTS

In view of the concerns about potential flow changes in the Mekong, it is not surprising that researchers and government agencies would attempt to develop predictions of the impacts of upstream flow regulation and water extraction activities on lower Mekong flows. One of the first attempts was that of Adamson (2001), who used a simple routing model to predict the macroimpacts of the proposed Yunnan reservoir cascade on downstream flows. He predicted that 20% regulation in China would produce a 50% increase in the average discharge in March, the minimum flow month, at Kratie in Cambodia. However, he found only very slight changes in wet season flows.

More substantial attempts to evaluate the potential effects of development on the river hydrology commenced in 2000 when the World Bank, under its GEF facility, funded the Mekong River Commission (MRC) Water Utilization Project (WUP), incorporating three components: a basin modeling package, development of rules for water utilization and institutional strengthening of the MRC secretariat, and the national Mekong committees (World Bank, 2000). The total funding for the project was USD $11 million, and the project was due for completion by December 2007. Its achievements have been modest.

The basin modeling component was apparently conceived as a model of everything. It was to include quantity and quality of surface water and groundwater, and to "incorporate components to allow the direct assessment of transboundary impacts on ecological, social and economic resources and conditions" (World Bank, 2000). The project was intended to develop a permanent modeling capacity within the MRC secretariat and the national Mekong committees so that the models could be applied and updated, but this aim has not been achieved, and the groundwater, water quality, and other components were never developed and were never possible due to lack of data.

The primary outcome of the modeling component of the WUP was a suite of linked hydrological models: a catchment runoff simulation model, a basin flows simulation model, and a hydrodynamic model. The catchment runoff model is based on the SWAT software developed by the US Department of Agriculture and is used to estimate flows to the other models. In the work discussed here, catchment runoff flows were the same for all scenarios. The basin flows simulation uses the IQQM software. It routes sub-basin flows through the river system and allows for diversions, consumptions, and dams and other control structures. This model was used for the river reach from the Chinese border to Kratie in Cambodia, where the river flows in a well-defined channel and floodplain inundation is minimal. Below Kratie, the package uses ISIS software to allow for modeling of tidal influences, the flow reversal of the Tonle Sap River, and salinity intrusions in the delta. The inputs to the ISIS model are the outputs from the IQQM model, so there will be a multiplication of errors as one moves downstream.

The models suffer most of the shortcomings that apply to deterministic quantitative models of complex systems (Pilkey and Pilkey-Jarvis, 2007), particularly when they are used in developing countries. The hydrological data on which the models were based contained large gaps and significant errors, which were corrected statistically, but the model has never been fully calibrated or validated. Following a review by an expert panel in August 2003 (MRC, 2004b), there was limited calibration of the models and Podger and co-workers (Podger et al., 2004) conducted further calibrations based on mass balances.

As a consequence, the models cannot be reliably used to make accurate quantitative predictions about future flows in the river, but they can be used for scenario testing. That is, given a particular flow in the river, what would be the consequences of various possible actions such as extractions of water for irrigation or the construction and operation of a dam. Even here, the predicted effects should be viewed as indicative (large, moderate or small), rather than precise numerical predictions.

The modeling of the effects of development of water resources incorporates a range of assumptions, apart from the assumptions included in the scenario development. These include assumptions about crop water use, evaporative losses, transmission losses, and dam operation rules. For some of these, there was data available from elsewhere that could provide a basis for deciding on values to incorporate in the models. However, the largest dams proposed in the Mekong basin are those in China, and although we know they will operate essentially as single purpose hydropower dams, the modelers had no further information about the guidelines or rules under which they would operate.

3. MEKONG DEVELOPMENT SCENARIOS

The Basin Development Plan Group at the MRC Secretariat devised a range of development scenarios that could be used to test the possible impacts of development on Mekong River flows. The scenarios were intended to be realistic, but not real. They were based on estimates of maximum development possible within a 20-year time frame within a series of sectors that impinge directly on water resources, including hydropower and irrigation. The scenarios were not intended as development plans in any sense, but were intended to allow testing of the sensitivity of river discharge and discharge patterns to various types of development within the basin.

The development scenarios became caught up in conflicts within the MRC and between the secretariat and the national Mekong committees. As a result, the CEO of the MRC, Dr Olivier Cogels, would not allow them to be publicly circulated, or the results discussed. Subsequently, several of the scenarios were used by a team from the World Bank, together with some additional scenarios they developed independently, and these were released publicly through the MRC website in 2004.

The outputs of five of the development scenarios initially modeled were used as the basis of discussion documents circulated internally by the MRC (MRC, 2005a,b) and, in part, released publicly by the World Bank (Podger et al., 2004). The internal MRC documents were used to support a project intended to assess the environmental and social consequences of impacts of flow changes that could result from the three development scenarios which resulted in the greatest hydrological changes.

Each of the modeled scenarios was compared with the baseline situation, as it was believed to exist in the year 2000. This specified a benchmark climate, based on that occurring in the basin between 1985 and 2000, land use, domestic and industrial demand, estimated area of irrigated crops, and existing hydropower dams and storage structures. The purpose of specifying a baseline condition was to allow the assessment of modeled change against a fixed baseline to provide a standard basis for comparison.

Each of the five modeled scenarios considered by the World Bank team had a different combination of development changes, including hydropower dams, irrigated areas, intensities of domestic use, and basin diversions (Table 16.1). The China dams scenario included the existing Manwan and Dachaoshan dams, and the Jinhong dam, which is under construction, and is of similar size to the others, as well as the two largest dams on the Mekong proposed by China: Xiaowan (which is under

TABLE 16.1 Summary of key parameters for modeled scenarios

Scenario	Domestic and industrial usage (mcm)	Irrigated areas (×'000 ha)	Hydropower dams' active storage volume (mcm)		Embankment area (×'000 ha)	Basin diversions (mcm)	
			lower Mekong basin	China		Intra	Inter
Baseline	1620	7422	6185	–	0	0	0
China dams	1620	7422	6185	22,700	0	0	0
Low development	3109	8316	12,433	10,300	0	0	0
Levees	3109	8316	12,433	10,300	130	0	0
Irrigated agriculture	4194	11,349	12,433	10,300	0	2200	3262
High development	4194	11,349	26,778	22,700	0	2200	3262

construction) and Nhaozhadu. The low development scenario included a minimum level of development based on population growth to 2020, including growth in domestic consumption, some growth in irrigation usage, and the impact of Xiaowan dam since it was under construction. The levees scenario was similar to the low development scenario, but with the assumption of 130,000 ha isolated from the Cambodian floodplain. The irrigated agriculture scenario included substantial increases in consumptive water use associated with irrigation expansion to the presumed limits of irrigable land in the lower basin. In other respects, it matched the low development scenario. The high development scenario included the water use of the irrigated agriculture scenario, the dams assumed in the China Dams scenario, and several more proposed tributary dams in the lower Mekong basin and a mainstream dam in Cambodia. Full details of all modeled flows can be found in Podger et al. (2004), so they will be discussed only in broad terms here.

The internal MRC documents considered four development scenarios, excluding the levees scenario included in Table 16.1 (MRC, 2004a,b). These documents were developed to inform a number of river and social scientists about potential hydrological changes as input to their own considerations about possible environmental and social consequences. Thus, the scenarios themselves included no consideration of any possible consequences.

The World Bank document (Podger et al., 2004), on the other hand, included a number of indicators to demonstrate the potential impacts of the development scenarios. These encompassed indicators of flow, Tonle Sap Lake and the floodplain, the delta, irrigation production, and hydropower. Some of these indicators, such as amount of hydropower generated, are relatively straightforward; others, such as those for fish impact are overly simplistic.

One of the key hydrological indicators included by Podger et al. (2004) was the wet season water level, because it provided an indicator of inundation. Inundation was thought to have ecological significance, affecting area of fish habitat for example, as well as social significance as an indicator of floods. On the floodplain, the period and extent of inundation were modeled as potentially important ecological factors. The extent of flooding indicates the area of wet season fish habitat, while the period of inundation would relate to how long fish had the habitat available, thus the length of the growth season for those fish whose larvae grow on the inundated floodplain.

Dry season discharges were also an indicator because of their potential importance to irrigation and salinity issues, but they also have ecological significance. It is during the dry season that farmers have most need for irrigation water, and low water levels require greater pumping costs; if levels are too low, there may be insufficient water to meet requirements. In addition if flows were very low, there may be insufficient water to dilute saline runoff known to occur in parts of northeast Thailand. Finally, the dry season is the season when many fish breed in the mainstream. Flows that are too low or too high can reduce available breeding habitat.

Flows were modeled for a series of sites along the river to indicate how patterns changed downstream, and also to allow assessment of potential changes at locations that are physically very different. Above Khone Falls, major concerns are flow impacts on the fishery and riverine ecosystems, water levels for navigation, especially in the dry season, and potential for flooding of river bank cities, including Vientiane, in the wet season. At Khone Falls and the Siphandone (Four Thousand Islands) area, maintenance of river morphology is an issue. Below Kratie, in Cambodia, the extent of floodplain inundation, the reversal of the Tonle Sap River, and the seasonal pattern of inundation around the Tonle Sap Great Lake are important issues. In the delta—where in February, March, and April, the estimated amount of water abstracted for irrigation (MRC, 2003b) would amount to 60%, 45%, and 40%, respectively, of the flow at Phnom Penh—adequate dry season flows for irrigation and reduction of saline intrusion are major issues.

4. PREDICTED HYDROLOGICAL CONSEQUENCES OF DEVELOPMENT

Agriculture constitutes the major potential extractive use of water in river systems. Irrigation is the greatest single human consumptive use of water. Compared with the volume of water used for domestic purposes such as drinking and washing, or for industrial purposes, water use for irrigation is usually three to four times the total of the other two. For example, Falkenmark and Rockstrom (2004) estimated the highest global per capita annual domestic and industrial water consumption occurred in the USA in the 1990s and was about 370 m^3, while the global average per capita water consumption for producing food was about 1200 m^3.

Thus, one of the major concerns investigated through scenario development and hydrological modeling was the potential for future irrigation development to impact on Mekong flows. The Basin Development Plan at MRC commissioned a consultant report into water use for agriculture in the lower Mekong basin (MRC, 2003b). The potential for irrigation is not unlimited in any area, since it requires suitable flat land, suitable soils, and suitable crops that can provide sufficient return on the investment.

Within the lower Mekong, the potential for extensive irrigation is limited by the availability of suitable land and soils. Total areas potentially suitable upstream of Kratie were identified as 1.9×10^6 ha in Thailand, 1.0×10^6 ha in Lao PDR, 4.1×10^6 ha in Cambodia, and 4.4×10^6 ha in Viet Nam. However, much of the irrigable land in the basin in Viet Nam and Thailand has already been developed, so the countries with the largest potential areas remaining are Cambodia (2.7×10^6 ha) and Lao PDR (0.7×10^6 ha) (MRC, 2003b).

Agricultural development has the potential to appreciably reduce dry season flows, especially during December (Podger *et al.*, 2004). The changes would be most pronounced downstream of Nong Khai, because the major areas with potential for irrigation development are in northeastern Thailand and Cambodia. The Kong-Chi-Mun plan, if it were to proceed, would divert irrigation water from the vicinity of Nong Khai, and irrigation in Cambodia would necessarily occur on the floodplain downstream of Kratie, and probably mainly downstream of Phnom Penh. Flow reductions, which Podger *et al.* (2004) predicted could be in the range of 200-300 $m^3 s^{-1}$ or of the order of 7% of the flow at the border between Viet Nam and Cambodia, could be offset by increases in dry season flows resulting from the Chinese dams.

Although the impact of increased irrigation is to reduce dry season flows, the World Bank team, in modeling the flow impact of irrigation, included increased hydropower development which increased dry season flows more than irrigation reduced them. This was not unreasonable since the most influential dams, such as Xiaowan, are already under construction; however, it precludes an analysis of the impacts of irrigation development alone.

As a result, the modeled flow changes for the irrigation development scenario show increasing rather than decreasing dry season flows. Irrigation and other diversions tend to counteract the increased dry season flows resulting from the operation of hydropower dams, but not sufficiently to eliminate their impact completely. The increase in dry season flows is most pronounced in upstream sites (Fig. 16.2), but even at Nong Khai, the increase is of the order of 400 $m^3 s^{-1}$ in March. The increase at Pakse is only half of that, and the increase is reduced to about 50 $m^3 s^{-1}$ by Kratie.

Wet season water levels are reduced under the irrigation scenario, which would reduce flooding, but once again the changes are most pronounced upstream, where flooding is least problematic (Fig. 16.3). The maximum

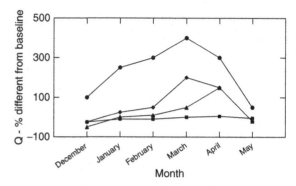

FIGURE 16.2 The percentage difference between modeled dry season flows and baseline flows under an irrigation development scenario for Nong Khai, Thailand (circles); Pakse, Lao PDR (diamonds); Kratie, Cambodia (triangles); and Tan Chau, Viet Nam (squares).

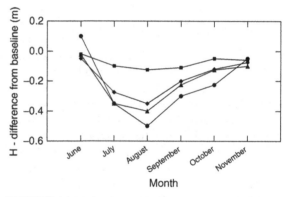

FIGURE 16.3 Modeled reductions in wet season water levels (in meters) under an irrigation development scenario at Nong Khai, Thailand (circles); Pakse, Lao PDR (diamonds); Kratie, Cambodia (triangles); and Tan Chau, Viet Nam (squares).

reduction was about 0.5 m at Nong Khai, reducing to about 0.13 m by Tan Chau, which, given the sensitivity of the model, is probably a null result. Overall, the difference in water-level range (Fig. 16.4), that is, the annual difference between maximum water levels and minimum water levels, resulting under the irrigation scenario from increasing dry season flows and reduced wet season flows, would be quite small, substantially less than the errors of the model.

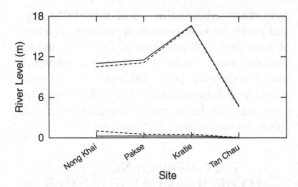

FIGURE 16.4 Modeled and baseline maximum and minimum water levels based on the agricultural development scenario (which also includes hydropower dams) for four sites along the Mekong main stream. Solid lines indicate baseline conditions and dashed lines indicate modeled changes.

The major finding from the modeling exercise was that the largest hydrological changes were generated by the Chinese dams. Since the dams are intended solely for hydropower generation, they will not result in a substantial reduction in annual discharge, but they will redistribute flows from the wet season to the dry season (Fig. 16.5). The precise extent of the redistribution depends on the operating rules for the dams, which are not known, but the modeled

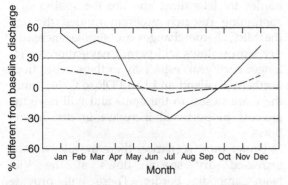

FIGURE 16.5 Percentage difference from baseline discharge at Chiang Saen in Thailand (solid line) and Kratie in Cambodia (dashed line) resulting from the modeled impacts of the construction of the first two large mainstream dams in China: Xiaowan and Nhaozhadu.

flow changes assume that the operating rules for the Chinese dams will be similar to those for existing dams in the Mekong system such as those in Lao PDR, Thailand, and Viet Nam.

Hydrological changes would decrease downstream. Changes in flow were modeled for Chiang Saen in Thailand and Kratie in Cambodia based on the predicted impact of Nuozhadu, Xiaowan, Manwan, and Dachaoshan reservoirs, two of which are constructed and one of which is under construction. At Chiang Saen, the most upstream hydrological monitoring site in the lower Mekong, the changes are proportionately larger than at Kratie in Cambodia, the furthest downstream site for which the flow model can be applied (Fig. 16.5). Sites between these two would display intermediate levels of flow change.

In addition to the increases in dry season flows and decreases in wet season flows predicted by the model, Adamson (2001) and MRC (2005a) predict that the onset of the wet season floods would be delayed as a result of upstream flow regulation. A common operating procedure for a hydropower dam is to run the water level down during the dry season and then retain the early wet season flows to refill the dam. Operators are reluctant to let the first wet season flows pass in case there is a wet season drought, and they have no later opportunity to refill the reservoir. The delay in the onset of the wet season flood would be most pronounced at sites close to the Chinese border, decreasing downstream as the influence of the large Lao tributaries becomes more pronounced.

Other scenarios generally had less pronounced impacts on the modeled results than the China dams and irrigated agriculture scenarios. In some cases, this is because the development is less intense—fewer dams and less water extraction for agriculture and domestic or industrial use. For the high development scenario, it is because some of the effects of the upstream dams are ameliorated by the downstream extractions and vice versa.

The effects of development on flood season water levels is generally less pronounced than the impact on dry season flows, mainly because the flood flows are so much larger. Podger *et al.* (2004) predicted a reduction of 0.2-0.8 m at Kratie in the period from July to September, and a smaller decrease of less than 0.3 m at Tan Chau in Viet Nam. The size of the change in water level depends partly on the shape of the basin at the prediction site, and at Tan Chau it is considerably wider than at Kratie.

The embankments scenario indicated that isolation of large areas of floodplain in Cambodia could influence flood heights at Tan Chau. Construction of flood embankments or other structures that prevent water moving on to the floodplain reduces floodplain storage, forcing water to move more rapidly downstream and raising the level of water in floodplain areas that are inundated. Large-scale construction of embankments to permit wet season irrigation in the delta is almost certainly contributing to a worsening of flood impacts in the Vietnamese delta, and any construction of similar structures in Cambodia would exacerbate the situation.

One impact of both the high development and China dams scenarios is to reduce the range of wet season inundation. A rising of the dry season water level increases the area permanently flooded, while a reduction in wet season flood height reduces the area seasonally inundated. In effect the seasonal inundation area is squeezed from both ends. Modeling showed extensive areas inundated for up to a month less under high development scenarios, but the total change in area for most inundation categories (e.g., area inundated with a depth of >0.5 m for 1-2 months) changed by only a maximum of 5%.

The impact on areas of wet season inundation is one aspect where predictive power is notably weak. Firstly, the predictions rely on the outcome of the upstream flow model, which already has errors. Secondly, the hydrological data for this region of the basin is weakest partly because of the physical nature of the environment, with frequent overbank flows and tidal influences, and partly because records are poorer as a result of historical political instability in Cambodia. Thirdly, small errors in the digital elevation model will have large influences on the predicted inundation area and few of those working with the data have confidence that the digital elevation model has only small errors.

5. CONSEQUENCES OF HYDROLOGICAL CHANGES

The models predict the hydrological consequences of the various development scenarios, but in order to make rational decisions about preferred futures, the people of the basin need to know the ecological, social, and economic consequences of the hydrological changes.

Both the MRC models and the earlier work of Adamson (2001) broadly concur on the hydrological consequences. There is no doubt that the large dams proposed and under construction in China will impact the river flows in all the downstream countries right to the delta. They will reduce wet season flows far less than they will increase dry season flows, and may delay the onset of the wet season flows.

The consequences of these changes are far harder to determine, and are the subject of a continuing research program funded through the MRC. Some changes are simple: increased dry season flows will permit navigation for far longer each year, especially in the reaches from northern Thailand and Lao PDR to China, since these are closest to the dams and will have the greatest proportional increases in dry season flows.

Under most scenarios, there are likely to be increased dry season flows at the Viet Nam-Cambodia border. These will provide greater security of irrigation water supplies, and potentially reduce the risk of sea water intrusion into the streams of the delta. This should favor rice irrigation which requires fresh water, but disadvantage shrimp aquaculture which requires brackish water (Kam et al., 2001).

Other mooted changes can be ruled out. The dams will not cause any substantial decrease in flooding in the downstream countries; the reduction in wet season river height will be far too small. The impacts of floods will continue to increase because of alienation of the floodplain raising flood heights, and increased development on the floodplain.

The greatest area of concern is the potential for impact on the fishery, especially the fishery in Cambodia. This provides the major animal protein source for people of the basin, and has a dollar value estimated at more than USD $6 billion (Hortle, Chapter 9).

Podger et al. (2004) developed a "feeding opportunity index" (FOI) as a first crude measure of possible fishery impact. The index was tested by visual graphical correlation with the Cambodia Dai fish data reported by Hortle et al. (2004): no statistical test was applied. They commented that the plot shows that the FOI increases with the catch, which they claim is considered to be a good indicator of total production of migratory fish in the region, but, they suggested, the FOI should be a more robust indicator of production. There are several difficulties with these claims.

The Cambodian Dai fishery is the only fishery in the Mekong for which there is reliable annual catch data. It constitutes a very small part of the total (<2%) and a very specialized part of the Mekong catch. Few biologists would consider the Dai fishery to be a good indicator of the total fishery, rather it is used because it is the only available data. The data used by Podger et al. was that from 1996-2000, because the hydrological models are based on the data from 15 years prior to 2000. The dai fish catch did track maximum water levels in Tonle Sap Lake very well from 1995-1996 to 2000-2001, but thereafter the relationship broke down (Hortle et al., 2004). The r^2 values for the regression of the dai catch on the depth of the flood

maintained for 31 days from 1995-1996 to 2000-2001 was 0.83, indicating that flood depth could explain 83% of the variation in fish catch; however if the full data set is used, from 1995-1996 to 2004-2005, the r^2 value drops to 0.425, indicating that flood depth explains only 43% of the variation. Had several more years data been included in the Podger et al. (2004) analysis, the relationship between the FOI and the catch would have deteriorated similarly.

Regardless of the relationship of the FOI to the Mekong fish catch, the impact of flow changes on fisheries is likely to come from effects other than alterations in floodplain inundation. In particular, the dry season is a critical time for migration and spawning for many Mekong fish species, and it is dry season flows which will be most altered. Baran (2006) identified flow modifications that increase dry season flows and those that delay the onset of the wet season flood as those most likely to deleteriously impact Mekong fish.

The impact on floodplain inundation and Tonle Sap Great Lake is more problematic. The elevation data for the floodplain is poor, and location in the floodplain of structures, such as elevated roads, make it difficult to predict how water movement will change. There are also disputes about existing water level data. Podger et al. (2004) and Campbell et al. (2006) both estimated that the change in floodplain inundation would amount to an approximate 10% decrease in seasonally inundated area around Tonle Sap Great Lake. However, this number should be viewed as speculative.

The change in floodplain inundation will not occur in the absence of other floodplain changes. Road construction, including the ADB southern coastal corridor project road, is continuing (Pinoypress, 2007) on the floodplain and will alter inundation patterns. Clearing of floodplain forest is also continuing and will degrade the quality of the inundated area, and increased population density in the area will degrade water quality.

Regardless of the size of the change in flood-plain inundation and degradation, most fisheries biologists expect that the types of hydrological changes predicted by the models would be sufficient alone to cause a major decline in the Mekong fishery.

6. CONCLUSIONS

The lower Mekong basin is a region with a large population and great poverty (Hook *et al.*, 2003). The people of the basin, through their governments, need to find development strategies and pathways which can equitably reduce poverty. At the same time, China is looking to use hydropower development on the upper Mekong, the Lancang River, as a means of promoting development and poverty alleviation in southwestern China.

Scenario modeling such as that discussed here provides one tool that enables governments and civil society to consider the potential outcomes of a range of development strategies before they are commenced. Those promoting large development schemes have always been outspoken about promoting the benefits. Scenario planning enables stakeholders to see more clearly the costs involved, and to identify the winners and the losers.

Within the Mekong basin, as in many other places, project proponents have often loudly denied that their projects would negatively impact other river users. In the Mekong the governor of Yunnan Province, Xu Rongkai, has denied that Chinese Dams would impact downstream countries (Xinhua, 2002), and even Olivier Cogels, CEO of the MRC, has argued that "the overall downstream impact of hydropower dams on the Lancang River in China is often exaggerated in the public opinion" (Cogels, 2007).

Breaking down the culture of denial and suppression of information is an essential first step toward competent and equitable resource management. Until information is made freely available, and problems, costs, and winners and losers are identified, there cannot be a sensible public debate either within or between the countries that share the basin. As Hirsch (2004) noted in relation to fisheries information in the Mekong, "the public good is likely to be best served by transparent, inclusive and accountable knowledge production."

These scenarios demonstrate that, even though there may be substantial benefits to host countries of projects such as irrigation schemes and hydropower dams in the Mekong, there will also be substantial impacts. Often the impacts will occur most severely in countries that will gain no financial benefit from the projects. Of the six Mekong countries, China and Laos have the greatest potential to benefit economically from hydropower and other development scenarios on the river, while Cambodia has the most to lose.

These scenarios should become the starting point for national and international discussion and negotiation on water resource development along the Mekong. A reduction in poverty in any Mekong country will benefit its neighbors through increased trade opportunities and a reduction of cross-border tension. But should developments that benefit one country lead to impoverishment in another; neighboring countries will also suffer through loss of markets and also through the flow of poor migrants seeking new economic opportunities.

ACKNOWLEDGMENTS

Scenario development and modeling as basin planning tools in the Mekong were initiated by Robyn Johnston and Steve Carson who argued long and passionately, and correctly, that they are the key tool if the people of the Mekong countries are to make informed decisions about their futures. Neither the World Bank Report nor this chapter would have been possible without their vision and persistence. Figure 16.5 is based on data supplied by Richard Beecham, who, together with Geoff Podger, did much of the modeling on which this chapter depends. The hydrological content of many of the MRC reports which is also critical to this chapter was produced by Peter Adamson.

References

Adamson, P. (2001). Hydrological perspectives of the Lower Mekong. *International Water Power and Dam Construction*, 2001:16-21.

Asia Times Online. (2002). *Mekong's dams wreak havoc on rural poor*. www.atimes.com/se-asia/DD10Ae04.html. Accessed 30 July 2009.

Bangkok Post. (2005). Green weed hopes wither. Bangkok Post, 11 June 2005.

Bangkok Post. (2006). Going against the current. Bangkok Post, Outlook, 13 March 2006.

Bangkok Post. (2007). Cambodia raps Laos over Mekong dams. http://www.bangkokpost.com/breaking_news/breakingnews.php?id=123623. Accessed 16 November 2007.

Baran, E. (2006). "Fish Migration Triggers in the Lower Mekong Basin and Other Tropical Freshwater Systems,". MRC Technical Paper No. 14. Mekong River Commission, Vientiane, 56pp.

Cambodianonline. (2004). Environmentalist warns of damage from Chinese damming. March 18, 2004. http://www.cambodianonline.net/homerivers.htm. Accessed 26 April 2007.

Campbell, I. C. (2007). Perceptions, data and river management—Lessons from the Mekong River. *Water Resources Research* 43, doi:10.1029/2006WR005130.

Campbell, I. C., Poole, C., Giesen, W., and Valbo-Jorgensen, J. (2006). Species diversity and ecology of the Tonle Sap Great Lake, Cambodia. *Aquatic Sciences* 68, 355-373.

Cogels O. (2007). Letter to Bangkok Post. January 9, 2007.

Douglas, M. M., Bunn, S. E., and Davies, P. M. (2005). River and wetland food webs in Australia's wet-dry tropics: General principles and implications for management. *Marine and Freshwater Research* 56, 329-342.

Falkenmark, M. and Rockstrom, J. (2004). "Balancing Water for Humans and Nature," Earthscan, London, 247pp.

Hirsch, P. (2004). Politics of fisheries knowledge in the Mekong River Basin. *In* "Proceedings of the Second International Symposium on the Management of Large Rivers for Fisheries," Volume II, Welcomme, R. and Petr, T. (eds.), pp 91-101. FAO Regional Office for Asia and the Pacific, Bangkok.

Hook, J, Novak, S., and Johnston, R. (2003). "Social Atlas of the Lower Mekong Basin," Mekong River Commission, Phnom Penh, 154pp.

Hori, H. (2000). "The Mekong. Environment and Development," United Nations University Press, Tokyo, 398pp.

Hortle, K. G., Ngor, P., Hem, R., and Lieng, S. (2004). Trends in the Cambodian dai fishery: Floods and fishing pressure. *Catch and Culture* 10(1), 7-9.

Junk, W. J. and Wantzen, K. M. (2004). The flood pulse concept: New aspects, approaches and applications—An update. In "Proceedings of the Second International Symposium on the Management of Large Rivers for Fisheries," Volume II, Welcomme, R. and Petr, T. (eds.), pp 117-140. FAO Regional Office for Asia and the Pacific, Bangkok.

Junk, W. J., Bailey, P. B., and Sparkes, R. E. (1989). The flood-pulse concept in river-floodplain-systems. *Canadian Special Publications for Fisheries and Aquatic Sciences* 106, 110-127.

Kam, S. P., Hoanh, C. T., Tuong, T. P., Khem, N. T., Dung, L. C., Phong, N. D., Barr, J., and Beb, D. C. (2001). Managing water and land resources under conflicting demands of shrimp and rice production for sustainable livelihoods in the Mekong River Delta, Vietnam. http://www.ciat.cgiar.org/inrm/workshop2001/docs/titles/1–2CPpaaerSPKam.pdf. Accessed 22 February 2008.

McMahon, T. A., Finlayson, B. L., Haines, A. T., and Srikanthan, R. (1992). "Global Runoff. Continental Comparisons of Annual Flows and Peak Discharges," Catena Verlag, Cremlingen, Germany, 166pp.

MRC. (2003a). "State of the Basin Report 2003," Mekong River Commission, Phnom Penh, 300pp.

MRC. (2003b). "Water Used for Agriculture in the Lower Mekong Basin," BDP Report 017. MRC Secretariat, Phnom Penh, 53pp.

MRC. (2004a). "Dams on the Upper Mekong in China and the Present Flows in the Lower Mekong Basin,". MRC Technical paper and Media Release. 26 March 2004. Mekong River Commission, Phnom Penh, 3pp.

MRC. (2004b). "Decision Support Framework. Water Utilization Project Component A: Final Report." Volume 11: Technical Reference Report. DSF 620 SWAT and IQQM Models. Mekong River Commission, Phnom Penh. 132pp.

MRC. (2005a). "Integrated Basin Flow Management," Phase 2. May 2005 Field Trip. Study Reach Notes. Draft 28 April 2005. Mekong River Commission, Vientiane, 77pp.

MRC. (2005b). "Integrated basin flow management," Report No. 6. Technical Workshop and Field Visit. Draft 17 June 2005. Mekong River Commission, Vientiane.

Nation. (2007). Mekong commission blasted over river dams. http://www.nationmultimedia.com/2007/11/14/national/national_30055997.php. Accessed 14 November 2007.

Osborne, M. (2004). "River at Risk. The Mekong and the Water Politics of China and Southeast Asia," Lowy Institute for International Policy, Paper No. 2. Double Bay, Australia, 56pp.

Pearce, F. (2004). Chinese dams blamed for Mekong's bizarre flow. New Scientist, 25 March 2004.

Pilkey, O. H. and Pilkey-Jarvis, L. (2007). "Useless Arithmetic. Why Environmental Scientists Can't Predict the Future," Columbia University Press, New York, 230pp.

Pinoypress. (2007). Korea, Australia and ADB provide $165.5 M for Viet Nam, Cambodia road improvements. http://www.pinoypress.net/2007/12/06/korea-australia-and-adb-provide-1655m-for-viet-nam-cambodia-road-improvements/. Accessed 20 December 2007.

Podger, G, Beecham, R., Blackmore, D., Perry, C., and Stein, R. (2004). "Modelled Observations of Development Scenarios in the Lower Mekong Basin," World Bank, Vientiane, 122pp.

Prateechaikul, V. (2003). Thaksin taps into Northeast dream. Bangkok Post, 28 July 2003.

Samabuddhi, K. (2004). Chinese dams upstream blamed for drastic decline in fish stock. Bangkok Post, 10 March 2004.

Sioli, H. (ed.). (1984). "The Amazon: Limnology and Landscape Ecology," Dr W. Junk Publishers, Dordrecht, Netherlands, 763pp.

Vidal, J. (2004). Dammed and dying: The Mekong and its communities face a bleak future. The Guardian, London, 25 March 2004.

WCMC. (2007). Freshwater biodiversity: A preliminary global assessment. World Conservation Monitoring Centre. http://www.unep-wcmc.org/infomation_services/publications/freshwater/4.htm. Accessed 29 August 2007.

World Bank. (2000). "Project Appraisal Document on a Proposed Grant from the Global Environment Facility in the Amount of SDR 8 Million (US$11 Million Equivalent) to the Mekong River Commission for a Water Utilization project," Report No.: 19625-EAP. Rural Development and Natural Resources Sector Unit. East Asia and Pacific Region. The World Bank, 72pp.

Xinhua. (2002). Hydraulic works not detrimental. Xinhua News Agency Release, September 7, 2002.

17

The Challenges for Mekong River Management

Ian Campbell

Principal Scientist, River Health, GHD, 180 Lonsdale Street, Melbourne 3000, Australia & Adjunct Research Associate, School of Biological Sciences, Monash University, 3800 Australia

1. INTRODUCTION

The Mekong River basin faces a multitude of problems. Some of these are historical, but, even with the welcome cessation of armed conflict in the region, there is increasing conflict over resource development. This conflict frequently occurs within countries, often between powerful development-focused agencies and the people dependent on the natural environment for subsistence, but increasingly conflict is appearing between countries, as evidenced by the recent angry outburst by Cambodia officials about proposed water resources developments in Lao PDR (Bangkok Post, 2007a). With a large number of mainstream dams on the lower Mekong now back on the agenda (Nette, 2008), the frequency and intensity of international conflict on the Mekong is likely to rise rapidly.

If conflict is to be avoided, river basin resource management and governance issues must be addressed. From the early 1950s, the four lower Mekong countries, Cambodia, Lao PDR, Thailand, and Viet Nam have been working to implement a regional approach to water resources development and management. This approach was formally adopted through the establishment of the Mekong Committee, which emphasized water resources development, and later through the establishment in 1995 of the Mekong River Commission (MRC), under an agreement with a much greater emphasis on sustainable water resource management.

More than 10 years after it was established, and after the expenditure of more than USD $100 million, of mostly donor money from developed countries, it is worthwhile to consider the contribution of the MRC to resource management in the basin. The organization is widely cited as a model of river basin management in the developing world (e.g., Varis, 2004), and has won an international award for its achievements. But some in civil society groups have claimed that the MRC is a failure, and have requested donors to withdraw their support (Bangkok Post, 2007b). In this regard, the revived proposals for mainstream dams on the lower Mekong present MRC with its greatest challenge (Nette, 2008).

What are the management issues in the Mekong basin, what has the MRC achieved, and what are its failings? Can the model be improved, or should it be abandoned?

2. THE NEED FOR MANAGEMENT

The Mekong is a large international river, providing a variety of benefits to a range of individuals, communities, and countries. But uses can conflict, and aspects of the river that benefit some, cause problems for others. There are a variety of issues that impact riparian states, and users of the river, and different stakeholders often have different perceptions of environmental and management issues (Campbell, 2007a). Several of the larger issues that will impact the Mekong are outlined below. Several are regional and beyond the capacity of a river basin management organization to manage alone, although they will impact the basin and

its management. Others fall well within the ambit of a river basin management organization.

2.1. Regional Issues

2.1.1. Population Growth

Like almost every other river basin on earth, the Mekong River basin is faced with increasing pressure from the growing population of humans living within it (Hook et al., 2003). Unless development is undertaken far more wisely than in most other places, the environmental impact per person will also increase. The outbreak of peace after decades of war, and improvements in health and nutrition that are slowly occurring, have contributed to high rates of population increase and populations heavily skewed toward younger people, ensuring that the population increases will continue for some time. Present population in the basin is estimated at about 72 million, with estimates that it is growing at between 1.1% and 1.5% annually (Pech and Sunada, 2008).

2.1.2. Poverty Alleviation and Development

The population of much of the Mekong basin, and the region within which it is located, is very poor. Annual GDP per capita in Cambodia was only $1361 in 1999, while that in Laos was $1471. Viet Nam was better off, with a GDP per capita of $1860, and Thailand is comparatively well off at $6000 (Hook et al., 2003). However, much of the area of Thailand within the Mekong basin, the northeast region of Isaan, is poor with incomes substantially lower than the national average, and per capita GDP between $1250 and $1500.

In view of the combination of rapid population growth and poverty, it is not surprising that national governments, and international agencies, have strongly emphasized the need for development both nationally and within the Mekong basin. Development is seen both as a means of poverty alleviation and as a means of raising national prestige. Thus, it has a political dimension that has tended to emphasize large projects. However, development does not necessarily lead to economic growth, and economic growth is not synonymous with poverty alleviation.

The relationship between economic growth and poverty reduction is complex, confounded by other factors, and difficult to elucidate partly because of the poor quality of the data. Kraay (2005) argues that sustained poverty reduction is impossible without sustained economic growth. However, Edward (2006) suggests that, although the poor do benefit from economic growth, it is much better for the rich. Globally, the poor benefit less in both absolute and relative terms than wealthier people, so that "relying on growth to reduce poverty is rather inefficient." Simms (2008) noted that it would take an additional $166 worth of global growth to generate $1 extra for those living on less than $1 per day.

Development in the Mekong could take a variety of trajectories (Hewett, 2007), with the extremes of the models being large-scale infrastructure projects, often funded by multinational development agencies and implemented by national government agencies, and small-scale, community-based projects often funded by NGOs. Öjendal (2000; pp. 87-90), for example, contrasts "mainstream" water resources development, which he identifies with large-scale dams and associated infrastructure, with "alternative" development approaches, which he identifies as "decentralized, participatory, small-scale, people-centered and non-state biased."

The development trajectory that has received the greatest publicity in the Mekong, and which has been most promoted by the governments and at times by the MRC and its predecessors (e.g., Cogels, 2005, 2006), has been via large infrastructure projects. That should not be surprising. For decades, large-scale water resources development projects, dams, have

been promoted as socially and economically beneficial for developing countries. It is only recently, following the World Commission on Dams report (World Commission on Dams, 2000), that this approach has been seriously questioned.

Large infrastructure projects may have a positive impact on economic growth (Munnell, 1992), but that is not necessarily the case. Large-scale water resources projects were one of the first applications of cost-benefit analyses because of concerns that some may not be contributing to a net economic benefit (Buss and Yancer, 1999; Margolis, 1959; Prest and Turvey, 1965). However, large infrastructure projects provide the best opportunities for large-scale corruption. Tanzi and Davoodi (1998) demonstrated a strong positive correlation between corruption indices and the levels of public investment in infrastructure, and a negative correlation between corruption indices and the quality and productivity of the infrastructure.

In any event, regardless of the trajectory, development will present environmental challenges. In the Mekong basin, the impacts of large-scale developments proposed and initiated by national governments are the main subject of concern. Obvious examples include the dams on the mainstream in China (China Daily, 2002), on the Nam Theun and at Khone Falls in Lao PDR, on the Se San in Viet Nam, and the existing Pak Mun Dam (Roberts, 1993), and proposed projects such as the Kong-Chi-Mun Project in Thailand (Hori, 2000).

Small-scale, decentralized, people-centered development would not necessarily be more efficient or environmentally friendly (e.g., see Wilder and Lankao, 2006). Growing population in the Mekong will inevitably increase pressure on natural resources. Even with extensive programs to intensify agriculture (producing more and higher value crops using less water), to develop transport infrastructure (so that crops can be delivered to markets more cheaply and in better condition), and to develop the skills base in the population (so that fewer people are dependant of agriculture for their income), 20 years are needed before pressure on the natural resource base—forests and the fishery—begins to decline.

There is already evidence of the cumulative impacts of many small-scale interventions within the Mekong. For example, MRC (2003) found a trend of increasing low flows and decreasing flood season flows presumably due to the construction of thousands of small dams and rice fields within the basin. These retain wet season flows and release them, either directly or through groundwater, during the dry season. The conversion of wetlands and forested areas to rice fields, the removal of forest and wooded areas through firewood harvesting, and the decline of giant fish species (Poulsen *et al.*, 2002) all appear to be unintended consequences of the small-scale actions of many local communities.

2.1.3. Governance

As in many developing regions, poor governance and corruption are substantial impediments to good resource management, and effective development and poverty alleviation in the Mekong. In a global context, this point has been has been repeatedly stressed in recent studies (Easterley, 2006; Kraay, 2005). Diversion of aid funding and government revenues from potentially productive activities to bank accounts in safe havens, informal taxation on the poor and powerless, and the warping of national priorities toward development projects that offer the greatest potential for personal enrichment, continue to be significant obstacles to poverty alleviation and economic growth in the Mekong region as they are in many other places. Transparency International rankings of the four lower Mekong countries in 2008 were 80 for Thailand, 121 for Viet Nam, 151 for Laos, and 166 for Cambodia, where 1 is perceived as the least corrupt country (Denmark) and 180 as the most corrupt (Somalia) (Transparency International, 2008).

Clearly, a river basin management organization cannot solve management issues such as these. Nevertheless, it can play a role in alerting national governments to the consequences of population growth, and it can try to ensure that decision makers and civil society are at least aware of the economic, social, and environmental costs of large infrastructure projects, and not just the benefits, which are always loudly proclaimed by the potential benefactors.

2.1.4. Political Commitment

For any transboundary river basin, effective management depends on the political commitment of the national participants. For success, all players must realize that good basin management is ultimately in all of their best interests, even though it will place constraints on their activities and their sovereignty. In fact, all international agreements place some constraints on the sovereignty of the signatories, with the understanding that the freedoms potentially gained will outweigh those lost.

In the case of the Mekong, there is considerable uncertainty about the commitment of the member states to the 1995 Agreement, and particularly the commitment of Thailand. The Thais have, in recent times, refused to sign up to rules for water utilization unless they were specified as "procedures," although the 1995 Agreement specifies that the countries will develop "rules." The Thais have argued that agreeing to "rules" will infringe their sovereignty. Many past and present staff members of the MRC secretariat have suggested that Thailand is playing an intentionally obstructive role in the organization, and that this is a deliberate policy of the national government.

2.2. Issues Specific to the Basin

2.2.1. Fishery

There is no doubt that the Mekong fishery is under pressure. While there is no evidence of a decline in overall catch, there is strong evidence of a decline in large species (Mattson et al., 2002) as well as a decline in catch per unit effort (Allan et al., 2005).

There are a number of obvious sources of pressure on the fishery. An increasing population of fishers, as a result of the general population increase in the basin, has intensified the harvest. In addition, the fishers have increasingly sophisticated fishing equipment. Nylon monofilament gill nets are widely and very cheaply available.

There are two key, but not unique, features of the Mekong fishery. The life cycles of many of the fish species are tied to the Mekong flood pulse, with reproduction of many species occurring in the dry season when water velocities are low and eggs and larvae are less likely to be washed out to sea. Secondly, many of the fish species migrate as part of their life cycles, either from the river to the flood plain or along the river. An appreciable proportion of the migrations are over long distances, encompassing the waters of several countries (Poulsen et al., 2002).

A consequence of the combination of pressures and key features is that the future of the fishery is not in the hands of a single nation. If the fishery of the lower Mekong is to be maintained, it must be managed through the cooperative actions of all the Mekong countries. China, Lao, and Viet Nam are the main countries exerting influence on the flow regime or with potential to do so. Cambodia, Lao, and Thailand are the countries with the greatest number of fishers and markets for the fish.

The Mekong fishery is critical to the well-being of the people of the basin. Aquatic animals provide a high proportion (40-80%) of the animal protein consumed by the people of the lower Mekong basin (MRC, 2003), contributing essential nutrients such as calcium (Jensen, 2001). Over 70 million people rely on it for food, and for many of those it is a major part of their livelihood. As a result, any substantial decline in the fishery has the potential to severely impact the people of the basin.

2.2.2. Environmental Flows

Changes to natural flow regimes are now recognized as a major human impact to rivers throughout the world (Tharme, 2003). Arguably, changes to the flow regime are the single greatest environmental change confronting the lower Mekong. Alterations to mainstream flows as a consequence of dams under construction, planned, or under consideration in China, Cambodia, and Thailand, and changes to tributary flows cause by dams under construction or planned in Lao and Viet Nam, could have widespread ecological and social impacts. Some of the consequences are discussed elsewhere in this volume.

One feature common to all of the proposed large dam developments in the Mekong is that the country which would gain the benefits from the proposal would not be the only country that would bear the ecological and social costs. Indeed in many cases, almost all the environmental and social costs will be borne by citizens of other countries, usually those downstream. Thus, as with the fishery, the construction and operation of dams has basin-wide implications, and should not be seen as a purely national concern.

2.2.3. Floodplain Alienation

After alteration to the flow regime, alienation of the floodplain (consequences of constructing flood protection levees, and elevated roads, which restrict or prevent flood waters from inundating the floodplain during high flows) is one of the most serious environmental changes confronting the people of the lower Mekong. The purpose of the construction is to protect the area behind the levee from the consequences of flooding, or in the case of elevated roads, to ensure the road is useable during the flood season.

Although the building of elevated roads has incidental impact, the creation of levees is of major concern. One negative consequence of flood plain alienation is exacerbated flooding elsewhere in the basin. Floodplains can receive and store an appreciable volume of the flood flows of the river. In the Mekong, it has been estimated that the Cambodian floodplains can store up to 87 km^3 (Fujii *et al.*, 2003). As the water passes downstream, water returns from the floodplain to the channel, extending the period of high water downstream, but reducing the maximum flood peak water level. Construction of levee banks along the river reduces the area of floodplain available to the river, and thus the available water storage volume. Consequently the water level in the river must be higher, since the same volume of water is present. Downstream floods therefore tend to have higher water levels over a shorter period.

Floodplain alienation may exacerbate flooding upstream as well as downstream of levee banks. Where areas of land on the floodplain are protected by levee banks, the levees may act as a partial dam, particularly if they occupy a significant proportion of the width of the floodplain. Thus water retained on the floodplain upstream may be deeper, flood a larger area, and be retained for longer because the levees act as a "choke point" in the river's flow.

Within the Mekong, there have been growing concerns about the impact of levee banks in the delta and on the border between Cambodia and Viet Nam. Within the Vietnamese area of the delta, as part of the move toward irrigated rice, and away from reliance on rising floodwaters for rice cultivation (MRC, 2003), large areas have been converted to polders by building levee banks so that the inflow of water during the flood season can be more tightly regulated. That process has alienated large areas that were formerly flooded, which may be raising flood heights in other areas of the delta.

Along the border between Cambodia and Viet Nam in the delta, there have been claims by Cambodians that Viet Nam has constructed a levee bank to hold back the Mekong flood.

Cambodian villagers claim that this has resulted in more severe flooding on their side of the border. The Vietnamese claim that they have only constructed a road to facilitate border security, and that flooding is being exacerbated by forest clearing in Cambodia. Whether or not the road was constructed with the intention of acting as a levee, it certainly does, and the poldering within Viet Nam must also be acting to create a choke point on the river, thus exacerbating flooding in both Cambodia and Viet Nam.

2.2.4. *Water Quality*

While not an immediate threat, over the longer term deterioration in water quality will become a serious management issue in the Mekong. Already serious deterioration in water quality has occurred in the upper Mekong and in the Mekong delta (Campbell, 2007a; Lu *et al.*, this volume; Ongley, Chapter 12, this volume). There are several obvious potential sources of water pollution. One is growth and industrialization of cities and towns within the basin, the second is spillages from navigation, the third is broad-scale land use change.

The Mekong contains a relatively small number of large cities along its banks: Can Tho in Viet Nam, Phnom Penh in Cambodia, Ubon Rachathani and Khon Kaen in Thailand, Vientiane and Luang Prabang in Lao PDR, and Jinhong in China are the most notable. Of these, Phnom Penh, with a population variously estimated at about half a million people, is the largest. None of the cities is heavily industrialized, with Khon Kaen and Jinhong probably the most industrial; consequently, sewage rather than toxic waste is the most substantial waste disposal problem.

Fortunately, many of the larger towns on the main stream dispose of sewage effluent in natural wetland systems rather than directly to the river. This is the case for Vientiane, Pakse, and Phnom Penh, for example. As a consequence, the discharged sewage is substantially treated before reaching the river. For example, waste from Vientiane finally flows into the river some 50 km downstream of the city, having flowed through That Luang swamp and a variety of small tributary streams. Unfortunately in all three cities, there is increasing encroachment into the swamps as the cities grow, so their waste treatment capacity is being increasingly lost at the same time that the waste volume and complexity are increasing. The benefits of free sewage treatment, a public good, are rarely factored into cost-benefit analyses on swamp drainage.

The second potential source of increased future water pollution is spillages and accidents from navigation. There are two sections of the river where this is of particular significance: from Phnom Penh south to the South China Sea and from Chiang Saen north to Jinhong. Downstream of Phnom Penh has been a traditional trade route with ocean-going ships bringing goods upstream to the Cambodian capital. It has been declining in national significance partly because of difficulties bringing ships through Vietnamese territory, and partly because it is quicker for most ships to dock at Sihanoukville and for goods to then be transported by road to Phnom Penh.

The upper Mekong navigation channel project has highlighted potential growth in navigation in the upper section of the river, and in many ways, this is of more concern. The channel here is far more dangerous, currents are faster, and the channel is rocky and less regular than the downstream section. A variety of cargo is transported, including petroleum products, with little or no regulation of vessels or cargoes, and no spillage emergency plans or cleanup equipment in place for the accidents that will occur.

The final source of potential increase in water pollution in the Mekong is through changing land use. The clearing of forests for farming is thought by many to have increased sedimentation in the river, although there is little direct evidence for this effect. However as farming becomes more intensive, with increasing use of fertilizers and pesticides, and as aquaculture expands with an increase in nitrogenous wastes discharged into the river, it is inevitable that diffuse pollution will become an increasing problem for the Mekong, as it has for many other rivers. As with other rivers, managing diffuse pollution will require effective land use management, which in turn requires educating and working with landowners.

2.2.5. Navigation

Navigation in the Mekong is an important issue (MRC, 2003; Osborne, 2007). There are as yet no comprehensive agreements about vessel passage, cargo regulation, or even a basin-wide system of channel markers and navigation procedures, although there has been agreement among lower Mekong countries. While navigation was restricted to small local boats, this was not a problem, but with more and more larger ships, and even suggestions by Chinese officials that the river should be made navigable from China to the sea, the lack of basin-wide navigation rules is an increasing impediment. Furthermore, although there was a bilateral agreement on waterway transport between Cambodia and Viet Nam in 1998, and a subsequent agreement on the transit of goods in 2000, these agreements have yet to be put into full operation (MRC, 2003; Podger *et al.*, 2004), and river transport from the sea to Phnom Penh is, as a result, still inhibited.

Navigation in the Mekong remains controversial. A 1993 report (cited by ESCAP, 1995) proposed a three-stage development plan for navigation in the upper Mekong. In 1995, ESCAP produced an initial environmental examination report on the plan (ESCAP, 1995), which identified a lack of information on the effects of the development on the fishery as a major gap in environmental knowledge. An article by an official from the Yunnan Provincial Navigation Bureau in 2000 elaborated on the earlier plan with a four-stage proposal, culminating in a "navigation channel ... capable of passage of a barge train of one tugboat and four 600 t barges ... through the navigation channel of 6 riparian countries ... to the South Sea" (Liu Daqing, 2000). In 2001, Chinese transportation officials, with cooperation firm governments of Myanmar, Laos, and Thailand, produced an environmental impact assessment of the first stage of the project. (Joint Experts Group, 2001). The document was severely criticized as inadequate by independent reviewers engaged by the MRC, who pointed out that, among other shortcomings, it did not meet the legislated requirements for environmental impact assessments of either Laos or Thailand (Cabrera, 2003). The project itself attracted strong negative publicity in the press in both Thailand and Cambodia (Cabrera, 2003; Cambodia Daily, 2002), leading to prominent Thai politicians publicly demanding that it be reviewed (Bangkok Post, 2002).

In meetings of the dialog partners of the MRC (Myanmar and China), Chinese delegates have insisted that the report by Liu Daqing had no official status and that China has no plans for navigation improvements beyond those identified in the 2001 document. However, whether or not Liu's paper had official status, the 1993 report was certainly an official document, and at the same meeting, delegates from Myanmar requested that China proceed with the next stage of the project, so they certainly believed China was still intending a multistage project.

2.2.6. Flooding

Although there is a widespread perception that flooding in the Mekong basin has become worse, there is no hydrological data to support that belief (Campbell, 2007a). Nevertheless, the consequences of flooding have become worse as the size of the population and the value of the infrastructure on the floodplain have increased (MRC, 2003).

In the future, the consequences of flooding for people are likely to become worse, especially in Cambodia and the delta. Populations and infrastructure in flood-prone areas will increase, and the floodplain alienation previously discussed is likely to cause higher flood levels in those areas which do flood. Construction of dams in China and Laos will reduce the flood peaks of small and medium floods, but modeling suggests that any reduction will be insignificant because the wet season flows are so large (Podger et al., 2004). Climate change, resulting in sea level rises, could also exacerbate flooding in the low-lying delta. Lebel et al. (undated) suggested that present policies and practices in the Mekong, rather than reducing risks, shift them to vulnerable and disadvantaged groups.

The flood management and mitigation program operated through the MRC has focused on flood warnings and flood responses. Attempts to prevent floods are now seen as both impractical as well as undesirable because of the recognition of the environmental benefits of floods (MRC, 2003). Provision of flood warnings is seen as an important part of reducing the human toll of flooding, and MRC has played an important role in developing flood forecasting models and in providing early warning systems for mainstream and major tributary floods; however, these are circulated to national government agencies, and the lack of good communication systems makes it difficult to circulate the information effectively to local communities who would gain the greatest benefit.

3. THE CHALLENGE OF MANAGEMENT

International transboundary rivers provide all the challenges of river management faced for rivers within a single country, as well as an additional complexity arising from the interactions between the national players (Campbell, 2007b).

The challenges of river and river basin management are formidable, and various aspects of the problem have been discussed at length elsewhere (e.g., Boon *et al.*, 2000; Harper and Ferguson, 1995; Heathcote, 1998; Hooper, 2005; Reimold, 1998). Key challenges are firstly the development of an appreciation among all the stakeholders that the river and its basin form a single complex integrated system where actions of one stakeholder can, and will, impinge on other stakeholders. The second challenge is to develop a common vision for the basin among the various river basin management players, and the third to improve coordination between them.

Within transboundary rivers, the difficulties of developing a shared vision and coordinating between players are exacerbated by national and political barriers, which often, as in the Mekong, include language barriers as well. Within the six Mekong basin countries, there is a range of political systems. Myanmar is almost a closed country, whereas Thailand is extremely open. The political systems are also undergoing fairly rapid change, with Viet Nam and China moving increasingly toward their own forms of market economies, Thailand having recently had a political coup and currently struggling to develop a rule of law and democratic government, and Myanmar affected by political unrest.

Obtaining stakeholder input from some key groups, such as subsistence fishers, can be very difficult. The Mekong countries are all poor, and many of the people most dependant on the river have relatively low levels of literacy and almost no political influence. Those with political power and influence often have visions of the future of the basin which takes little account of the needs and aspirations of other stakeholders.

Below, I identify some basic needs for managing river basins, their applicability to the Mekong situation, and the extent to which they are currently being met.

3.1. Management Needs

In order to manage the Mekong, as with any large river, there are some readily identified basic requirements. These are the fundamental needs of those whose job is to manage or coordinate the management of the system.

3.1.1. Well-Defined Objectives

Barrow (1998) echoed the suggestion of Le Moigne *et al.* (1994) that the river basin development and planning and management agencies need to focus on well-defined objectives, which take into consideration the demands on the basin and the resources available. Furthermore, it is important that the objectives are clear, consistent, and not contradictory.

3.1.2. Information and Technical Capacity

Neither the Mekong, nor any other river basin, can be managed without information. The stakeholders and decision makers need reliable information on river flows, water quality, fishery, river ecology, the people of the basin, and much more. It is not possible to have a sensible debate or to develop rational policy in the absence of reliable information.

That implies an agency or agencies with the technical capacity to collect reliable data and make it accessible to the stakeholder community. The data collection agencies must be trusted by the stakeholders to produce reliable data, and to release all the relevant data that are available in any particular discussion. If the range of stakeholders is broad, there may also be a requirement to interpret the data that are released.

Basin management will work best when stakeholders have confidence that the full range of development options is open for discussion, and their advantages and disadvantages openly identified. Environmental decisions are in the end value judgments, although they should be informed by technical decisions (Campbell, 2007b). Therefore, providing information about the options and their costs and benefits will not remove the need for debate, or lead to scientific decision making as some have wished, but it will provide a more informed basis for value judgments to be made and debated.

3.1.3. Information Sharing

The technical information collected and assessed for decision making must be shared between stakeholders. In the case of large international basins such as the Mekong it is important, for example, that all the countries have access to hydrological and rainfall data, as well as data on water quality and the fishery. The data should be available to national government line agencies, but also to broader civil society including researchers inside and outside the basin, NGOs, and various interest groups such as those representing fishers and farmers. Access to data and information can reduce mutual suspicion and ensure that all groups participating in resource management discussions are potentially operating from the same technical understanding.

In international basins, countries need to be aware of plans by their neighbors, and should have an opportunity to influence decisions that may impact on them. For example, within Europe there is a formal process by which this can occur through the Espoo Convention on transboundary environmental impact assessment (ECE, 1991), but there is no similar agreement in southeast Asia, even between the member countries of the MRC.

3.1.4. Cooperation between Stakeholders

Management of an international transboundary river like the Mekong requires cooperation between the various national players, but also between players at other levels of government. Many of the management problems in large river basins can only be addressed through the agreement, cooperation, and support of local communities. They may require national or provincial level legislation and regulations as well, but in the end they succeed or fail based on the attitude of the local people.

Fisheries management in a system like the Mekong cannot succeed unless the fishers understand and support the management goals and regulatory mechanisms. Unlike large ocean fisheries (which governments have conspicuously failed to manage anyway; see, e.g., Pilkey and Pilkey, 2007) the fishery in inland waters of developing countries involves millions of participants using simple low-cost equipment, primarily gill nets, and traps. Such a system cannot be controlled by policing alone. Instead, local communities must see the value in protecting the fishery, utilizing it sustainably, and enforce their own regulatory processes. This is a major aim of those promoting community based fishery management.

A second example is the management of diffuse pollution from urban and especially rural runoff. Management techniques include maintaining appropriate buffer zones along streams, changing cultivation procedures to reduce the times and areas where soil is exposed, and use of natural or artificial wetlands to treat runoff before it enters the river. Those sorts of management practices require the cooperation of farmers and other landowners, and the administrations of small towns. They cannot be forced on people. A buffer zone can be destroyed within hours if farmers see it as a waste of space, and once again enforcement alone will never be sufficient to maintain good practice.

4. THE MRC AND MEKONG MANAGEMENT

MRC has been praised (Varis, 2004) and criticized (Chenoweth *et al.*, 2001) as a basin management organization. Barrow (1998) reviewed river basin development planning and management globally and concluded that the results to that time had been disappointing. He compiled a list of suggestions for improvement of basin organizations including a focus on well-defined objectives, ensuring developments are mutually beneficial to all countries involved, being realistic in terms of resources available, and having regulatory mechanisms, trained staff, and satisfactory organizational structures.

4.1. Well-Defined Set of Objectives?

MRC attempts to be both a development agency and a basin management agency. The 1995 agreement includes both an undertaking to "promote, support, cooperate and coordinate in the development of sustainable benefits" and an undertaking to "protect the environment and the ecological balance." The two roles are incompatible, a difficulty that has been discussed by Hirsch and Jensen (2006) and the MRC Review (2007). A consequence has been that the organization has swung from one role to the other, identifying itself as a basin management organization under some chief executive officers, and as a development agency under others. Both Hirsch and Jensen (2006) and the MRC Review (2007) apparently accepted the frequent statements by Oliver Cogels, the CEO at the time, that his MRC as a development facilitation organization model was the preferred model of the member governments. But, while it was clear that there were a number of government officials from the member countries who were enthusiastic supporters of the model, the strongest proponent of the model was Cogels himself, who, shortly after his arrival at the Commission espoused projects such as the Senegal River development under the Organisation pour la Mise en Valeur du Fleuve Sénégal (OMVS) as models for the MRC to emulate. Cogels contract was not extended when his term was completed, and neither his successor nor his predecessor espoused the "MRC as development facilitator" model. It may have been

members of the Joint Committee had been carried away by claims that the MRC could attract tens of millions of dollars of development money for large scale water resources infrastructure developments and were later disappointed.

However, it is clear that the MRC cannot succeed in fulfilling its mandate until the ambiguity and contradictions in the mandate are resolved. Hirsch and Jensen (2006) provide a list of possible functions of river basin organizations which could be relevant to the MRC, noting that while some are complementary others are contradictory. Either the 1995 agreement will need to be renegotiated clarifying the roles of the organization, or the council will need to pass clarifying resolutions if the organization is to be able to maintain a sensible, consistent, long-term focus.

4.2. Source of Information and Technical Capacity?

The role of the MRC as a source of information and technical capacity, and in the collection, analysis and dissemination of data has waxed and waned. Often this has been in parallel with its identification as a development agency or a basin management organization. During development agency phases the organization has tended to be far less open, restricting the release of data, which did not support the development options that it favored, while at other times it has been far more open.

Various commentators have had very different perceptions of the technical capacity of the MRC. Chenoweth (1999; Chenoweth et al., 2001) was highly critical of the MRC's data collection, particularly the water quality and hydrological data which he states was of such poor quality it could not be used for management purposes (Chenoweth et al., 2001). Unfortunately much of his criticism was out of date, uninformed or both. The database was in excellent condition, a number of quality assurance procedures and data validation exercises were in place, and the data were being used for management.

Data from the MRC water quality database has now been used a number of times for management. It was used by the MRC for the State of the Basin Report (MRC, 2003), is being used in support of the MRC environmental flows work, and has provided the basis for several papers that evaluate water quality issues within the basin (Campbell, 2007a; Ongley, Chapter 12, this volume). The hydrological database has likewise been extensively used for basin flow modeling under the MRC WUP project as by the World Bank team (Podger et al., 2004; Campbell, Chapter 16, this volume). While the databases certainly have weaknesses, they are not substantially poorer in quality, or less useful, than many in government agencies in developed countries.

In contrast, Hirsch and Jensen (2006) were extremely complementary about the technical capacity of MRC. They commented that "donor funds to the MRC have largely been well spent on the development of state-of-the-art scientific knowledge and technical capacity," although they later commented on a "brain drain" that presumably had depleted that capacity to some extent. The MRC Review (2007) felt that the secretariat had no difficulty in finding expatriate staff but more difficulty locating riparian expertise. In particular, they suggested that riparian staff with sectoral expertise could be found, but that they rarely had integrative skills in fields such as Integrated Water Resources Management (IWRM).

My experience was that that the MRC secretariat struggled to maintain an adequate level of technical expertise. The organization has great difficulty in attracting appropriately skilled expatriate staff for technical areas, largely due to the annual employment contracts and the cultural constraints on appointments. Recruiting highly qualified technical staff from the member countries is even more difficult because there is a limited pool of technically skilled people in the riparian countries, and the need to balance recruitment across all four countries and recruit through the National Mekong Committees (NMCs) are additional constraints. Riparian staff are also restricted to a maximum 6 year appointment causing high turnover.

4.3. Promoting Information Sharing

One of the roles of the MRC is to promote information exchange between member states. The intention is apparently to allow potentially impacted countries to become aware of developments that may impact on them sufficiently early to allow discussion and comment, and "to address and resolve issues and problems that may arise ... in an amicable, timely and good neighborly manner" (MRC, 2004). Under the 1995 agreement, member countries are required to notify the joint committee of any intrabasin uses and interbasin diversions of water on the tributaries, and notification, prior consultation, and even prior agreement of the other member states are required for diversions and uses on the main stream.

The requirement for notification on tributaries has been of little value. Most of the notification documents have been cursory at best. For example, the notification of Buon Kuop hydropower station, a 280 MW facility constructed on a tributary of the Sre Pok River by Viet Nam, was submitted on 19 December 2003, with construction of the project due to commence in the same month (Dao Trong Tu, 2003). The notification comprised a total of nine pages, and the complete environmental assessment was four paragraphs. Most other notifications have been similar, with the exception of the Nam Theun II dam project for which the entire EIA was provided. The Joint Committee has not permitted the secretariat to pass comment on the notifications.

There is a great reluctance to share data among the Mekong countries. Even hydrological data, the most fundamental to river management, is not readily shared (Affeltranger, 2005). China will share high-flow data to assist flood forecasting, but not low-flow data, with the downstream countries. The agreement on data sharing amongst the MRC member countries under the water utilization project has had no detectable impact on information flow.

Information produced by the secretariat has been useful in promoting discussion within the basin and between players in the basin and the outside world. Documents such as the social atlas (Hook *et al.*, 2003), fisheries technical reports, and the State of the Basin Report (MRC, 2003) are now widely used as source documents by national government agencies, NGOs, and others. So, overall the organization has increased information sharing within the basin, but the increase has been disappointing.

4.4. Developing Stakeholder Cooperation

It is difficult to assess the contributions that the MRC and its predecessors have made to international cooperation in the region, but they do not appear substantial. The World Bank concluded of the Mekong Committee that "more than a decade of effort on the Indicative Plan had not stimulated cooperation between the countries" (Kirmani and Le Moigne, 1997), but that work was conducted at a time of political upheaval in the region with the American-Viet Nam war in progress and related armed conflict in Cambodia and Laos. However, the recent unusual public attack on the Lao PDR at a donor meeting by NMC officials from Cambodia (Bangkok Post, 2007a) indicates that progress since that time is limited.

The MRC secretariat engages with provincial or local government through the NMCs, a process seen as ineffective by some commentators (Satoru, 2005). With only small staffs, the NMCs do not have the resources for lengthy and detailed consultation processes. NMCs have fewer language difficulties in communicating with local stakeholders, but become a filter between stakeholders and the secretariat.

The MRC has established a dialogue with a range of civil society groups. Some of the larger groups, such as IUCN and WWF, are invited as observers to Joint Committee and Council meetings. The MRC also had regular meetings with Oxfam America while it was based in Phnom Penh, as well as with other local groups such as the Se San villagers. A common difficulty has been a belief by some civil society groups that the MRC can act as a basin police force, preventing member governments from proceeding with projects that may be deleteriously impacting local people. In fact, because the MRC is made up of the member governments it has no such power. Recently a large number of small civil society groups from the region publicly questioned the value of the MRC, and called on donors to withdraw support if the MRC fails to protect the river from unsound projects (Bangkok Post, 2007c; Nation, 2007).

Hirsch and Jensen (2006) were critical of the secretariat for not engaging more with universities and research organizations in the region. They note that seven universities have Mekong-related centers but are rarely drawn upon. That comment is largely misinformed. The MRC secretariat has actively engaged or attempted to engage relevant academics, but regrettably there is no established research culture within universities in the member countries, so the responses are very limited. Between 2001 and 2005, the MRC Environment Programme received only two proposals for research support from riparian nationals working in universities, both were supported. During the same period academics from at least eight other universities and research institutes were contracted to conduct consultancy work, and all told that the program would be receptive to relevant research proposals, but none were received.

5. CONCLUSIONS

The Mekong River faces a range of serious environmental challenges. Many of these are transboundary. In some cases the actions of one country will seriously impact downstream countries, as is the case with large dam construction. In other cases the resource is shared between all countries, and poor management in one will affect utilization in others, as is the case for the fisheries.

There is an urgent need for an effective river management agency to coordinate and drive management of the Mekong. It will need clear objectives, and should be a source of high-quality technical capacity and reliable data. It must develop credible scenarios to allow the riparian countries and other stakeholders to discuss and debate possible futures for the Mekong, and how they may be achieved.

There is no other organization within the Mekong basin that can fulfill a river basin management role, but the MRC cannot fulfill that role and also act as a development agency. The two roles are incompatible. Development agencies are seen, rightly or not, as selecting the information they release to support their preferred development options, and are mistrusted as a result.

The contributions of the MRC to sustainable management of the Mekong have been intermittent and limited. On the positive side is it has played a critical role in alerting governments and people in the region to the value and importance of the Mekong fishery. It has also played an important role in collecting and analyzing essential environmental data, including data on flows, water quality, and ecological health. However, it has failed to develop the high level of technical expertise, and sufficient impartiality, to contribute consistently and substantially to the public debate about water resource management issues in the basin. Crippled by political interference, it has failed

to release the development scenarios necessary to promote debate between the countries and stakeholders about the Mekong they want in the future.

There is also an urgent need for a forum within which political issues surrounding river basin management and development can be discussed and resolved. The MRC council and Joint Committee have not yet lived up to their promise as adequate fora for resolving intergovernmental disagreements about water resources issues in the basin. Consistently governments have avoided bringing difficult issues to the MRC and have dealt with them bilaterally. Examples include the discussions between Cambodia and Viet Nam over the impacts of Yali Dam, the signing of the Upper Mekong Navigation Agreement between Thailand, Lao, China, and Myanmar, and the impacts of Nam Theun II dam on the river. These failings have led donors and civil society groups to question the value of the MRC.

The two roles, technical river basin manager and political forum, do not necessarily need to be conducted by a single organization, and it may be preferable that they are not. Technical information should inform political negotiations, and political debates should influence the sort of technical data collected and the kinds of analyses produced, but not the outcome. To retain relevance in the future the MRC needs to clarify its role, ensure it has an administrative structure in the secretariat that can deliver that role, and reform its recruitment and employment mechanisms so that the best people possible are employed to do the job.

ACKNOWLEDGMENTS

The ideas in this chapter arise from numerous discussions over many years with people from the Mekong basin, people who have worked with other river management organizations, and past and present staff of the MRC. My former colleagues in the Environment Department at MRC, Anders Thuren and Hans Guttman were influential as were Robin Johnston, Chris Barlow, and John Metzger who worked in other parts of the MRC, and Peter Clews who worked as a consultant.

References

Affeltranger, B. (2005). Data sharing and sustainability of river basin organizations. The Mekong River Commission as case study. Communication presented *In* "First Global International Studies Conference," Istanbul, August 2005. http://www.afes-press-books.de/pdf/Istanbul/affeltranger-paper.pdf. Accessed 23 January 2008.

Allan, J. D., Abell, R., Hogan, Z., Revenga, C., Taylor, B. W., Welcomme, R. L., and Winemiller, K. (2005). Overfishing of inland waters. *Bioscience* **55**, 1041-1051.

Bangkok Post. (2002). Chavalit urges reef plan review. Bangkok Post, 19 July 2002, 1p.

Bangkok Post. (2007a). Cambodia raps Laos over Mekong dams. Bangkok Post, 16 November 2007. http://www.bangkokpost.com/breaking_news/breakingnews.php?id=123623. Accessed 16 November 2007.

Bangkok Post. (2007b). Civic groups call Mekong Commission a failure. http://www.bangkokpost.com/breaking_news / breakingnews.php?id=123544. Accessed 14 November 2007.

Bangkok Post. (2007c). NGOs against Mekong dams. http://www.bangkokpost.com/News/13Nov2007_news13.php. Accessed 13 November 2007.

Barrow, C. J. (1998). River basin development planning and management: A critical review. *World Development* **26**, 171-186.

Boon, P. J., Davies, B. R., and Petts, G. E. (eds.). (2000). "Global Perspectives on River Conservation. Science, Policy and Practice," John Wiley and Sons, Chichester, UK, 548pp.

Buss, T. F. and Yancer, L. C. (1999). Cost-benefit analysis: A normative perspective. *Economic Development Quarterly* **13**, 29-37.

Cabrera, J. (2003). The rape of a river. Bangkok Post, Perspective, 5 January 2003, 6pp.

Cambodia Daily. (2002). Chinese river plans may threaten Cambodia. The Cambodia Daily 2 April 2002, 1p.

Campbell, I. C. (2007a). Perceptions, data and river management—Lessons from the Mekong River. *Water Resources Research* **43**, doi:10.1029/2006WR005130.

Campbell, I. C. (2007b). The management of large rivers: Technical and political challenges. *In* Large Rivers," Gupta, A. (ed.). John Wiley, New York.

Chenoweth, J. (1999). Effective multi-jurisdictional river basin management. Data collection and exchange in the Murray-Darling and the Mekong River basins. *Water International* **24**, 368-376.

Chenoweth, J. L., Malano, H. M., and Bird, J. F. (2001). Integrated basin management in multi-jurisdictional river basins: The case of the Mekong River Basin. *Water Resources Development* **17**, 365-377.

China Daily. (2002). Xiaowan Dam, a reservoir for progress. http://www.china.org.cn/english/environment/42990.htm. 16 September 2002. Accessed 22 March 2007.

Cogels, O. (2005). River commission takes on development role in the Lower Mekong Basin. *Mekong Update and Dialogue* **8**(2), 2-3.

Cogels, O. (2006). Opening address to Asia 2006. *In* "The International Symposium on Water Resources and Renewable Energy Development in Asia." www.mrcmekong.org/MRC_news/speeches/30-nov-06_open_htm. Accessed 10 January 2006.

Dao Trong Tu. (2003). Memorandum on notification on the Buon Kuop Hydropower Project from Viet Nam. 22 December 2003.

Easterley, W. R. (2006). "The White Man's Burden. Why the West's Efforts to Aid the Rest Have Done So Much Ill and So Little Good," The Penguin Press, New York, 436pp.

ECE. (1991). "Convention on Environmental Impact Assessment in a Transboundary Context. Done at Espoo Finland 1991." United Nations Economic Commission on Europe, Geneva.

Edward, P. (2006). Examining inequality: Who really benefits from global growth? *World Development* **34**, 1667-1695.

ESCAP. (1995). "Environmental Impact Assessment for Inland Water Transport Development Projects in the Upper Mekong Subregion," Initial Environmental examination report. United Nations Economic and Social Commission for Asia and the Pacific, Bangkok, 49pp.

Fujii, H., Garsdal, H., Ward, P., Ishii, M., Morishita, K., and Boivin, T. (2003). Hydrological roles of the Cambodian floodplain of the Mekong River. *International Journal of River Basin Management* **1**, 1-14.

Harper, D. M. and Ferguson, A. J. D. (eds.). (1995). "The Ecological Basis of River Basin Management," John Wiley and Sons, Chichester, UK, 614pp.

Heathcote, I. W. (1998). "Integrated Watershed Management Principles and Practice," John Wiley and Sons, New York, 414pp.

Hewett, A. (2007). Plenty of aid, but is it for a "greater" Mekong? www.onlineopinion.com.au/view.asp?article=6488. Accessed 30 July 2009.

Hirsch, P. and Jensen, K. M. (2006). "National Interests and Transboundary Water Governance in the Mekong," Australian Mekong Resource Centre in collaboration with DANIDA, Sydney, Australia, 171pp.

Hook, J., Novak, S., and Johnston, R. (2003). "Social Atlas of the Lower Mekong Basin," Mekong River Commission, Phnom Penh, 154pp.

Hooper, B. P. (2005). "Integrated River Basin Governance. Learning from International Experience," IWA Publishing, London, 306pp.

Hori, H. (2000). "The Mekong. Environment and Development," United Nations University Press, Tokyo, 398pp.

Jensen, J. G. (2001). Traditional fish products: The milk of southeast Asia. *Catch and Culture* 6, 1-16.

Joint Experts Group. (2001). Report on Environmental Impact Assessment. The Navigation Channel Improvement Project of the Lancang-Mekong River from China-Myanmar Boundary Marker 243 to Ban Houei Sai of Laos, The Joint Experts Group on EIA of China, Laos, Myanmar, and Thailand, 124pp September 2001.

Kirmani, S. and Le Moigne, G. (1997). "Fostering Riparian Cooperation in International River Basins. The World Bank at its Best in Development Diplomacy," World Bank Technical Paper 335. World Bank, Washington DC, 42pp.

Kraay, A. (2005). Aid, growth and poverty. *In* "Presentation for IMF Seminar on Foreign Aid and Macroeconomic Management," Maputo, Mozambique. www.imf.org/external/np/seminars/eng/2005/famm/pdf/kraay.pdf. Accessed 30 July, 2009.

Lebel, L., Sinh, B. T., Garden, P., Hien, B. V., Subsin, N., Tuan, L. A., and Vinh, N. T. P. (undated). Risk reduction or redistribution? Flood management in the Mekong region. http://www.mpowernet.org/download_pub-doc.php?doc=3706. Accessed 9 October 2007.

Le Moigne, G., Subramanian, A., Mei Xie, and Giltner, S. (eds.). (1994). "A Guide to the Formulation of Water Resources Strategy," World Bank Technical Paper 263. World Bank, Washington DC, 102pp.

Liu, D. (2000). Provisional plan for development of navigation and regulation of navigation channel on the Lancang-Mekong River. Paper presented *In* "Economic and Social Commission for Asia and the Pacific Subregional Workshop on Technological Development of Inland Water Transport Infrastructure," Kunming, China, 12pp 11-15 December 2000.

Margolis, J. (1959). The economic evaluation of federal water resource development. *The American Economic Review* 49, 96-111.

Mattson, N. S., Kongpen, B., Naruepon, S., Nguyen, T., and Ouk, V. (2002). "Cambodian Giant Fish Species: On Their Management and Biology," MRC Technical Paper No. 3. Mekong River Commission, Phnom Penh, 29pp.

MRC. (2003). "The State of the Basin Report 2003," Mekong River Commission, Phnom Penh, 300pp.

MRC. (2004). "Agreement on the Cooperation for the Sustainable Development of the Mekong River Basin 5 April 1995," Mekong River Commission, Vientiane, 19.

MRC Review. (2007). Independent organisational, financial and institutional review of the Mekong River Commission secretariat and the Natioal Mekong Committees. Final Report, January 2007. http://www.mrcmekong.org/download/free_download/Financial_and_Institutional_Review.pdf. Accessed 24 October 2008.

Munnell, A. H. (1992). Policy Watch. Infrastructure investment and economic growth. *Journal of Economic Perspectives* 6, 189-198.

Nation. (2007). Mekong commission blasted over river dams. http://www.nationmultimedia.com/2007/11/14/national/national_30055997.php. Accessed 14 November 2007.

Nette, A. (2008). Mekong Commission fends off credibility charges. http://ipsnews.net/news.asp?idnews=42321. Accessed 15 February 2009.

Öjendal, J. (2000). Sharing the Good. Modes of managing water resources in the Lower Mekong River Basin, Ph.D. Dissertation. Göteborg University, Sweden, 235pp.

Osborne, M. (2007). "The Water Politics of China and Southeast Asia II. Rivers, Dams, Cargo Boats and the Environment," Lowy Institute for International Policy. Perspectives. Lowy Institute, Sydney, Australia, 30pp.

Pech, S. and Sunada, K. (2008). Population growth and natural-resources pressure in the Mekong River Basin. *Ambio* 37(3), 219-225.

Pilkey, O. H. and Pilkey-Jarvis, L. (2007). "Useless Arithmetic. Why Environmental Scientists Can't Predict the Future," Columbia University Press, New York, 230pp.

Podger, G., Beecham, R., Blackmore, D., Perry, C., and Stein, R. (2004). "World Bank. Mekong Regional Water Resources Assistance Strategy.Modelled Observations on Development Scenarios in the Lower Mekong Basin," World Bank, Vientiane, 126pp.

Poulsen, A. F., Ouch, P., Sintavong, V., Ubolratana, S., and Nguyen, T. T. (2002). "Fish Migrations of the Lower Mekong River Basin: Implications for Development Planning and Environmental Management," MRC Technical Paper No. 8. Mekong River Commission, Phnom Penh, 62pp.

Prest, A. R. and Turvey, R. (1965). Cost-benefit analysis: A survey. *The Economic Journal* 75, 683-735.

Reimold, R. J. (ed.). (1998). "Watershed Management. Practice, Policies and Coordination," McGraw Hill, New York, 391pp.

Roberts, T. (1993). Just another dammed river? Negative impacts of Pak Mun dam on the fishes of the Mekong Basin. *Natural History Bulletin of the Siam Society* 41, 105-133.

Satoru, M. (2005). The Mekong region: Incorporating the views of regional civil society. *In* "The State of the Environment in Asia. 2005/2006," Awaji, T. and Teranishi, S. (eds.), pp 131-148. Japan Environmental Council, Springer-Verlag, Tokyo, 385.

Simms, A. (2008). Trickle-down myth. *New Scientist* **200** (2678), 49.

Tanzi, V. and Davoodi, H. (1998). Corruption, public investment and growth. *In* "The Welfare State, Public Investment and Growth," Shibata, H. and Ihori, T. (eds.), pp 41-60. Springer-Verlag, Tokyo, 334pp.

Tharme, R. (2003). A global perspective on environmental flow assessments: Emerging trends in the development and application of environmental flow methodologies in rivers. *River Research and Applications* **19**, 397-441.

Transparency International. (2008). 2008 Corruption Perceptions Index. http://www.transparency.org/news_room/in_focus/2008/cpi2008/cpi_2008_table. Accessed 16 October 2008.

Varis, O. (2004). Basin organization models: The case of the Mekong River. *In* "Water Governance in West Africa," Niasse, M., Ly, I., Iza, A., and Varis, O. (eds.), pp 223-236. IUCN International Law Series, Bonn, 247pp.

Wilder, M. and Lankao, P. R. (2006). Paradoxes of decentralization: Water reform and social implications in Mexico. *World Development* **34**, 1977-1995.

World Commission on Dams. (2000). "Dams and Development: A New Framework for Decision Making," Earthscan Publications, London, UK, 404pp.

Index

Note: Page numbers followed by "f" indicate figures, "t" indicate tables.

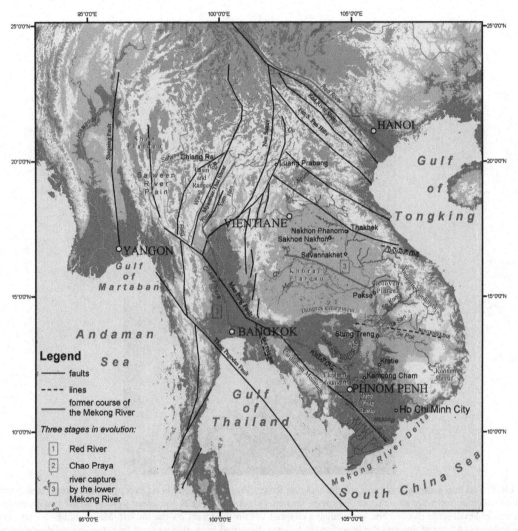

PLATE 1 Simplified tectonic map of Indochina with information drawn from various sources cited in the text but chiefly from Mouret (1994). The Indochina block sits between the South China Block to the north-east and the Central-Sunda Plate to the south-west. The boundaries between these three blocks are, respectively, the Red River Shear and the Danang Line in the north-east and the Rovieng, Pursat, and Sakeo Lines in the south-west. The possible former courses of the Mekong River are discussed in the text. (See Fig. 1 in Chapter 2, p. 14.)

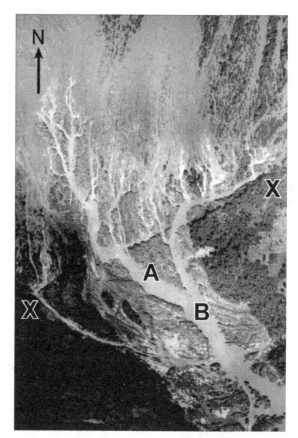

PLATE 2 Vertical view of anabranch of the Mekong River close to Kon Phapheng in Siphandone area, Laos. Structural lineations, aligned from top to center of the image and curving to the right in the lower portion of the image, control the alignment of small-scale rivulets and the avulsive channel within the forest on the far left. Major channels may conform to the lineation (A) or cut through the lineation (B) but are disrupted by a distinct fault line running obliquely across the image (X to X). From air photograph, 1993, 1:15,000. Horizontal field of view ~1.5 km. (See Fig. 2 in Chapter 2, p. 15.)

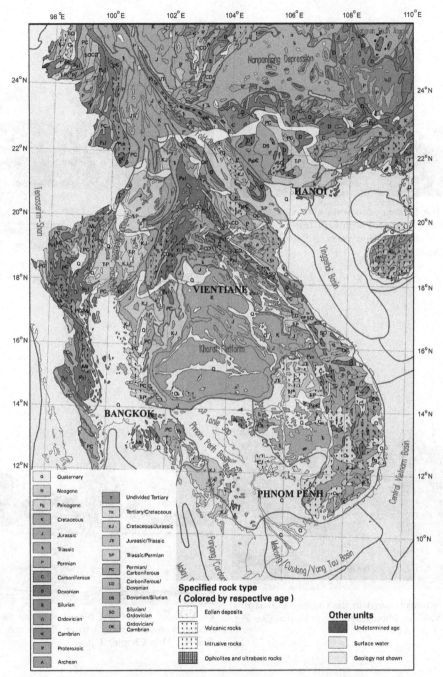

PLATE 3 Geological map of Indochina. Boundaries of the main geological provinces are shown as red lines. Scale: 1:5,000,000. Redrawn from Steinshouer *et al.* (1997). (See Fig. 3 in Chapter 2, p. 19.)

PLATE 4 The Mekong downstream of Chiang Saen, Thailand. The river is in rock with very little accommodation space for sediment. Photograph: Avijit Gupta. (See Fig. 2 in Chapter 3, p. 31.)

PLATE 5 Upper Mekong Valley near Er Lake. Photograph: Avijit Gupta. (See Fig. 6 in Chapter 3, p. 37.)

PLATE 6 Sediment stored in the channel and narrow valley of the Mekong. Photograph: Avijit Gupta. (See Fig. 10 in Chapter 3, p. 43.)

PLATE 7 Rock-cored islands showing sediment accumulation around a rocky barrier. Photograph: Avijit Gupta. (See Fig. 11 in Chapter 3, p. 48.)

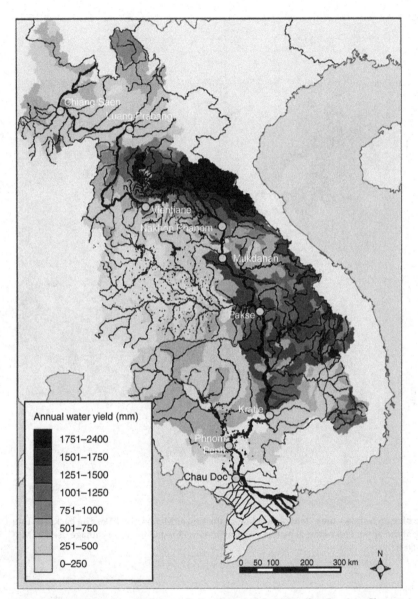

Annual water yield (mm)

■	1751–2400
■	1501–1750
■	1251–1500
■	1001–1250
■	751–1000
■	501–750
■	251–500
□	0–250

0 50 100   200   300 km

N

PLATE 8   Annual water yield of the Lower Mekong Basin (from MRC, 2008). (See Fig. 3 in Chapter 4, p. 58.)

PLATE 9 The blue area shows the extent of flooding in the Lower Mekong in 2006. The backflow into Tonle Sap is clear, as is the flooding at the upper end of the delta (Source: Dartmouth Flood Observatory image used in MRC, 2007). (See Fig. 5 in Chapter 4, p. 60.)

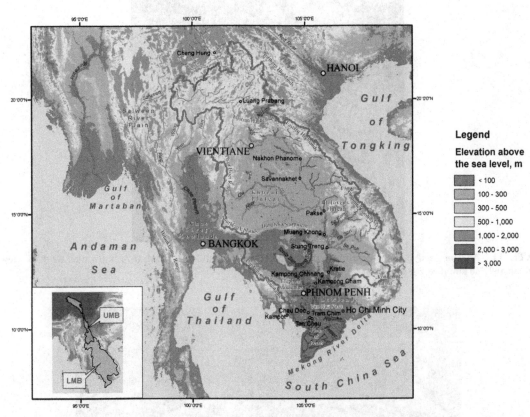

PLATE 10 Mekong River Basin: topography and major towns and cities along the Mekong River. (See Fig. 1 in Chapter 5, p. 79.)

PLATE 11 Anastomosed bedrock-controlled fluvial network—Siphandone. Landsat-7, horizontal field of view is approximately 40 km. (See Fig. 8 in Chapter 5, p. 89.)

PLATE 12 Oblique aerial view of 4000 islands reach of Mekong River (large island is approximately 900 m wide at widest point). Photograph: Stuart Chape. (See Fig. 9 in Chapter 5, p. 89.)

PLATE 13 Satellite image of the Mekong River during flood season in vicinity of Kampong Cham. Landsat-7: field of view approximately 45 × 45 km. (See Fig. 11 in Chapter 5, p. 91.)

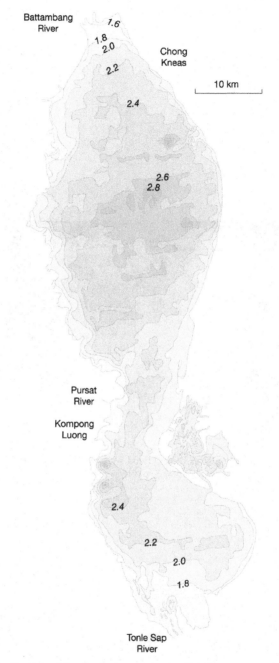

PLATE 14 Bathymetry of the Tonlé Sap lake. Contours are in meters for dry-season water level. (See Fig. 13 in Chapter 5, p. 93.)

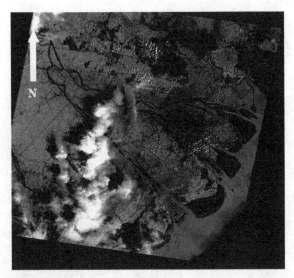

PLATE 15 Satellite image of the Mekong River delta—Landsat-7: 185 × 185 km. (See Fig. 14 in Chapter 5, p. 94.)

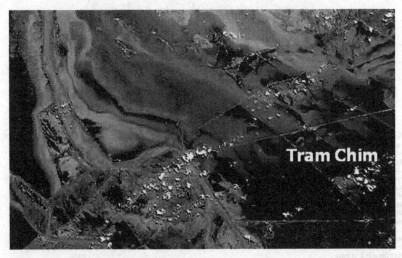

PLATE 16 Example of distributary channel and floodwater flow over the Plain of Reeds; 20 July 2000—Horizontal field of view approximately 25 km. (See Fig. 15 in Chapter 5, p. 95.)

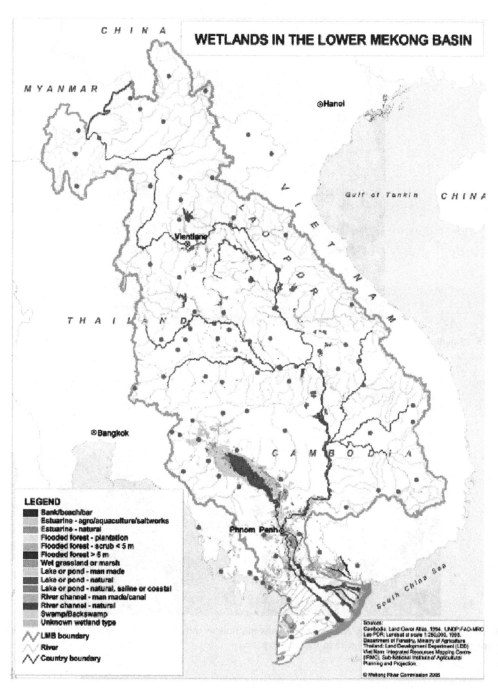

PLATE 17 Distribution of wetlands (courtesy of MRC). (See Fig. 17 in Chapter 5, p. 97.)

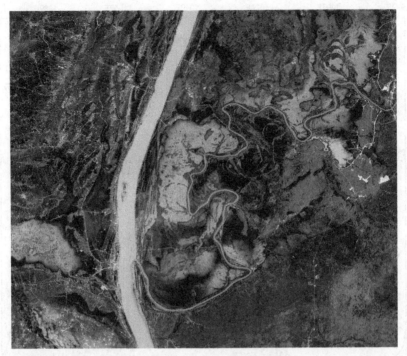

PLATE 18 Wetland inundation at tributary confluence with the Mekong River. Flow top to bottom of image. Landsat-7: field of view approximately 30 × 30 km. (See Fig. 18 in Chapter 5, p. 98.)

PLATE 19 Inundation of floodplain to left of riparian levée. Photograph: Joe Garrison, Garrison Photographic, Phnom Penh. (See Fig. 21 in Chapter 5, p. 102.)

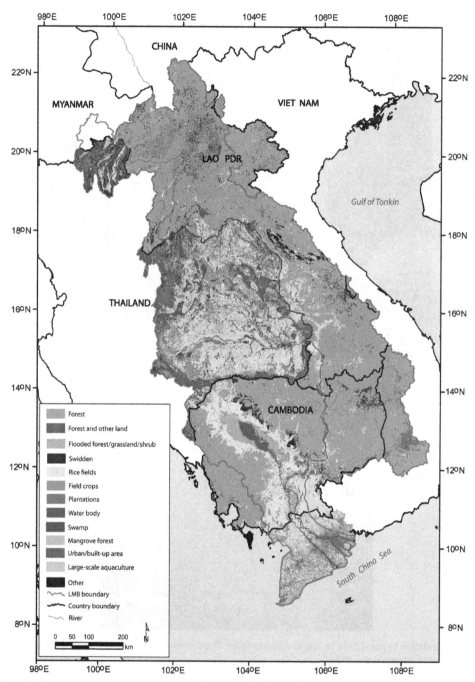

Legend:
- Forest
- Forest and other land
- Flooded forest/grassland/shrub
- Swidden
- Rice fields
- Field crops
- Plantations
- Water body
- Swamp
- Mangrove forest
- Urban/built-up area
- Large-scale aquaculture
- Other
- LMB boundary
- Country boundary
- River

PLATE 20   Land-use in the lower Mekong basin. (See Fig. 1 in Chapter 9, p. 215.)

PLATE 21 Aerial photograph of flooded forest vegetation in the Battambang river area. Photograph: Joe Garrison, Garrison Photographic, Phnom Penh. (See Fig. 8 in Chapter 10, p. 263.)

PLATE 22 Fish traps at the mouth of the Battambang River in 2004. Photograph: Ian Campbell. (See Fig. 9 in Chapter 10, p. 265.)

PLATE 23 Irrawaddy dolphin calf from Koh Pidau Pool, Kratie Province, Mekong River. (See Fig. 2 in Chapter 15, p. 367.)

PLATE 24 An Irrawaddy dolphin from Kampi Pool, Kratie Province, Mekong River. The Irrawaddy dolphin is a charismatic mega-vertebrate species that is ideal as a flagship species in the Mekong River. Photograph: Yim Sak Sang. (See Fig. 7 in Chapter 15, p. 384.)

Printed in the United States
By Bookmasters